Ergebnisse der Mathematik und ihrer Grenzgebiete

Volume 59

3. Folge

A Series of Modern Surveys in Mathematics

For further volumes:
www.springer.com/series/728

Frédéric Paugam

Towards the Mathematics of Quantum Field Theory

Springer

Frédéric Paugam
Institut de Mathématiques de Jussieu
Université Pierre et Marie Curie
Paris, France

ISSN 0071-1136 ISSN 2197-5655 (electronic)
Ergebnisse der Mathematik und ihrer Grenzgebiete. 3. Folge / A Series of Modern Surveys in Mathematics
ISBN 978-3-319-04563-4 ISBN 978-3-319-04564-1 (eBook)
DOI 10.1007/978-3-319-04564-1
Springer Cham Heidelberg New York Dordrecht London

Library of Congress Control Number: 2014933388

Mathematics Subject Classification: 81-02, 81T70, 81T13, 58A03, 58A20, 18-02, 18G55, 81T20, 81T17

Printed on acid-free paper

Springer is part of Springer Science+Business Media (www.springer.com)

To S., J. and P.

Preface

This book is an applied pure mathematics textbook on quantum field theory. Its aim is to introduce mathematicians (and, in particular, graduate students) to the mathematical methods of theoretical and experimental quantum field theory, with an emphasis on

coordinate-free presentations

of the mathematical objects in play, but also of the mathematical theories underlying those mathematical objects. The main objective of this presentation is to promote interactions between mathematicians and physicists, by supplying a common and flexible language, hopefully for the good of both communities, even if the mathematical one is our primary target.

We propose here a reference book that gives a coherent and systematic mathematical toolbox for the coordinate-free treatment of classical and quantum field theories. Giving such a complete presentation, made accessible to graduate students and researchers through a compact formalism, is an original contribution to the mathematical literature, despite the fact that there already exist good specialized references on some parts of the subject, cited in the bulk of our book.

We don't pretend to exhaustivity, because it is unattainable for such a wide subject. We have aimed to produce a book that attains the optimal unity and generality that could be hoped for within our working capacities.

Moreover, we have always tried, with the aim of teaching in mind, to favor general methods that apply to all of the numerous examples treated in this book. This allows us to avoid unworthy repetitions, which would have prevented us from covering so much material in a single volume. The mathematician's approach of going from the general to the particular case also allows us to give optimal and precise hypotheses and results on theories that are usually treated in the physics literature through the study of particular examples.

We also describe a large class of examples of variational problems from classical, quantum and theoretical physics, directly in our language. Finally, we give an overview of various modern quantization and renormalization procedures, together with a mathematical formalization of them that is coherent with our general formalism.

The main problem that we had to face when we started this project was the vast zoology of types of structures that appear in the mathematical formalization of quantum field theory. Here are the specifications that we used to arrive at a locally finitely presented book.

Our viewpoint of physics is very naive and reductionist: a physical theory is given by a family of experiments and a model, i.e.

a mathematical machinery,

that explains these experiments. The main constraint on such a theory is that one must be able to discuss it with colleagues and students, so that they test it, correct it and improve it with later developments. We will mostly be interested in models, and refer to the physics literature for the description of the corresponding experiments. Some of the models we present are theoretical extrapolations of existing physical models. They are either hoped to be tested at some point, or only serve to uncover and prove nice mathematical theorems.

There are various ways to describe a mathematical machinery, and all of them are based on some common metamathematical (i.e. natural) language.

In the set theoretical language, one has the *axiom of choice* which provides a tool to prove

existence results.

In the categorical language (see Lawvere [Law05, Law06] and Ehresmann [Ehr81]), one has *Yoneda's lemma*, which allows us to prove general

unicity statements.

The rough idea of categorical mathematics is to define general mathematical objects and theories not as sets with additional structures, but as objects of a (possibly higher) category (in a metamathematical, i.e., linguistic sense), with a particular universal mapping property, or equivalently, as representing objects for a given (generalized) functor.

To describe all the constructions needed in this book in a unified and compact setting, we decided to identify (higher) categorical logic and (higher) category theory.

It is known that distilled axioms are pretty indigestible. We strongly suggest the reader follow the complicated alembics of Chaps. 1 and 2 simultaneously with respective sections of more wholesome following chapters.[1]

We don't aim at a complete account of all these theories, but we will present enough of each of them to make our description of physical models mathematically clear and consistent. There are many other ways to approach the mathematics of quantum field theory. All of them have advantages on those used here. We have chosen these approaches simply because we are more comfortable with them, and because we don't see another way of covering all of the material presented in this book.

Paris, France Frédéric Paugam

[1]Exercise: find another occurrence of this paragraph in the mathematical literature.

Acknowledgements

This work started with the author's coorganization of a two year workshop at Jussieu's mathematical institute (2008–2010), entitled "mathématiques de la physique moderne". The author's notes were first formally shaped for an advanced course at Instituto Nacional de Matemática Pura e Aplicada (IMPA, Rio de Janeiro) in 2010, and completely rewritten for two advanced courses at Université Pierre et Marie Curie (UPMC, Paris) in 2011 and 2012. They were completed during an advanced course at Université du Luxembourg (UL, Luxembourg) in 2012.

This course was given at IMPA with the financial support of UPMC, on infrastructures of the IMPA/UPMC contract for scientific exchanges. It was also given in two consecutive years at UPMC's advanced classes, and as an intensive one month course at UL, with the financial support of the Luxembourg laboratory of mathematics. I thank IMPA for giving me the opportunity to teach this course and the IMPA/Paris 6 contract for its support (and in particular, their organizers H. Rosenberg and S. Sorin). I thank the University of Paris 6, Jussieu's Mathematical Institute (particularly the "Operator algebras" and "Algebraic Analysis" groups), IMPA and the University of Luxembourg for providing stimulating atmospheres and excellent working conditions. I also thank the administrative staff of the institute, and in particular N. Fournaiseau, I. Houareau-Sarazin and Z. Zadvat. I thank N. Poncin and G. Bonavolonta for their interest in my work and N. Poncin for inviting me to give this course in Luxembourg. I thank my students both at IMPA, UPMC and Luxembourg for their questions and comments and for giving me the opportunity to improve the presentation. Among them, special thanks are due to G. Bonavolonta, L. Continanza, T. Covolo, T. Drummond, C. Grellois, J. Guere, A. Heleodoro, D. Khudaverdyan, D. Lejay, T. Lemanissier, and H. Pugh.

I thank the members of the workshop "mathématiques de la physique moderne",[2] which lasted at UPMC (Paris) from February 2008 to January 2010, for their participation, and in particular those who gave talks: M. Bermudez, L. Boutet de Monvel, P. Cartier, A. Connes, A. Dabholkar, B. Delamotte, G. Esposito-Farèse, O. Gabriel,

[2] Workshop's archives: http://www.math.jussieu.fr/~fpaugam/GdT-physique/GdT-physique.html.

G. Godefroy, F. Helein, B. Jacob, K. Noui, L. Vitagliano, W. van Suijlekom, J. Zinn-Justin, and J.-B. Zuber. A special thanks is due to Miguel Bermúdez for uncountable great and enthusiastic discussions about mathematics and physics and for his important involvement in the workshop's activities.

I thank the members of the workshops "renormalisation et algèbres de factorisation",[3] which took place at UPMC (Paris) from September 2010 to January 2012, and particularly the coorganizers D. Calaque and G. Ginot. They certainly had a great influence on my homotopical inclinations. These two workshops were based on Costello's book [Cos11], Costello–Gwilliam's book [CG10] and Francis–Gaitsgory's article [FG11]. I had great email and verbal exchanges with these authors concerning their work. I also had inspiring discussions with N. Rozenblyum on a preliminary version of his ongoing project [Roz12]. The original Beilinson–Drinfeld book [BD04] was also an important source of inspiration. I thank the members and organizers of the Glanon workshop "Théorie des champs, avec un parfum dérivé" for inspiring discussions and courses.

I thank the contributors of the nlab website, and in particular J. Baez, U. Schreiber, M. Shulman, and M. Weber, for their inspiring contributions on higher category theory. My book was clearly influenced by the nPOV.

I also benefitted greatly from the courses of A. Connes at collège de France, L. Vitagliano at the diffiety school, O. Gwilliam at the Glanon mathematical meeting, and from discussions with M. Anel, M. Bächtold, G. Barnich, D. Bennequin, D. Ben-Zvi, M. Bermudez, L. Boutet de Monvel, C. Brouder, H. Bursztyn, D. Calaque, A. Cattaneo, P. Cartier, M. Chaperon, L. Charles, S. Chemla, D.-C. Cisinski, K. Costello, Dang Nguyen V., F. Deglise, E. De Faria, W. De Mello, P. Dingoyan, J.-P. Dos Santos, T. Drummond, M. Dubois-Violette, M. Egeileh, A. Fabretti, G. Felder, E. Ferrand, J. Francis, A. Gabrielov, D. Gaitsgory, H. Gies, G. Ginot, F. Helein, G. Henkin, V. Humilière, P. Iglesias-Zemmour, A. Joyal, B. Keller, Y. Kosmann-Schwarzbach, J. Krasilsh'chik, L. Lazzarini, N. Lerner, M. Livernet, F. Loeser, J. Lurie, G. Maltsiniotis, J. Millès, V. Minerbe, I. Moerdijk, G. Morando, J. Oesterlé, F. Petit, K. Rejzner, M. Romagny, M. Rosso, N. Rozenblyum, F. Schaetz, P. Schapira, U. Schreiber, M. Shulman, G. Skandalis, J. Stasheff, M. Stienon, J. Tapia, B. Toen, W. van Suijlekom, M. Vaquié, A. Verbovetsky, G. Vezzosi, A. Vinogradov, L. Vitagliano, M. Weber, A. Weinstein, and J. Zinn-Justin.

[3]Workshop's archives: http://www.math.jussieu.fr/~fpaugam/documents/GdT-renormalisation-factorisation.html.

Contents

Chapter 1
Introduction

We will now give an elementary presentation, based on simple examples, of our general approach to the fulfillment of our commitments, exposed in the preface. For references and historical background on the notions introduced here we refer to the heart of the book, and to the end of this introduction for a (non-exhaustive) list of the author's original contributions.

1.1 Variational Calculus and Parametrized Geometry

The notion of field is very general. From the mathematician's viewpoint, it means a function

$$x : M \longrightarrow C$$

between two spaces that represents the motion of a physical object in a configuration space C, with parameter space M. Here the notion of space must be taken in a widely generalized sense, because even electrons have a mathematical formalization as functions between spaces of a non-classical kind, i.e., which are not modeled on subsets of the real affine space $\mathbb{R}^n$.

One may then base classical mechanics on variational calculus, following Lagrange and Hamilton. This is done by using Maupertuis' principle of least action, which says that the true physical trajectories of a system may be chosen among fields as those that minimize (or extremize) a given functional $S(x)$, called the action functional.

Let us first illustrate this very general notion with the simple example of Newtonian mechanics in the plane. Let $M = [0, 1]$ be a time parameter interval, and $C = \mathbb{R}^2$ be the configuration space. In this case, a field is simply a function

$$x : [0, 1] \longrightarrow \mathbb{R}^2$$

that represents the motion of a particle (i.e., a punctual object) in the plane (see Fig. 1.1).

F. Paugam, *Towards the Mathematics of Quantum Field Theory*,
Ergebnisse der Mathematik und ihrer Grenzgebiete. 3. Folge / A Series of Modern
Surveys in Mathematics 59, DOI 10.1007/978-3-319-04564-1_1,

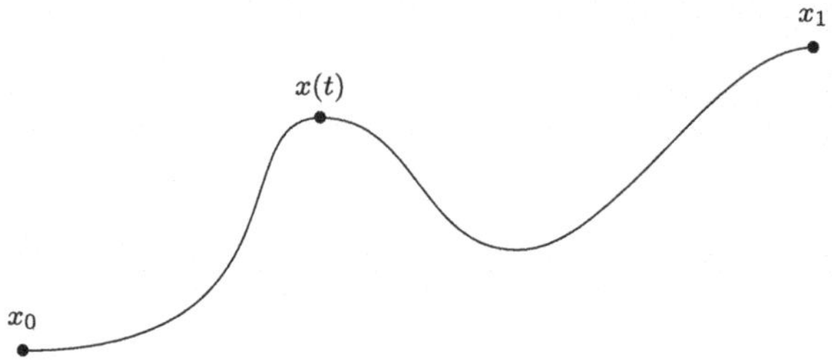

Fig. 1.1 Motion of a particle in the plane

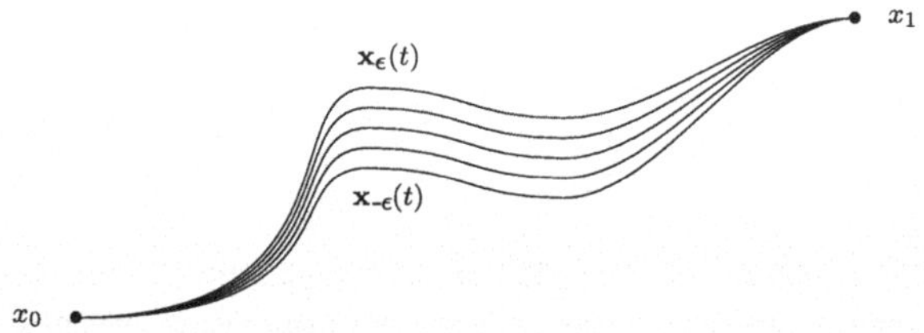

Fig. 1.2 Path in the space of trajectories with fixed end-points

The action functional of the particle in free motion is given by

$$S(x) := \int_M \frac{1}{2} m \left\| \partial_t x(t) \right\|^2 dt,$$

where m is a real number encoding the mass of the particle. We will suppose given the additional datum of some starting and ending points $x_0, x_1 \in C$ for the trajectories, and of the corresponding speed vectors $\mathbf{v}_0$ and $\mathbf{v}_1$.

With this additional datum, one may compute the physical trajectories by computing the infinitesimal variation of the action functional S along a small path

$$\begin{aligned} \mathbf{x} :]-\epsilon, \epsilon[\times [0,1] &\to \mathbb{R}^2 \\ (\lambda, t) &\mapsto \mathbf{x}_\lambda(t) \end{aligned}$$

in the space of fields at a field $x = \mathbf{x}_0$, with fixed starting and ending points (see Fig. 1.2), by the formula

$$\frac{\delta S}{\delta \mathbf{x}}(x) := \partial_\lambda S(\mathbf{x}_\lambda)|_{\lambda=0}.$$

By applying an integration by parts, one gets

$$\frac{\delta S}{\delta \mathbf{x}}(x) = \int_M \langle -m \partial_t^2 x, \partial_\lambda \mathbf{x} \rangle dt,$$

and this expression is zero for all variations $\mathbf{x}$ of x if and only if Newton's law for the motion of a free particle

$$m \partial_t^2 x = 0$$

is satisfied. This equation may be solved using the given initial condition $(x_0, \mathbf{v}_0)$, or the given final condition $(x_1, \mathbf{v}_1)$. In both cases, the model tells us, with no surprise,

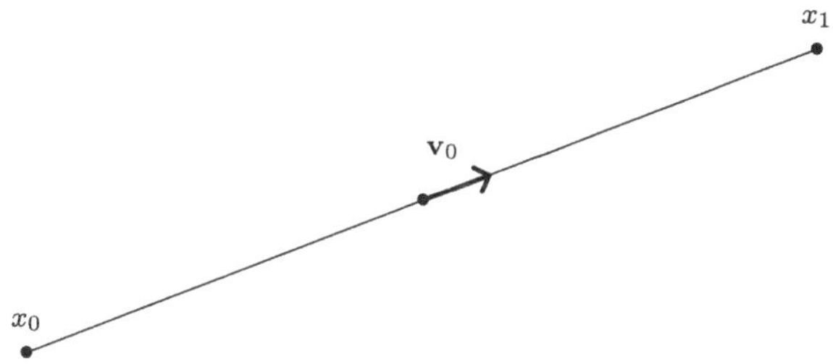

Fig. 1.3 The free particle moving on a *straight line* at constant speed

that the free particle is simply moving on a straight line at constant speed, starting at the given point, with the given fixed speed (see Fig. 1.3). We may identify the space of trajectories with the space of pairs $(x_0, \mathbf{v}_0)$ forming the initial conditions for the corresponding differential equation. This space of pairs is the cotangent space

$$T^*C \cong \mathbb{R}^2 \times \mathbb{R}^2$$

of the configuration space.

We remark that the above computation of the equations of motion for the given action functional does not really involve the datum of a topology on the space H of fields with fixed starting and ending point. One only needs to know the value of the action $S : H \to \mathbb{R}$ on a family $\{\mathbf{x}_\lambda\}$ of fields parametrized by the small interval $]-\epsilon, \epsilon[$. One may even easily define the value of S on such a family, whenever λ is a point in an open subset U of $\mathbb{R}^n$ for some n. The awkward idea of replacing the simple parameter interval $U =]-\epsilon, \epsilon[$ by any such open set will allow us to think of the space H of fields as a space of the same nature as the spaces M and C of parameters and configurations. For $U \subset \mathbb{R}^n$, let us denote by $M(U)$, $C(U)$ and $\mathbb{R}(U)$ the sets of smooth maps from U to M, C and $\mathbb{R}$. One may think of these sets as sets of points of M, C and $\mathbb{R}$ parametrized by U. Let us also define the space of histories by setting

$$\begin{aligned} H(U) := \big\{ & x : U \times M \to C \text{ smooth, such that} \\ & x(u, 0) = x_0 \text{ and } x(u, 1) = x_1 \text{ for all } u \in U \big\}. \end{aligned}$$

These are points of the space of fields parametrized by U, with given fixed starting and ending points. We give an example with parameter λ in the open unit disc $U = D(0, 1) \subset \mathbb{R}^2$ in Fig. 1.4. Note that these sets are all compatible with smooth changes of parameters $V \to U$, because of the stability of smoothness by composition. One may then define a smooth functional

$$S : H \to \mathbb{R}$$

as the datum of a family of set functions

$$S_U : H(U) \to \mathbb{R}(U),$$

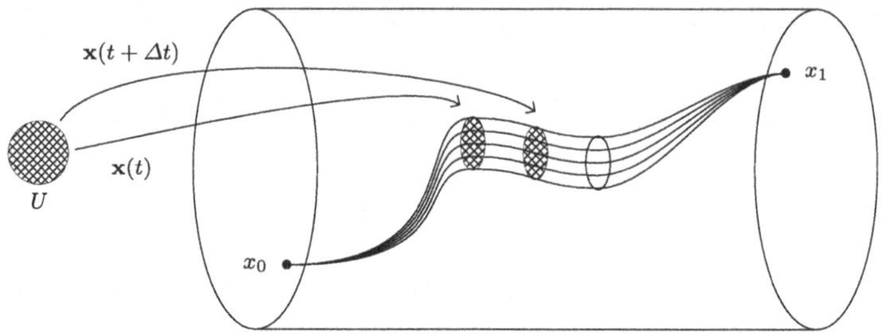

Fig. 1.4 A point in the space of trajectories parametrized by the unit disc

which are compatible with smooth changes of parameters. In modern mathematical language, M, C, H and $\mathbb{R}$ defined as above are functors from the category of smooth open subsets in real affine spaces to sets, and S is a natural transformation.

The main advantage of the above interpretation of variational calculus is that it allows us to work with a much more general notion of space, and these generalized kinds of spaces are necessary to understand the computations of quantum field theory from a completely geometrical viewpoint, generalizing what we explained above. One may call this geometry

parametrized geometry.

The combination of parametrized geometry with the usual differential geometric tools, such as manifolds, involves a refinement of the functorial notion of parametrized space by the notion of a sheaf, which allows us to define a smooth manifold as the parametrized space obtained by pasting some basic open subset $U \subset \mathbb{R}^n$, viewed as basic parametrized spaces, along smooth maps.

1.2 Motion with Interactions and the Geometry of Differential Equations

We continue to consider the motion of a Newtonian particle in the plane, given by a map

$$x : M \to C,$$

with $M = [0, 1]$ and $C = \mathbb{R}^2$, however we now suppose that it is not moving freely, but is interacting with its environment, as described in Fig. 1.5. Since Newton, the mathematical formalization of this environment has been given by a function

$$V : C \to \mathbb{R}$$

called the potential energy density. The associated force field is given by the gradient vector field

$$\mathbf{F} = -\overrightarrow{\text{grad}}(V).$$

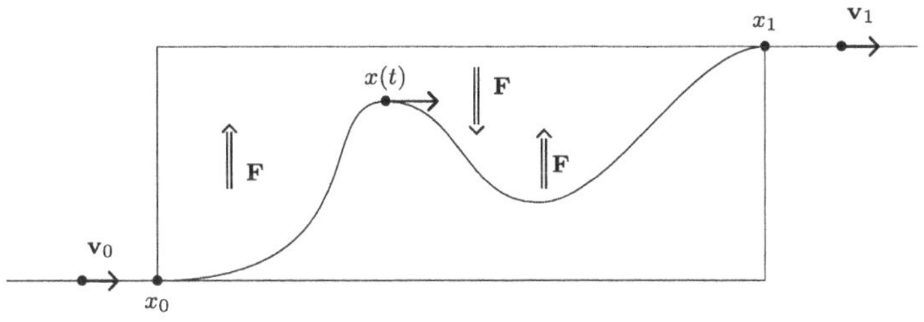

Fig. 1.5 Motion of a particle in a potential

The action functional for this interacting particle is given by a functional

$$S(x) := \int_M L\big(x(t), \partial_t x(t)\big)dt,$$

where the function

$$L(x, \partial_t x) := \frac{1}{2}m\|\partial_t x\|^2 + V(x)$$

is called the Lagrangian density. Functionals S of the above form are called local functionals. The associated equations of motion are again given by Newton's law

$$m\partial_t^2 x = \mathbf{F}(x).$$

As before, we suppose that the initial and final conditions $(x_0, \mathbf{v}_0)$ and $(x_1, \mathbf{v}_1)$ are fixed. We may split the action functional $S = S_{free} + S_{int}$ into a sum of the free (kinetic) action functional

$$S_{free}(x) = \int_M \frac{1}{2}m\big\|\partial_t x(t)\big\|^2 dt,$$

whose equations of motion were already considered in the previous section, and the interaction functional

$$S_{int}(x) = \int_M V\big(x(t)\big)dt.$$

We may then represent the motion of our particle as being free before and after the experiment, and being influenced by the potential within the experiment's spacetime box.

Following Lagrange, it will be convenient, along the way, to develop a more algebraic version of variational calculus corresponding to a differential calculus for local functionals.

In the above example, one simply considers the manifold $\mathrm{Jet}(M, C)$ with coordinates (t, x_i) given by formal derivatives of the variables x in $C = \mathbb{R}^2$, indexed by integers $i \in \mathbb{N}$. One may view the Lagrangian density as a function $L : \mathrm{Jet}(M, C) \to \mathbb{R}$

defined by

$$L(t, x_0, x_1) := \frac{1}{2}m\|x_1\|^2 + V(x_0).$$

The only additional datum needed to formulate variational calculus in purely algebraic terms is the differential relation between the family of formal coordinates x_i, given by

$$\partial_t x_i = x_{i+1}.$$

This universal differential relation is encoded in a connection ∇ on the bundle $\mathrm{Jet}(M, C) \to M$, called the Cartan connection. As the action of ∂_t on the jet bundle functions is given by the total derivative operator $\frac{d}{dt}$, the differential calculus underlying variational calculus is completely encoded by the above differential structure on the jet bundle.

There is also a parametrized geometry underlying this differential calculus that is morally a geometry of formal solution spaces to non-linear partial differential equations, called $\mathcal{D}$-geometry. Its parameter spaces are essentially given by jet bundles $(\mathrm{Jet}(U, U \times V), \nabla)$ of morphisms $\pi : U \times V \to U$ (for $U \subset \mathbb{R}^n$ and $V \subset \mathbb{R}^m$ open subsets), together with their Cartan connection. This geometry of differential equations is often used by physicists to study classical and quantum systems, and in particular their local symmetries and conservation laws.

1.3 Measurable Densities and Functorial Analysis

We continue to study the motion of a particle $x : [0, 1] \to \mathbb{R}^2$ in the plane, but we shall change the action functional S by replacing it with a parametrization-invariant version, given by the formula

$$S(x) := \int_M \frac{1}{2}m\left\|\partial_t x(t)\right\|dt$$

which involves the norm, i.e., the square root of the metric on $\mathbb{R}^2$. From the viewpoint of parametrized geometry, this expression makes sense when we replace the trajectory x by a smooth family

$$x : [0, 1] \times U \to \mathbb{R}^2$$

parametrized by an arbitrary open set $U \subset \mathbb{R}^n$. However, since the square root is a non-smooth function at 0, the function

$$\begin{aligned} U &\longrightarrow \mathbb{R} \\ \lambda &\longmapsto S(x_\lambda) \end{aligned}$$

will not be smooth in the parameter in general if the particle has zero speed at some point of the trajectory. We thus have to restrict the domain of definition of the functional S to a parametrized subspace $D_S \subset H$, called its domain of definition.

More generally, we could take any Lebesgue integrable density $L : [0, 1] \times T\mathbb{R}^2 \to \mathbb{R}$ and force the formula

$$S(x_\lambda) := \int_M L\big(t, x_\lambda(t), \partial_t x_\lambda(t)\big)dt$$

to define a smooth function of the auxiliary parameter $\lambda \in U$. The domain of definition for this functional is easily computed using Lebesgue's domination criterion for the derivative of an integral with parameters. It is given by the subspace $D_S \subset H$ with parametrized points

$$D_S(U) := \big\{x_\lambda |\ \text{locally on } U, \exists g_i \in L^1\big([0, 1]\big),\ \big|\partial_\lambda^i L\big(t, x_\lambda(t), \partial_t x_\lambda(t)\big)\big| \le g_i(t)\big\}.$$

The above discussion shows that it is necessary to generalize the notion of functional introduced in parametrized geometry by authorizing partially defined functions, thus obtaining more flexibility and a better compatibility with the usual tools of functional analysis. This extension of functional analysis is called

functorial analysis.

1.4 Differential Calculus and Functional Geometry

It was already remarked by Weil that one may formulate differential calculus in an algebraic way, by using algebras with nilpotent elements, such as $\mathcal{C}^\infty(\mathbb{R})/(\epsilon^2)$ for example, where $\epsilon = \mathrm{id} : \mathbb{R} \to \mathbb{R}$ is the standard coordinate function on $\mathbb{R}$. One may think of this algebra as the algebra of coordinate functions on an infinitesimal thickening $\vec{\bullet}$ of the point $\bullet$, defined by adding a universal tangent vector to it. Indeed, if M is a manifold, the choice of a tangent vector $\mathbf{v}$ to M at a point $m \in M$ is equivalent to the choice of a derivation

$$D : \mathcal{C}^\infty(M) \to \mathbb{R}$$

compatible with the evaluation algebra morphism

$$\begin{aligned} \mathrm{ev}_m : \mathcal{C}^\infty(M) &\to \mathbb{R} \\ f &\mapsto f(m), \end{aligned}$$

meaning that for all $f, g \in \mathcal{C}^\infty(M)$, one has

$$D(fg) = f(m)D(g) + D(f)g(m).$$

The datum of such a derivation is also equivalent to that of the algebra morphism

$$\begin{aligned} h_D : \mathcal{C}^\infty(M) &\to \mathcal{C}^\infty(\mathbb{R})/\big(\epsilon^2\big) \\ f &\mapsto f(m) + \epsilon D(f). \end{aligned}$$

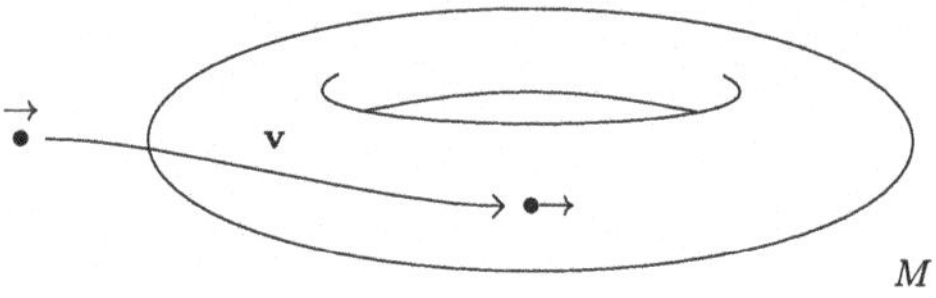

Fig. 1.6 Infinitesimally parametrized points are tangent vectors

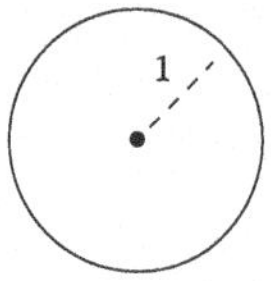

Fig. 1.7 The unit circle

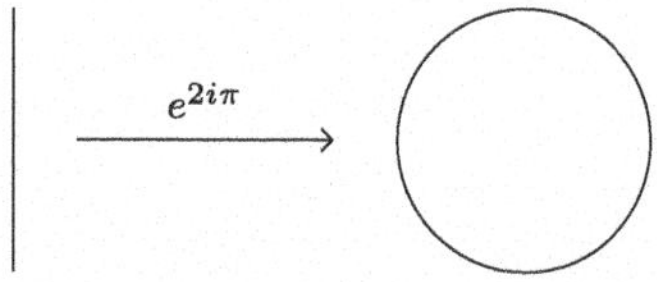

Fig. 1.8 The exponential map

We may interpret this algebra morphism as a morphism $\mathbf{v} : \vec{\bullet} \to M$, i.e., as a point of M together with a tangent vector on it, as described in Fig. 1.6.

This last interpretation of tangent vectors admits a very general formulation which works essentially without change in all of the parametrized settings discussed in this introduction.

To make this general formulation work, say, in the smooth setting, one first has to replace the parameters $U \subset \mathbb{R}^n$ by their algebras $\mathcal{C}^\infty(U)$ of functions, which makes no essential difference because of the equivalence between the datum of U and the datum of $\mathcal{C}^\infty(U)$.

Then, one needs to allow more general algebras as parameters. For example, we would like to consider the nilpotent algebra $\mathcal{C}^\infty(\mathbb{R})/(\epsilon^2)$ as an admissible parameter. More generally, algebras of formal solutions to equations, such as $\mathcal{C}^\infty(U)/(f_1, \dots, f_n)$, may also be useful. Note that using such algebras is also a very natural thing to do in classical differential geometry, where one is interested in varieties defined by equations. For example, the algebra of smooth functions on the real circle S^1, pictured in Fig. 1.7, may be easily defined as the quotient algebra

$$\mathcal{C}^\infty(S^1) := \mathcal{C}^\infty(\mathbb{C})/(|z|^2 - 1) \cong \mathcal{C}^\infty(\mathbb{R}^2)/(x^2 + y^2 - 1).$$

One may use the exponential map $e^{2i\pi} : \mathbb{R} \to S^1$, drawn in Fig. 1.8, or the implicit function theorem, pictured in Fig. 1.9, to define S^1 as a smooth manifold. However it is less comfortable to work with charts here than to work directly with the above algebra. Indeed, $S^1 \subset \mathbb{C}$ is naturally defined by the closed algebraic condition $|z|^2 = 1$, and most of its properties may be actually derived from the immediate study of this equation. However, if we think of such quotients as the usual real algebras, given by sets, together with compatible multiplication and addition operations,

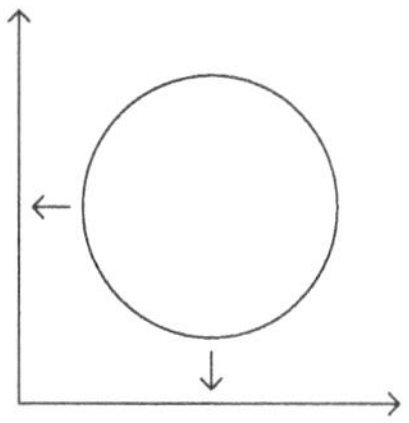

Fig. 1.9 The implicit function theorem

we run into the following difficulty: the natural product map

$$\mathcal{C}^\infty(U) \otimes \mathcal{C}^\infty(V) \longrightarrow \mathcal{C}^\infty(U \times V)$$

is not an isomorphism because some functions on the right may only be described as infinite converging series of tensors.

This problem may be resolved by using a completed topological tensor product. However, this also has the drawback of introducing unnatural complications, which, furthermore, are incompatible with generalizations to the other kinds of parametrized geometries discussed in this introduction. There is however a very elegant solution to this problem of completing the category of algebras of the form $\mathcal{C}^\infty(U)$, for $U \subset \mathbb{R}^n$, in a way that forces the product maps

$$\mathcal{C}^\infty(U) \otimes \mathcal{C}^\infty(V) \longrightarrow \mathcal{C}^\infty(U \times V)$$

to be isomorphisms. One does that by observing that the functor

$$\underline{\mathcal{C}}^\infty(U) : \mathbb{R}^m \supset V \mapsto \mathcal{C}^\infty(U, V)$$

of smooth functions on U with values in open sets of arbitrary real affine spaces commutes with transversal fiber products of open sets, i.e., satisfies

$$\underline{\mathcal{C}}^\infty\left(U, V \overset{t}{\underset{W}{\times}} V'\right) \cong \underline{\mathcal{C}}^\infty(U, V) \underset{\underline{\mathcal{C}}^\infty(U,W)}{\times} \underline{\mathcal{C}}^\infty(U, V').$$

One then defines a more general smooth algebra to be a set-valued functor

$$\mathbb{R}^m \supset V \mapsto A(V)$$

satisfying the above condition of commutation with transversal fiber products. This process of generalizing spaces by generalizing functions is called

functional geometry.

When conveniently combined with parametrized geometry, it furnishes a workable definition of a vector field on a space of fields and, more generally, of differential calculus on parametrized spaces.

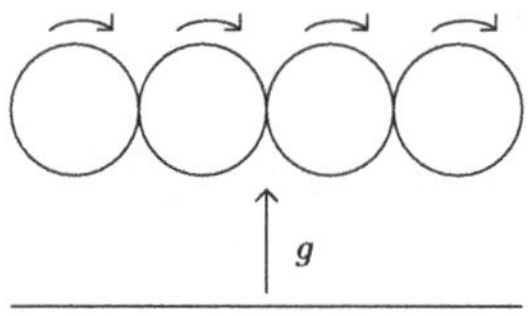

Fig. 1.10 Gauge symmetries with values in S^1

1.5 Motion with Symmetries and Homotopical Geometry

Before further discussing the quantization problem, we have to deal with systems that have some given gauge symmetries, which cause the Cauchy problem for the equations of motion to be ill-posed. To obtain a well defined notion of quantization for such a system, one needs a canonical way of breaking the symmetry, in order to retrieve a well-posed Cauchy problem with nice initial conditions. This method is usually called the gauge fixing procedure.

Here is how things appear in Yang-Mills gauge theory, which is at the heart of modern particle physics. Let M be a manifold (such as the affine space $\mathbb{R}^2$ we used before) equipped with a metric, and let G be a compact Lie group with Lie algebra $\mathfrak{g}$, equipped with an invariant pairing. Let P be a principal G-bundle over M. It is given by a (locally trivial) fiber bundle $P \to M$ together with an action of the group bundle $G_M := G \times M$ that is simply transitive, giving locally an isomorphism $G_M \cong P$.

A principal G-connection on P is morally given by a G-equivariant parallel transport of sections of P along infinitesimally closed points in M. This may also be formalized as a G-invariant differential 1-form

$$A \in \Omega^1(P, \mathfrak{g})^G$$

on P satisfying an additional non-degeneracy condition. The corresponding action functional is then given by

$$S(A) := \int_M \frac{1}{4} \langle F_A \wedge *F_A \rangle,$$

where

$$F_A := dA + [A \wedge A] \in \Omega^2(P, \mathfrak{g})^G$$

is the curvature of the connection and $*$ is the Hodge $*$-operator on differential forms associated to the given metric g on M. The solutions of the corresponding equations of motion on a compact manifold M contain flat connections.

Let us consider the situation $M = \mathbb{R}$ and P a principal S^1-bundle. The gauge symmetries, pictured in Fig. 1.10, are then given by functions $g : M \to S^1$ and a principal connection A on the trivial bundle $P = S^1 \times \mathbb{R}$ is given by a parallel transport along $\mathbb{R}$ for points in the circle fibers S^1 of P, as described in Fig. 1.11.

There is no reason to choose a particular principal bundle P, and one may also think of the action functional as a function $S(P, A)$ on isomorphism classes of principal bundles with connection. Homotopical geometry is a setting where one may

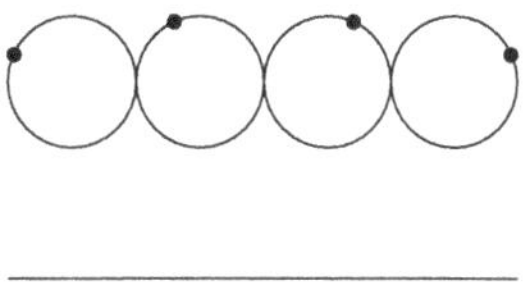

Fig. 1.11 Parallel transport along a principal connection

naturally interpret the bundle P as a field, i.e., a function

$$P : M \longrightarrow C = BG,$$

where BG is the classifying space of principal G-bundles. Bundles with connections also have a classifying space BG_{conn}, so that pairs (P, A) composed of a principal bundle and a principal G-connection may also be regarded as fields

$$(P, A) : M \longrightarrow C = BG_{conn},$$

with values in this classifying space. It is not possible to define spaces such as BG in the classical parametrized setting: one must enhance the sets of parametrized points to groupoids. Moreover, if we want to work with higher principal bundles (also called gerbes), associated to cohomology classes in $H^n(X, G)$ (where G is now assumed to be a commutative Lie group, like S^1 above), we also need to work with higher groupoids. This general enhancement of equivariant geometry under group actions is given by the setting of homotopical geometry. The space of fields of Yang-Mills theory is given by the space $\underline{\mathrm{Hom}}(M, BG_{conn})$ whose points parametrized by an open subset $U \subset \mathbb{R}^n$ are given by the groupoid of isomorphism classes of pairs (P, A) parametrized by U.

The advantage of formulating gauge theory in this setting is that it allows a clear treatment of gauge symmetries, which are manifest in the above definition of Yang-Mills theory. It also gives a more geometric intuition of the obstructions to quantizing a system with symmetries, called anomalies in physics, and (differential) group and Lie algebra cohomology in mathematics.

1.6 Deformation Theory and Derived Geometry

Many problems of physics can be formulated as deformation problems. As we already explained when we discussed functional geometry, differential calculus itself can be formulated in terms of infinitesimal thickenings and infinitesimal deformations of points, corresponding to tangent vectors. We have also seen that the geometrical formulation of theories with symmetries, like Yang-Mills theory, involve the use of an equivariant geometry, called homotopical geometry. The definition of differential invariants of equivariant spaces also involves the use of derived geometrical parameters. Perturbative quantization, as we will see, is also a deformation problem from classical field theory to quantum field theory.

Derived geometry is a parametrized geometry, where one replaces the usual open parameters $U \subset \mathbb{R}^n$ by some homotopical generalizations $\widetilde{U} \to \mathbb{R}^n$, obtained

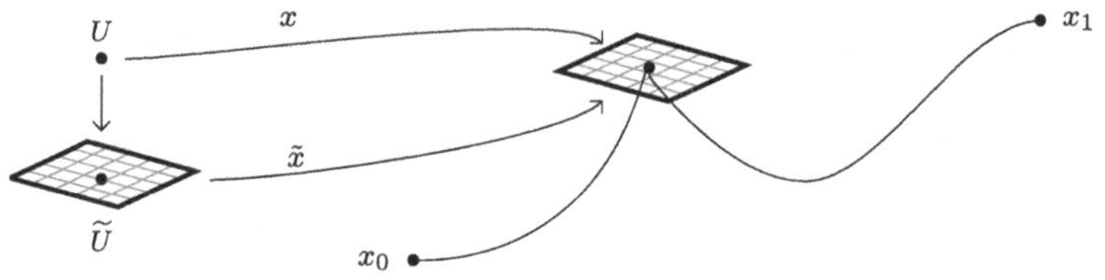

Fig. 1.12 Trajectories parametrized by homotopical coordinates

Fig. 1.13 The singular cross

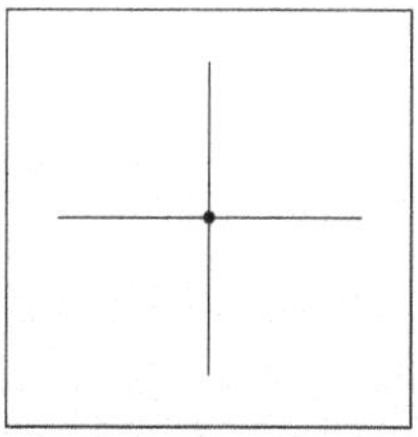

Fig. 1.14 First order infinitesimal thickening of the singular cross

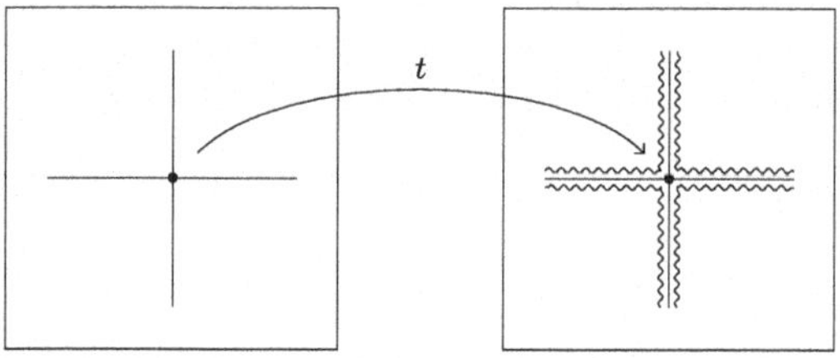

by adding to the real coordinates $u \in U$ higher homotopical coordinates $\tilde{u} \in \tilde{U}$, as pictured in Fig. 1.12. The coordinate algebra on $\tilde{U}$ is a simplicial smooth algebra $\mathcal{C}^\infty(\tilde{U})$ such that $\pi_0(\mathcal{C}^\infty(\tilde{U})) = \mathcal{C}^\infty(U)$. These parameters are used to measure obstructions to deformations, i.e., non-smoothness properties. To be precise, coordinate functions on these homotopical parameter spaces are obtained by generalizing smooth algebras (used to formalize infinitesimal calculus in functional geometry). This generalization is obtained by considering functors $\mathbb{R}^m \supset U \mapsto A(U)$ on open subsets of arbitrary real affine spaces, with values in simplicial sets, and commuting (homotopically) with transversal fiber products of open sets.

To illustrate the general relation between deformation theory and derived geometry in simple terms, let us consider the singular cross, drawn in Fig. 1.13, given by the union X of the two main axes in the plane, with polynomial equation $xy = 0$ in $\mathbb{R}[x, y]$. This equation is singular at the origin. This singularity can be explained by the fact that the coordinate $\mathbb{R}$-algebra $\mathbb{R}[x, y]/(xy)$ on X is not smooth, because the universal point $\mathrm{id} : X \to X$ of X cannot be deformed along the infinitesimal thickening $t : X \to T$, with coordinate algebra $\mathbb{R}[x, y]/(xy)^2$, pictured in Fig. 1.14. Such a deformation is defined as a factorization of the map $\mathrm{id} : X \to X$ as a sequence

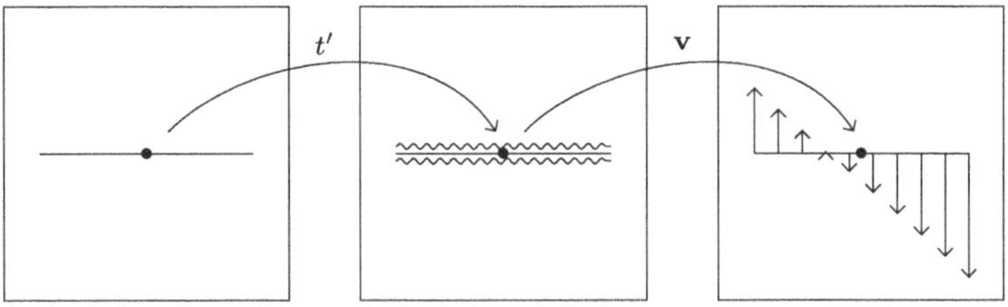

Fig. 1.15 A global vector field as an infinitesimally parametrized point

$X \xrightarrow{t} T \to X$, which would correspond to a factorization

$$\mathbb{R}[x,y]/(xy) \longrightarrow \mathbb{R}[x,y]/(xy)^2 \xrightarrow{t^*} \mathbb{R}[x,y]/(xy) \quad (\text{composite} = \mathrm{id})$$

at the level of coordinate algebras. But such a factorization does not exist. The obstruction to this existence problem lies in the cohomology group $\mathrm{Ext}^1(\mathbb{L}_X, \mathcal{I})$, where $\mathcal{I} = (xy)$ is the square zero ideal in $\mathbb{R}[x,y]/(xy)^2$, equipped with its canonical module structure over $\mathcal{O}_X = \mathbb{R}[x,y]/(xy)$, and $\mathbb{L}_X$ is a differential graded module over $\mathcal{O}_X$ called the cotangent complex.

In the case of the simpler equation $y = 0$ of the horizontal axis L, we don't have this problem, because the choice of a deformation of the universal point of L along its order two thickening $t' : L \to T'$ is given by a factorization

$$\mathbb{R}[x,y]/(y) \longrightarrow \mathbb{R}[x,y]/(y)^2 \xrightarrow{(t')^*} \mathbb{R}[x,y]/(y), \quad (\text{composite} = \mathrm{id})$$

corresponding to the choice of a global vector field on L (drawn in Fig. 1.15 with a rotation of 90 degrees), given by a morphism $\mathbf{v} : x \mapsto x + \mathbf{v}(x).y$, with $\mathbf{v}(x) \in \mathbb{R}[x]$. The cotangent complex $\mathbb{L}_L$ of this smooth space is concentrated in degree zero, and is equal there to the module of differential forms Ω^1_L. The space of deformations of the identity point $\mathrm{id} : L \to L$ along the thickening $L \to T'$ is then given by the space

$$\mathrm{Hom}_L\big(T', L\big) \cong \mathrm{Ext}^0(\mathbb{L}_L, \mathcal{O}_L)$$

of sections of the thickening map $t' : L \to T'$. So in the smooth case, one can interpret deformations of points in a given space in terms of degree zero cohomology of the cotangent complex.

The aim of derived geometry is also to interpret the space of obstruction classes $[c] \in \mathrm{Ext}^1(\mathbb{L}_X, \mathcal{I})$ to the deformation of the singular cross X as a space of infinitesimal extensions

$$\mathrm{Hom}_L(T, X) \cong \mathrm{Ext}^1(\mathbb{L}_X, \mathcal{I}_X).$$

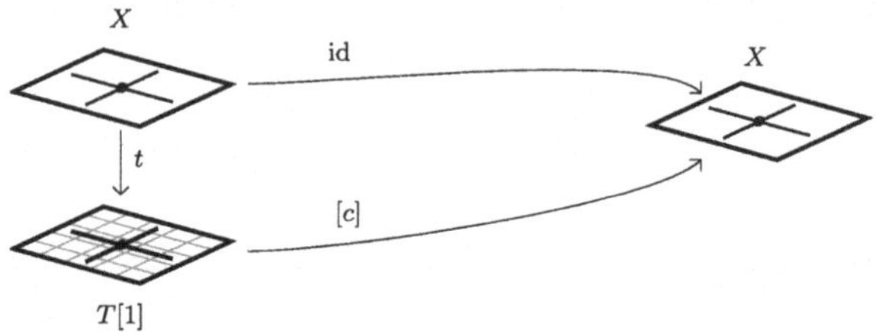

Fig. 1.16 Obstruction to deformations as homotopically parametrized points

One may also think of it as a way to define canonical formal (or more precisely, homotopical) tubular neighborhoods of singular subspaces. This can only be done by using an infinitesimal extension $t : X \to T[1] = X(\mathcal{I}[1])$, pictured in Fig. 1.16, with coordinate ring the nilpotent extension

$$\mathrm{Sym}_{\mathcal{O}_X}(\mathcal{I}[1]) := \mathbb{R}[x,y]/(xy) \oplus (xy)/(xy)^2[1],$$

with cohomological infinitesimal variable in a module of cohomological degree -1, and zero differential.

Another important motivation for working with derived geometrical spaces is given by the fact that many spaces considered in physics are given by non-transverse intersections of subspaces $L_1, L_2 \subset X$ in some given space X. For example, the space of trajectories for Yang-Mills theory is given by the intersection

$$\mathrm{Im}(dS) \cap H \subset T^*H$$

of the image of the differential of the action functional with the zero section inside the cotangent bundle T^*H. Morally, this intersection is non-transverse, and one would like to take a small generic perturbation of $\mathrm{Im}(dS)$ (by which we mean a kind of formal tubular neighborhood of this subspace inside T^*H) to retrieve a transverse intersection.

Let us illustrate this idea of a small generic perturbation in a simpler setting. Let us consider the intersection, in the plane, of the horizontal axis $y = 0$ and the line $y = \lambda x$ of slope $\lambda \geq 0$. If we vary $\lambda > 0$ the line rotates, as pictured in Fig. 1.17, but the intersection remains the origin, and the dimension of the intersection is zero there, because the coordinate ring on an intersection is given by the tensor product

$$\mathbb{R}[x,y]/(y, y-\lambda x) = \mathbb{R}[x,y]/(y) \otimes_{\mathbb{R}[x,y]} \mathbb{R}[x,y]/(y-\lambda x),$$

and is isomorphic to $\mathbb{R}$, i.e., to the coordinate algebra of functions on a point. We may also compute this dimension of the point by differential methods: its cotangent space is also of dimension zero. However, if λ tends to zero, the two lines coincide, and their intersection is no longer transversal. The intersection is then a line, not a point, and the intersection dimension is 1. We thus have bad permanence properties for the intersection dimension in parametrized families. But the derived tensor product

$$\mathbb{R}[x,y]/(y) \otimes^{\mathbb{L}}_{\mathbb{R}[x,y]} \mathbb{R}[x,y]/(y)$$

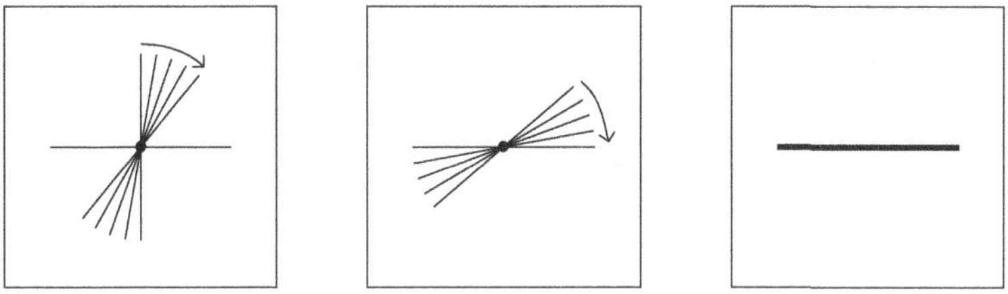

Fig. 1.17 Generic deformation of a line

may be thought of as a coordinate algebra for the derived intersection of the line with itself. This tensor product is computed by replacing one of the two sides of the tensor product by the differential graded $\mathbb{R}[x, y]$-algebra

$$\mathrm{Sym}_{dg-\mathbb{R}[x,y]}\big(\mathbb{R}[x, y][1] \xrightarrow{\cdot y} \mathbb{R}[x, y]\big)$$

that is a cofibrant resolution of the algebra $\mathbb{R}[x, y]/(y)$ (we may omit the symmetric algebra operation, because the corresponding differential graded module is already canonically a differential graded algebra). This results in a differential graded ring over $\mathbb{R}[x, y]/(y)$ that has cohomology only in degrees 0 and 1. The cotangent complex of this space, i.e., its complex of differential forms, has cohomology of dimension 1 in degree zero, and 1 in degree 1. The Euler characteristic of its cohomology thus returns a derived intersection of virtual dimension 0 at the origin. One may think of this derived intersection as a kind of generic perturbation of the non-transverse intersection, obtained by moving one of the lines back to a line of non-zero slope.

So we will keep in mind that the optimal notion of formal generic perturbation, which has good permanence properties in families, and which makes all reasonable intersections transverse, is given by the derived intersection. In the Yang-Mills setting, this corresponds to a derived mapping space

$$\mathbb{R}\underline{\mathrm{Hom}}(M, BG_{flat})$$

encoding derived principal bundles with flat connections. As mentioned earlier, its parameters are given by homotopical infinitesimal extensions $U \to \tilde{U}$ of open subsets $U \to \mathbb{R}^n$ that encode obstructions to deforming the field variables (P, A). One may show that this derived space is naturally equipped with a local symplectic structure, making it an analog of the space T^*C of initial conditions for the free particle mechanics, which has the advantage of keeping the gauge symmetries of the problem manifest.

The definition of a well-posed Cauchy problem for the equations of motion is usually given by physicists through the additional choice of a so-called Lagrangian subspace L in the (graded) derived critical space, called the gauge fixing condition.

Physicists also give another way to define a gauge fixed Cauchy problem, based on a generalization of Noether's theorem, which roughly states that the relations between the equations of motion, called Noether identities, are in one to one correspondence with local symmetries of the system. This general method, called the Batalin-Vilkovisky formalism, essentially yields, in the Yang-Mills case, the same result as the above alluded to derived critical space of the action functional on the stacky space of gauge field histories (P, A).

1.7 Fermionic Particles and Super-Geometry

One way of mathematically interpreting Dirac's quantization of the electron is to consider that this quantum particle has a classical counterpart whose time parameter anticommutes with itself. Let us first explain how one may give a mathematical description of such a strange object.

We have already roughly defined the notion of space parametrized by smooth open subsets of affine spaces. Since such an open subset $U \subset \mathbb{R}^n$ is determined, up to smooth isomorphism, by its algebra of real valued smooth functions $\mathcal{C}^\infty(U)$, we may define a more general notion of parametrized geometry by using more general types of algebras. Let us consider Grassmann algebras of the form $\mathcal{C}^\infty(U, \wedge^*\mathbb{R}^m)$ for $U \subset \mathbb{R}^n$ smooth. These associative algebras are not commutative but fulfill a super-commutation rule, given by

$$fg = (-1)^{|f|\cdot|g|} gf,$$

where $|f|$ and $|g|$ denote the degree in the anticommuting variable. We will think of these algebras as coordinates $\mathcal{C}^\infty(U^{n|m})$ on some spaces $U^{n|m}$, these being parameters for a new type of geometry called parametrized super-geometry.

This geometry generalizes parametrized smooth geometry in the following sense: we may extend the parametrized spaces $M = [0, 1]$, $C = \mathbb{R}^2$ and H to the category of super-opens $U^{n|m}$ by setting, for example

$$\mathbb{R}^2(U^{n|m}) = \mathbb{R}^{2|0}(U^{n|m}) := \mathcal{C}^\infty(U, \wedge^{2*}\mathbb{R}^m)^2,$$

and more generally

$$M(U^{n|m}) := \mathrm{Hom}_{\mathrm{ALG}}(\mathcal{C}^\infty(M), \mathcal{C}^\infty(U^{n|m})).$$

We shall not explain here why the above extensions are natural, but will illustrate these new spaces by describing the classical counterpart of the fermionic particle.

The parameter space for the electronic particle is the odd line $M = \mathbb{R}^{0|1}$, with coordinate functions the polynomial algebra $\mathbb{R}[\theta]$ in an anticommuting parameter. So we consider the space of maps

$$x : \mathbb{R}^{0|1} \to C,$$

where $C = \mathbb{R}^2$ is the plane, as before. The natural inclusion $\{*\} \subset \mathbb{R}^{0|1}$ of the point $\{*\} = \mathbb{R}^{0|0}$ into the odd line induces a bijection of sets

$$\mathrm{Hom}(\mathbb{R}^{0|1}, C) \xrightarrow{\sim} \mathrm{Hom}(\{*\}, C) = C,$$

so that the electronic particle looks like an inert object, because its trajectories do not depend on the parameter θ.

However, working with the parametrized super-space $\underline{\mathrm{Hom}}(\mathbb{R}^{0|1}, C)$ of maps from $\mathbb{R}^{0|1}$ to C, whose points parametrized by $U^{n|m}$ are defined by

$$\underline{\mathrm{Hom}}\big(\mathbb{R}^{0|1}, C\big)\big(U^{n|m}\big) := \mathrm{Hom}\big(\mathbb{R}^{0|1} \times U^{n|m}, C\big),$$

one may easily prove that this space is a non-trivial super-space, with algebra of coordinates the superalgebra

$$\Omega^*(C) = \mathcal{C}^\infty\big(\mathbb{R}^2, \wedge^*\big(\mathbb{R}^2\big)^*\big)$$

of differential forms on $C = \mathbb{R}^2$. This superspace is called the odd tangent bundle of C, and denoted by $T[1]C$. The given fixed metric on C identifies it with the odd cotangent bundle of C, denoted by $T^*[1]C$, whose algebra of coordinates is the algebra of multivector fields. We will also think of it as the space of initial conditions for the equations of motion.

1.8 Quantization of the Free Classical and Fermionic Particles

For the two classical systems described above let us now define a simplified notion of quantization, that is very close to the one originally used by Dirac.

Recall that we have defined two kinds of spaces of trajectories in $C = \mathbb{R}^2$: the system of the classical particle is given by maps $x : [0, 1] \to C$, with space of initial conditions for the equations of motion given by T^*C, and the fermionic particle is given by maps $x : \mathbb{R}^{0|1} \to C$, with space of initial conditions for the equations of motion given by the odd cotangent bundle $T^*[1]C$. Both spaces are equipped with a symplectic structure, allowing us to give a static formulation of the mechanics of the system, called Hamiltonian mechanics.

A canonical quantization of a system with space of initial conditions P (chosen among T^*C and $T^*[1]C$) is given by a filtered algebra (A, F^*) (meaning an algebra A with a collection of imbricated subspaces $F^0 \subset \cdots \subset F^n \subset \cdots \subset A$ compatible with its product operation), such that there is an isomorphism of the associated graded algebra

$$\mathrm{gr}^F A := \oplus_{n \geq 0} F^{n+1}/F^n \xrightarrow{\sim} \mathcal{C}^\infty(P).$$

It is easy to see that the so-called Weyl algebra of differential operators

$$\mathcal{D}_C = \left\{\sum_n a_n(t)\partial_t,\ a_n \text{ smooth on C}\right\}$$

on $C = \mathbb{R}^2$, with its filtration by the degree in ∂_t, is a canonical quantization of the algebra of functions $\mathcal{C}_p^\infty(T^*C)$ on T^*C that are polynomial in the fiber coordinate. This algebra has a natural representation on the space $\mathcal{C}^\infty(C, \mathbb{R})$ of real valued functions on C.

The canonical quantization of the algebra $\mathcal{C}^\infty(T^*[1]C)$ is given by the Clifford algebra $\mathrm{Cliff}(T^*C, g)$, where g is (opposite to) the standard metric on $\mathbb{R}^2$. This algebra has a natural representation on the odd Hilbert space $H = L^2(C, S)$, where S is the spinorial representation of the Clifford algebra, regarded as a super-vector space lying in odd degree.

One may refine the above notion of canonical quantization to obtain in addition operators $(\Delta + m^2)$ and $(\not{D} + m)$ on the above super-Hilbert spaces, which give information on the associated quantum equations of motion.

The models of modern particle physics are based on a second quantization procedure which views the above quantum operators as the classical equations of motion for some new field configurations, with new parameter space M' the original configuration space C of punctual mechanics, and new configuration spaces C' the supermanifolds $\mathbb{R}$ and S. The spaces of field configurations are thus given by parametrized spaces of maps

$$\varphi : M' \to \mathbb{R} \quad \text{and} \quad \psi : M' \to S,$$

called bosonic and fermionic fields. These constructions show that it is very helpful to have a formalism for classical and quantum field theory that allows a uniform treatment of all these examples.

1.9 Quantizing Interactions: The Functional Integral

We continue to discuss the basic example of the motion of a particle

$$x : M \to C,$$

with time parameter in $M = [0, 1]$, configuration space the affine plane $C = \mathbb{R}^2$, and interacting Lagrangian density

$$L(x, \partial_t x) := \frac{1}{2} m \|\partial_t x\|^2 + V(x).$$

Recall that the equations of motion are given by

$$m\partial_t^2 x = -\overrightarrow{\mathrm{grad}}(V)(x).$$

The Cauchy theorem for differential equations tells us that the space of solutions to the above equation is, again, isomorphic to the cotangent space T^*C which encodes the initial conditions. One may very well quantize this space as above, but the associated quantum equations of motion are not as simple as the operator $\Delta + m^2$ that we previously used: one must take into account interactions. This is where the perturbative renormalization method comes in: one will treat the theory as depending on a small (or formal) coefficient λ, called the coupling constant, that one may put in front of the potential, yielding a term $\lambda V(x)$ in the Lagrangian density.

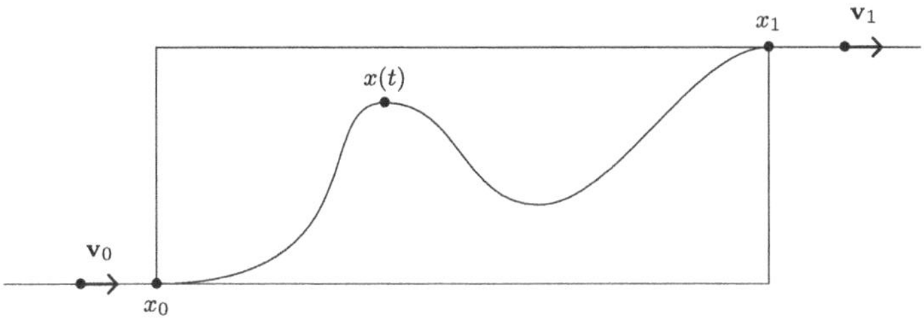

Fig. 1.18 Motion of a particle in a Newtonian potential

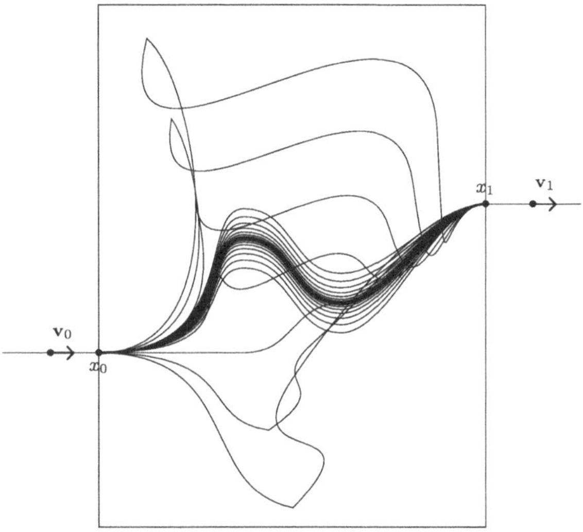

Fig. 1.19 Quantum trajectories concentrate around classical solutions

One may give a very intuitive presentation of this perturbative quantization method by using Feynman's functional integral approach to quantization. Recall from Fig. 1.18 the shape of the classical motion of a particle in a Newtonian potential V. Feynman's interpretation of a quantum process is that the quantum particle actually takes all possible paths between the incoming and the outgoing configurations, but that the probability of observing it on a given path concentrates around the classical solutions to the equations of motion, as pictured in Fig. 1.19.

Note that the widely spread idea that Feynman's functional integral does not exist is false. What is true, is that this mathematical device is not an integral in the naive sense of functional analysis, meaning a continuous positive functional

$$\int_H : \mathcal{C}^0(H) \to \mathbb{R},$$

where $H \subset \mathrm{Hom}(M, C)$ is the topological space of field histories. This would not even make any sense in the case of electrons, which are fermionic particles, because of their anti-commuting coordinates.

In all known cases, though, one may use functorial analysis to describe a space $\mathcal{O}(H)$ of (possibly formal) functionals on H together with a well defined functional

$$\langle - \rangle_H : \mathcal{O}(H) \to \mathbb{R}[[\hbar]]$$

which formally satisfies properties similar to the average value of a normalized Gaussian integration measure on H of the form

$$d\mu_S = \frac{e^{\frac{i}{\hbar}S(x)} d\mu(x)}{\int_H e^{\frac{i}{\hbar}S(x)} d\mu(x)}.$$

The definition of this integral is even harder to give if one is working with the higher dimensional parameter space $M = [0,1] \times \mathbb{R}^3$ given by Lorentzian space-time, and with configuration space, say, $C = \mathbb{R}$. Field histories are then given by real valued functions

$$\varphi : M \to \mathbb{R},$$

and, M being non-compact, one is forced to work with a very big infinite dimensional analog of a Gaussian integral.

A very simple way to develop an intuitive understanding of the renormalization problem is to formulate, following Feynman's thesis, the problem of defining the functional integral as the limit of a family of finite dimensional Lebesgue integrals, by choosing a compact box $B \subset M$, which represents the laboratory, and a lattice $\Lambda \subset M$, which represents the grid of the experimental device. The intersection $M' = B \cap \Lambda \subset M$ is a finite set, so that the mean value $\langle - \rangle_H$ is well defined on the space of restricted field configurations $\varphi_{|M'} : M' \to \mathbb{R}$ which is identified with the finite dimensional real vector space $\mathbb{R}^{M'}$. The renormalization problem then comes from the fact that, if $f \in \mathcal{O}(H)$ is a reasonable functional with finite dimensional mean value

$$\langle f \rangle_H(B, \Lambda) := \langle f \rangle_{\mathbb{R}^{M'}},$$

the limit

$$\langle f \rangle_H := \lim_{\substack{B \to M \\ \Lambda \to M}} \langle f \rangle_H(B, \Lambda)$$

does not exist when the box size increases up to M and the lattice step tends to zero. The whole point of renormalization theory is to give a refined way of taking this limit, by eliminating, inductively on the degree of the formal power series involved, all the infinities that appear in it.

1.10 The Geometry of Perturbative Quantization

The main advantage of formulating classical field theory in terms of parametrized geometry is that it allows to view the various perturbative quantization methods that

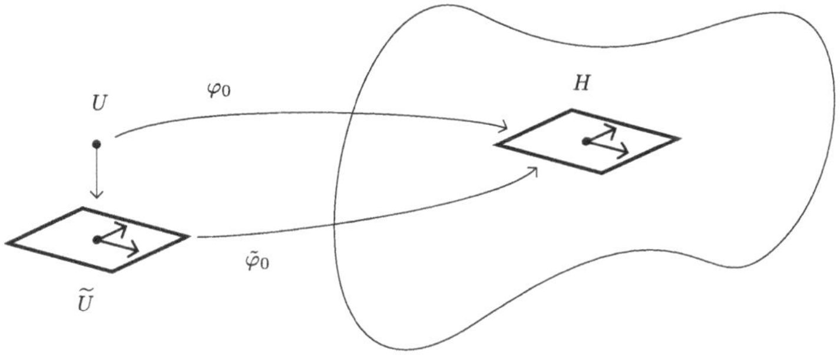

Fig. 1.20 A two dimensional infinitesimal point

appear in physics geometrically, working directly on the space of field configurations. This way of formulating quantization, which is close to DeWitt's covariant field theory, has the advantage of allowing a general treatment of theories with symmetries.

The basic idea is to consider a given solution $\varphi_0 : M \to C$ of the equations of motion, called the background field, and to work with the formal completion $\widehat{\mathcal{O}}_{\varphi_0}(H)$ of the algebra of functionals on the space H of field histories (one may also work with the derived critical space of the action functional here) around this background field. To get an idea of the type of algebra that one may obtain by this formal completion process, consider a vectorial field theory on a manifold M, with configuration space $C = E \to M$ a vector bundle and space of histories $H = \underline{\Gamma}(M, E)$ its space of sections. If φ_0 is the zero field, the formal completion $\widehat{\mathcal{O}}_{\varphi_0}(H)$ may be described explicitly by using distributional symmetric formal power series:

$$\widehat{\mathcal{O}}_{\varphi_0}(H) = \prod_{n\geq 0} \mathrm{Hom}_{\mathbb{R}}\big(\Gamma\big(M^n, E^{\boxtimes n}\big), \mathbb{R}\big)^{S_n}.$$

One may think of these formal power series as forming the algebra of functions on the restriction of the space $\underline{\Gamma}(M, E)$ to infinitesimal thickenings $\bullet \to \widetilde{T}$ of the point $\varphi_0 = 0$ in H, given by algebras of the form $C^\infty(\mathbb{R}^n)/(x_1, \ldots, x_n)^m$. A drawing for the case $n = 2$ and $m = 2$ is given in Fig. 1.20.

In the case of a theory with symmetries, e.g., Yang-Mills theory, we take the formal completion

$$\widehat{\mathrm{Crit}^h(S_{YM})}_{(P_0, A_0)}$$

of the (flat part of the) derived critical space $\mathrm{Crit}^h(S_{YM}) = \mathbb{R}\mathrm{Hom}(M, BG_{flat})$ at a given background flat gauge field (P_0, A_0). This formal completion, pictured in Fig. 1.21, is defined as the restriction of the derived critical space to the category of nilpotent homotopical thickenings $\tilde{T}$ of the point (P_0, A_0). The formal derived space is completely determined by its tangent (homotopical) local Lie algebra $\mathfrak{g}$, so that the formal completion of the functionals of the classical field theory is completely

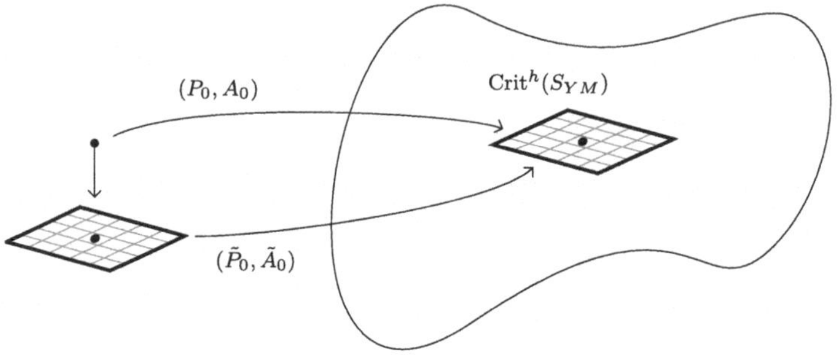

Fig. 1.21 The formal completion of the derived critical space at a given homotopical point

determined as the Chevalley-Eilenberg algebra

$$\mathcal{O}\big(\widehat{\mathrm{Crit}^h(S_{YM})_{(P_0,A_0)}}\big) = \mathrm{CE}^*(\mathfrak{g})$$

of this Lie algebra (formulated in the $\mathcal{D}$-geometrical setting, to make $\mathfrak{g}$ finite dimensional in the differential sense). There is a local (i.e., differentially structured) Poisson bracket on the derived critical space $\mathrm{Crit}^h(S_{YM})$, and this induces an invariant pairing on $\mathfrak{g}$, i.e., a Poisson structure on $\mathrm{CE}^*(\mathfrak{g})$.

One then formulates and solves various different (but essentially equivalent) renormalization problems to define a perturbative functional integral (i.e., a renormalized Gaussian average)

$$\langle - \rangle_H : \widehat{\mathcal{O}}_{\varphi_0}(H) \to \mathbb{R}[\![\hbar]\!].$$

Parametrized geometry is also the natural setting to formalize a non-perturbative renormalization problem, in terms of non-formal functionals in $\mathcal{O}(H)$.

Finally, one may give a very algebraic formulation of the perturbative quantization problem, that is a quantum analog of Lagrange's algebraic formulation of variational calculus. This involves the use of the Ran space $\mathrm{Ran}(M)$ of configurations of points in the parameter space for trajectories, which gives a geometric way of encoding formal power series of elements in $\mathcal{O}(H)$, once for all background fields $\varphi_0 : M \to C$. This space is also a convenient setting for the geometry of multi-jets, which provides a tool for the simultaneous study of Taylor series of functions at different points.

This space may be defined as the parametrized space whose points with values in an open set $U \subset \mathbb{R}^n$ are given by

$$\mathrm{Ran}(M)(U) = \big\{[S] : S \times U \to M \text{ smooth, with } S \text{ a finite set}\big\}.$$

It is morally the union of all finite powers M^S of M along all diagonal maps (and in particular, along symmetries induced by bijections of S). One may also view this space as a useful way of encoding notions that are compatible with the collapsing of families of points in M to one given point (see Fig. 1.22).

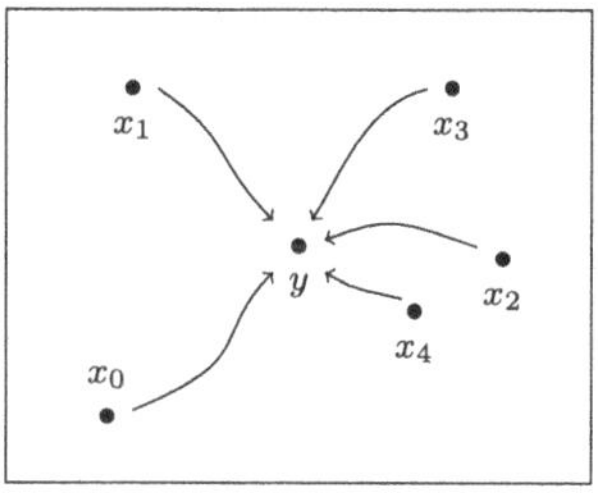

Fig. 1.22 The collapsing of a family of points in a manifold

The corresponding parametrized geometry is given by the $\mathcal{D}$-geometry of factorization spaces on the Ran space. The quantization problem may be formalized in the following way: if X is a derived $\mathcal{D}$-space over M (for example, the derived critical space $\mathrm{Crit}^h(S)$ of a given action functional, given by the Euler-Lagrange system of partial differential equations), one may associate to it a derived $\mathcal{D}$-space $X^{(Ran)}$ (by a kind of homotopical multi-jet construction) over $\mathrm{Ran}(M)$, with an additional factorization property. Its algebra of functions is called a commutative factorization algebra. One may try to deform this commutative structure to a non-commutative factorization algebra. In most cases, this is not possible in the non-derived setting, as one can see in the example of quantum group deformations: there are no such deformations in a strict set theoretical context, because of the classical Heckman-Hilton argument of algebraic topology, which says that two compatible monoid operations on a set are equal and commutative. However, if one works with a homotopical version of factorization spaces, whose algebras of functions are commutative homotopical algebras, the quantization problem is better behaved. This forces us to view the Ran space as a derived space, whose points are parametrized by homotopical smooth algebras. The most natural thing to do is to work directly with the differential graded category of modules over the space $X^{(Ran)}$, which forms a factorization category, whose deformations are called categorical factorization quantizations. This gives an optimal setting in which one can state a general quantization problem, giving a kind of quantum version of Lagrange's analytic formulation of mechanics. Note that this formulation cannot be obtained without the use of homotopical methods.

1.11 Doctrines and Theories

We now briefly describe the general setting that allowed us to treat all the constructions described in this introduction in a unified setting, basing our approach on (higher) categorical logic.

1.11.1 The Categorical Approach to Geometry

We first motivate our systematic use of categories by explaining how it lies at the basis of parametrized (see Sect. 1.1) and functional (see Sect. 1.4) geometry.

Suppose given a category LEGOS of simple building blocks for a given geometry (e.g., $U \subset \mathbb{R}^n$ smooth open subset). Usually, this category has the following drawbacks:

- It doesn't have enough limits (e.g., spaces of solutions to equations, fiber products), and not enough colimits (e.g., quotients).
- It doesn't contain "spaces of maps", i.e., functional spaces $\underline{\mathrm{Hom}}(U, V)$.

The solution to this problem chosen by Grothendieck, Lawvere and Ehresmann is to use the Yoneda embeddings

$$\mathrm{LEGOS} \hookrightarrow \underline{\mathrm{Hom}}\big(\mathrm{LEGOS}^{op}, \mathrm{SETS}\big), \qquad \mathrm{LEGOS}^{op} \hookrightarrow \underline{\mathrm{Hom}}(\mathrm{LEGOS}, \mathrm{SETS})$$

and to restrict the class of functors via the condition of commutation to limits (e.g., sheaves in parametrized geometry, sketches in functional geometry).

If LEGOS is the category of smooth open subsets $U \subset \mathbb{R}^n$, we may view

1. *functional geometry* as the definition of generalized algebras, regarded as *functors of functions*
$$A : \mathrm{LEGOS} \to \mathrm{SETS}$$
that commute with prescribed limits (e.g., transversal pullbacks), e.g.
$$\underline{\mathcal{C}}^{\infty}(M) : U \mapsto \mathcal{C}^{\infty}(M, U).$$
2. *parametrized geometry* as the definition of generalized spaces, regarded as *functors of points*
$$X : \mathrm{LEGOS}^{op} \to \mathrm{SETS}$$
that commute with prescribed limits (nerves of coverings), e.g.,
$$\underline{M} : U \mapsto \mathcal{C}^{\infty}(U, M).$$

One then has, by definition, the right tensor product for generalized algebras

$$\underline{\mathcal{C}}^{\infty}(U \times V) \cong \underline{\mathcal{C}}^{\infty}(U) \otimes \underline{\mathcal{C}}^{\infty}(V),$$

and the right pasting for generalized spaces

$$\underline{U \coprod_{U \cap V} V} \cong \underline{U} \coprod_{\underline{U \cap V}} \underline{V}.$$

The main interest of this categorical approach is that it generalizes easily to the setting of higher categories, which we have chosen to express obstruction theoretic statements using generalized universal properties. The category SETS is the category $0-\text{CAT}$ of 0-categories, which one can replace by the ∞-category ${}^{\infty}\text{GRPD}$ of ∞-groupoids (for homotopical geometrical generalizations) or by other higher categories, e.g., the 2-category CAT of (1-)categories. This allows a smooth treatment of obstruction theories (homology, homotopy). It also allows us to give natural definition of higher categorical algebraic objects (e.g., monoidal categories).

1.11.2 Higher Categories and Categorical Logic

Suppose given a notion of n-category for $n \geq 0$. All theories, types of theories and their models (also called semantics) that will be considered in this book can be classified, in a coordinate-free fashion, by the following definition of a doctrine.

Definition 1.11.1 *Let $n \geq 1$ be an integer. A* doctrine *is an $(n+1)$-category $\mathcal{D}$. A* theory *of type $\mathcal{D}$ is an object $\mathcal{C}$ of $\mathcal{D}$. A* model *for a theory of type $\mathcal{D}$ in another theory of the same type is an object*

$$M : \mathcal{C}_1 \to \mathcal{C}_2$$

of the n-category $\underline{\text{Mor}}(\mathcal{C}_1, \mathcal{C}_2)$.

The coordinate definition of a given theory is usually called its *syntax*, and the study of its models is called its *semantics*. We will often describe theories using their syntax, but work with their semantics in a coordinate-free, higher categorical fashion.

Note that the doctrines we use are usually given by $(n+1)$-categories of n-categories with additional *structures* and *properties*. One may define additional structures via a (possibly inductive) sketch-like construction (see Chaps. 2 and 3), and additional properties via the existence of representing objects for some functors. We decided to use the above general identification between higher category theory and categorical logic in order to obtain the optimal pedagogical presentation, but the proper formulation of a categorical analog of the axiom of choice (an existence axiom for models of certain theories), called the *doctrinal axiom of choice* would necessarily involve a refined language, necessary to define the finiteness notion of the *coherence* of a theory. However, as this book is applications-oriented, we didn't want to burry its results under six feet of formalism.[1]

The types of doctrines used in our presentation of quantum field theory are given, for example, by:

- The 2-category of categories with finite products (or more generally finite limits) with product preserving functors between them and natural transformations, whose theories are called algebraic theories. These will be used to formalize infinitesimals in smooth differential calculus.

[1] Did we?

- The 2-category of (nerves of) sites (categories with Grothendieck topology), whose models are called sheaves, or spaces. These will be used to formalize differential geometry on spaces of fields.
- The 2-category of monoidal (resp. symmetric monoidal, resp. multi-) categories with monoidal functors and monoidal natural transformations between them, whose theories are called PROs (resp. PROPs, resp. multicategories). These will be used to formalize fermionic differential calculus, local functional calculus and homotopical Poisson reduction of general gauge theories.
- The 2-category of Quillen's model categories with Quillen adjunctions between them. These will be used to formalize geometrically homotopical Poisson reduction of general gauge theories and higher gauge field theories.
- The ${}^{\infty}2$-category of ∞-categories, which are useful to formalize deformation quantization of gauge theories; the difficulties of passing from ${}^{\infty}1$- to ${}^{\infty}2$-categories are not greater than those that have to be overcome to pass to ${}^{\infty}n$-categories, and the general theory is very useful for making dévissages and classifying theories in the spirit of the Baez and Dolan's categorification program.
- We will also encounter n-doctrines related to topological quantum field theories.
- There is no good reason to restrict to doctrines for $n = 1$, because we may also be interested in studying the relations between doctrines, their geometries, and so on. These may only be studied in higher doctrines.

1.12 What's New in this Book?

We now give a (non-exhaustive) list of some original tools, ideas and results that are developed in this book.

We may roughly say that the bundle composed of Chaps. 2, 3, 4, 12, 13, 20, 22 and 24 (see also Sect. 9.11) contains substantial contributions of the author to the mathematical formalization of classical and quantum field theory. The other chapters contain results that are gathered in the mathematical and physical literature, but which were often in a form that made them inaccessible to graduate students in mathematics, so that our presentation is at least pedagogically original. An important aspect of the pedagogical and bibliographical content of this work was the cautious choice (and the concise but exhaustive description) of the optimal mathematical tools required for the presentation of our results.

Here is a more precise description of some of our main scientific contributions:

- We describe in Sect. 2.1 the main tools of higher category theory in a purely abstract setting, independent of the model chosen to describe higher categories. This allows us to shorten the usual presentation of these tools, and to concentrate on their applications. Some of the notions we introduced here are not yet in the literature.
- To give a conceptual presentation of the various structures used in this book, we introduced in Sect. 1.2 the general doctrinal techniques for higher categorical logic, and developed a systematical categorification process, called the doctrine

machine (see Definition 9.11.2), which automatically gives the right definitions for both higher categorical structures and algebraic and geometric structures, in both classical (see Chap. 3) and homotopical (see Chap. 10) settings.

- In particular, our definition of monoidal and symmetric monoidal higher categories in Sects. 2.2 and 2.3 is obtained by directly generalizing the categorical logical interpretation of monoids and commutative monoids. This is equivalent to other approaches (in particular, to those of Mac Lane and Segal), but looks much more natural. It also has the advantage of having a direct generalization to the higher categorification of other sketch-like theories (e.g., finitely presented or finite limit theories).
- The idea of parametrized and functional geometry (see Chap. 3), was already present in the literature, for example in the work of Grothendieck and Souriau's schools, and in the synthetic geometry community, grounded by Lawvere's categorical approach to dynamics. However, its systematic use in the formalization of the known physical approaches to quantum field theory is new. Our constructions are directly adapted to spaces with boundary and corners (see Example 3.2.4), in contrast to the above references, which mostly work with closed manifolds. We also treat in the same setting super-symmetric field theories (see Sect. 3.3.3), and the geometry of non-algebraic partial differential equations (see Sect. 12.5).
- The proper generalization of classical functional analysis to parametrized and functional geometry, called functorial analysis (see Chap. 4), is also a new idea that is a particularly fruitful way of combining homotopical and super-manifold methods with more classical methods of analysis. There is not really another way to solve this important problem of the formalization of (non-perturbative) quantum field theory computations.
- The systematic definition of differential invariants using categorical methods, grounded in Quillen's work, has been extended in our work (see Sect. 2.5). By adopting this approach, we don't have to repeat the definition of differential invariants in the various geometrical settings we use. The application of these methods to define jet spaces with boundaries and jet spaces in smooth super-geometry seems to be new.
- We have introduced the notion of monoidally enriched differential geometry (see Sect. 3.3.3) to cover various types of functional geometries present in this book, such as super-geometry, graded geometry and relatively algebraic $\mathcal{D}$-geometry.
- We have developed the new notion of a smooth algebra with corners (see Example 3.2.4), that gives, via the associated differential and jet calculus, a natural setting for the study of non-topological higher dimensional quantum field theories with boundaries. This also opens the door to interesting research directions in the differential geometry of higher and derived stacks with corners, and to their factorization quantizations (see Chap. 24).
- We have given a non-abelian cohomological interpretation of Cartan's formalism in general relativity (see Sects. 7.4 and 14.4.2). This opens the possibility to study global properties of moduli stacks of Cartan geometries.
- We have formalized supersymmetric local differential calculus purely in terms of $\mathcal{D}$-modules (see Chap. 12), which gives a better way to interpret its analogy with classical local differential calculus.

- Our formalization of $\mathcal{D}$-geometry over the Ran space (see Sects. 12.3, 12.5, and 24.4) in smooth and analytic geometry is also new.
- Our formalization of the classical Batalin-Vilkovisky formalism and gauge fixing procedure (see Chap. 13), presented in a coordinate-free language, which also works for supersymmetric theories, with optimal finiteness hypotheses, is also a new mathematical result. This part of the theory, which forms the technical heart of our contributions, expands and improves on the two publications [Pau11a] and [Pau11b], by adding a new and precise geometric formalization of the gauge fixing procedure for local field theories.
- The mathematical formalization of non-perturbative quantum field theory (see Chap. 20) using functorial analysis and the Wilson-Wetterich non-perturbative renormalization group, compatible with our formalization of gauge theories, is also new.
- Our formalization of the Epstein-Glaser causal renormalization method (see Chap. 22) in terms of functional geometry and functorial analysis improves on the purely functional analytic approach used in the literature, by giving a clear meaning to the notion of a space of graded or differential graded fields adapted to this method.
- Our general formalization of the categorical quantization problem for factorization spaces in Sect. 24.6 is also new.
- The use of the analytic derived Ran space, and of microlocal algebraic analysis in the geometric study of factorization algebras (see Remark 24.2.3), is also an original contribution which opens the door to a causal treatment, à la Epstein-Glaser, of the quantization problem for factorization algebras on Lorentzian spacetimes, which is an interesting issue of the theory, already studied in the physics literature in particular cases, for example by Hollands and Wald.

1.13 What's Not in this Book?

First, we have omitted a lot of the many applications of quantum field theoretic tools in mathematics.

Moreover, this books essentially treats the formalization of

quantitative aspects

of quantum field theories, the main aim being to give mathematicians direct access to the standard physics textbooks in this subject.

A very interesting mathematical research direction that we have not treated at all in this book is the study of

qualitative aspects

of quantum field theories, that one may putatively call the theory of quantum dynamical systems.

The starting point for the theory of dynamical systems may be said to be Poincaré's paper on the three body problem: he transformed a mistake into interesting mathematical developments. The entrance into the theory of differential dynamical systems that is the closest to the spirit of our book would be Thom's transversality theory (grounded in [Tho56]), expanded to multijets by Mather [Mat70] (see also [Cha01] for a presentation of the subject and [GG73], Sect. II.4 for proofs). Note that these transversality methods are based on codimension considerations for subspaces of infinite dimensional spaces of functions (called spaces of fields in our book) defined by partial differential equations. The better permanence properties of transversality in derived geometrical spaces may give new tools to mathematicians to refine their knowledge of classical transversality methods in dynamics.

The qualitative study of non-perturbative quantum field theory, through the Wilson-Wetterich non-perturbative (also sometimes called functional) renormalization group method, is another interesting open mathematical problem that has, to the author's knowledge, not yet been studied in any depth by pure mathematicians.

Our precise mathematical formalization of the theory via the methods of functorial analysis, which complete, but are strictly compatible with the functional analytic methods, may be useful to interested researchers.

References

[Cha01] Chaperon, M.: Jets, transersalité, singularités:petite introduction aux grandes idées de René Thom (2001, preprint)

[GG73] Golubitsky, M., Guillemin, V.: Stable Mappings and Their Singularities. Graduate Texts in Mathematics, vol. 14, p. 209. Springer, New York (1973)

[Mat70] Mather, J.N.: Stability of C^∞ mappings. V. Transversality. Adv. Math. **4**, 301–336 (1970)

[Pau11a] Paugam, F.: Histories and observables in covariant field theory. J. Geom. Phys. **61**(9), 1675–1702 (2011). doi:10.1016/j.geomphys.2010.11.002

[Pau11b] Paugam, F.: Homotopical Poisson reduction of gauge theories. In: Mathematical Foundations of Quantum Field Theory and Perturbative String Theory. Proc. Sympos. Pure Math., vol. 83, pp. 131–158. Am. Math. Soc., Providence (2011)

[Tho56] Thom, R.: Un lemme sur les applications différentiables. Bol. Soc. Mat. Mexicana **1**, 59–71 (1956)

Part I
Mathematical Preliminaries

Chapter 2
A Categorical Toolbox

The reader can skip this chapter at first reading and refer to it when needed. We give here a very short account of the tools from (higher) category theory that will be used throughout this book. Some of these are not so standard outside of the category theory community, so we shall record them for later reference in other chapters. Mac Lane's book [ML98], Grothendieck's seminar [AGV73] and Kashiwara-Schapira's book [KS06] are also excellent complementary references for classical categorical results. We refer to Chap. 9 for a quick presentation of the tools of higher category theory used in our book, with a documented reference list. More information may be found in Baez and Dolan's groundbreaking article [BD98], in Simpson's textbook [Sim10], in Cheng and Lauda's introduction [CL04], in Lurie's reference books [Lur09c] and [Lur09d], and in Joyal's book [Joy11] (see also [Rez09, Mal10] and [Ara12]).

In this chapter, we work with a broad notion of *sets* (also called *classes*). This leads to problems only when one works with limits indexed by very big classes. We will use standard techniques, when needed, to overcome this problem. The reader interested in technical details can refer to the theory of Grothendieck universes [AGV73].

2.1 Higher Categories, Doctrines and Theories

In this section, by quite directly generalizing the usual categorical methods, we describe the basic constructions of higher category theory which will be used throughout this book. Readers who are only interested in 1-categories should simply replace n by 1 in the following.

Definition 2.1.1 A category $\mathcal{C}$ is given by the following data:

1. a class $\mathrm{Ob}(\mathcal{C})$ called the *objects* of $\mathcal{C}$;
2. for each pair of objects X, Y, a class $\mathrm{Hom}(X, Y)$ called the class of *morphisms*;
3. for each object X a morphism $\mathrm{id}_X \in \mathrm{Hom}(X, X)$ called the *identity morphism*;

F. Paugam, *Towards the Mathematics of Quantum Field Theory*,
Ergebnisse der Mathematik und ihrer Grenzgebiete. 3. Folge / A Series of Modern Surveys in Mathematics 59, DOI 10.1007/978-3-319-04564-1_2,

4. for each triple of objects X, Y, Z, a composition law for morphisms

$$\circ : \mathrm{Hom}(X, Y) \times \mathrm{Hom}(Y, Z) \to \mathrm{Hom}(X, Z).$$

One assumes moreover that this composition law is associative, i.e., $f \circ (g \circ h) = (f \circ g) \circ h$ and that the identity is a unit, i.e., $f \circ \mathrm{id} = f$ and $\mathrm{id} \circ f = f$.

Example 2.1.1 Here are some well known examples of categories:

1. SETS whose objects are sets and morphisms are maps.
2. GRP whose objects are groups and morphisms are group morphisms.
3. TOP whose objects are topological spaces and morphisms are continuous maps.

Definition 2.1.2 A *functor* $F : \mathcal{C}_1 \to \mathcal{C}_2$ between two categories is given by the following data:

1. for each object X of $\mathcal{C}_1$, an object $F(X)$ of $\mathcal{C}_2$;
2. for each morphism $f : X \to Y$ of $\mathcal{C}_1$, a morphism $F(f) : F(X) \to F(Y)$ of $\mathcal{C}_2$.

One assumes moreover that F is compatible with identities and composition, i.e.,

$$F(\mathrm{id}) = \mathrm{id} \quad \text{and} \quad F(f \circ g) = F(f) \circ F(g).$$

A *natural transformation* $\Phi : F \Rightarrow G$ between two functors $F, G : \mathcal{C}_1 \to \mathcal{C}_2$ is given by the datum, for every object X of $\mathcal{C}_1$, of a morphism $\Phi_X : F(X) \to G(X)$, such that for every morphism $f : X \to Y$ in $\mathcal{C}_1$, the following diagram commutes:

$$\begin{array}{ccc} F(X) & \xrightarrow{\Phi_X} & G(X) \\ {\scriptstyle F(f)}\downarrow & & \downarrow{\scriptstyle G(f)} \\ F(Y) & \xrightarrow[\Phi_Y]{} & G(Y) \end{array}$$

Example 2.1.2 Important examples of functors are given by forgetful functors, such as the functor GRP $\to$ SETS that sends a group to the underlying set, or the functor TOP $\to$ SETS that sends a topological space to the underlying set. One may also cite the free group functor SETS $\to$ GRP.

We now give a simple definition of the notion of n-category, following Simpson's book [Sim10]. One can find a finer axiomatization in [Sim01] and [BS11]. We will give a precise definition of higher categories in Sect. 9.10.

Definition 2.1.3 A theory of n-categories is assumed to have a notion of sum and product. A 0-category is a set. An n-*category* $\mathcal{A}$ is given by the following data:

1. (OB) a set $\mathrm{Ob}(\mathcal{A})$ of *objects*, compatible with sums and products of n-categories.

2. (MOR) for each pair (x, y) of objects, an $(n-1)$-category $\underline{\mathrm{Mor}}(x, y)$. One defines

$$\underline{\mathrm{Mor}}(\mathcal{A}) := \coprod_{x,y\in \mathrm{Ob}(\mathcal{A})} \underline{\mathrm{Mor}}(x, y)$$

and by induction i-morphisms in a higher category $\underline{\mathrm{Mor}}^i(\mathcal{A}) := \underline{\mathrm{Mor}}(\underline{\mathrm{Mor}}(\cdots(\mathcal{A})))$ for $0 \le i \le n$. One has $\underline{\mathrm{Mor}}^0(\mathcal{A}) := \mathcal{A}$ and $\underline{\mathrm{Mor}}^n(\mathcal{A})$ is a set. Let $\mathrm{Mor}(\mathcal{A}) = \mathrm{Ob}(\underline{\mathrm{Mor}}(\mathcal{A}))$. By construction, one has source and target maps

$$s_i, t_i : \mathrm{Mor}^i(\mathcal{A}) \to \mathrm{Mor}^{i-1}(\mathcal{A})$$

that satisfy

$$s_i s_{i+1} = s_i t_{i+1}, \qquad t_i s_{i+1} = t_i t_{i+1}.$$

3. (ID) for each $x \in \mathrm{Ob}(\mathcal{A})$, there should be a natural element $1_x \in \mathrm{Mor}(x, x)$. This gives morphisms

$$e_i : \mathrm{Mor}^i(\mathcal{A}) \to \mathrm{Mor}^{i+1}(\mathcal{A})$$

such that $s_{i+1}e_i(u) = u$ and $t_{i+1}e_i(u) = u$.
4. (EQUIV) on each set $\mathrm{Mor}^i(\mathcal{A})$, there is an equivalence relation $\sim$ compatible with the source and target maps. The induced equivalence relation on $\mathrm{Mor}^i(x, y)$ is also denoted by $\sim$.
5. (COMP) for any $0 < i \le n$ and any three $i-1$-morphisms u, v and w sharing the same source and target, there is a well-defined composition map

$$\left(\mathrm{Mor}^i(u, v)/\sim\right) \times \left(\mathrm{Mor}^i(v, w)/\sim\right) \to \left(\mathrm{Mor}^i(u, w)/\sim\right)$$

which is associative and has the classes of identity morphisms as left and right units.
6. (EQC) Equivalence and composition are compatible: if $0 \le i < n$ and $u, v \in \mathrm{Mor}^i(u, v)$ share the same source and target, then $u \sim v$ if and only if there exists an $f \in \mathrm{Mor}^{i+1}(u, v)$ and $g \in \mathrm{Mor}^{i+1}(v, u)$ such that $f \circ g \sim 1_u$ and $g \circ f \sim 1_v$. This allows one to define the category $\tau_{\le 1}\underline{\mathrm{Mor}}^i(u, v)$ with objects $\mathrm{Mor}^i(u, v)$ and morphisms between $w, z \in \mathrm{Mor}^i(u, v)$ given by equivalence classes in $\mathrm{Mor}^{i+1}(w, z)/\sim$.

An n-category is called *strict* if its composition laws all lift on morphisms (meaning that all compositions maps are well defined and strictly associative). We will also sometimes call an n-category *weak* if it is not strict.

We will give in Sect. 9.10 a sketch of a proof of the following theorem.

Theorem 2.1.1 *In the ZFC axiomatic formulation of set theory, there exists a theory of n-categories, for all n. The class of n-categories forms an $(n+1)$-category n*CAT.

Proof We refer to Simpson's book [Sim10], Rezk's paper [Rez09] and Ara's paper [Ara12] for three different (but essentially equivalent) constructions of a theory of n-categories. □

Simpson and Rezk actually define homotopical versions of higher categories, called ${}^{\infty}n$-categories. There is also a unicity result for these theories, due to Barwick and Schommer-Pries [BS11]. All the higher categorical definitions of this section are also adapted to this homotopical setting, without changes. The reader is invited to specialize to the case where $n = 1$ to obtain the usual notions of category theory.

The notion of n-category allows us to define a very general notion of type of structure (in the sense of categorical logic), that we call a doctrine.

Definition 2.1.4 A *doctrine* is an $(n+1)$-category $\mathcal{D}$. A *theory* of type $\mathcal{D}$ is an object $\mathcal{C}$ of $\mathcal{D}$. A *model* for a theory of type $\mathcal{D}$ in another theory of the same type is an object

$$M : \mathcal{C}_1 \to \mathcal{C}_2$$

of the n-category $\underline{\mathrm{Mor}}_{\mathcal{D}}(\mathcal{C}_1, \mathcal{C}_2)$.

The above definition is an extension of Lawvere's terminology [Law06], which was used in a special case. In the literature, a theory (or more precisely its generating data) is also called a language and a syntax and the study of its models is called its semantics. The idea of categorical logics is to define theories using a language and a syntax, but also to study their models using categorical methods, yielding tools which enable us to compare theories and treat them in a coordinate-free and invariant (or, as physicists would say, covariant) way. Relating the above general notion to the explicit presentations of doctrines used by Lawvere would involve a general notion of *structures and properties* on n-categories (e.g., of an n-monad in an n-category), which is beyond the scope of these notes, however the notion will be clear in each of the specific examples we shall describe.

Definition 2.1.5 Let $\mathcal{D}$ be a doctrine. A *presentation* of $\mathcal{D}$ is the datum of a higher doctrine $\mathcal{E}$ and of two theories $\mathcal{C}_1$ and $\mathcal{C}_2$ of type $\mathcal{D}$ such that there is an equivalence of doctrines

$$\mathcal{D} \cong \underline{\mathrm{Mor}}_{\mathcal{E}}(\mathcal{C}_1, \mathcal{C}_2)$$

identifying theories of type $\mathcal{D}$ with models of the theory $\mathcal{C}_1$ in the theory $\mathcal{C}_2$.

We now define the notion of the tensor product of theories, which provides a very useful tool enabling us to define new theories from existing theories.

Definition 2.1.6 Let $\mathcal{D} \subset n\mathrm{CAT}$ be a doctrine of higher categories with additional structures. We say that $\mathcal{D}$ has *internal homomorphisms* if, for two theories $\mathcal{C}_1$ and $\mathcal{C}_2$, the higher category $\underline{\mathrm{Mor}}(\mathcal{C}_1, \mathcal{C}_2)$ is naturally equipped with the structure of an object of $\mathcal{D}$. If $\mathcal{D}$ is a doctrine with internal homomorphisms, and $\mathcal{C}_1$ and $\mathcal{C}_2$ are two theories, their *Lawvere tensor product* is the theory $\mathcal{C}_1 \otimes \mathcal{C}_2$ such that for all theories $\mathcal{C}_3$, one has a natural equivalence

$$\underline{\mathrm{Mor}}\big(\mathcal{C}_1, \underline{\mathrm{Mor}}(\mathcal{C}_2, \mathcal{C}_3)\big) \cong \underline{\mathrm{Mor}}(\mathcal{C}_1 \otimes \mathcal{C}_2, \mathcal{C}_3).$$

One may also use ${}^{\infty}n$-categories to study the homotopy theory of doctrines and theories (structures up to homotopy), which plays an important role in physics. The corresponding objects will be called ∞-doctrines and ∞-theories. This may also be adapted to the homotopical setting. We refer to Sect. 9.11 for more details.

Many of the doctrines that we will use are given by 2-categories very close to the 2-category CAT of categories, which we now describe concretely.

Definition 2.1.7 The 2-category CAT is the strict 2-category whose:

1. objects are categories;
2. morphisms are functors;
3. 2-morphisms are natural transformations.

One of the most important tools of category theory is the notion of adjunction. We first give a 2-categorical definition of this notion.

Definition 2.1.8 Let $\mathcal{E}$ be a 2-category. An *adjoint pair* in $\mathcal{E}$ is a pair

$$F : \mathcal{D} \leftrightarrows \mathcal{C} : G$$

of morphisms, together with a pair of morphisms

$$i : 1_{\mathcal{C}} \to F \circ G \quad \text{and} \quad e : G \circ F \to 1_{\mathcal{D}},$$

satisfying the triangular equalities

$$(Gi) \circ (eG) = 1_G \quad \text{and} \quad (Fe) \circ (iF) = 1_F.$$

We now define a notion of adjunction between functors that is better adapted to homotopical generalizations. For $n = 1$, it is compatible with the above definition when $\mathcal{E} = \mathrm{CAT}$ is the 2-category of categories.

Definition 2.1.9 Let $\mathcal{C}$ and $\mathcal{D}$ be two n-categories. A *pair of adjoint functors* between $\mathcal{C}$ and $\mathcal{D}$ is a pair

$$F : \mathcal{D} \leftrightarrows \mathcal{C} : G$$

of functors together with a pair of morphisms of functors

$$i : 1_{\mathcal{C}} \to F \circ G \quad \text{and} \quad e : G \circ F \to 1_{\mathcal{D}},$$

such that for all objects $X \in \mathcal{C}$ and $Y \in \mathcal{D}$, the induced morphisms

$$\underline{\mathrm{Mor}}_{\mathcal{C}}(FY, X) \leftrightarrows \underline{\mathrm{Mor}}_{\mathcal{D}}(Y, GX)$$

are equivalences of $(n-1)$-categories inverse to each other.

Example 2.1.3 The pair

$$F : \text{GRP} \leftrightarrows \text{SETS} : G$$

given by the forgetful functor F and the free group functor G is an adjoint pair in CAT.

We now define the notion of left and right Kan extensions, which are useful higher categorical analogs of the notion of universal object in a category. These are useful to formulate limits and colimits by their universal properties, and to discuss localization of categories and derived functors, which are necessary tools used in the formulation of homotopical and homological algebra.

Definition 2.1.10 Let $\mathcal{D}$ be an $(n+1)$-category for $n \geq 1$, $F : C \to D$ and $f : C \to C'$ be morphisms between two objects of $\mathcal{D}$. A (local) *right Kan extension* of F along f, if it exists, is a morphism

$$\text{Ran}_f F = f_* F : C' \to D$$

equipped with a 2-morphism $a : \text{Ran}_f F \circ f \Rightarrow F$ that induces, for all $G \in \text{Mor}^1(C', D)$, an equivalence of $(n-1)$-categories

$$\begin{aligned} \underline{\text{Mor}}_{\underline{\text{Mor}}(C',D)}(G, \text{Ran}_f F) &\xrightarrow{\sim} \underline{\text{Mor}}_{\underline{\text{Mor}}(C,D)}\big(f^* G, F\big) \\ c &\mapsto a \star c, \end{aligned}$$

where $f^* : \underline{\text{Mor}}(C', D) \to \underline{\text{Mor}}(C, D)$ denotes composition by f. Similarly, one defines a left Kan extension as a morphism

$$\text{Lan}_f F = f_! F : C' \to D$$

equipped with a natural equivalence

$$\underline{\text{Mor}}_{\underline{\text{Mor}}(C,D)}\big(F, f^*(-)\big) \cong \underline{\text{Mor}}_{\underline{\text{Mor}}(C',D)}(\text{Lan}_f F, -).$$

Remark 2.1.1 The notion of Kan extension may also be defined in a weaker "almost cartesian closed" setting, where $\mathcal{D}$ is an n-category whose morphism $(n-1)$-categories $\underline{\text{Mor}}_{\mathcal{D}}(C, D)$ naturally form n-categories. This will be useful when simplifying the treatment of ${}^{\infty}1$-Kan extensions, avoiding having to pass through the setting of ${}^{\infty}2$-categories, by using the fact that functors between two ${}^{\infty}1$-categories also form an ${}^{\infty}1$-category. This method is used by Lurie in [Lur09d], in the setting of quasi-categories.

More concretely, a right Kan extension is a pair $(\mathrm{Ran}_f F, a)$ composed of a morphism $\mathrm{Ran}_f F : C' \to D$ and a 2-morphism $a : \mathrm{Ran}_f F \circ f \Rightarrow F$ giving a diagram

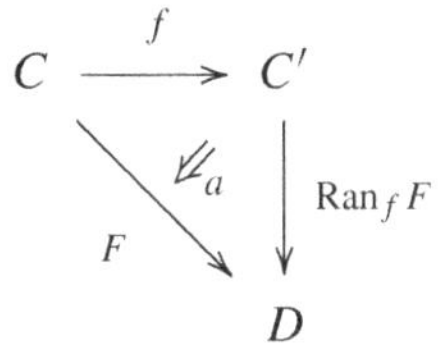

such that for every pair (G, b) composed of a morphism $G : C' \to D$ and of a 2-morphism $b : G \circ f \Rightarrow F$, there exists a 2-morphism $c : G \Rightarrow \mathrm{Ran}_f F$, such that the equivalence $b \sim a \circ c$, illustrated by the following diagrams

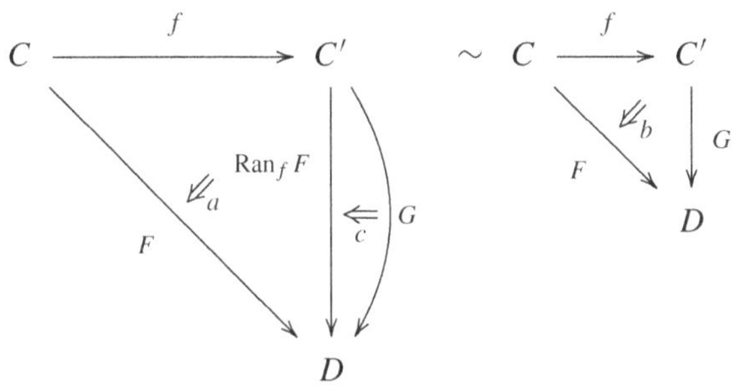

is true up to a coherent system of higher equivalences.

We now define the notion of limits and colimits in n-categories, which will allow us to define interesting doctrines.

Definition 2.1.11 Let $\mathcal{D}$ be an $(n+1)$-category for $n \geq 1$, with a terminal object $\mathbb{1}$, defined by the fact that for all object $\mathcal{C}$ of $\mathcal{D}$, one has

$$\underline{\mathrm{Mor}}(\mathcal{C}, \mathbb{1}) \cong \mathbb{1}_{n\mathrm{CAT}},$$

where $\mathbb{1}_{n\mathrm{CAT}}$ is the n-category with only one k-morphism for all k. Let I and $\mathcal{C}$ be two objects of $\mathcal{D}$. A *diagram* in $\mathcal{C}$ is a morphism $F : I \to \mathcal{C}$. Let $f = 1 : I \to \mathbb{1}$ be the terminal morphism in $\mathcal{D}$. The *limit* of F is defined as the right Kan extension

$$\lim F := \mathrm{Ran}_1 F$$

and the *colimit* of F is defined as the left Kan extension

$$\mathrm{colim}\, F := \mathrm{Lan}_1 F.$$

Here is a more concrete description of this notion, in the case where $\mathcal{D} = n\mathrm{CAT}$ is the $(n+1)$-category of n-categories. Given a diagram in $\mathcal{C}$ and an object N of $\mathcal{C}$, we denote by $c_N : I \to \mathcal{C}$ the constant functor given by $c_N(X) = N$. The functor $N \mapsto c_N$ is identified with the composition

$$1^* : \underline{\mathrm{Mor}}(\mathbb{1}, \mathcal{C}) \to \underline{\mathrm{Mor}}(I, \mathcal{C})$$

with the final morphism $1 : I \to \mathbb{1}$.

Definition 2.1.12 A *cone* from N to F (resp. *cocone* from F to N) is a 2-morphism $\psi : c_N \to F$ (resp. $\psi : F \to c_N$).

A limit $L = \lim F$ for a functor $F : I \to \mathcal{C}$ is a universal cone, meaning a cone $\psi_L : c_L \to F$ such that for every other cone $\psi : c_N \to F$, there exists a factorization

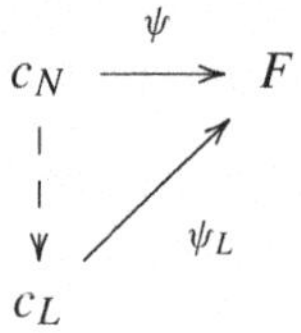

that is "unique up to higher equivalence", meaning that for all N, composition with ψ_L is an equivalence of $(n-1)$-categories

$$\psi_L^* : \underline{\mathrm{Mor}}_{\underline{\mathrm{Mor}}(I,\mathcal{C})}(c_N, c_L) \xrightarrow{\sim} \underline{\mathrm{Mor}}_{\underline{\mathrm{Mor}}(I,\mathcal{C})}(c_N, F).$$

In the case $n = 1$ of usual categories, the above map is simply a bijection between sets, meaning that the above factorization is unique.

Inverting the direction of arrows, one identifies the notion of the colimit $L = \operatorname{colim} F$ with that of the universal cocone.

More concretely, in the case of categories, i.e., for $n = 1$, a cone is a family of morphisms $\psi_X : N \to F(X)$ indexed by objects in I such that for every morphism $f : X \to Y$ in I, the following diagram

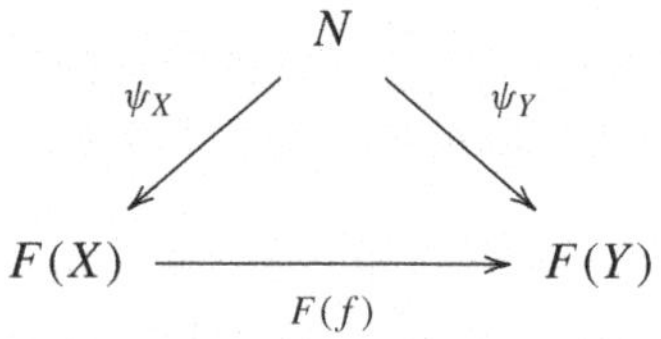

commutes. A limit $L = \lim F$ for a functor $F : I \to \mathcal{C}$ is a cone such that for every cone ψ from N to F, there is a unique morphism $N \to L$ that makes the diagram

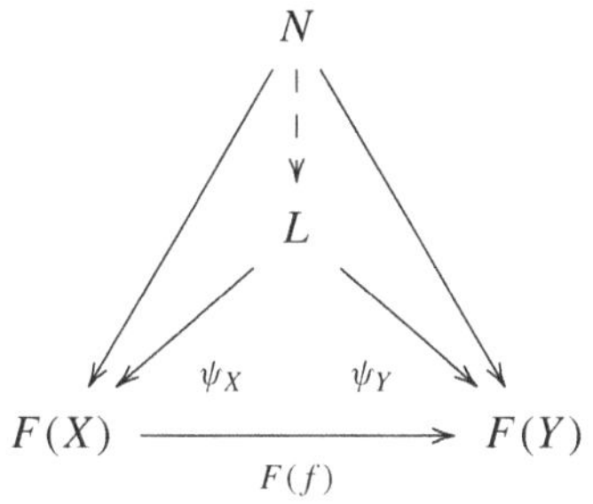

commute.

Example 2.1.4 Limits (resp. colimits) indexed by discrete categories are identified with products (resp. coproducts).

We will call a limit or colimit in the $(n+1)$-category of n-categories *small* if it is indexed by an n-category whose $(n-1)$-truncation (given by identifying isomorphic $(n-1)$-morphisms) is an object of the n-category $(n-1) - \mathrm{CAT}$ of small $(n-1)$-categories, defined by induction from the category of small sets.

Proposition 2.1.1 *Let $\mathcal{C}$ be an $(n+1)$-category. For every object X of $\mathcal{C}$, the functor*

$$\underline{\mathrm{Mor}}(X, -) : \mathcal{C} \to n\mathrm{CAT}$$

commutes with all small limits.

Proof This follows from the definition of limits. □

Definition 2.1.13 Let $\mathcal{C}$ be an n-category. Its Yoneda dual n-categories are defined as

$$\mathcal{C}^{\wedge} := \underline{\mathrm{Mor}}_{n\mathrm{CAT}}\big(\mathcal{C}, (n-1)\mathrm{CAT}\big) \quad \text{and} \quad \mathcal{C}^{\vee} := \underline{\mathrm{Mor}}_{n\mathrm{CAT}}\big(\mathcal{C}^{op}, (n-1)\mathrm{CAT}\big).$$

If $X \in \mathcal{C}$, we denote by $h^X := \underline{\mathrm{Mor}}(X, -) \in \mathcal{C}^{\wedge}$ and $h_X := \underline{\mathrm{Mor}}(-, X) \in \mathcal{C}^{\vee}$ the associated functors.

The main interest of the Yoneda dual n-category is that it contains all small limits and colimits, so that it is very similar to the n-category $(n-1)\mathrm{CAT}$. One may think of it as a kind of limit or colimit completion of the given category.

Proposition 2.1.2 *Let $\mathcal{C}$ be an n-category. Then its Yoneda duals*

$$\mathcal{C}^{\wedge} := \underline{\mathrm{Mor}}_{n\mathrm{CAT}}\big(\mathcal{C}, (n-1)\mathrm{CAT}\big) \quad \textit{and} \quad \mathcal{C}^{\vee} := \underline{\mathrm{Mor}}_{n\mathrm{CAT}}\big(\mathcal{C}^{op}, (n-1)\mathrm{CAT}\big)$$

have all limits and colimits. Moreover, the embedding $\mathcal{C} \to \mathcal{C}^{\wedge}$ preserves limits and the embedding $\mathcal{C} \to (\mathcal{C}^{\vee})^{op}$ preserves colimits.

Proof The result follows from the definition of limits and colimits. □

Definition 2.1.14 Let $\mathcal{C}_1$ and $\mathcal{C}_2$ be two n-categories. A functor $F : \mathcal{C}_1 \to \mathcal{C}_2$ is *fully faithful* if for every pair (X, Y) of objects of $\mathcal{C}_1$, the corresponding morphism

$$\mathrm{Mor}(F) : \underline{\mathrm{Mor}}_{\mathcal{C}_1}(X, Y) \to \underline{\mathrm{Mor}}_{\mathcal{C}_2}\big(F(X), F(Y)\big)$$

is an equivalence of $(n-1)$-categories.

We now give a formulation of the higher Yoneda lemma, which is a reasonable hypothesis to impose on the formalism. The case $n = 1$ is the usual Yoneda lemma.

Hypothesis 1 (Yoneda's Lemma) *Let $\mathcal{C}$ be an n-category. The functors*

$$\begin{array}{ccc} \mathcal{C}^{op} \to \mathcal{C}^{\wedge} & & \mathcal{C} \to \mathcal{C}^{\vee} \\ & \text{and} & \\ X \mapsto h^X & & X \mapsto h_X \end{array}$$

are fully faithful, meaning that the morphisms of $(n-1)$-categories

$$\underline{\mathrm{Mor}}_{\mathcal{C}}(X, Y) \to \underline{\mathrm{Mor}}_{\mathcal{C}^{\wedge}}\big(h^Y, h^X\big) \quad \text{and} \quad \underline{\mathrm{Mor}}_{\mathcal{C}}(X, Y) \to \underline{\mathrm{Mor}}_{\mathcal{C}^{\vee}}(h_X, h_Y)$$

are equivalences.

Yoneda's lemma play a crucial role in the definition of objects of a category by their universal properties. It means that an object X is uniquely determined by the corresponding functors h_X and h^X. We now illustrate this fact by a few examples.

Example 2.1.5 The product of two objects X and Y in a category $\mathcal{C}$ is an object $X \times Y$ such that there is a functorial isomorphism

$$\mathrm{Hom}(-, X \times Y) \cong \mathrm{Hom}(-, X) \times \mathrm{Hom}(-, Y).$$

The coproduct of two objects X and Y in a category $\mathcal{C}$ is an object $X \coprod Y$ such that there is a functorial isomorphism

$$\mathrm{Hom}\Big(X \coprod Y, -\Big) \cong \mathrm{Hom}(X, -) \times \mathrm{Hom}(Y, -).$$

For example, the sum (resp. product) of two sets is their disjoint union (resp. their product). The sum of two abstract groups is their so-called free product. The sum of two commutative unitary rings is their tensor product.

Example 2.1.6 The fiber product of two morphisms $f : X \to S$ and $g : Y \to S$ is an object $X \times_S Y$ such that for all objects Z, there is a natural bijection

$$\mathrm{Hom}(Z, X \times_S Y) \cong \{(h, k) \in \mathrm{Hom}(Z, X) \times \mathrm{Hom}(Z, Y) | f \circ h = g \circ k\}.$$

The definition of amalgamated sum is obtained by reversing all arrows. For example, in the category TOP of topological spaces, if $a < b < c < d$ denote four real numbers, one has

$$]a, c[\times_{]a,d[}]b, d[=]b, c[,$$

so that the fiber product corresponds to the intersection operation, and

$$]a, c[\coprod_{]b,c[}]b, d[=]a, d[,$$

so that the amalgamated sum correspond to the pasting operation.

The first example of a doctrine is given by the notion of an n-algebraic theory (see Lawvere's thesis [Law04] for the case $n = 1$). The generalization to higher dimension is useful for the study of monoidal categories (see Sects. 2.2 and 2.3).

Definition 2.1.15 The doctrine of *finite product n-categories* nFPCAT $\subset$ nCAT is the $(n + 1)$-category:

1. whose objects are n-categories $\mathcal{T}$ with finite products (and in particular, a final object, given by the empty product);
2. whose morphisms are product preserving functors.

If $\mathcal{T} \in$ FPCAT is a theory, the set of generators of the monoid of isomorphism classes of its objects is called its set of *sorts*, and its morphisms are also called its *operations*. An *algebraic theory* is a theory $\mathcal{T} \in$ FPCAT with only one sort, i.e., whose objects are all of the form $x^{\times r}$ for a fixed object $x \in \mathcal{T}$ and $r \geq 0$. We denote by FALG the doctrine of algebraic theories.

Example 2.1.7 The category $\mathcal{T}_{\mathrm{GRP}}$ opposite to that of finitely generated free groups with group morphisms between them has finite products, and is called the algebraic theory of groups. If G is a group, one may define a model $G : (\mathcal{T}_{\mathrm{GRP}}, \times) \to (\mathrm{SETS}, \times)$ by setting, if F_n denotes the free group on n elements,

$$G(F_n) = \mathrm{Hom}_{\mathrm{GRP}}(F_n, G) \cong G^n.$$

This gives an equivalence between the category of models of the theory of groups (with natural transformations as morphisms) and the category of groups. The theory $\mathcal{T}_{\mathrm{GRP}}$ is single-sorted. The theory of rings (given by the category opposite to that of free finitely generated rings) is also single sorted.

Example 2.1.8 More generally, given a set theoretic one sorted algebraic structure (rings, modules over them, etc....) that admits a free construction, the corresponding theory $\mathcal{T}$ is given by the opposite category to the category of free finitely generated objects, and models

$$M : (\mathcal{T}, \times) \to (\text{SETS}, \times)$$

correspond exactly to the given type of structure on the set $M(\{1\})$. An interesting example is given by the algebraic theory $\text{AFF}_{\mathcal{C}^\infty}$ of smooth rings, whose objects are smooth affine spaces $\mathbb{R}^n$ for varying n and whose morphisms $f : \mathbb{R}^n \to \mathbb{R}^m$ are smooth functions. A model of this theory

$$A : (\text{AFF}_{\mathcal{C}^\infty}, \times) \to (\text{SETS}, \times)$$

is called a smooth ring. These will be useful for the formulation of infinitesimal calculus on functional spaces.

Example 2.1.9 The 2-sorted theory $\mathcal{T}_{\text{RINGS},\text{MOD}}$ of commutative rings and modules is the category with finite products opposite to the category whose objects are pairs (A, M) composed of a free finitely generated commutative ring and a free module on it, and whose morphisms are pairs $(f, \varphi) : (A, M) \to (B, N)$ composed of a ring morphism $f : A \to B$ and a module morphism $\varphi : M \to f^*N :=_A N$.

We now define the notion of a higher sketch. Sketches were first introduced by Ehresmann in [Ehr82]. This is a natural generalization of the notion of an algebraic theory. Many theories we will use are given by (higher) sketches. We refer to Barr-Wells [BW05] for an overview of sketches. Note that Ehresmann's original definition uses graphs with partial compositions, not categories. This definition is more elegant since it really gives a kind of generators and relations presentation of categories, but generalizing it to higher dimensions is not that easy, so that we leave this charming exercise to the interested reader.

Definition 2.1.16 Let $(\mathfrak{L}, \mathfrak{C}) = (\{I\}, \{J\})$ be two classes of small n-categories, called categories of indices. The doctrine of $(\mathfrak{L}, \mathfrak{C})$-*sketches* is the $(n+1)$-category:

1. whose objects are triples $(\mathcal{T}, \mathcal{L}, \mathcal{C})$, called *sketches*, composed of n-categories $\mathcal{T}$, equipped with a class $\mathcal{L}$ of I-indexed cones for $I \in \mathfrak{L}$ and a class $\mathcal{C}$ of J-indexed cocones for $J \in \mathfrak{C}$;
2. whose morphisms are functors that respect the given classes of cones and cocones.

A sketch $(\mathcal{T}, \mathcal{C}, \mathcal{L})$ is called a *full sketch* if $\mathcal{C}$ and $\mathcal{L}$ are composed of colimit cocones and limit cones indexed by categories $I \in \mathfrak{L}$ and $J \in \mathfrak{C}$.

Remark 2.1.2 Note that some doctrines may be described by sketch theories, with additional *properties*. For example, the theory of cartesian closed categories (resp. that of elementary topoi) is described by theories for the finite limit sketch, with the additional property of having an inner homomorphism object (resp. a power-object, analogous to the set of parts of a given set). The general notion of doctrine encompases these generalizations, and many others.

Example 2.1.10

1. The doctrine FPCAT of finite product theories coincides with the doctrine of $(\mathfrak{L}, \mathfrak{C})$-sketches, with $\mathfrak{L}$ the class of all finite discrete categories (whose only morphisms are identities) and $\mathfrak{C} = \emptyset$.
2. The doctrine FPALG of finitely presented algebraic theories is defined as the doctrine of $(\mathfrak{L}, \mathfrak{C})$-sketches with $\mathfrak{L}$ the class of all finite discrete categories and $\mathfrak{C}$ the class of finite categories. An example of a full sketch for this doctrine is given by the category of finitely presented algebras (coequalizers of two morphisms between free algebras) for a given algebraic theory $\mathcal{T}$. For example, finitely presented groups or finitely presented commutative unital $\mathbb{R}$-algebras are theories of this kind.
3. The doctrine FLALG of finite limit theories is defined by setting $\mathfrak{L}$ to be the class of finite categories and $\mathfrak{C} = \emptyset$. We will use the following example of finite limit theory in our formalization of geometry on spaces of fields. Let $\mathcal{T} = \text{OPEN}_{\mathcal{C}^\infty}$ be the category of smooth open subsets $U \subset \mathbb{R}^n$ for varying n with smooth maps between them. Let $\mathcal{C} = \{.\}$ be the point (final object) and $\mathcal{L}$ be the class of transversal pullback diagrams. Recall that these are pullback diagrams

$$\begin{array}{ccc} U \times_V W & \longrightarrow & V \\ \downarrow & & \downarrow g \\ U & \xrightarrow{f} & W \end{array}$$

such that for all $x \in U$ and $y \in V$ with $f(x) = g(y)$, the images of tangent spaces are transversal, i.e., $D_x f(T_x U) \cup D_y g(T_y V)$ generate $T_{f(x)} W$. This transversality condition means that the fiber product is a smooth open (because of the implicit function theorem). A model

$$A : \left(\text{OPEN}_{\mathcal{C}^\infty}, \{.\}, - \overset{t}{\underset{-}{\times}} -\right) \to \left(\text{SETS}, \{.\}, - \underset{-}{\times} -\right)$$

of this finite limit theory gives a smooth algebra

$$A_{|\text{AFF}_{\mathcal{C}^\infty}} : \text{AFF}_{\mathcal{C}^\infty} \to \text{SETS}$$

so that one can think of it as a refinement of the notion of a smooth algebra. We will see in Proposition 3.2.1 that both objects are actually equivalent.

2.2 Monoidal Categories

We will start by defining the most general notion of a monoidal category in a doctrinal fashion, and then give more concrete definitions of particular examples.

Definition 2.2.1 Let $\mathcal{D}$ be the doctrine given by the $(n+2)$-category of $(n+1)$-categories with finite products. Let $(\mathcal{T}_{\text{MON}}, \times)$ be the finitary algebraic theory of monoids, given by the category with finite products opposite to that of finitely generated free monoids. The theory $\mathcal{E}_k$ is defined as the tensor product of $\mathcal{D}$-theories

$$\mathcal{E}_k := \mathcal{T}_{\text{MON}}^{\otimes k}$$

in the sense of Definition 2.1.6. A model of the theory $\mathcal{E}_k$ with values in $(n\text{CAT}, \times)$ is called a *k-tuply monoidal n-category*. A 1-monoidal n-category is also called a *monoidal n-category*.

Concretely, a monoidal category is simply a product preserving morphism

$$(\mathcal{C}, \otimes) : (\mathcal{T}_{\text{MON}}, \times) \to (\text{CAT}, \times)$$

with values in the (weak) 2-category of categories. It is thus the categorical analog of a monoid.

Note that our definition of a monoidal higher category differs a little from the one usually used in the literature, but both are equivalent, by Mac Lane's coherence theorem [ML98]. The following theorem was first stated by Baez and Dolan in [BD98]. A complete proof can be found in Lurie's book [Lur09c], Sect. 5.1.2.

Theorem 2.2.1 *A k-monoidal n-category is equivalent to an $(n+k)$-category with only one morphism in degrees smaller than k.*

The following result was stated by Baez and Dolan in [BD98] and is fully proved by Simpson in [Sim10], Chap. 23.

Theorem 2.2.2 *A k-tuply monoidal n-category for $k > n+1$ is equivalent to a $k+1$-tuply monoidal n-category.*

Note that this doctrinal viewpoint looks quite abstract, but it has the advantage of having a direct homotopical generalization, in contrast to the practice of writing all of the diagrams explicitly. It is moreover very easy to generalize this higher categorification method to other kinds of sketch-like theories (e.g., finite product, finite limit or finite presentation theories). We now give Mac Lane's compact presentation, which is based on his coherence theorem. This theorem says roughly that fixing one associativity constraint for the three term tensor products suffices to uniquely fix all others, if a coherence among these constraints, called the pentagonal axiom, is satisfied. The next theorem may also be taken as a definition, if one wants to avoid the use of higher categories.

Theorem 2.2.3 *A* monoidal category *is given by a tuple*

$$(\mathcal{C}, \otimes) = \big(\mathcal{C}, \otimes, \mathbb{1}, \mathrm{un}^r, \mathrm{un}^l, \mathrm{as}\big)$$

composed of:

(a) *a category* $\mathcal{C}$;
(b) *a bifunctor* $\otimes : \mathcal{C} \times \mathcal{C} \to \mathcal{C}$;
(c) *an object* $\mathbb{1}$ *of* $\mathcal{C}$ *called the unit object;*
(d) *for each object* A *of* $\mathcal{C}$, *two unity isomorphisms*

$$\mathrm{un}^r_A : A \otimes \mathbb{1} \xrightarrow{\sim} A \quad \textit{and} \quad \mathrm{un}^l_A : \mathbb{1} \otimes A \xrightarrow{\sim} A;$$

(e) *for each triple* (A, B, C) *of objects of* $\mathcal{C}$, *an associativity isomorphism*

$$\mathrm{as}_{A,B,C} : A \otimes (B \otimes C) \xrightarrow{\sim} (A \otimes B) \otimes C,$$

that are assumed to make the following diagrams commutative:

(i) *pentagonal axiom for associativity isomorphisms:*

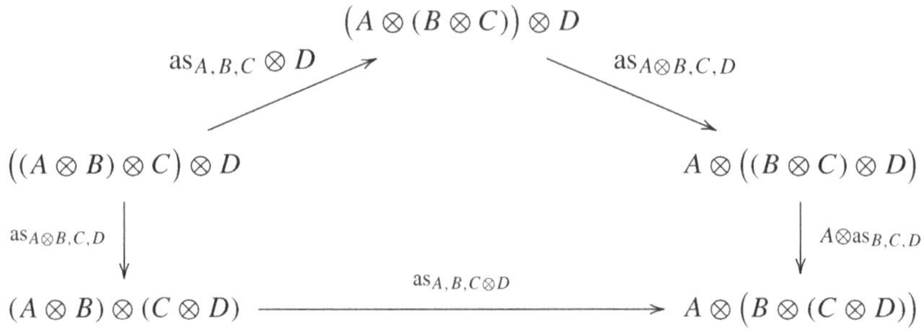

(ii) *compatibility of unity and associativity isomorphisms:*

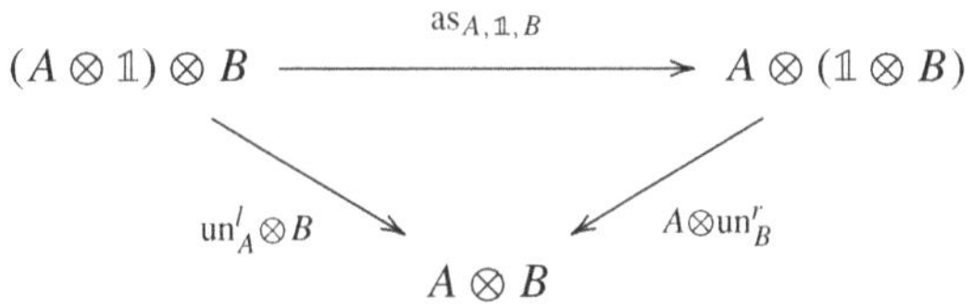

Proof We refer to Lurie's book [Lur09c], Sect. 5.1.2 and Mac Lane's book [ML98], Sect. VII.2. □

Definition 2.2.2 A monoidal category $(\mathcal{C}, \otimes)$ is:

1. *closed* if it has internal homomorphisms, i.e., if for every pair (B, C) of objects of $\mathcal{C}$, the functor
$$A \mapsto \mathrm{Hom}(A \otimes B, C)$$
is representable by an object $\underline{\mathrm{Hom}}(B, C)$ of $\mathcal{C}$;
2. *strict* if the associativity and unity transformations are equalities.

Example 2.2.1

1. Any category with finite products is a monoidal category with monoidal structure given by the product.
2. The category $\text{MOD}(K)$ of modules over a commutative unital ring K with its ordinary tensor product is a closed monoidal category. If one works with the standard construction of the tensor product as a quotient

$$M \otimes_K N := K^{(M \times N)} / \sim_{bil},$$

of the free module on the product by the bilinearity relations, the associativity isomorphisms are not equalities. Instead, they are given by the corresponding canonical isomorphisms, uniquely determined by the universal property of tensor products.
3. The category $(\text{CAT}, \times)$ of categories with their product is monoidal.
4. The category $(\text{Endof}(\mathcal{C}), \circ)$ of endofunctors $F : \mathcal{C} \to \mathcal{C}$ of a given category, with its monoidal structure given by composition, is a monoidal category.

Definition 2.2.3 Let $(\mathcal{C}, \otimes)$ be a monoidal category. A *monoid* in $\mathcal{C}$ is a triple $(A, \mu, 1)$ composed of:

1. an object A of $\mathcal{C}$;
2. a multiplication morphism $\mu : A \otimes A \to A$;
3. a unit morphism $1 : \mathbb{1} \to A$,

such that for each object V of $\mathcal{C}$, the above maps satisfy the usual associativity and unit axioms with respect to the given associativity and unity isomorphisms in $\mathcal{C}$. A comonoid in $(\mathcal{C}, \otimes)$ is a monoid in the opposite category $(\mathcal{C}^{op}, \otimes)$. A (left) *module* over a monoid $(A, \mu, 1)$ in $\mathcal{C}$ is a pair (M, μ_M) composed of an object M of $\mathcal{C}$ and a morphism $\mu_M : A \otimes M \to M$ that is compatible with the multiplication μ on A. A comodule over a comonoid is a module over the corresponding monoid in the opposite category.

Example 2.2.2

1. In the monoidal category $(\text{SETS}, \times)$, a monoid is simply an ordinary monoid.
2. In the monoidal category $(\text{MOD}(K), \otimes)$ for K a commutative unital ring, a monoid A is simply a (not necessarily commutative) K-algebra. A module over A in $(\text{MOD}(K), \otimes)$ is then an ordinary module over A. A comonoid A is a coalgebra and a comodule over it is a comodule in the usual sense.
3. In the monoidal category $(\text{Endof}(\mathcal{C}), \circ)$ of endofunctors of a category $\mathcal{C}$, a monoid is called a monad (or a triple) and a comonoid is called a comonad. If

$$F : \mathcal{D} \leftrightarrows \mathcal{C} : G$$

is an adjunction, the endofunctor $A = F \circ G : \mathcal{C} \to \mathcal{C}$ is equipped with natural unit $1 : \text{Id}_{\mathcal{C}} \to A$ and multiplication morphisms $\mu : A \circ A \to A$ that makes it a monad. The functor $B = G \circ F : \mathcal{D} \to \mathcal{D}$ is equipped with natural counit $1 : B \to \text{Id}_{\mathcal{C}}$ and comultiplication morphisms $\Delta : B \to B \circ B$ that makes it a comonad.

Definition 2.2.4 The doctrine MONCAT is defined as the 2-category:

1. whose objects are monoidal categories;
2. whose morphisms are monoidal functors, i.e., functors $F : \mathcal{C}_1 \to \mathcal{C}_2$ equipped with natural isomorphisms

$$u : \mathbb{1}_{\mathcal{D}} \to F(\mathbb{1}_{\mathcal{C}}) \quad \text{and} \quad \mu_{X,Y} : F(X) \otimes F(Y) \to F(X \otimes Y)$$

   that satisfy natural compatibility conditions with respect to the monoidal structures;
3. whose 2-morphisms are monoidal natural transformations, i.e., natural transformations $\Phi : F \Rightarrow G$ between monoidal functors compatible with the additional structure.

The monoidal analog of Lawvere's algebraic theories (see Definition 2.1.15) is given by the following notion.

Definition 2.2.5 Theories for the doctrine MONCAT given by strict monoidal categories whose objects are all of the form $x^{\otimes n}$ for a fixed object x and $n \geq 0$ are called *linear theories*.

Example 2.2.3 The simplest non-trivial linear theory is the strict monoidal category $(\Delta, +)$ of finite ordered sets (finite ordinals) with increasing maps between them and monoidal structure given by the sum of ordinals. A model

$$M : (\Delta, +) \to (\mathcal{C}, \otimes)$$

in a monoidal category is simply a monoid in $(\mathcal{C}, \otimes)$.

One can easily define a notion of a multilinear theory as a linear theory with many generators. We refer to Leinster's book for a complete account of these theories [Lei04].

Definition 2.2.6 A *PRO* is a triple $(\mathcal{C}, \otimes, S)$ composed of a strict monoidal category $(\mathcal{C}, \otimes)$ and a set S such that:

1. the monoid of objects of $\mathcal{C}$ is isomorphic to the free monoid on S;
2. for any elements $b_1, \ldots, b_m$ and $a_1, \ldots, a_n$ of S, the natural map

$$\coprod_{a_1^1, \ldots, a_n^{k_n}} \left(\mathrm{Hom}_{\mathcal{C}}\left(a_1^1 \otimes \cdots \otimes a_1^{k_1}, a_1\right) \times \cdots \times \mathrm{Hom}_{\mathcal{C}}\left(a_n^1 \otimes \cdots \otimes a_n^{k_n}, a_n\right)\right)$$
$$\to \mathrm{Hom}_{\mathcal{C}}(b_1 \otimes \cdots \otimes b_m, a_1 \otimes \cdots \otimes a_n)$$

   is a bijection, where the union is over all $n, k_1, \ldots, k_n \in \mathbb{N}$ and $a_i^j \in S$ such that there is an equality of formal sequences

$$\left(a_1^1, \ldots, a_1^{k_1}, \ldots, a_n^1, \ldots, a_n^{k_n}\right) = (b_1, \ldots, b_m).$$

A *morphism of PROs* is a pair (F, f) composed of a monoidal functor $F : (\mathcal{C}_1, \otimes) \to (\mathcal{C}_2, \otimes)$ and a map of sets $f : S_1 \to S_2$ such that the map on monoids of objects is the result of applying the free monoid functor to f.

One usually thinks of a PRO (also called a *multicategory*) as a category $\mathcal{C}$ equipped with families of multilinear operations $\mathrm{Hom}_{\mathcal{C}}(a_1, \ldots, a_n; b)$ with composition maps for multilinear operations satisfying natural associativity and unit conditions.

Definition 2.2.7 A category $\mathcal{C}$ *enriched over a monoidal category* $(\mathcal{V}, \otimes)$ is given by the following data:

1. a class $\mathrm{Ob}(\mathcal{C})$ of objects;
2. for each pair (X, Y) of objects, an object $\mathrm{Hom}(X, Y)$ in $\mathcal{V}$;
3. for each object X, a morphism $\mathrm{id}_X : \mathbb{1} \to \mathrm{Hom}(X, X)$ in $\mathcal{V}$;
4. for each triple (X, Y, Z) of objects, a composition morphism

$$\mathrm{Hom}(X, Y) \otimes \mathrm{Hom}(Y, Z) \to \mathrm{Hom}(X, Z)$$

with the usual associativity and unit conditions for composition (with respect to the given monoidal structure).

Example 2.2.4

1. A monoidal category with internal homomorphisms is enriched over itself. In particular, the monoidal category $(\mathrm{MOD}(K), \otimes)$, for K a commutative unital ring, is enriched over itself.
2. A category enriched over the monoidal category $(\mathrm{CAT}, \times)$ of categories is a strict 2-category. More generally, a category enriched over the monoidal category $(\mathit{Str}n\mathrm{CAT}, \times)$ of strict n-categories is a strict $(n+1)$-category.

2.3 Symmetric Monoidal Categories

We present here a definition of symmetric monoidal higher categories that is equivalent to the one that is traditionally used (generalizing Segal's theory of Γ-spaces, as in [Lur09c]). The equivalence, due to Shulman, results from the fact that Γ is the free semi-cartesian symmetric monoidal theory, and that its cartesian closure is the theory $\mathcal{T}_{\mathrm{CMON}}$ that we use. The clear advantage of our approach is that it allows us to define other kinds of categorified algebraic structures, based on other sketch theories (e.g., finite product theories, finite limit theories and finitely generated theories).

Definition 2.3.1 Let $\mathcal{D}$ be the doctrine given by the $(n+2)$-category of $(n+1)$-categories with finite products. Let $\mathcal{E}_\infty := (\mathcal{T}_{\mathrm{CMON}}, \times)$ be the finitary algebraic theory of commutative unital monoids, given by the category with finite products opposite to that of finitely generated free commutative unital monoids. A *symmetric monoidal n-category* is a model for the theory $\mathcal{E}_\infty$ in the $(n+1)$-category $n\mathrm{CAT}$.

We use the general Definition 2.2.1 to define braided monoidal categories.

Definition 2.3.2 A *braided monoidal category* is a 2-tuply monoidal category.

The following result is very similar to the result in the setting of monoidal structures described in Theorem 2.2.2.

Proposition 2.3.1 *There is an equivalence of* $(n+1)$*-categories between symmetric monoidal* n*-categories and* k*-tuply monoidal categories for* $k > n+1$. *In particular, a* 3*-tuply monoidal category is a symmetric monoidal category.*

Proof See the proof of loc. cit. for references. □

We now give a theorem that may be used as a more concrete definition.

Theorem 2.3.1 *A* braided monoidal category *is a tuple*

$$(\mathcal{C}, \otimes) = \left(\mathcal{C}, \otimes, \mathbb{1}, \mathrm{un}^r, \mathrm{un}^l, \mathrm{as}, \mathrm{com}\right)$$

such that $(\mathcal{C}, \otimes, \mathbb{1}, \mathrm{un}^r, \mathrm{un}^l, \mathrm{as})$ *is a monoidal category, equipped with the following additional datum:*

(f) *for each pair* (A, B) *of objects of* $\mathcal{C}$, *a commutation isomorphism (also called a braiding)*

$$\mathrm{com}_{A,B} : A \otimes B \xrightarrow{\sim} B \otimes A,$$

that is assumed to make the following diagrams commutative:

(iii) *hexagonal axiom for compatibility between the commutation and the associativity isomorphisms:*

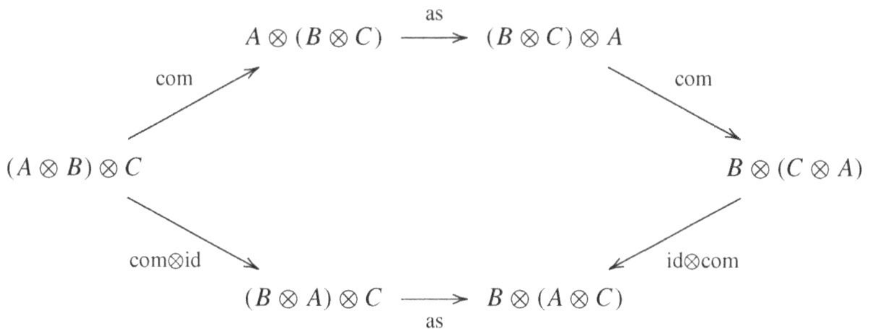

(iv) *compatibility between units and commutation:*

$$\mathrm{un}^r \circ \mathrm{com} = \mathrm{un}^l.$$

A symmetric monoidal category *is a braided monoidal category that further satisfies*

(v) *the idempotency of the commutation isomorphism*:

$$\mathrm{com}_{A,B} \circ \mathrm{com}_{B,A} = \mathrm{id}_A.$$

Proof We refer to Lurie's book [Lur09c], Sect. 5.1.2. □

We now introduce an important finiteness condition on an object of a monoidal category: that of having a monoidal dual.

Definition 2.3.3 Let $(\mathcal{C}, \otimes)$ be a monoidal category. A *dual pair* in $(\mathcal{C}, \otimes)$ is given by a pair $(A, A^\vee)$ of objects of $\mathcal{C}$ and two morphisms

$$i : \mathbb{1} \to A^\vee \otimes A \quad \text{and} \quad e : A \otimes A^\vee \to \mathbb{1}$$

satisfying the triangle equalities, meaning that the composite maps

$$A \xrightarrow{A \otimes i} A \otimes A^\vee \otimes A \xrightarrow{e \otimes A} A$$

$$A^\vee \xrightarrow{i \otimes A^\vee} A^\vee \otimes A \otimes A^\vee \xrightarrow{A^\vee \otimes e} A^\vee$$

are the identities. One then calls $A^\vee$ a right dual of A and A a left dual of $A^\vee$. A monoidal category whose objects have both left and right duals is called *rigid*.

A dual pair can be viewed as an adjoint pair in the sense of Definition 2.1.8, in the "delooped" 2-category $B(\mathcal{C}, \otimes)$, with only one object, morphisms given by objects of $\mathcal{C}$ (whose composition is given by the tensor product) and 2-morphisms given by morphisms in $\mathcal{C}$ (with their usual composition).

Remark 2.3.1 Rigid braided monoidal categories have a well defined notion of trace for endomorphisms. Indeed, if $(A, A^\vee)$ is a dual pair with unit $i : \mathbb{1} \to A^\vee \otimes A$ and counit $e : A \otimes A^\vee \to \mathbb{1}$, one can define a trace map

$$\mathrm{Tr} : \mathrm{End}(A) \to \mathrm{End}(\mathbb{1})$$

by the composition

$$\mathrm{Tr}(f) := \mathbb{1} \xrightarrow{i} A^\vee \otimes A \xrightarrow{A^\vee \otimes f} A^\vee \otimes A \xrightarrow{\mathrm{com}} A \otimes A^\vee \xrightarrow{e} \mathbb{1}.$$

Example 2.3.1

1. Any category with finite products is symmetric monoidal with $\otimes = \times$.

2. The category $(\text{MOD}(K), \otimes)$ of modules over a commutative unital ring is closed symmetric monoidal with monoidal structure given by the usual tensor product of modules and the canonical associativity, unit and commutativity constraints. These constraints are constructed by applying the universal property

$$\text{Bil}(M, N; K) \cong \text{Hom}(M \otimes N, K)$$

of the tensor product. The internal homomorphism objects $\underline{\text{Hom}}(M, N)$ are given by the natural module structure on the set $\text{Hom}_{\text{MOD}(K)}(M, N)$ of module morphisms.
3. The category $\text{MOD}_{fd}(K)$ of finite dimensional modules over a commutative field K is a closed and rigid symmetric monoidal category, i.e., it has duals and internal homomorphisms. The trace of an endomorphism $f : M \to M$ in the monoidal sense gives its usual trace in $\text{End}(\mathbb{1}) = K$.
4. The category $\text{Cob}(n)$ of oriented $(n-1)$-dimensional manifolds with morphisms given by oriented cobordisms $\partial N = M \coprod \bar{M}'$ between them (with N an n-dimensional manifold with boundary) is a rigid monoidal category, with duality given by the orientation reversing operation $M^\vee := \bar{M}$.

An important motivation for introducing the formalism of symmetric monoidal categories is given by the following example, which plays a crucial role in physics.

Example 2.3.2 The categories $\text{MOD}_g(K)$ and $\text{MOD}_s(K)$ of $\mathbb{Z}$-graded and $\mathbb{Z}/2$-graded modules over a commutative unital ring K are symmetric monoidal categories. Objects of $\text{MOD}_g(K)$ (resp. $\text{MOD}_s(K)$) are direct sums

$$V = \oplus_{k\in\mathbb{Z}} V^k \quad \left(\text{resp. } V = \oplus_{k\in\mathbb{Z}/2} V^k\right)$$

of K-modules and morphisms between them are linear maps respecting the grading. If $a \in V^k$ is a homogeneous element of a graded module V, we denote by $\deg(a) := k$ its degree. The tensor product of two graded modules V and W is defined as the graded module given by the usual tensor product of the underlying modules, equipped with the grading

$$(V \otimes W)^k = \oplus_{i+j=k} V^i \otimes W^j.$$

The tensor product is associative with unit $\mathbb{1} = K$ in degree 0, and with the usual associativity isomorphisms of K-modules. The main difference with the tensor category $(\text{MOD}(K), \otimes)$ of usual modules is given by the non-trivial commutation isomorphisms

$$\text{com}_{V,W} : V \otimes W \to W \otimes V$$

defined by extending by linearity the rule

$$v \otimes w \mapsto (-1)^{\deg(v)\deg(w)} w \otimes v.$$

One thus obtains symmetric monoidal categories $(\mathrm{MOD}_g(K), \otimes)$ and $(\mathrm{MOD}_s(K), \otimes)$ that are moreover closed, i.e., have internal homomorphisms. Once one has fixed the symmetric monoidal category $(\mathrm{MOD}_g, \otimes, \mathrm{as}, \mathrm{un}, \mathrm{com})$, one gets the following canonical additional constructions for free. The internal homomorphism object in $(\mathrm{MOD}_g, \otimes)$ is given by the usual module

$$\underline{\mathrm{Hom}}(V, W) = \underline{\mathrm{Hom}}_{\mathrm{MOD}(K)}(V, W)$$

of all linear maps, equipped with the grading

$$\underline{\mathrm{Hom}}_{\mathrm{MOD}_g(K)}(V, W) = \oplus_{n \in \mathbb{Z}} \mathrm{Hom}^n(V, W)$$

where the degree n component $\mathrm{Hom}^n(V, W)$ is the set of all linear maps $f : V \to W$ such that $f(V^k) \subset W^{k+n}$. There is a natural evaluation morphism on the underlying modules

$$\mathrm{ev} : \underline{\mathrm{Hom}}_{\mathrm{MOD}(K)}(M, N) \otimes_{\mathrm{MOD}(K)} M \to N$$

and one only has to check that it is compatible with the grading to show that, for every $U \in \mathrm{MOD}_g(K)$, there is a natural bijection

$$\mathrm{Hom}_{\mathrm{MOD}_g(K)}\big(U, \underline{\mathrm{Hom}}(V, W)\big) \cong \mathrm{Hom}_{\mathrm{MOD}_g(K)}(U \otimes V, W).$$

One can check that the tensor product of two internal homomorphisms $f : V \to W$ and $f' : V \to W'$ is given by the Koszul sign rule on homogeneous components:

$$(f \otimes g)(v \otimes w) = (-1)^{\deg(g)\deg(v)} f(v) \otimes g(w).$$

Definition 2.3.4 If $(\mathcal{C}, \otimes)$ is symmetric monoidal, a *commutative monoid* in $\mathcal{C}$ is a monoid $(A, \mu, 1)$ that satisfies the commutativity condition

$$\mu \circ \mathrm{com}_{A,A} = \mu.$$

We denote by $\mathrm{ALG}_{\mathcal{C}}$ (resp. $\mathrm{CALG}_{\mathcal{C}}$) the category of monoids (resp. commutative monoids) in $\mathcal{C}$. These will also be called *algebras* and *commutative algebras* in $\mathcal{C}$.

Example 2.3.3

1. Commutative monoids in the symmetric monoidal category $(\mathrm{SETS}, \times)$ are simply commutative monoids.
2. Commutative monoids in the symmetric monoidal category $(\mathrm{MOD}(\mathbb{Z}), \otimes)$ of $\mathbb{Z}$-modules with their usual tensor product are ordinary commutative unital rings.
3. Commutative monoids in the symmetric monoidal category $(\mathrm{MOD}_g(K), \otimes)$ of graded K-modules are called graded commutative K-algebras. Commutative algebras in $(\mathrm{MOD}_s(K), \otimes)$ are called (commutative) superalgebras over K. One can find more on superalgebras in [DM99]. Recall that the commutativity condition for the multiplication of a superalgebra uses the commutation isomorphism

of the tensor category $(\mathrm{MOD}_s(K), \otimes)$, so that it actually means a graded commutativity: (A, μ) is commutative if $\mu \circ \mathrm{com}_{A,A} = \mu$, which means

$$a \cdot b = (-1)^{\deg(a)\deg(b)} b \cdot a.$$

If $(\mathcal{C}, \otimes)$ is symmetric monoidal, A is a commutative monoid in $\mathcal{C}$, and M is a left A-module, one can put on M a right A-module structure $\mu^r : M \otimes A \to M$ defined by $\mu^r_M := \mu^l_M \circ \mathrm{com}_{M,A}$. We will implicitly use this right A-module structure in the forthcoming.

The left and right module structures are really different if the commutation isomorphism is non-trivial, e.g., in the category $\mathrm{MOD}_g(K)$ of graded modules. This is a key point of the formalism of symmetric monoidal categories: the sign nightmares are hidden in the monoidal structure, so that one can compute with generalized algebras as if they were just ordinary algebras.

Definition 2.3.5 Let $(\mathcal{C}, \otimes)$ be a symmetric monoidal category with small colimits and $V \in \mathcal{C}$. The *symmetric algebra* of V is the free commutative monoid in $\mathcal{C}$ defined by

$$\mathrm{Sym}(V) := \coprod_{n \geq 0} V^{\otimes n}/S_n.$$

If $F : \mathrm{CALG}_{\mathcal{C}} \to \mathcal{C}$ is the forgetful functor, for every commutative monoid B in $\mathcal{C}$, one has a natural isomorphism

$$\mathrm{Hom}_{\mathcal{C}}\big(V, F(B)\big) \cong \mathrm{Hom}_{\mathrm{CALG}_{\mathcal{C}}}\big(\mathrm{Sym}(V), B\big).$$

Definition 2.3.6 The doctrine SMONCAT is defined as the 2-category:

1. whose objects are symmetric monoidal categories;
2. whose morphisms are monoidal functors that are compatible with the commutativity constraint;
3. whose 2-morphisms are monoidal natural transformations.

Example 2.3.4 The symmetric monoidal category $\mathrm{Cob}(n)$ of cobordisms, whose objects are oriented $(n-1)$-dimensional manifolds, morphisms are oriented cobordisms and monoidal structure is the disjoint union, can be regarded as a theory for the doctrine SMONCAT, called the theory of topological quantum field theories. Its models

$$M : \Big(\mathrm{Cob}(n), \coprod\Big) \to (\mathcal{C}, \otimes)$$

are called (Atiyah-type) topological quantum field theories. Its higher categorical extension $\mathrm{Cob}_{\infty_n}(n)$, as described by Baez-Dolan and refined by Lurie, is in some sense very simple because it is the free symmetric monoidal ${}^{\infty}n$-category with fully dualizable objects on one object (the point), as shown by Lurie in [Lur09] (full dualizability is a higher categorical analog of monoidal dualizability).

The analog of Lawvere's algebraic theories (see Definition 2.1.15) in the doctrine SMONCAT is given by the notion of PROP. We refer to [Val04] and [Mer10] for a full account of this theory.

Definition 2.3.7 A *PROP* is a strict symmetric monoidal category whose objects are all of the form $x^{\otimes n}$ for a single object x and $n \geq 0$. A *wheeled PROP* is a rigid PROP, i.e., a strict symmetric monoidal category with duals generated by a single object x.

There is a nice combinatorial way to encode PROPs, given by the notion of a properad, that clarifies the fact that PROPs are theories of multilinear operations with multiple input and multiple output. Among them, theories with simple outputs are called operads. We refer to [Val04] for a full account of properad theory and to [LV12] and [Fre09] for a full account of operad theory.

Let $(\mathcal{V}, \otimes)$ be a symmetric monoidal category with finite sums.

Definition 2.3.8 An $\mathbb{S}$-*bimodule* in $\mathcal{V}$ is a sum $\mathcal{M} = \coprod_{n,m\geq 0} \mathcal{M}(n, m)$ whose generic element $\mathcal{M}(n, m)$ is both a left $\mathbb{S}_n$-module and a right $\mathbb{S}_m$-module in $\mathcal{V}$. An $\mathbb{S}$-*module* is a sum $\mathcal{M} = \coprod_{n\geq 0} \mathcal{M}(n)$ whose generic element $\mathcal{M}(n)$ is a left $\mathbb{S}_n$-module.

Definition 2.3.9 A *properad* (resp. *wheeled properad*) $\mathcal{P}$ in $\mathcal{V}$ is a PROP (resp. a wheeled PROP) $(\mathcal{C}, \otimes)$ enriched over $(\mathcal{V}, \otimes)$. Its underlying $\mathbb{S}$-bimodule is

$$\mathcal{P}(n, m) := \underline{\mathrm{Hom}}(n, m).$$

An *operad* in $\mathcal{V}$ is a PROP $(\mathcal{C}, \otimes)$ enriched over $(\mathcal{V}, \otimes)$ with finite sums whose $\mathbb{S}$-bimodule is of the form

$$\mathcal{P}(n, m) = \oplus_{n_1+\cdots+n_m=n} \underline{\mathrm{Hom}}(n_1, 1) \otimes \cdots \otimes \underline{\mathrm{Hom}}(n_m, 1).$$

If $\mathcal{P}$ is an operad, we denote

$$\mathcal{P}(n) := \mathcal{P}(n, 1).$$

We denote by PROPERADS$(\mathcal{V}, \otimes)$, WPROPERADS$(\mathcal{V}, \otimes)$ and OPERADS$(\mathcal{V}, \otimes)$ the categories of properads, wheeled properads, and operads in $\mathcal{V}$.

Definition 2.3.10 An *algebra over a properad* (resp. *over an operad*) in $\mathcal{V}$ is a model

$$M : (\mathcal{P}, \otimes, \mathbb{1}) \to (\mathcal{V}, \otimes, \mathbb{1})$$

in the doctrine SMONCAT$_{\mathcal{V}}$ of symmetric monoidal categories enriched over $\mathcal{V}$. More concretely:

1. If one defines the *endomorphism properad* of an object $M \in \mathcal{V}$ by

$$\mathcal{E}nd(M)(n, m) := \underline{\mathrm{Hom}}\big(M^{\otimes n}, M^{\otimes n}\big),$$

an algebra over a properad $\mathcal{P}$ is the datum of a family of $\mathbb{S}_n \times \mathbb{S}_m$-equivariant morphisms

$$\mathcal{P}(n,m) \to \mathcal{E}nd(M)(n,m)$$

which is compatible with tensor products and composition of morphisms.

2. If one defines the *endomorphism operad* of an object $M \in \mathcal{V}$ by

$$\mathcal{E}nd(M)(n) := \underline{\mathrm{Hom}}\big(M^{\otimes n}, M\big),$$

an algebra over an operad $\mathcal{P}$ is the datum of a family of $\mathbb{S}_n$-equivariant morphisms

$$\mathcal{P}(n) \to \mathcal{E}nd(M)(n)$$

which is compatible with tensor products and composition of morphisms.

Example 2.3.5

- The operad Comu in $(\mathrm{VECT}_k, \otimes)$ given by $\mathrm{Comu}(n) = k = \mathbb{1}_{\mathrm{VECT}_k}$ is the commutative operad, whose algebras are commutative (unital) algebras in VECT_k. More generally, representations of Comu in a symmetric monoidal category $(\mathcal{C}, \otimes)$ are algebras (i.e., commutative monoids) in $(\mathcal{C}, \otimes)$.
- The operad Assu in $(\mathcal{C}, \otimes)$ is given by $\mathrm{Assu}(n) = \mathbb{1}[\mathbb{S}_n]$. Algebras over Assu are associative unital algebras (i.e., non-commutative monoids) in $(\mathcal{C}, \otimes)$. The morphism Assu $\to$ Comu sends $\sigma \in \mathbb{S}_n$ to 1.
- The operad Ass is given by $\mathrm{Ass}(0) = 0$ and $\mathrm{Ass}(n) = \mathbb{1}[\mathbb{S}_n]$ for $n \geq 1$. Its algebras are non-unital associative algebras.
- The operad Lie $\subset$ Ass is given by $\mathrm{Lie}(n) = \mathrm{Ind}_{\mathbb{Z}/n\mathbb{Z}}^{S_n}(V)$, where V is the representation of $\mathbb{Z}/n\mathbb{Z}$ given by its action on a primitive n-th root of unity. The operad LIE has as algebras Lie algebras (not easy to prove; see [LV12], Sect. 13.2 for more details and references).

One can also define operads by generators and relations, starting with $\mathbb{S}$-modules and using the free operad construction. There is also a purely combinatorial description of properads in terms of $\mathbb{S}$-bimodules. We only give this description for operads, referring to [Val04] for the case of properads.

Every $\mathbb{S}$-module M defines an endofunctor $\tilde{M} : \mathcal{V} \to \mathcal{V}$, called the Schur functor, by the formula

$$\tilde{M}(V) := \coprod_{n \geq 0} M(n) \otimes_{\mathbb{S}_n} V^{\otimes n}.$$

The composite of two $\mathbb{S}$-modules M and N is the $\mathbb{S}$-module

$$M \circ N = \oplus_{n \geq 0} M(n) \otimes_{\mathbb{1}[S_n]} N^{\otimes n}$$

and the unit $\mathbb{S}$-module is given by $M(0) = \mathbb{1}$.

Proposition 2.3.2 *The composition product gives $\mathbb{S}$-modules a monoidal category structure*

$$\big(\mathbb{S} - \text{MOD}(\mathcal{V}), \circ, \mathbb{1}\big).$$

The Schur functor is a strict monoidal functor

$$\big(\mathbb{S} - \text{MOD}(\mathcal{V}), \circ\big) \to \big(\text{Endof}(\mathcal{V}), \circ\big)$$
$$M \mapsto \tilde{M}$$

Note that an operad naturally defines a monad on $\mathcal{V}$, because the Schur functor is strictly monoidal.

Proposition 2.3.3 *The functor*

$$\text{OPERADS}(\mathcal{V}, \otimes) \to \text{MON}(\mathbb{S} - \text{MOD}, \circ, \mathbb{1})$$
$$(\mathcal{P}, \otimes) \mapsto \big(\mathcal{P}(n)\big)_{n \geq 0}$$

from the category of operads to the category of monoids in the monoidal category of $\mathbb{S}$-modules has a left adjoint.

Using the above adjunction, we will also call a monoid in the monoidal category of $\mathbb{S}$-modules an operad.

2.4 Grothendieck Topologies

The category of sheaves for a given Grothendieck topology on a category $\mathcal{C}$ is essentially defined to improve on the category of presheaves (i.e., the Yoneda dual $\mathcal{C}^\vee$). Recall that the Yoneda embedding

$$\mathcal{C} \to \mathcal{C}^\vee := \underline{\text{Hom}}\big(\mathcal{C}^{op}, \text{SETS}\big)$$

is fully faithful and sends all limits to limits. The problem is that it does not respect colimits. Consider the example of the category OPEN_X of open subsets of a given topological space X. If U and V are two open subsets, their union can be described as the colimit

$$U \cup V = U \coprod_{U \times_X V} V.$$

If $U \mapsto \underline{U} := \text{Hom}(-, U)$ denotes the Yoneda embedding for $\mathcal{C} = \text{OPEN}_X$, one usually has a map

$$\underline{U} \coprod_{\underline{U \times_X V}} \underline{V} \to \underline{U \cup V}$$

in the category $\mathcal{C}^\vee$, but this is not an isomorphism. If one works in the category $\text{SH}(\text{OPEN}_X, \tau)$ of sheaves on X, this becomes an isomorphism.

Definition 2.4.1 Let $\mathcal{C}$ be a category with finite limits. A *Grothendieck topology* τ on $\mathcal{C}$ is the datum of families of morphisms $\{f_i : U_i \to U\}$, called *covering families*, and denoted Cov_U, such that the following holds:

1. (Identity) The identity map $\{\text{id} : U \to U\}$ belongs to Cov_U.
2. (Refinement) If $\{f_i : U_i \to U\}$ belongs to Cov_U and $\{g_{i,j} : U_{i,j} \to U_i\}$ belong to Cov_{U_i}, then the composed covering family $\{f_i \circ g_{i,j} : U_{i,j} \to U\}$ belongs to Cov_U.
3. (Base change) If $\{f_i : U_i \to U\}$ belongs to Cov_U and $f : V \to U$ is a morphism, then $\{f_i \times_U f : U_i \times_U V \to V\}$ belongs to Cov_V.
4. (Local nature) If $\{f_i : U_i \to U\}$ belongs to Cov_U and $\{f_j : V_j \to U\}$ is a small family of arbitrary morphisms such that $f_i \times_U f_j : U_i \times_U V_j \to U_i$ belongs to Cov_{U_i}, then $\{f_j\}$ belongs to Cov_U.

A category $\mathcal{C}$ equipped with a Grothendieck topology τ is called a *site*.

We remark that, for this definition to make sense one needs the fiber products that appear in it to exist in the given category. A more flexible and general definition in terms of sieves can be found in [KS06].

Suppose that we are working in the category $\mathcal{C} = \text{OPEN}_X$ of open subsets of a given topological space X, with inclusion morphisms and its ordinary topology. The base change axiom then says that a covering of $U \subset X$ induces a covering of its open subsets. The local character means that families of coverings of elements of a given covering induce a (refined) covering.

Definition 2.4.2 Let $(\mathcal{C}, \tau)$ be a category with Grothendieck topology. A functor $X : \mathcal{C}^{op} \to \text{SETS}$ is called a *sheaf* if the sequence

$$X(U) \longrightarrow \prod_i X(U_i) \rightrightarrows \prod_{i,j} X(U_i \times_U U_j)$$

is exact for every covering family $\{f_i : U_i \to U\}$. The category of sheaves is denoted by $\text{SH}(\mathcal{C}, \tau)$.

One can think of a sheaf in this sense as being analogous to the family of continuous functions on a topological space. A continuous function on an open set is uniquely defined by a family of continuous functions on the open subsets of a given covering, whose values are equal on their intersections.

Since sheaves are given by a limit preservation property, they can be viewed as models of a sketch (see Definition 2.1.16) with underlying category $\mathcal{C}^{op}$.

From now on, one further supposes that the topology is sub-canonical, meaning that for all $U \in \mathcal{C}$, $\text{Hom}(-, U) : \mathcal{C}^{op} \to \text{SETS}$ is a sheaf for the given topology.

We now describe in full generality the problem solved by Grothendieck topologies.

Definition 2.4.3 Let $(\mathcal{C}, \tau)$ be a site. The *nerves of coverings* associated to τ in $\mathcal{C}$ are defined by

$$N(\{U_i \to U\}) = \{N_j = U_{i_1} \times_U \cdots \times_U U_{i_n}\} \to U.$$

Theorem 2.4.1 *Let $(\mathcal{C}, \tau)$ be a site with a sub-canonical topology. The Yoneda functor*

$$\mathcal{C} \to \mathrm{SH}(\mathcal{C}, \tau)$$

is a fully faithful embedding that preserves general limits, and sends the nerves

$$N(\{U_i \to U\}) = \{N_j = U_{i_1} \times_U \cdots \times_U U_{i_n}\} \to U$$

of covering families to colimits.

This means that, contrary to the Yoneda embedding $\mathcal{C} \to \mathcal{C}^\vee$, the embedding of $\mathcal{C}$ in the category $\mathrm{SH}(\mathcal{C}, \tau)$ preserves coverings. This will be essential for the definition of "varieties" modeled on objects in $\mathcal{C}$, by pasting of representable functors along open coverings.

2.5 Categorical Infinitesimal Calculus

In this section, we use Quillen's tangent category [Qui70] to give a purely categorical definition of the main tools of differential calculus. The main advantage of this abstract approach is that it is very concise and allows us to deal at once with all the generalized geometric situations we will use.

Note that one can generalize this approach directly to the higher categorical setting of doctrines and theories, by replacing everywhere the category $\mathcal{C}$ by an ${}^{\infty}n$-category, and the category of abelian groups in $\mathcal{C}/A$ by the stabilization $\mathrm{Stab}(\mathcal{C}/A)$ (see Sect. 9.7). We refer to Sect. 10.6 for an example of a concrete application of this general idea to differential calculus in homotopical geometry. This can be useful when studying the deformation theory of higher categories with additional structures (which play an important role in homological mirror symmetry, for example; see [Kon95]).

In the following, we work with categories that do not necessarily admit finite limits.

Definition 2.5.1 Let $\mathcal{C}$ be a category with final object $\underline{0}$.

1. An *abelian group object* in $\mathcal{C}$ is a triple $(A, \mu, 0)$ composed of an object A of $\mathcal{C}$, a multiplication morphism $\mu : A \times A \to A$ and an identity morphism $0 : \underline{0} \to A$ that define a functor

$$\mathrm{Hom}(-, A) : \mathcal{C} \to \mathrm{GRAB}$$

from $\mathcal{C}$ to the category of abelian groups.

2. A *torsor over an abelian group object* $(A, \mu, 0)$ is a pair (B, ρ) composed of an object B of $\mathcal{C}$ and an action morphism $\rho : A \times B \to B$ that induce an action of $\mathrm{Hom}(-, A) : \mathcal{C} \to \text{GRAB}$ on $\mathrm{Hom}(-, B) : \mathcal{C} \to \text{SETS}$ and such that the morphism

$$(\rho, \mathrm{id}) : \mathrm{Hom}(-, A) \times \mathrm{Hom}(-, B) \to \mathrm{Hom}(-, B) \times \mathrm{Hom}(-, B)$$

is an isomorphism.

The categorical definition of square zero extensions can be found on the nlab contributive website.

Definition 2.5.2 Let I be the category with two objects and a morphism between them. Let $\mathcal{C}$ be a category.

1. The *arrow category* $[I, \mathcal{C}]$ is the category whose objects are morphisms $B \to A$ in $\mathcal{C}$ and whose morphisms are given by commutative squares.
2. The *tangent category* $T\mathcal{C}$ is the category of abelian group objects in the arrow category $[I, \mathcal{C}]$, whose objects are triples

$$(B \to A, \mu : B \times_A B \to B, \underline{0} : A \to B),$$

and whose morphisms are commutative squares

$$\begin{array}{ccc} B & \longrightarrow & B' \\ \downarrow & & \downarrow \\ A & \xrightarrow{f} & A' \end{array}$$

such that the induced morphism $B \to f^*B' := A \times_{A'} B'$ is a morphism of abelian group objects over A. The fiber of $T\mathcal{C}$ at A is denoted $\text{MOD}(A)$ and is called the category of modules over A.
3. A left adjoint to the domain functor

$$\mathrm{dom} : T\mathcal{C} \to \mathcal{C}$$
$$[B \to A] \mapsto B$$

is called a *cotangent functor* and is denoted $\Omega^1 : \mathcal{C} \to T\mathcal{C}$.

Example 2.5.1 If $\mathcal{C} = \text{RINGS}$ is the category of commutative unital rings, we will see in Theorem 3.3.1 that its tangent category $T\mathcal{C}$ is equivalent to the category MOD of pairs (A, M), composed of a ring A and an A-module M. The corresponding arrow in $\mathcal{C}/A$ is given by the square zero extension

$$p : A \oplus M \to A.$$

In this setting, the value of the cotangent functor on $A \in \mathcal{C}$ is simply the module of Kähler forms, given by the square zero extension

$$A \oplus I/I^2 \cong (A \otimes A)/I^2 \to A,$$

with zero section induced by $p_1^* : A \to A \otimes A, a \mapsto a \otimes 1$.

Example 2.5.2 The category $\mathcal{C} = \text{OPEN}_{C^\infty}^{op}$ does not have many pullbacks, meaning that it is hard to take amalgamated sums of smooth opens. This implies that for every $A \in \mathcal{C}$, there are not many abelian group objects $p : B \to A$ in $\mathcal{C}/A$. We will now show that the only such object is given by the identity map $\text{id}_A : A \to A$. Indeed, suppose given an abelian cogroup object $p^* : V \to U$. This means that one has a section $0^* : U \to V$, so that $p^* : V \to U$ is injective. Moreover, it is open and closed because 0^* is a continuous section of $V \to U$. This means that V is a connected component of U. Let $U = V \coprod W$. In this case, the amalgamated sum exists and is given by

$$U \coprod_V U \cong V \coprod W \coprod V.$$

The comultiplication $\mu^* : U \to U \coprod_V U$ has a section $\text{id}_U \coprod_V 0^*$, so that it is an open and closed map, and sends U to a connected component of the above disjoint sum. This implies that $U = V$. We conclude that the tangent category $T\mathcal{C}$ is equal to $\mathcal{C}$, and that the cotangent functor $\Omega^1 : \mathcal{C} \to T\mathcal{C}$ is just the identity functor. This shows why the use of generalized algebras is necessary to define an interesting categorical differential calculus in smooth geometry.

Definition 2.5.3 Let $\mathcal{C}$ be a category. Let $M \to A$ be a module over A. A section $D : A \to M$ is called an M-valued *derivation*. We denote by $\text{Der}(A, M)$ the set of M-valued derivations.

By definition, if there exists a cotangent module Ω_A^1, one has

$$\text{Der}(A, M) \cong \text{Hom}_{\text{MOD}(A)}\big(\Omega_A^1, M\big).$$

We now refine the tangent category construction to define higher infinitesimal thickenings.

Definition 2.5.4 Let $\mathcal{C}$ be a category.

1. The *first thickening category* $\text{Th}^1\mathcal{C}$ is the category of torsors

$$(C \to A, \rho : B \times_A C \to C)$$

in the arrow category $[I, \mathcal{C}]$ over objects $(B \to A, \mu, \underline{0})$ of $T\mathcal{C}$.

2. The *n-th thickening category* $\mathrm{Th}^n\mathcal{C}$ is the category of objects $D \to A$ in the arrow category $[I, \mathcal{C}]$ such that there exists a sequence

$$D = C_n \to C_{n-1} \to \cdots \to C_0 = A$$

with all $C_{i+1} \to C_i$ in $\mathrm{Th}^1\mathcal{C}$.
3. Given $A \in \mathcal{C}$, a left adjoint to the forgetful functor

$$\mathrm{Th}^n\mathcal{C}/A \to \mathcal{C}/A$$

$$[B \to A] \mapsto [B \to A]$$

is called an *infinitesimal thickening functor* and is denoted $\mathrm{Th}^n : \mathcal{C}/A \to \mathrm{Th}^n\mathcal{C}/A$.
4. In particular, the functor

$$\mathrm{Jet}^n := \mathrm{Th}^n \circ \Delta^* : \mathcal{C} \to \mathrm{Th}^n\mathcal{C},$$

where $\Delta^*(A) := [A \coprod A \to A]$ is the codiagonal map, is called the *jet functor*.

Proposition 2.5.1 *Let $\mathcal{C}$ be a category with pullbacks and $A \in \mathcal{C}$. There is a natural isomorphism*

$$\mathrm{Jet}^1 A \xrightarrow{\sim} \Omega^1_A$$

in $\mathrm{Th}^1\mathcal{C}/A$.

Proof By definition, $[\mathrm{Jet}^1 A \to A]$ is the infinitesimal thickening of the codiagonal map $[A \coprod A \to A]$. In particular, it has two retractions, denoted $d_0, d_1 : A \to \mathrm{Jet}^1 A$. One can consider d_0 as a trivialization of the torsor $\mathrm{Jet}^1 A \to A$ over an abelian group object, giving $\mathrm{Jet}^1 A \to A$ the structure of an abelian group object over A, equipped with a section $d_1 : A \to \mathrm{Jet}^1 A$. With this structure, it satisfies the universal property of the abelian group object $[\Omega^1_A \to A]$, so that both are isomorphic. □

Definition 2.5.5 Let (LEGOS, τ) be a site and $X \in \mathrm{SH}(\mathrm{LEGOS}, \tau)$ be a space. Suppose that there exists a thickening functor Th^k on $\mathrm{SH}(\mathrm{LEGOS}, \tau)^{op}$. Let $C \to M$ be a morphism of spaces. The relative k-th jet space $\mathrm{Jet}^k(C/M) \to M$ is defined as the space

$$\mathrm{Jet}^k(C/M) := C \times_M \mathrm{Jet}^k(M).$$

We now define the notion of a Grothendieck connection, which is a kind of infinitesimal pasting (i.e., descent) datum. Let (LEGOS, τ) be a site and $X \in \mathrm{SH}(\mathrm{LEGOS}, \tau)$ be a space. Suppose that there exists a first thickening functor Th^1 on $\mathrm{SH}(\mathrm{LEGOS}, \tau)^{op}$. Denote by $X^{(1)} := \mathrm{Th}^1(\Delta_{X^2})$ the first infinitesimal thickening of $\Delta_{X^2} : X \to X \times X$ and $X^{(1)}_3 := \mathrm{Th}^1(\Delta_{X^3})$ the first infinitesimal thickening of $\Delta_{X^3} : X \to X \times X \times X$. Let $p_1, p_2 : X^2 \to X$ be the two natural projections and let $p_{1,2}, p_{2,3}, p_{1,3} : X^3 \to X \times X$ the natural projections. We will also denote by $p_1, p_2, p_3 : X^3 \to X$ the natural projections.

Definition 2.5.6 Let $C \to X$ be a space morphism. A *Grothendieck connection* on $C \to X$ is the datum of an isomorphism

$$A : p_1^* C_{|X^{(1)}} \xrightarrow{\sim} p_2^* C_{|X^{(1)}}$$

over $X^{(1)}$ that reduces to the identity on X by pullback along $X \to X^{(1)}$. Given a Grothendieck connection on $C \to X$, one can pull it back to $X_3^{(1)}$ via the projections $p_{i,j}$, obtaining isomorphisms

$$A_{i,j} : p_i^* C_{|X_3^{(1)}} \xrightarrow{\sim} p_j^* C_{|X_3^{(1)}}.$$

The *curvature of the connection* $A : p_1^* C_{|X^{(1)}} \xrightarrow{\sim} p_2^* C_{|X^{(1)}}$ is the isomorphism

$$F_A := A_{1,3}^{-1} \circ A_{2,3} \circ A_{1,2} : p_1^* C_{|X_3^{(1)}} \xrightarrow{\sim} p_1^* C_{|X_3^{(1)}}.$$

An *integrable connection* is a connection whose curvature F_A is the identity morphism.

This notion of connection actually generalizes to any kind of geometrical object on X that has a natural notion of pull-back, for example a linear connection on a vector bundle, a G-equivariant connection on a G-bundle, etc....

Example 2.5.3 Let $X = \mathbb{R}$ and $C = \mathbb{R}^2$ be equipped with their polynomial algebras of functions $\mathcal{O}_X = \mathbb{R}[x]$ and $\mathcal{O}_C = \mathbb{R}[x, u]$. We work in the category $\mathcal{C}$ of $\mathbb{R}$-algebras. One then has

$$\mathrm{Jet}^1 \mathcal{O}_X := \mathbb{R}[x, x_0]/(x - x_0)^2,$$

$$\mathcal{O}_{p_1^* C} = \mathbb{R}[x, x_0, u]/(x - x_0)^2, \qquad \mathcal{O}_{p_2^* C} = \mathbb{R}[x_0, x, u]/(x - x_0)^2.$$

A connection on C must be the identity on X, i.e., modulo $(x - x_0)$. The trivial Grothendieck connection on $C \to X$ is given by the isomorphism

$$\mathrm{id} : p_1^* C \xrightarrow{\sim} p_2^* C.$$

If $A \in (x - x_0)/(x - x_0)^2$ is an element (that represents a differential form on X), then $\epsilon : u \mapsto u + A$ is also a Grothendieck connection on C, whose inverse isomorphism is $u \mapsto u - A$.

Definition 2.5.7 Let (LEGOS, τ) be a site with amalgamated sums, $\mathcal{C}$ be the opposite category LEGOS^{op}, and $X \in \mathrm{SH}(\mathrm{LEGOS}, \tau)$ be a sheaf. A morphism $U \to T$ in LEGOS that is an object of $\mathrm{Th}^n(\mathcal{C})$ for some n is called an *infinitesimal thickening*. One says that X is:

1. *formally smooth* (resp. *formally unramified*, resp. *formally étale*) if for every infinitesimal thickening $U \to T$, the natural map

$$X(T) \to X(U)$$

is surjective (resp. injective, resp. bijective);
2. *locally finitely presented* if X commutes with directed limits;
3. *smooth* (resp. unramified, resp. étale) if it is locally finitely presented and formally smooth (resp. formally unramified, resp. formally étale).

Definition 2.5.8 Let (LEGOS, τ) be a site with amalgamated sums and $X \in \text{SH}(\text{LEGOS}, \tau)$ be a sheaf. The *de Rham space* of X is the sheaf associated to the presheaf whose points with values in $T \in \text{LEGOS}$ are defined by

$$X_{DR}(T) := X(T)/\sim_{inf}$$

where $x \sim_{inf} y : T \to X$ if there exists an infinitesimal thickening $U \to T$ such that $x_{|U} = y_{|U}$. We denote by $p_{DR} : X \to X_{DR}$ the natural projection map.

Remark 2.5.1 We may also define the de Rham space as the quotient of X by the action of the groupoid $\Pi_1^{inf}(X)$ of pairs of points $(x, y) \in X^2(T)$ that are infinitesimally close, i.e., such that there exists a thickening $S \to T$ with $x_{|U} = y_{|U}$. This is a subgroupoid of the groupoid $\text{Pairs}(X)$ whose arrows are pairs of points in X, with composition

$$(x, y) \circ (y, z) = (x, z).$$

If X is a manifold, both groupoids are related to the fundamental groupoid $\Pi_1(X)$ of paths in X up to homotopy. An action of the groupoid $\text{Pairs}(X)$ on a bundle $E \to X$ is equivalent to a trivialization of E. An action of the groupoid $\Pi_1(X)$ on E is a parallel transport for the elements of the fibers of E along (homotopy classes of) paths in X. The groupoid $\Pi_1^{inf}(X)$ is intermediary between these two groupoids: an action of it on a bundle $E \to X$ is equivalent to a parallel transport for elements of the fibers of E along infinitesimal paths in X. More concretely, it is an integrable connection. The relation with the fundamental groupoid is given by the fact that a connection on a path is simply a differential equation, and that solving the corresponding Cauchy problem allows the definition of a parallel transport along paths. If one has a parallel transport along paths, one also morally gets, by passing to the infinitesimal limit, a connection.

Remark 2.5.2 One may also define the de Rham space as the space $\pi_0(\Pi_\infty^{inf}(X))$ of connected components of the ∞-groupoid of infinitesimal simplices in X, whose degree n component is

$$\Pi_\infty^{inf}(X)_n(U) := \big\{(t, x_i) \in T(U) \times X^n(T),\ t : U \to T \text{ thickening}$$
$$\forall i, j,\ x_{i|U} = x_{j|U}\big\}.$$

We refer to Sect. 10.3 for more details. If

$$X \to X^{(1)} \to X \times X$$

is the infinitesimal thickening of the diagonal to order 1, we may regard it as a point in $\Pi_\infty^{inf}(X)_2(U)$ for $U = X$, because if we let $T = X^{(1)}$, we have that the natural map $\gamma : T \to X^2$ may be interpreted as a pair $(x_1, x_2) \in X^2(T)$ of points of X with values in T that are equal on U, the natural map $U \to T$ being a first order thickening. One interprets γ as an infinitesimal path between the points x_1 and x_2.

Proposition 2.5.2 *If a Grothendieck connection A on $B \to X$ is integrable, the isomorphism A can be extended to an action of $\Pi_\infty^{inf}(X)$ on B, so that it gives a space $B' \to X_{DR}$ whose pullback along $\pi : X \to X_{DR}$ is B.*

Definition 2.5.9 Let (LEGOS, τ) be a site with amalgamated sums and $X \in \text{SH}(\text{LEGOS}, \tau)$ be a sheaf.

1. A *crystal of space* on X is a space morphism $Z \to X_{DR}$.
2. A crystal of space $Z \to X_{DR}$ is called a *crystal of varieties* over X if its pullback $Z \times_{X_{DR}} X \longrightarrow X$ is a variety (in the sense of Definition 3.1.1) in $\text{VAR}(\text{LEGOS}, \tau)$.
3. A *crystal of module* $\mathcal{M}$ over X is a module over the space X_{DR}, i.e., an object $X_{DR} \to \mathcal{M}$ of the tangent category $T(\text{SH}(\text{LEGOS}, \tau)^{op})$.

The category of crystals of varieties (resp. crystals of modules) over X is denoted by VAR_{cris}/X (resp. $\text{MOD}_{cris}(X)$).

Proposition 2.5.3 *Let (LEGOS, τ) be a site with amalgamated sums. Let X be a formally smooth space in $\text{SH}(\text{LEGOS}, \tau)$. There is a natural adjunction*

$$p_{DR}^* : \text{MOD}_{cris}(X) \leftrightarrows \text{MOD}(X) : p_{DR,*}$$

between modules on X and crystals of modules on X. More generally, there is a natural adjunction

$$p_{DR}^* : \text{Sp}/X_{DR} \leftrightarrows \text{Sp}/X : p_{DR,*}.$$

One denotes by $\text{Jet}(C/X) \to X_{DR}$ the crystal of space over X given by $p_{DR,*}(C)$, and also (by extension), the corresponding space $\text{Jet}(C/X) = p_{DR}^*(p_{DR,*}C) \to X$.

References

[AGV73] Artin, M., Grothendieck, A., Verdier, J.-L.: Théorie des Topos et Cohomologie Étale des Schémas. Tome 3, Lecture Notes in Mathematics, vol. 305, p. 640. Springer, Berlin (1973). Séminaire de Géométrie Algébrique du Bois-Marie 1963–1964 (SGA 4), Dirigé par M. Artin, A. Grothendieck et J.L. Verdier. Avec la collaboration de P. Deligne et B. Saint-Donat

[Ara12] Ara, D.: Higher quasi-categories vs higher Rezk spaces. arXiv e-prints (2012). arXiv: 1206.4354

[BD98] Baez, J.C., Dolan, J.: Categorification. arXiv Mathematics e-prints (1998). arXiv: math/9802029

[BS11] Barwick, C., Schommer-Pries, C.: On the unicity of the homotopy theory of higher categories. arXiv e-prints (2011). arXiv:1112.0040

[BW05] Barr, M., Wells, C.: Toposes, Triples and Theories. Repr. Theory Appl. Categ. (2005). 12:x+288. Corrected reprint of the 1985 original [MR0771116]

[CL04] Cheng, E., Lauda, A.: Higher dimensional categories: an illustrated guidebook (2004, preprint)

[DM99] Deligne, P., Morgan, J.W.: Notes on supersymmetry (following Joseph Bernstein). In: Quantum Fields and Strings: A Course for Mathematicians, Vol. 1, 2 (Princeton, NJ, 1996/1997), pp. 41–97. Am. Math. Soc., Providence (1999)

[Ehr82] Ehresmann, C.: Œuvres complètes et commentées. IV-2. Esquisses et structures monoïdales fermées. Cah. Topol. Géom. Différ. **23**(suppl. 2), 407–823 (1983). 1982. Edited and with commentary in English by Andrée Charles Ehresmann

[Fre09] Fresse, B.: Modules over Operads and Functors. Lecture Notes in Mathematics, vol. 1967, p. 308. Springer, Berlin (2009). doi:10.1007/978-3-540-89056-0. ISBN 978-3-540-89055-3

[Joy11] Joyal, A.: Notes on quasi-categories (2011, preprint)

[Kon95] Kontsevich, M.: Homological algebra of mirror symmetry. In: Proceedings of the International Congress of Mathematicians, Vol. 1, 2 (Zürich, 1994), pp. 120–139. Birkhäuser, Basel (1995)

[KS06] Kashiwara, M., Schapira, P.: Categories and Sheaves. Grundlehren der Mathematischen Wissenschaften [Fundamental Principles of Mathematical Sciences], vol. 332, p. 497. Springer, Berlin (2006). ISBN 978-3-540-27949-5; 3-540-27949-0

[Law04] Lawvere, F.W.: Functorial semantics of algebraic theories and some algebraic problems in the context of functorial semantics of algebraic theories. Theory Appl. Categ. **5**, 1–121 (2004). Reprinted from Proc. Natl. Acad. Sci. USA **50**, 869–872 (1963) [MR0158921] and Reports of the Midwest Category Seminar. II, pp. 41–61, Springer, Berlin (1968) [MR0231882]

[Law06] Lawvere, F.W.: Adjointness in foundations. Theory Appl. Categ. **16**, 1–16 (2006). Reprinted from Dialectica **23** (1969)

[Lei04] Leinster, T.: Higher Operads, Higher Categories. Cambridge University Press, Cambridge (2004)

[Lur09] Lurie, J.: On the classification of topological field theories. arXiv e-prints (2009). arXiv: 0905.0465

[Lur09c] Lurie, J.: Higher algebra (2009, preprint)

[Lur09d] Lurie, J.: Higher Topos Theory. Annals of Mathematics Studies, vol. 170, p. 925. Princeton University Press, Princeton (2009). ISBN 978-0-691-14049-0; 0-691-14049-9

[LV12] Loday, J.-L., Vallette, B.: Algebraic Operads. Grundlehren der Mathematischen Wissenschaften [Fundamental Principles of Mathematical Sciences], vol. 346. Springer, Heidelberg (2012). ISBN 978-3-642-30361-6. xxiv+634 pp.

[Mal10] Maltsiniotis, G.: Grothendieck ∞-groupoids, and still another definition of ∞-categories. arXiv e-prints (2010). arXiv:1009.2331

[Mer10] Merkulov, S.A.: Wheeled pro(p)file of Batalin-Vilkovisky formalism. Commun. Math. Phys. **295**, 585–638 (2010). doi:10.1007/s00220-010-0987-x. 0804.2481

[ML98] Mac Lane, S.: Categories for the Working Mathematician, 2nd edn. Graduate Texts in Mathematics, vol. 5, p. 314. Springer, New York (1998). ISBN 0-387-98403-8

[Qui70] Quillen, D.: On the (co-) homology of commutative rings. In: Applications of Categorical Algebra (Proc. Sympos. Pure Math., Vol. XVII, New York, 1968), pp. 65–87. Am. Math. Soc., Providence (1970)

[Rez09] Rezk, C.: A cartesian presentation of weak n-categories. arXiv e-prints (2009). arXiv: 0901.3602

[Sim01] Simpson, C.: Some properties of the theory of n-categories. arXiv Mathematics e-prints (2001). arXiv:math/0110273

[Sim10] Simpson, C.T.: Homotopy theory of higher categories. arXiv e-prints (2010). arXiv: 1001.4071

[Val04] Vallette, B.: Dualite de Koszul des PROPs. arXiv Mathematics e-prints (2004). arXiv: math/0405057

Chapter 3
Parametrized and Functional Differential Geometry

The aim of this chapter is to present flexible tools to do differential geometry on spaces of fields, i.e., on spaces of "functions" between generalized "spaces". We choose here a viewpoint that is directly adapted to the homotopical generalizations of Chap. 10.

Let us describe our general approach in the particular case of differential geometry. Let

$$\mathrm{LEGOS} = \mathrm{OPEN}_{\mathcal{C}^\infty}$$

be the category of building blocks for differential geometry, given by open subsets $V \subset \mathbb{R}^n$ for varying n with smooth maps between them. Note that this category has two main drawbacks:

1. It doesn't contain enough limits (e.g., solutions spaces for equations and fiber products), and not enough colimits (e.g., quotients).
2. It doesn't contain "spaces of fields", i.e., functional spaces $\underline{\mathrm{Hom}}(U, V)$.

The solution to this problem, introduced by Grothendieck in algebraic geometry and Lawvere and Ehresmann in categorical logic, is to use the Yoneda embeddings

$$\mathrm{LEGOS} \hookrightarrow \underline{\mathrm{Hom}}\big(\mathrm{LEGOS}^{op}, \mathrm{SETS}\big) \quad \text{and} \quad \mathrm{LEGOS}^{op} \hookrightarrow \underline{\mathrm{Hom}}(\mathrm{LEGOS}, \mathrm{SETS})$$

into functor categories that have all the desired limits, colimits and mapping spaces, and to restrict the class of functors (by some limit-commutation conditions) to have some permanence properties for the geometric building blocks in the bigger categories. This gives essentially two viewpoints in the modern approach to differential geometry: the *parametrized* and the *functional* viewpoint. Let $U \subset \mathbb{R}^n$ be an open subset.

1. The parametrized approach is based on the study of the *functor of points*

$$\begin{aligned} \underline{U} : \mathrm{LEGOS}^{op} &\to \mathrm{SETS} \\ V &\mapsto \mathrm{Hom}(V, U) \end{aligned}$$

   of U parametrized by other geometric spaces V (e.g., a point, an interval, etc. ...);

F. Paugam, *Towards the Mathematics of Quantum Field Theory*,
Ergebnisse der Mathematik und ihrer Grenzgebiete. 3. Folge / A Series of Modern
Surveys in Mathematics 59, DOI 10.1007/978-3-319-04564-1_3,

2. The functional approach is based on the study of the *functor of functions*

$$\begin{aligned} \mathcal{O}(U) : \text{LEGOS} &\to \text{SETS} \\ V &\mapsto \text{Hom}(U, V) \end{aligned}$$

   on U with values in other geometric spaces V (e.g., the real or complex numbers).

Note that these two approaches are equivalent if we vary U and V symmetrically. The main idea is the same in both settings. One generalizes spaces by using contravariant functors

$$X : \text{LEGOS}^{op} \to \text{SETS}$$

that commute with some prescribed limit cones (e.g., those given by nerves of coverings; this gives sheaves on Grothendieck sites), yielding a sensible notion of

pasting of building blocks

in the category of generalized spaces. One generalizes algebras of functions by using covariant functors

$$\mathcal{A} : \text{LEGOS} \to \text{SETS}$$

that commute with some prescribed limit cones (e.g., those given by finite products or transversal fiber products; this gives generalized algebras in the sense of Lawvere and Dubuc), yielding a sensible notion of

intersection of building blocks or *solution space for an equation.*

Both these constructions are naturally formalized in the setting of doctrines described in Sect. 2.1, and in particular, using sketches (see Definition 2.1.16). This presentation of geometry has the great advantage of allowing a direct homotopical generalization (see Chap. 10) to treat obstruction theoretic questions geometrically.

The parametrized approach has been systematically developed by Grothendieck and his school (see [Gro60] and [AGV73]) in algebraic geometry. In differential geometry, it is called the diffeological approach, and was developed by Chen [Che77], Souriau [Sou97] and Iglesias Zeimour [IZ99], with applications to classical mechanics (see [BH11] for a survey).

The functional approach is widely used in analysis, with deep roots in functional analytic methods. It has also been developed on more categorical and logical grounds following Lawvere's thesis [Law04]. A particular combination of the functional and parametrized approach in classical differential geometry is called synthetic differential geometry. It has been developed following Lawvere's ideas [Law79] by Moerdijk-Reyes in [MR91] and Kock [Koc06]. It has also been used in analytic geometry by Dubuc, Taubin and Zilber (see [DT83, DZ94] and [Zil90]), in a way that is closer to our presentation.

Our aim is to have flexible tools to do differential geometry on spaces of fields (i.e., spaces of "functions" between "spaces", in a generalized sense). The parametrized approach gives a well behaved notion of differential form, and the functional approach gives a well behaved notion of vector field. We will use a combination of the parametrized and functional approach that is optimal for our purpose of studying the mathematics of classical and quantum field theories with geometrical tools.

We give here an axiomatic presentation that incorporates all the examples of geometric settings that will be used in this book.

3.1 Parametrized Geometry

We first recall some results about sites from Sect. 2.4.

Let $(\textsc{Legos}, \tau)$ be a site (a category with a Grothendieck topology) whose objects, called legos, will be the basic building blocks for our generalized spaces. Let $\textsc{Legos}^{\vee} := \underline{\mathrm{Hom}}(\textsc{Legos}^{op}, \textsc{Sets})$ be the Yoneda dual category. One has a fully faithful embedding

$$\begin{aligned} \textsc{Legos} &\to \textsc{Legos}^{\vee} \\ X &\mapsto \underline{X} \end{aligned}$$

where

$$\underline{X} := \mathrm{Hom}(-, X) : \textsc{Legos}^{op} \to \textsc{Sets}$$

is the contravariant functor naturally associated to X. The Yoneda dual category has all limits (e.g. fiber products) and colimits (e.g. quotients) and also internal homomorphisms, so that it could be a nice category in which to do geometry. However, the natural embedding $\textsc{Legos} \to \textsc{Legos}^{\vee}$ does not preserve finite colimits given by pastings of open subspaces. For example, if $\textsc{Legos}$ is the category $\textsc{Open}_X$ of open subsets of a given topological space X, and one considers a union

$$U = U_1 \cup U_2 = U_1 \coprod_{U_1 \cap U_2} U_2 \subset X$$

of open subsets of X, the natural morphism in $\textsc{Legos}^{\vee}$ given by

$$\underline{U_1} \coprod_{\underline{U_1 \times_X U_2}} \underline{U_2} \to \underline{U} := \underline{U_1 \coprod_{U_1 \cap U_2} U_2}$$

is usually *not* an isomorphism. The notion of sheaf for a Grothendieck topology was introduced precisely to solve this problem: the above natural morphism, with colimits taken in the category $\textsc{Sh}(\textsc{Legos}, \tau)$, is indeed an isomorphism.

We will thus work with generalized spaces given by sheaves on Grothendieck sites. We now define varieties as generalized spaces that can be obtained by pasting legos.

Definition 3.1.1 Let (LEGOS, τ) be a site and let $f : X \to Y$ be a morphism of sheaves on (LEGOS, τ).

- If X and Y are representable in LEGOS, one says that f is *open* if it is injective on points and it is the image of a morphism

$$\coprod_i f_i : \coprod U_i \to Y$$

for $\{U_i \xrightarrow{f_i} Y\}$ a covering family for τ.
- In general, one says that f is *open* if it is injective on points and it is universally open, meaning that, for every $x : \underline{U} \to Y$, with $\underline{U}$ representable, the fiber

$$f \times_Y x : X \times_Y \underline{U} \to \underline{U}$$

is isomorphic to the embedding $\underline{W} \subset \underline{U}$ of an open subset of the representable space $\underline{U}$.
- A *variety* is a space X that has an open covering by representable spaces, meaning that there exists a family $f_i : \underline{U_i} \to X$ of open embeddings with $U_i \in \text{LEGOS}$ such that the map

$$\coprod_i f_i : \coprod_i \underline{U_i} \to X$$

is an epimorphism of sheaves.

We denote by $\text{VAR}(\text{LEGOS}, \tau) \subset \text{SH}(\text{LEGOS}, \tau)$ the category of varieties.

Definition 3.1.2 For the following sites (LEGOS, τ), the corresponding varieties are called:

- $\mathcal{C}^k$*-manifolds* for $k \leq \infty$, if LEGOS is the category $\text{OPEN}_{\mathcal{C}^k}$ of open subsets of $\mathbb{R}^n$ for varying n with $\mathcal{C}^k$-maps between them, and τ is the usual topology;
- $\mathcal{C}^k$*-manifolds with corners* for $k \leq \infty$ (see Example 3.2.4 for a refined notion that uses maps respecting the corner stratification), if LEGOS is the category $\text{OPEN}_{\mathcal{C}^k_+}$ of open subsets of $\mathbb{R}^n_+$ for varying n with smooth maps between them (recall that for $U \subset \mathbb{R}^n_+$, a map $f : U \to \mathbb{R}$ is $\mathcal{C}^k$ if there exists an open $\tilde{U} \subset \mathbb{R}^n$ with $U = \tilde{U} \cap \mathbb{R}^n_+$ and a $\mathcal{C}^k$-function $\tilde{f} : \tilde{U} \to \mathbb{R}$ such that $\tilde{f}_{|U} = f$), and τ is the usual topology;
- *real analytic manifolds* if LEGOS is the category $\text{OPEN}_{an,\mathbb{R}}$ of open subsets of $\mathbb{R}^n$ with real analytic maps between them and τ_{an} is the usual analytic topology generated by subsets of the form

$$U_{f,g} := \{x \in U; |f(x)| < |g(x)|\}$$

for $f, g \in \mathcal{O}(U)$;
- *complex analytic manifolds* if LEGOS is the category $\text{STEIN}_{\mathbb{C}}$ of Stein open subsets U of $\mathbb{C}^n$ with analytic maps between them and τ_{an} the usual analytic topology defined as in the real analytic case;

- *schemes* if LEGOS is the category $\text{AFF}_{\mathbb{Z}} := \text{CALG}_{\mathbb{Z}}^{op}$ of affine schemes (opposite the category of commutative unital rings), equipped with the topology generated by the localization maps $\text{Spec}(A[f^{-1}]) \to \text{Spec}(A)$;
- *algebraic varieties* (schemes of finite type) if LEGOS is the category $\text{AFF}_{\mathbb{Z}}^{ft}$ dual to the category $\text{CALG}_{\mathbb{Z}}^{ft}$ of $\mathbb{Z}$-algebras of finite type (quotients $A = \mathbb{Z}[X_1, \ldots, X_n]/I$ of some polynomial algebra with finitely many variables).

To be able to define differential invariants using a categorical setting for infinitesimal thickenings as in Sect. 2.5, one needs at least to have the following condition.

Definition 3.1.3 A Grothendieck site (LEGOS, τ) is called *differentially convenient* if LEGOS has finite coproducts and an initial object.

In the above examples, the only differentially convenient sites are the categories of schemes. Indeed, the pasting of two open subsets U and V of $\mathbb{R}^n$ (or $\mathbb{C}^n$) for $n > 0$ along a point can't be described by an open subset of $\mathbb{R}^m$ for any $m \geq 0$. An example of an infinitesimal thickening in the setting of schemes is given by the embedding

$$\text{Spec}(\mathbb{R}) \to \text{Spec}\big(\mathbb{R}[x]/(x^2)\big)$$

of a point in its first infinitesimal neighborhood. Its importance for the algebraic formulation of differential calculus was emphasized by Weil. We will see later in this section how to improve on the above examples of sites to obtain infinitesimals in all the situations of Definition 3.1.2.

The most standard example of a generalized geometrical setting, with spaces defined as sheaves on a site, is given by the theory of diffeology, which was developed by J.W. Smith [Smi66] and K.T. Chen [Che77], and used in the physical setting by J.-M. Souriau to explain the geometric methods used by physicists to study variational problems (see [Sou97] and [IZ99] for references and historical background, and [BH11] for an overview).

Definition 3.1.4 Let $\text{OPEN}_{\mathcal{C}^\infty}$ be the category of open subsets of $\mathbb{R}^n$ for varying n with smooth maps between them and τ be the usual topology on open subsets. The category of *diffeologies* (also called *smooth spaces*) is the category $\text{SH}(\text{OPEN}_{\mathcal{C}^\infty}, \tau)$.

Definition 3.1.5 An *ordinary smooth manifold* is a topological space X and a family of open embeddings $f_i : U_i \to X$, with $U_i \subset \mathbb{R}^n$ open subsets, such that for every pair (U_i, U_j), the continuous map

$$f_{i|U_i \cap U_j}^{-1} \circ f_{j|U_i \cap U_j} : U_i \cap f_i^{-1}\big(f_j(U_j)\big) \to U_i \cap f_j^{-1}\big(f_i(U_j)\big)$$

is a smooth map (i.e., a morphism in $\text{OPEN}_{\mathcal{C}^\infty}$). The family $(f_i : U_i \to X)_{i \in I}$ is called an *atlas*. An atlas which is maximal with respect to atlas inclusion is called a *maximal atlas*. A morphism of manifolds is a continuous map that induces a morphism of maximal atlases (in the obvious sense). We denote by $\text{VAR}_{\mathcal{C}^\infty}^{ordinary}$ the category of ordinary smooth manifolds.

We now check, on a simple example, that our notion of variety corresponds to the usual one.

Theorem 3.1.1 *The map*

$$\begin{aligned}\mathrm{VAR}_{\mathcal{C}^\infty}^{ordinary} &\to \mathrm{SH}(\mathrm{OPEN}_{\mathcal{C}^\infty}, \tau)\\ M &\mapsto \left[U \mapsto \mathrm{Hom}_{\mathrm{VAR}_{\mathcal{C}^\infty}^{ordinary}}(U, M)\right]\end{aligned}$$

is a fully faithful embedding of smooth varieties in diffeologies and it induces an equivalence of categories

$$\mathrm{VAR}_{\mathcal{C}^\infty}^{ordinary} \xrightarrow{\sim} \mathrm{VAR}(\mathrm{OPEN}_{\mathcal{C}^\infty}, \tau).$$

Proof There is a natural Grothendieck topology τ on the category $\mathrm{VAR}_{\mathcal{C}^\infty}^{ordinary}$ with covering families given by open morphisms $f_i : U_i \to M$ that cover M with $U_i \in$ LEGOS. We can thus use the natural fully faithful embedding

$$\mathrm{VAR}_{\mathcal{C}^\infty}^{ordinary} \to \mathrm{SH}\left(\mathrm{VAR}_{\mathcal{C}^\infty}^{ordinary}, \tau\right).$$

If $U_i \to M$ is a covering family, one defines its nerve system $N(\{U_i \to M\})$ to be the inductive system of all finite fiber products $N_j = U_{i_1} \times_M \cdots \times_M U_{i_n}$. Note that all its objects are legos. The inductive limit $\varinjlim N_j$ is equal to M, by definition of our notion of variety. There is a natural functor

$$\mathrm{SH}\left(\mathrm{VAR}_{\mathcal{C}^\infty}^{ordinary}, \tau\right) \to \mathrm{SH}(\mathrm{OPEN}_{\mathcal{C}^\infty}, \tau)$$

sending X to its evaluation on legos. If $Y \in \mathrm{SH}(\mathrm{VAR}_{\mathcal{C}^\infty}^{ordinary}, \tau)$ is a sheaf, for every ordinary manifold $X = \varinjlim_j N_j$ given by the limit of the nerve of a covering, one has

$$\begin{aligned}Y(X) &= \mathrm{Hom}_{\mathrm{SH}_{\mathrm{VAR}}}(X, Y) = \mathrm{Hom}_{\mathrm{SH}_{\mathrm{VAR}}}\left(\varinjlim_j N_j, Y\right)\\ &= \varprojlim_j \mathrm{Hom}_{\mathrm{VAR}}(N_j, Y) = \varprojlim_j Y(N_j)\end{aligned}$$

so that $Y(X)$ is determined by the values $Y(N_j)$ of Y on some legos. Applying the same reasoning to morphisms, one concludes that the above functor is fully faithful. Note that in both categories, varieties are defined as inductive limits of nerves for the same topology with same covering opens in $\mathrm{OPEN}_{\mathcal{C}^\infty}$. One thus gets an equivalence

$$\mathrm{VAR}\left(\mathrm{VAR}_{\mathcal{C}^\infty}^{ordinary}, \tau\right) \xrightarrow{\sim} \mathrm{VAR}(\mathrm{OPEN}_{\mathcal{C}^\infty}, \tau).$$

To prove the theorem, it remains to show that the natural functor

$$\mathrm{VAR}_{\mathcal{C}^\infty}^{ordinary} \to \mathrm{VAR}\left(\mathrm{VAR}_{\mathcal{C}^\infty}^{ordinary}, \tau\right)$$

is essentially surjective, since it is already fully faithful. Let X be a variety in $\mathrm{VAR}(\mathrm{VAR}_{C^\infty}^{ordinary}, \tau)$. By hypothesis, we have a family $f_i : \underline{U_i} \hookrightarrow X$ of open embeddings of legos $U_i \in \mathrm{LEGOS}$ such that

$$\coprod_{i \in I} \underline{U_i} \to X$$

is a sheaf epimorphism. This means that $X = \varinjlim_i \underline{U_i}$ in the category of sheaves. By hypothesis, all maps $\underline{U_i} \to X$ are open, so that the maps of the nerve inductive system $N_j = N_j(\{U_i \to X\})$ of legos are open. Let $\tilde{X}$ be the topological space given by the colimit $\tilde{X} = \varinjlim N_j$ in the category TOP of topological spaces. One has open embeddings $U_i \subset X$ that are smoothly compatible, so that $\tilde{X}$ is a variety. In fact, $\tilde{X}$ is the inductive limit $\varinjlim N_j$ in the category $\mathrm{VAR}_{C^\infty}^{ordinary}$. The main properties of sheaves for a Grothendieck topology, given in Theorem 2.4.1, is that their colimits along nerves of coverings correspond to actual colimits in the original category. This shows that $\underline{\tilde{X}} \cong X$ and concludes the proof. □

We now recall, for comparison purposes, the notion of locally ringed space. Note that it allows the introduction of infinitesimal thickenings, that can be used for a categorical formulation of differential calculus. We refer to [Gro60] for more details. The main drawbacks of this category are the following:

- It is algebraic in nature (i.e., germs are local rings), so that it is not always well adapted to the functorial statement of non-algebraic equations of analysis.
- It does not have a notion of internal homomorphisms, i.e., the locally ringed space of morphisms between locally ringed spaces does not exist in general.

We will see in Sect. 3.2 another more natural path to the functorial definition of differential invariants.

Definition 3.1.6 A *locally ringed space* consists of a pair $(X, \mathcal{O}_X)$ composed of a topological space X and a sheaf of functions $\mathcal{O}_X$, with germs given by local rings. A *morphism of locally ringed spaces* is given by a pair

$$(f, f^\natural) : (X, \mathcal{O}_X) \to (Y, \mathcal{O}_Y)$$

with $f : X \to Y$ a continuous map and $f^\natural : f^{-1}\mathcal{O}_Y \to \mathcal{O}_X$ a morphism of sheaves of rings (meaning for each open $U \subset Y$, morphisms $f^\natural : \mathcal{O}_Y(U) \to \mathcal{O}_X(f^{-1}(U))$, compatible with restrictions) such that the corresponding morphism of germs is local (i.e. sends the maximal ideal of one ring into the maximal ideal of the other) at every point of X.

The category of locally ringed spaces contains all kinds of varieties, and has a special auto-duality property: their varieties are no more complicated than themselves.

Proposition 3.1.1 *Let* LRSPACES *be the category of locally ringed spaces. Equip* LRSPACES *with its standard topology* τ *given by open immersions compatible with the sheaves of functions.*

- *If* VAR $\subset$ LRSPACES *is a full subcategory with fiber products (but not necessarily stable by fiber products in* LRSPACES*) with a well defined restriction of the standard topology* τ*, the natural functor*

$$\text{VAR} \to \text{VAR}(\text{VAR}, \tau)$$

is fully faithful.
- *The categories of smooth varieties, schemes and algebraic varieties described in Definition* 3.1.2 *embed fully faithfully in the category of locally ringed spaces.*

Proof The first statement follows directly from the local nature of locally ringed spaces. The second can be checked locally, in which case it is clear because one has enough functions (the fact that germs are local is not easy to prove, though). □

We now define algebraic varieties relative to a symmetric monoidal category.

Definition 3.1.7 Let $(\mathcal{C}, \otimes)$ be a symmetric monoidal category and denote by

$$\text{LEGOS}_{\mathcal{C}} := \text{ALG}_{\mathcal{C}}^{op}$$

the category dual to commutative $\mathcal{C}$-algebras. Let A be a commutative algebra in $(\mathcal{C}, \otimes)$. The functor $\underline{\text{Spec}}(A) \in \text{LEGOS}^{\vee}$ given by

$$\begin{aligned} \underline{\text{Spec}}(A) : \text{LEGOS}_{\mathcal{C}}^{op} = \text{ALG}_{\mathcal{C}} &\to \text{SETS} \\ B &\mapsto \text{Hom}(A, B) \end{aligned}$$

is called the *spectrum* of A.

It is possible, following Toen and Vaquié in [TV09b], to define a Zariski topology on LEGOS by generalizing a known characterization of Zariski opens in ordinary schemes (monomorphisms of spaces that are flat and finitely presented).

Definition 3.1.8 An algebra morphism $f : A \to B$ (or the corresponding morphism $\underline{\text{Spec}}(B) \to \underline{\text{Spec}}(A)$) is called a *Zariski open* if it is a flat and finitely presented monomorphism:

1. (monomorphism) for every algebra C, $\underline{\text{Spec}}(B)(C) \subset \underline{\text{Spec}}(A)(C)$;
2. (flat) the base change functor $- \otimes_A B : \text{MOD}(A) \to \text{MOD}(B)$ is left exact (i.e., commutes with finite limits); and
3. (finitely presented) if $\underline{\text{Spec}}_A(B)$ denotes $\underline{\text{Spec}}(B)$ restricted to A-algebras, $\underline{\text{Spec}}_A(B)$ commutes with filtered inductive limits.

A family of morphisms $\{f_i : \underline{\text{Spec}}(A_i) \to \underline{\text{Spec}}(A)\}_{i \in I}$ is called a *Zariski covering* if all its elements are Zariski open and moreover

4. there exists a finite subset $J \subset I$ such that the functor

$$\prod_{i \in J} - \otimes_A A_j : \mathrm{MOD}(A) \to \prod_{j \in J} \mathrm{MOD}(A_j)$$

is conservative (i.e., preserves isomorphisms).

The *Zariski topology* on $\mathrm{LEGOS}_{\mathcal{C}}$ is the Grothendieck topology generated by Zariski coverings. The corresponding varieties and spaces are called *schemes* and *spaces relative to* $(\mathcal{C}, \otimes)$ and their categories are denoted by $\mathrm{SCH}(\mathcal{C}, \otimes)$ and $\mathrm{SH}(\mathcal{C}, \otimes)$.

Example 3.1.1

1. Usual schemes are schemes relative to the symmetric monoidal category $(\mathrm{MOD}(\mathbb{Z}), \otimes)$ of abelian groups. This result was proved by Grothendieck in EGA [Gro67].
2. Super and graded schemes are schemes relative to the symmetric monoidal categories $(\mathrm{MOD}_s(\mathbb{Z}), \otimes)$ and $(\mathrm{MOD}_g(\mathbb{Z}), \otimes)$.
3. If $(X, \mathcal{O})$ is a locally ringed space, schemes relative to the symmetric monoidal category $(\mathrm{QCoh}(X), \otimes)$ of quasi-coherent sheaves on X (cokernels of morphisms of arbitrary free modules) are called relative schemes on X.

We now explain how to use contravariant constructions on a site (LEGOS, τ) to define contravariant constructions (e.g. differential forms) on spaces in $\mathrm{SH}(\mathrm{LEGOS}, \tau)$. These contravariant constructions apply to all the geometric sites (LEGOS, τ) we will use in this book.

Definition 3.1.9 Suppose given a category $\mathcal{C}$ and a site (LEGOS, τ). Let

$$\Omega : \mathrm{LEGOS}^{op} \to \mathcal{C}$$

be a sheaf for the topology τ with values in $\mathcal{C}$, which we will call a *differential geometric construction* on LEGOS. Let $X \in \mathrm{SH}(\mathrm{LEGOS}, \tau)$ be a space, and let LEGOS/X denote the category of morphisms $x : U_x \to X$ where $U_x \in \mathrm{LEGOS}$. The *differential geometric construction* $\Omega(X)$ *induced on* X is defined as the inverse limit in $\mathcal{C}$ (assumed to exist)

$$\Omega(X) := \varprojlim_{x \in \mathrm{LEGOS}/X} \Omega(U_x).$$

More generally, if $i : U \to X$ is an open morphism of spaces, one can define $\Omega(U)$. This gives (if it exists) a sheaf on X for the topology induced by τ

$$\Omega : \mathrm{SH}(\mathrm{LEGOS}, \tau)/X \to \mathcal{C}.$$

Example 3.1.2 Let $(\mathrm{LEGOS}, \tau) = (\mathrm{OPEN}_{\mathcal{C}^\infty}, \tau)$ be the usual topology on smooth legos and let

$$\Omega = \Omega^1 : \mathrm{OPEN}_{\mathcal{C}^\infty}^{op} \to \mathrm{VECT}_{\mathbb{R}}$$

be the sheaf of differential 1-forms. If $X : \text{OPEN}_{\mathcal{C}^\infty} \to \text{SETS}$ is a diffeology, a differential 1-form on X is the datum, for each $x : \underline{U} \to X$ of a differential form $[x^*\omega]$ on U, i.e., a section of $T^*U \cong U \times (\mathbb{R}^n)^* \to U$, such that if $f : x \to y$ is a morphism, i.e., a commutative diagram

$$\begin{array}{ccc} \underline{U_x} & \xrightarrow{x} & X \\ {\scriptstyle f}\downarrow & \nearrow_{y} & \\ \underline{U_y} & & \end{array},$$

one has

$$f^*[y^*\omega] = x^*\omega.$$

One can similarly define differential n-forms and the de Rham complex

$$(\Omega_X^*, d)$$

of the diffeology X. This example of course generalizes to any kind of spaces whose local models (i.e., LEGOS) have a well behaved (i.e., local and contravariant) notion of differential form. Note that the de Rham complex of a space of maps $\underline{\text{Hom}}(X, Y)$ between two smooth varieties has a nice description (up to quasi-isomorphism) in terms of iterated integrals, that is due to Chen [Che77] in the very particular case of the circle $X = S^1$, and much more recently to Ginot-Tradler-Zeinalian [GTZ09] in general. This description roughly says that there is a natural iterated integral quasi-isomorphism of differential graded algebras (see Sect. 9.8 for a definition of this notion)

$$\text{it}_X : \text{CH}_X(\Omega^*(M)) \longrightarrow \Omega^*(\underline{\text{Hom}}(X, M))$$

that gives a concrete combinatorial description of the de Rham cohomology of the mapping space in terms of the higher Hochschild homology (whose definition is beyond the scope of these notes) of differential forms on M.

Remark 3.1.1 One can go one step further by defining the notion of a locally trivial bundle on a space. Let GRPD be the 2-category of groupoids (i.e., categories with only invertible morphisms). The idea is to use the sheaf

$$\Omega = \text{BUN} : \text{LEGOS}^{op} \to \mathcal{C} = \text{GRPD}$$

with value in the 2-category of groupoids. A bundle B over a diffeology X is the following datum:

- for each point $x : U \to X$, a bundle $[x^*B] \to X$;
- for each morphism $f : x \to y$ of points, an isomorphism

  $$i_f : f^*[y^*B] \xrightarrow{\sim} [x^*B]$$

  of bundles on U;

- for each commutative triangle

$$T = \begin{array}{ccc} x & \xrightarrow{g \circ f} & z \\ {\scriptstyle f}\downarrow & \nearrow_{g} & \\ y & & \end{array}$$

  of points, an isomorphism

$$i_T : f^* i_g \xrightarrow{\sim} i_{g \circ f}$$

- and so on...

It is not so easy to write down all the commutative diagrams in play by hand and there is a nice mathematical theory which enables the proper treatment of this type of geometrical notion, called homotopical geometry. We will come back to this example in Chap. 10.

3.2 Functional Geometry

Recall that if X and Y are sets, one defines the X-indexed product of Y by

$$\prod_X Y := \mathrm{Hom}_{\textsc{Sets}}(X, Y).$$

In particular, $\prod_{\emptyset} X$ is always equal to the one point set $\{.\} = \mathrm{Hom}(\emptyset, X)$. More generally, an empty product in a given category is a final object.

Up to now, we have considered spaces defined as sheaves X on some Grothendieck site $(\textsc{Legos}, \tau)$, meaning contravariant functors

$$X : \textsc{Legos}^{op} \to \textsc{Sets}$$

that respect nerves of coverings in τ. In particular, a given lego U gives a space

$$\underline{U} := \mathrm{Hom}(-, U)$$

also called the functor of points of U. The main drawbacks of this approach are the following:

1. Classical categories of legos (e.g., the category $\textsc{Open}_{\mathcal{C}^\infty}$ of smooth open subsets of $\mathbb{R}^n$) are usually not rich enough to contain infinitesimal thickenings, which play a central role in the categorical approach to differential calculus, presented in Sect. 2.5. This has to do with the fact that the types of function algebras that we have on legos are not general enough to allow the definition of infinitesimal thickenings, so that the corresponding differential invariants (e.g., the tangent category) are trivial. See Example 2.5.2 for a more precise explanation.

2. Given a smooth function $f : U \to \mathbb{R}$ on an open subset of $\mathbb{R}^n$, one would like the zero set space $Z(f) \subset \underline{U}$, defined by the formula

$$\begin{aligned} Z(f) : \mathrm{OPEN}_{\mathcal{C}^\infty}^{op} &\to \mathrm{SETS} \\ V &\mapsto Z(f)(U) := \{x : V \to U,\, f \circ x = 0\}, \end{aligned}$$

to be representable by a lego whose algebra of functions ought to be something like $\mathcal{C}^\infty(U)/(f)$.

The most natural way to solve the above problems by enriching a category of legos was discovered by Lawvere in [Law04]. One can simply embed it in the opposite category $\overline{\mathrm{LEGOS}} = \mathrm{ALG}_{\mathrm{LEGOS}}^{op}$ of a category $\mathrm{ALG}_{\mathrm{LEGOS}}$ of covariant functors

$$A : \mathrm{LEGOS} \to \mathrm{SETS}$$

that respect some prescribed categorical structures (e.g., limits). Of course, one would like to have a natural Grothendieck topology $\bar{\tau}$ induced by τ and a fully faithful embedding of Grothendieck sites

$$(\mathrm{LEGOS}, \tau) \to (\overline{\mathrm{LEGOS}}, \bar{\tau}),$$

to retrieve the usual notion of varieties. We will use sketches (see Definition 2.1.16) to solve this problem of "adding infinitesimals and solution spaces to legos".

The following example is due to Lawvere [Law79], and was used by Dubuc [Dub81] and Moerdijk-Reyes [MR91] to describe generalized smooth spaces, called synthetic differential spaces.

Example 3.2.1 Here is an example of Lawvere's algebraic theory (see Definition 2.1.15) related to (generalized) differential geometry: For $k \leq \infty$, let $\mathrm{AFF}_{\mathcal{C}^k}$ be the category with objects the affine spaces $\mathbb{R}^n$ for varying $n \geq 0$ and with morphisms the $\mathcal{C}^k$-maps between them. It is a full finite product sketch, i.e., a category with finite products. A model

$$A : \mathrm{AFF}_{\mathcal{C}^k} \to \mathrm{SETS}$$

for this theory (i.e., a product preserving functor) is called a (finitary) $\mathcal{C}^k$*-algebra*. We denote by $\mathrm{ALG}_{\mathcal{C}^k}^{fin}$ the category of $\mathcal{C}^k$-algebras. One can think of a $\mathcal{C}^k$-algebra A as a set $A(\mathbb{R})$, equipped with operations

$$[f] : A(\mathbb{R})^n \to A(\mathbb{R})$$

associated with smooth functions $f : \mathbb{R}^n \to \mathbb{R}$. In particular, $A(\mathbb{R})$ is equipped with an $\mathbb{R}$-algebra structure induced by that on $\mathbb{R}$, since the operations

$$\begin{aligned} + : \mathbb{R} \times \mathbb{R} &\to \mathbb{R} \\ \times : \mathbb{R} \times \mathbb{R} &\to \mathbb{R} \\ 0 : \{.\} &\to \mathbb{R} \\ 1 : \{.\} &\to \mathbb{R} \end{aligned}$$

are all $\mathcal{C}^k$-maps. The affine space

$$\begin{aligned}\mathbb{A}^n : \text{ALG}^{fin}_{\mathcal{C}^k} &\to \text{SETS}\\ A &\mapsto A(\mathbb{R})^n\end{aligned}$$

is representable by the $\mathcal{C}^k$-algebra $\mathcal{C}^k(\mathbb{R}^n)$, that is the free $\mathcal{C}^k$-algebra on n generators. Moreover, if $f : \mathbb{R}^n \to \mathbb{R}$ is a $\mathcal{C}^k$-function, its space of zeroes

$$Z(f)(A) := \left\{x \in \mathbb{A}^n(A),\ f(x) = 0\right\}$$

is representable by the $\mathcal{C}^k$-algebra $\mathcal{C}^k(\mathbb{R}^n)/(f)$.

Remark 3.2.1 If $\text{OPEN}_{\mathcal{C}^k}$ denotes the category of open subsets in $\mathbb{R}^n$ for $n \geq 0$ with $\mathcal{C}^k$-maps between them, there is a fully faithful embedding

$$\begin{aligned}\mathcal{C}^k : \text{OPEN}^{op}_{\mathcal{C}^k} &\to \text{ALG}^{fin}_{\mathcal{C}^k}\\ U &\mapsto \left[\mathbb{R}^n \mapsto \text{Hom}_{\mathcal{C}^k}\left(U, \mathbb{R}^n\right)\right]\end{aligned}$$

that sends transversal fiber products (i.e., the products that are well defined in this category) to pushouts. This category is the one usually used by people working in synthetic differential geometry (see e.g., Moerdijk-Reyes [MR91]), because it contains algebras of the form $\mathcal{C}^\infty(\mathbb{R})/(\epsilon^2) \cong \mathbb{R}[\epsilon]/(\epsilon^2)$, that play the role of infinitesimal thickenings. If we denote by $\text{OPEN}_{\mathcal{C}^k_+}$ the category of open subset of $\mathbb{R}^n_+$ with smooth morphisms between them, there is also a fully faithful embedding

$$\mathcal{C}^k_+ : \text{OPEN}^{op}_{\mathcal{C}^k_+} \to \text{ALG}^{fin}_{\mathcal{C}^k}.$$

One can actually show that if $U \subset \mathbb{R}^n_+$ is an open subset, one has

$$\mathcal{C}^k_+(U) = \lim_{\tilde{U} \supset U} \mathcal{C}^k(\tilde{U})$$

where the limit is taken in $\text{ALG}^{fin}_{\mathcal{C}^k}$ and is indexed by open subsets $\tilde{U} \subset \mathbb{R}^n$ such that $U \subset \tilde{U} \cap \mathbb{R}^n_+$ (see [MR91], Chap. 1). One can also use finer notions of morphisms between open subset of $\mathbb{R}^n_+$ with corners. This give a corresponding theory of $\mathcal{C}^k$-ring with corners (see Example 3.2.4 for a refined discussion).

A drawback of the theory of smooth rings described in Example 3.2.1 is that it has no a priori topology and one has to fix a topology on the category of smooth rings (by using the Zariski topology on finitely generated rings, for example). We now give another way of defining a notion of generalized algebra that allows the direct definition of a Grothendieck topology on their opposite category. We are inspired here by Dubuc's work on synthetic analytic geometry, with Taubin and Zilber (see [DT83, DZ94, Zil90]).

Definition 3.2.1 A *differential geometric context* is a tuple

$$\mathbf{C} = (\text{LEGOS}, \tau, \mathfrak{L}, \mathfrak{C}, \mathcal{L}, \mathcal{C}, \text{ALG}_{\mathbf{C}})$$

composed of a Grothendieck site (LEGOS, τ), a category ALG, two classes $\mathfrak{L}$ and $\mathfrak{C}$ of small categories, and an $(\mathfrak{L}, \mathfrak{C})$-sketch (LEGOS, $\mathcal{L}, \mathcal{C}$). A model

$$A : (\text{LEGOS}, \mathcal{L}, \mathcal{C}) \to \text{SETS}$$

is called a *generalized algebra* (in the given differential geometric context). The category $\text{ALG}_{\mathbf{C}}$ is a sub-category of the category of generalized algebras that contains the image of LEGOS. If $U \in$ LEGOS, we denote by $\mathcal{O}(U)$ the corresponding generalized algebra of functions given by

$$\mathcal{O}(U)(V) := \text{Hom}(U, V).$$

A generalized algebra is called *local* if it sends covering nerves to colimits.

Definition 3.2.2 Let $\mathbf{C} = (\text{LEGOS}, \tau, \mathfrak{L}, \mathfrak{C}, \mathcal{L}, \mathcal{C}, \text{ALG}_{\mathbf{C}})$ be a differential geometric context. We denote by Spec(A) an object of $\overline{\text{LEGOS}} := \text{ALG}_{\mathbf{C}}^{op}$. Suppose that the families

$$\{\text{Spec}(A_j) \to \text{Spec}(A)\}_{j \in J}$$

whose pullback along any map Spec$(\mathcal{O}(U)) \to$ Spec(A), for $U \in$ LEGOS, is induced by a covering family $\{U_j \to U\}_{j \in J}$ in τ, form a Grothendieck topology $\bar{\tau}$ on $\overline{\text{LEGOS}}$. The differential geometric context is called *well adapted* if the natural functor

$$(\text{LEGOS}, \tau) \to (\overline{\text{LEGOS}}, \bar{\tau})$$

is a fully faithful embedding of sites compatible with the topology and induces an embedding

$$\text{VAR}(\text{LEGOS}, \tau) \to \text{VAR}(\overline{\text{LEGOS}}, \bar{\tau})$$

of the corresponding categories of varieties.

We now illustrate the above construction by defining a natural category of generalized $\mathcal{C}^k$, complex and real analytic rings, equipped with a Grothendieck topology.

Example 3.2.2 Let $\mathfrak{L}$ be the classes of finite (possibly empty) categories and $\mathfrak{C} = \emptyset$. Recall that an empty limit is equivalent to a final object. The doctrine of $(\mathfrak{L}, \mathfrak{C})$-sketches is called the doctrine of finite limit sketches. Let LEGOS be one of the following categories:

1. open subsets $U \subset \mathbb{R}^n$ with $\mathcal{C}^k$-maps between them, for varying n, and some fixed $0 < k \leq \infty$;
2. open subsets $U \subset \mathbb{R}^n$ with real analytic maps between them;

3. open subsets $U \subset \mathbb{C}^n$ (one may also restrict to Stein subsets) with complex analytic maps between them.

Equip LEGOS with its usual topology, the class $\mathcal{L}$ of transversal (possibly empty) pullbacks. The corresponding algebras are respectively called $\mathcal{C}^k$-rings, real analytic rings and complex analytic rings. There exists choices of sub-categories $\mathrm{ALG}_{\mathrm{C}}$ of algebras that give a well adapted differential geometric context (see [MR91] and [DZ94]). Let $(\overline{\mathrm{LEGOS}}, \bar{\tau})$ be the corresponding category of generalized affine spaces. The natural functor

$$\mathrm{LEGOS} \to \mathrm{SH}(\overline{\mathrm{LEGOS}}, \bar{\tau})$$

is a fully faithful embedding that extends to the corresponding category of varieties. The categories of sheaves thus constructed are called the category of $\mathcal{C}^k$-spaces and generalized real and complex analytic spaces respectively. The main advantages of the category of generalized spaces on ordinary varieties are the following:

1. Generalized algebras contain infinitesimal objects, like $\mathcal{C}^\infty(\mathbb{R})/(\epsilon^2)$ for example in the smooth case. This allows us to use the methods of categorical infinitesimal calculus (see Sect. 2.5) to define their differential invariants.
2. One can define the solution algebra $\mathcal{O}(U)/(f)$ for a given equation $f = 0$ with $f \in \mathcal{O}(U)$.
3. Generalized spaces contain spaces of functions $\underline{\mathrm{Hom}}(X, Y)$ between two given spaces. This is important since spaces of functions, called spaces of fields by physicists, are omnipresent in physics.

Remark 3.2.2 There are various refinements of the notion of an analytic ring and building blocks for synthetic analytic geometry. Their main interest is that they allow the treatment of local properties of analytic spaces, and not only of infinitesimal properties. These are based on Dubuc-Taubin-Zilber's notion of a local analytic ring [Zil90] and locally analytically ringed space [DT83]. These analytic schemes based on local analytic rings have the advantage on usual analytic rings of making germs of analytic functions at a given point "finitely generated" in the categorical sense (see [DZ94]). One may also use Stein opens as building blocks for analytic rings.

A nice result is that finitary smooth algebras are equivalent to smooth algebras. This result is no longer true in the holomorphic or algebraic case, and is deeply related to the fact that smooth manifolds have a partition of unity, which implies that the Zariski and usual topology are identified in this smooth setting.

Proposition 3.2.1 *Let* $\mathrm{ALG}_{\mathcal{C}^\infty}$ *be the category of smooth algebras, given by functors*

$$A : \left(\mathrm{OPEN}_{\mathcal{C}^\infty}, - \underset{-}{\overset{t}{\times}} -\right) \to \left(\mathrm{SETS}, - \underset{-}{\times} -\right)$$

that commute with transversal and empty fiber products. Let $\mathrm{ALG}^{fin}_{\mathcal{C}^\infty}$ *be the category of finitary smooth algebras, given by functors*

$$A : (\mathrm{AFF}_{\mathcal{C}^\infty}, \times) \to (\mathrm{SETS}, \times)$$

on the category of smooth affine spaces $\mathbb{R}^n$ *commuting with (possibly empty) products. The natural functor*

$$\mathrm{ALG}_{\mathcal{C}^\infty} \to \mathrm{ALG}^{fin}_{\mathcal{C}^\infty}$$

is an equivalence.

Proof This result follows from the following facts (see [BD86], Theorem 1.6):

1. The finitary smooth algebra of functions on an open subset $U \subset \mathbb{R}^n$ is naturally identified with the localization

$$\mathcal{C}^\infty(\mathbb{R}^n)[f^{-1}] := \mathcal{C}^\infty(\mathbb{R}^{n+1})/(x_{n+1}f - 1)$$

   of the algebra of functions on the affine space along a given smooth function $f : \mathbb{R}^n \to \mathbb{R}$.
2. For two finitely generated ideals $I, J \subset A$, one has

$$A/I \otimes_A A/J \cong A/(I, J).$$

3. The functor $\mathrm{OPEN}_{\mathcal{C}^\infty} \to \mathrm{ALG}^{fin}_{\mathcal{C}^\infty}$ sends transversal pullbacks to pushouts (follows from the implicit function theorem, see [MR91], Theorem 2.8).
4. If $A \in \mathrm{ALG}^{fin}_{\mathcal{C}^\infty}$ is a finitary smooth algebra and $U \overset{t}{\underset{V}{\times}} W$ is a transversal fiber product, one has

$$A\left(U \overset{t}{\underset{V}{\times}} W\right) \cong A(U) \times_{A(V)} A(W),$$

   where $A(U) := \mathrm{Hom}_{\mathrm{ALG}^{fin}_{\mathcal{C}^\infty}}(\mathcal{O}(U), A)$. This allows us to construct a left inverse to the above natural functor.
5. A $\mathcal{C}^\infty$-algebra A is uniquely determined by its value at $\mathbb{R}$. Indeed, for $U \subset \mathbb{R}^n$ an open subset (described as the complement of the set of zeroes of a function $f : U \to \mathbb{R}$), the following diagrams are transversal pullbacks:

$$\begin{array}{ccc} \mathbb{R}^* & \xrightarrow{(x,1/x)} & \mathbb{R}\times\mathbb{R} \\ \downarrow & & \downarrow \\ \{\cdot\} & \longrightarrow & \mathbb{R} \end{array} \qquad \begin{array}{ccc} U & \longrightarrow & \mathbb{R}^n \\ \downarrow & & \downarrow f \\ \mathbb{R}^* & \longrightarrow & \mathbb{R} \end{array}$$

□

We now describe various types of semianalytic spaces which play an important role in algebraic analysis. We are inspired here by non-Archimedean analytic geometry (see [Ber90] and [Hub96]), and by the theory of real semianalytic and subanalytic sets (see [Gab96, Łoj93, BM88]).

Example 3.2.3 Let LEGOS be one of the following categories:

1. Relatively compact complex rational domains $U \subset \mathbb{C}^n = \mathbb{A}^n_{\mathbb{C}}$ defined by finite intersections of domains

$$U_{f,g} := \{x \in \mathbb{C}^n, |f(x)| < |g(x)|\},$$

   for f and g analytic, with overconvergent functions (i.e., functions that are analytic on the closure $\bar{U}_{f,g}$).
2. Relatively compact rational domains $U \subset \mathbb{C}^n$ stable by complex conjugation σ (i.e., relatively compact domains U in the real Berkovich affine space $\mathbb{A}^n_{\mathbb{R}} = \mathbb{C}^n/\sigma$), with overconvergent real analytic functions on them.

One may also possibly use only Stein semianalytic subsets, depending on the specifications for the geometry to be constructed. Equip LEGOS with the Grothendieck topology generated by usual coverings by rational domains, and with the class of limit cones given by transversal pullbacks. The corresponding algebras are called (overconvergent) complex and real semianalytic algebras and the corresponding varieties are called (overconvergent) semianalytic spaces. These can be equipped with their natural Grothendieck topology, induced by the usual topology (and given by finite coverings). Note that complex subanalytic subsets, defined as projections of relatively compact constructible domains (in the algebra generated by rational domains for complement, union and intersection), along affine projections $\mathbb{A}^n_X := X \times \mathbb{C}^n \to X$ identify naturally with Gabrielov real subanalytic subsets[1] of the complex space X. To obtain a similar result in the real semianalytic setting, one needs to add étale coverings (working with projections of étale constructible subsets). Note also that the space of points of the semianalytic site on a given semianalytic space is something like a real or complex analytic version of the Zariski-Riemann space of a ring (see [HK94]). It may also be useful to work with spaces modeled on the Grothendieck site of subanalytic subsets, to relate our constructions to the methods of algebraic analysis, presented in Chap. 11.

One may also define a refined notion of smooth algebra with corners by the following process. We are inspired here by Joyce's notion of a manifold with corners of any dimension (see [Joy09]), but give a different approach.

Example 3.2.4 An open subset with corners $U \subset (\mathbb{R}_+)^n$ may be stratified by the depth of its points. The depth of a point $u = (u_1, \ldots, u_n)$ is the number of its com-

[1] Remark due to Gabrielov: use that $|f| = c$ is constructible and $\log |e^f| = \mathrm{Re}(f)$.

ponents that are nonzero. There is a stratification

$$U = \coprod_{k=0}^{n} D^k(U)$$

where $D^k(U) := \{x \in U, \text{depth}(x) = k\}$. Recall that the smooth structure on a subset of $(\mathbb{R}_+)^n$ is obtained by using smooth functions on an arbitrary $\tilde{U}$ around U, where $\tilde{U} \subset \mathbb{R}^n$ is open and $U = \tilde{U} \cap (\mathbb{R}_+)^n \subset \mathbb{R}^n$. We denote by $\mathrm{OPEN}_{C^\infty}^{strat}$ the category of stratified open subsets with corners, with smooth maps respecting the stratification, in the following sense: a point of depth k must be sent to a point of depth less than or equal to k. With this condition there are non-trivial maps $[0,1] \to \mathbb{R}$, and any non-trivial map $\mathbb{R} \to [0,1]$ sends $\mathbb{R}$ into $]0,1[\subset [0,1]$. One may define the notion of a transversal fiber product of stratified open sets, and use it to define the category $\mathrm{ALG}_{C^\infty}^{strat}$ of stratified smooth algebras. The topology on stratified open subsets is induced by the usual topology on $\mathbb{R}^n$, using the fact that a smooth open subset of $\mathbb{R}^n_+$ correspond to germs of open subsets of $\mathbb{R}^n$ around $\mathbb{R}^n_+$. This gives a corresponding category of stratified smooth spaces. One may also use the formalism of Sect. 2.5 to define partial differential equations on stratified spaces. Now if $S : \underline{\Gamma}(M, C) \to \mathbb{R}$ is a local action functional on the space of sections of a bundle C on a compact stratified space M, then we may use the stratified structure to explicitly compute the variation of the action, via a stratified integration by parts. The relation of this with the stratified jet formalism remains to be explored. Recall that throughout this book, we have chosen to use the notion of histories (see our viewpoint of Lagrangian variational problems in Definition 8.1.1) precisely to kill the boundary terms of the integration by parts, and avoid the above technicalities. This formalism could help to give a differential version of the notion of higher topological quantum field theory (see [Lur09]).

3.3 Differential Invariants

We now describe the concrete implementations of the tools of categorical infinitesimal calculus, presented in Sect. 2.5, in the differential geometric settings that will be in use in our description of physical systems.

3.3.1 Algebraic Differential Invariants

Here, we essentially follow Lychagin [Lyc93] and Toen–Vezzosi [TV08].

Definition 3.3.1 A symmetric monoidal category $(\mathcal{V}, \otimes)$ is called *pre-additive* if the initial and final object coincide (we denote it 0), the natural map

$$M \oplus N \to M \times N,$$

given by $(m, n) \mapsto (m, 0) \oplus (0, n)$ is an isomorphism, and the monoidal structure commutes with finite direct sums.

From now on, we denote by $(\mathcal{V}, \otimes)$ a pre-additive symmetric monoidal category. Let $\mathrm{ALG}_{\mathcal{V}}$ be the category of commutative monoids in $\mathcal{V}$. These are the basic building blocks of (generalized) algebraic varieties, as explained in Definition 3.1.8.

Lemma 3.3.1 *Let $A \in \mathrm{ALG}_{\mathcal{V}}$ be a commutative monoid in $\mathcal{V}$ and $M \in \mathrm{MOD}(A)$ be a module (in the usual sense of symmetric monoidal categories). The direct sum $A \oplus M$ is naturally equipped with a commutative monoid structure such that the projection $A \oplus M \to A$ is a monoid morphism.*

Proof The multiplication map on $A \oplus M$ is defined by the formula

$$\mu_{A \oplus M} : (A \oplus M) \otimes (A \oplus M) \cong A \otimes A \oplus A \otimes M \oplus M \otimes A \oplus M \otimes M \to A \oplus M$$

given by combining the four morphisms

$$\begin{aligned} &\mu \oplus 0 : A \otimes A \to A \oplus M, \\ &0 \oplus \mu_M^l : A \otimes M \to A \oplus M, \\ &\mu_M^r \oplus U : M \otimes A \to A \oplus M, \quad \text{and} \\ &0 : M \otimes M \to M. \end{aligned}$$

The unit $1 : \mathbb{1} \to A \oplus M$ is given by $1_A \oplus 0$. This makes $A \oplus M$ a commutative monoid in $\mathcal{V}$ such that the projection $A \oplus M \to A$ is a monoid morphism. □

Theorem 3.3.1 *Let $\mathcal{V} = \mathrm{MOD}(\mathbb{Z})$ be the monoidal category of abelian groups. Let $\mathrm{MOD}_{\mathcal{V}}$ be the category of pairs (A, M) composed of a commutative monoid A and a module M over it, with morphisms given by pairs $(f, f^\sharp) : (A, M) \to (B, N)$ of a monoid morphism $f : A \to B$ and an A-module morphism $f^\sharp : M \to f_*N$ (where f_*N is N regarded as an A-module).*

1. *The functor*

$$\begin{aligned} \mathrm{MOD}_{\mathcal{V}} &\to T\mathrm{ALG}_{\mathcal{V}} \\ (A, M) &\mapsto [A \oplus M \to A] \end{aligned}$$

that sends a module over a monoid to the corresponding square zero extension is an equivalence.

2. *The category $\mathrm{Th}^n \mathrm{ALG}_{\mathcal{V}}$ is the subcategory of $[I, \mathcal{C}]$ whose objects are quotient morphisms $A \to A/I$, with kernel a nilpotent ideal I (i.e. nilpotent sub A-module of A) of order $n+1$ (i.e., satisfying $I^{n+1} = 0$).*
3. *The jet functor $\mathrm{Jet}^n : \mathcal{C} \to \mathrm{Th}^n \mathcal{C}$ is given by the corresponding ordinary jet algebra:*

$$\mathrm{Jet}^n(A) = (A \otimes A)/I^{n+1}$$

where I is the kernel of the multiplication map $m : A \otimes A \to A$.

Proof We restrict to the case of usual algebras (monoids in $\text{MOD}(\mathbb{Z})$), which is due to Beck and Quillen (see the module section on the nlab contributive website), but the result may extend to more general (e.g. graded) settings. Let A be a ring and $p : B \to A$ be an abelian group object in its overcategory. The unit axiom gives a diagram

$$\begin{array}{ccc} A & \longrightarrow & B \\ & {}_{\mathrm{id}}\searrow \quad \swarrow_{p} & \\ & A & \end{array}$$

that makes B, as an abelian group, a direct sum $A \oplus M$, with $M = \ker(p)$, and if $a \in A$ and $m \in M$, we have $am \in M$. The product of $A \oplus M$ in the overcategory is given by $A \oplus M \oplus M$. The addition operation on the abelian group object is therefore a morphism

$$\begin{array}{ccc} A \oplus M \oplus M & \longrightarrow & A \oplus M \\ & {}_{\mathrm{id}}\searrow \quad \swarrow_{p} & \\ & A & \end{array}$$

The unit axiom on this operation together with the fact that the top morphism is a ring homomorphism imply that this morphism is $\mathrm{id} \oplus (\mathrm{id} + \mathrm{id})$, i.e., addition on M. In this way one obtains an A-module structure on M, such that $B = A \oplus M$ is equipped with its standard square zero extension ring structure. If $p : C/I^2 \to C/I = A$ is a quotient by a general square zero ideal, it may be regarded similarly as a torsor under the abelian group object given by the square zero extension $A \oplus I/I^2 \to A$. The remaining results follow from the definitions. □

Note that $\mathrm{Jet}^n A$ is naturally equipped with two A-algebra structures denoted by d_0 and d_1.

Definition 3.3.2 The A-module of *vector fields* is defined by

$$\Theta_A := \underline{\mathrm{Hom}}_A\big(\Omega^1_A, A\big).$$

The A-module of *differential operators of order n* is defined by

$$\mathcal{D}^n_A := \underline{\mathrm{Hom}}_A\big(\big(\mathrm{Jet}^n(A), d_1\big), A\big).$$

One defines a composition operation

$$\mathcal{D}^n_A \underset{\mathbb{1}}{\otimes} \mathcal{D}^n_A \to \mathcal{D}^n_A$$

by $D_1 \otimes D_2 \mapsto D_1 \circ D_2 := D_1 \circ d_0 \circ D_2 : \mathrm{Jet}^n(A) \to A$. The A-module of *differential operators* is the colimit

$$\mathcal{D}_A := \operatorname{colim}_n \mathcal{D}_A^n.$$

The intuition behind the definition of derivations in our general setting is the following: a homomorphism $D : A \to M$ in $\mathcal{C}$ will be called a derivation if

$$D(fg) = D(f)g + fD(g).$$

To be mathematically correct, we need to define derivations internally using the tensor structure. Here we suppose that our monoidal category is additive.

Definition 3.3.3 Let M be an A-module. The *internal derivation object* $\underline{\mathrm{Der}}(A, M)$ is the kernel of the inner Leibniz morphism (defined by adjunction of inner homomorphisms with the tensor product)

$$\begin{array}{llll} \underline{\text{Leibniz}}: & \underline{\mathrm{Hom}}(A, M) & \longrightarrow & \underline{\mathrm{Hom}}(A \otimes A, M) \\ & D & \longmapsto & \left(D \circ \mu,\ \mu_M^l \circ (\mathrm{id}_A \otimes D) + \mu_M^r \circ (D \otimes \mathrm{id}_A)\right) \end{array}$$

In the setting $\mathcal{V} = \mathrm{MOD}(\mathbb{Z})$, we have $\Theta_A = \underline{\mathrm{Der}}(A, A)$.

Note that for $\mathcal{C} = \mathrm{MOD}_g(K)$ the symmetric monoidal category of graded modules, the Leibniz condition can be expressed as a graded Leibniz rule by definition of the right A-module structure on M above. More precisely, an internal morphism $D : A \to M$ is a graded derivation if it satisfies the graded Leibniz rule

$$D(ab) = D(a)b + (-1)^{\deg(D)\deg(f)} a D(b).$$

The restriction of this construction to the category $\mathrm{MOD}_s(K)$ of super modules gives the notion of super-derivation.

We carefully inform the reader that if $A = \mathcal{C}^\infty(\mathbb{R})$ is the algebra of smooth function on $\mathbb{R}$, then in Ω_A^1, one has

$$d\exp(x) \neq \exp(x)dx.$$

This shows that in contrast to the notion of derivation, which is a purely algebraic notion depending only on the algebra of function, and not on the kind of geometry one is working with (smooth, analytic, algebraic, etc....), differential forms depend on the type of geometry in play. We refer to Sect. 3.3.2 for a solution to this problem.

We will now suppose that the monoidal category $\mathcal{V} = \mathrm{MOD}(K)$ is that of modules over a commutative unital (possibly graded) ring and denote by A a commutative monoid in $\mathcal{V}$.

The module Θ_A is endowed with a Lie bracket

$$[\cdot,\cdot]:\Theta_A\otimes\Theta_A\to\Theta_A$$

induced by the commutator of endomorphisms in $\underline{\mathrm{Hom}}(A,A)$. There is also a natural action

$$[\cdot,\cdot]:\Theta_A\otimes A\to A$$

which is compatible with the above bracket and induces for each $\partial\in\Theta_A$ a derivation $\partial:A\to A$. All the above data is called the A-Lie algebroid structure on Θ_A.

Recall from [BD04], 2.9.1, some basics on Lie algebroids.

Definition 3.3.4 A *Lie algebroid over A* is an A-module L equipped with a bracket

$$[\cdot,\cdot]:L\otimes L\to L \quad \text{and} \quad \text{an } A\text{-linear anchor map } \tau:L\to\Theta_A$$

that satisfies for every $x,y,z\in L$ and $a\in A$:

1. Jacobi's identity: $[x,[y,z]]=[[x,y],z]+[y,[x,z]]$;
2. anti-commutativity: $[x,y]=-[y,z]$;
3. compatibility of the anchor with the Lie bracket: $[x,ay]=\tau(x)(a)\cdot y+a[x,y]$.

An L-module is an A-module M with an action of L compatible with the L-action on A: for every $a\in A$, $m\in M$ and $x\in L$, one has:

1. $x(am)=x(a)m+a(xm)$;
2. $(ax)m=a(xm)$ (left module structure) or $(ax)m=x(am)$ (left module structure).

The tangent Lie algebroid Θ_A is clearly the final Lie A-algebroid.

Proposition 3.3.1 *If A is smooth (meaning that Ω^1_A is a projective A-module of finite type), then $\mathcal{D}_A$ (resp. $\mathcal{D}_A^{op}$) is the enveloping algebra of the Lie-A-algebroid Θ_A of vector fields: if B is an associative A-algebra in $(\mathcal{V},\otimes)$ equipped with an A-linear map $i:\Theta_A\to B$ (that induces the adjoint Θ_A-action on B) such that the inclusion $A\subset B$ is a morphism of left (resp. right) Θ_A-modules, there exists a unique morphism*

$$\mathcal{D}_A\to B$$
$$\left(\text{resp. } \mathcal{D}_A^{op}\to B\right)$$

extending $i:\Theta_A\to B$.

Definition 3.3.5 A (left)$\mathcal{D}_A$*-module* is an object of M in $\mathcal{V}$, equipped with a left multiplication

$$\mu^l_{\mathcal{D}}:\mathcal{D}_A\otimes M\to M$$

that is compatible with the multiplication in $\mathcal{D}_A$.

Proposition 3.3.2 *Suppose that A is smooth, so that $\mathcal{D}_A$ is the enveloping algebra of the A-algebroid Θ_A of vector fields. The category of left $\mathcal{D}_A$-modules is equipped with a symmetric monoidal structure* $(\mathrm{MOD}(\mathcal{D}_A), \otimes)$ *defined by*

$$\mathcal{M} \otimes \mathcal{N} := \mathcal{M} \otimes_A \mathcal{N}$$

with the $\mathcal{D}_A$-module structure induced by the action of derivations $\partial \in \Theta_A$ by

$$\partial(m \otimes n) = \partial(m) \otimes n + m \otimes \partial(n).$$

Proof The action above is compatible with the action of functions on vector fields and with the Lie bracket of vector fields, so that it extends to an action of $\mathcal{D}_A$ since A is smooth by Proposition 3.3.1. □

Definition 3.3.6 A *differential complex* is a graded A-module M equipped with a $\mathcal{D}_A^{op}$-linear morphism $d : M \otimes_A \mathcal{D}_A \to M \otimes_A \mathcal{D}_A[1]$. Let $(\mathrm{Diffmod}_{dg}(A), \otimes)$ be the symmetric monoidal category of differential complexes. The *algebra of differential forms* on A is the free algebra in $(\mathrm{Diffmod}_{dg}(A), \otimes)$ on the de Rham differential $d : A \to \Omega_A^1$, given by the symmetric algebra

$$\Omega_A^* := \mathrm{Sym}_{(\mathrm{Diffmod}_{dg}(A),\otimes)}\left(\left[\underset{0}{A} \xrightarrow{d} \underset{1}{\Omega_A^1}\right]\right).$$

We now recall the basic properties of differential forms.

Theorem 3.3.2 *The natural map*

$$\Theta_A \to \underline{\mathrm{Hom}}_A\left(\Omega_A^1, A\right)$$

extends to a morphism

$$i : \Theta_A \to \mathrm{Hom}_{\mathrm{Diffmod}_g(A)}\left(\Omega_A^*, \Omega_A^*[1]\right)$$

called the interior product map (its images are odd derivations of Ω_A^*). The natural map

$$\Theta_A \to \underline{\mathrm{Hom}}(A, A)$$

extends to a morphism

$$L : \Theta_A \to \mathrm{Hom}_{\mathrm{MOD}_g(A)}\left(\Omega_A^*, \Omega_A^*\right),$$

called the Lie derivative. One has "Cartan's formula"

$$L_X\omega = i_X(d\omega) + d(i_X\omega).$$

Proof One can consider the monoidal category $(\mathrm{Diffmod}_{tridg}(A), \otimes)$ composed of graded A-differential-modules M equipped with a triple of differential operators

$$\overset{L}{\circlearrowright} M \underset{i}{\overset{d}{\rightleftarrows}} M[1] .$$

Recall that the Lie derivative is defined by $L_X(f) = X(f)$ on functions $f \in A$ and by $(L_X\omega)(Y) = \omega([X, Y])$ on 1-forms $\omega \in \Omega^1_A$. One then has, for every $X \in \Theta_A$, an isomorphism

$$\left(\Omega^*_A, d, i_X, L_X\right) \cong \mathrm{Sym}_{(\mathrm{Diffmod}_{tridg}(A),\otimes)}\left(\left[\begin{array}{c} \overset{L_X}{\circlearrowright} \underset{0}{A} \underset{i_X}{\overset{d}{\rightleftarrows}} \underset{1}{\Omega^1_A} \overset{L_X}{\circlearrowleft} \end{array}\right]\right)$$

and the equality $i_X(df) = X(f)$ implies Cartan's formula by extension to the symmetric algebra. □

Example 3.3.1 In $(\mathcal{V}, \otimes) = (\mathrm{VECT}_{\mathbb{R}}, \otimes)$, for $A = \mathbb{R}[X_1, \ldots, X_n]$, the interior product of a $k + 1$-form with a vector field is a k-form given by

$$i_X\omega(X_1, \ldots, X_k) = \omega(X, X_1, \ldots, X_k).$$

It can be computed explicitly by using the fact that i_X is a derivation and that it is known on 1-forms. The Lie derivative can be computed explicitly using the Leibniz rule for the exterior differential

$$d(\omega \wedge \eta) = d(\omega) \wedge \eta + (-1)^{\deg(\eta)}\omega \wedge d(\eta),$$

the Leibniz rule for the interior product

$$i_X(\omega \wedge \eta) = i_X(\omega) \wedge \eta + (-1)^{\deg(\eta)}\omega \wedge i_X(\eta),$$

and Cartan's formula

$$L_X\omega = i_X(d\omega) + d(i_X\omega).$$

We now describe the natural geometrical setting for the functor of points approach to algebraic partial differential equations.

Definition 3.3.7 Let A be an algebra in $(\mathcal{V}, \otimes)$. Algebras in the monoidal category

$$\left(\textsc{Mod}(\mathcal{D}_A), \otimes\right)$$

are called $\mathcal{D}_A$*-algebras*, and their category is denoted by $\textsc{Alg}_{\mathcal{D}_A}$.

Recall from Beilinson and Drinfeld's book [BD04], Sect. 2.3.2, the existence of universal $\mathcal{D}$-algebras, called algebraic jet algebras. This way of looking at infinite jets shows that they are typically finite dimensional objects ($\mathcal{D}$-algebras of finite type), contrary to what one may think at first sight.

Proposition 3.3.3 *Let A be an algebra in $(\mathcal{V}, \otimes)$ and $\textsc{Alg}_{\mathcal{D}}$ be the category of commutative unital $\mathcal{D}_A$-algebras. The functor*

$$\mathrm{Forget}_{\mathcal{D}} : \textsc{Alg}_{\mathcal{D}} \to \textsc{Alg}(A)$$

that sends a $\mathcal{D}_M$-algebra to the underlying A-algebra has a right adjoint called the algebraic jet algebra $\mathrm{Jet}_{alg}(B)$.

Proof The algebra $\mathrm{Jet}_{alg}(B)$ is given by the quotient of the symmetric algebra

$$\mathrm{Sym}^*(\mathcal{D}_A \otimes_A B)$$

in the symmetric monoidal category $(\textsc{Mod}(\mathcal{D}), \otimes)$ by the ideal generated by the elements

$$\partial(1 \otimes r_1.1 \otimes r_2 - 1 \otimes r_1 r_2) \in \mathrm{Sym}^2(\mathcal{D} \otimes_A B) + \mathcal{D} \otimes_A B$$

and

$$\partial(1 \otimes 1_B - 1) \in \mathcal{D} \otimes B + A,$$

$r_i \in B$, $\partial \in \mathcal{D}$, 1_B the unit of B. □

Example 3.3.2 If $A = \mathbb{R}[X]$ is the ring of polynomial functions on $\mathbb{R}$ and $B = \mathbb{R}[x, u]$ is the algebra of coordinates on the trivial algebraic bundle $\pi : C = \mathbb{R}^2 \to \mathbb{R} = M$, one gets as jet algebra

$$\mathrm{Jet}(B) = \mathbb{R}[x, u_0, \ldots, u_n, \ldots]$$

the infinite polynomial algebra with action of ∂_x given by

$$\partial_x u_i = u_{i+1}.$$

It corresponds to the algebraic jet bundle $\pi_\infty : \mathrm{Jet}(C) \to M$. A section of the bundle $\pi : C \to M$ is a polynomial $P \in \mathbb{R}[x]$ and the jet map

$$\Gamma(M, C) \to \Gamma\big(M, \mathrm{Jet}(C)\big)$$

is given by sending $P(x) \in \Gamma(M, C) \cong \mathbb{R}[x]$ to its Taylor expansion

$$j_\infty P(x) \in \Gamma\big(M, \mathrm{Jet}(C)\big) \cong \mathbb{R}[x]^{\{u_0,\dots,u_n,\dots\}},$$

given by

$$(j_\infty P)(x) = \big(\partial_x^n P(x)\big)_{n\geq 0} = \sum_{n\geq 0} \frac{\partial_x^n P(x)}{n!}(X-x)^n.$$

Remark 3.3.1 One can apply the methods of algebraic differential calculus presented in this section directly to the symmetric monoidal category $(\mathrm{MOD}(\mathcal{O}_X), \otimes_{\mathcal{O}_X})$ of quasi-coherent modules over a variety $(X, \mathcal{O}_X)$ to define covariant differential invariants (vector fields, differential operators). However, algebraic methods don't give the right result in general for contravariant constructions (differential forms, jets). Indeed, in this way one only gets Kähler differentials and jets, which are insufficient for the study of smooth equations. This problem can be overcome in at least three ways:

1. One can do categorical infinitesimal calculus in a full subcategory of the category of locally ringed spaces.
2. One can restrict the functors of differential calculus (derivations, differential operators) to a category of geometric modules. This is the approach used by Nestruev [Nes03].
3. One can work with smooth algebras, as will be explained in Sect. 3.3.2; we will use this approach because it is optimal for our purposes.

3.3.2 Smooth Differential Invariants

Having studied the algebraic case in the previous section, we now show that the formalism of categorical infinitesimal calculus given in Sect. 2.5 gives back ordinary smooth differential calculus on smooth varieties. This is the purpose of the following theorem.

Theorem 3.3.3 *Let*

$$\begin{aligned} \mathcal{C}^\infty(M) : \mathrm{AFF}_{\mathcal{C}^\infty} &\to \mathrm{SETS} \\ \mathbb{R}^n &\mapsto \mathcal{C}^\infty\big(M, \mathbb{R}^n\big) \end{aligned}$$

be the smooth algebra of functions on a smooth variety M.

1. *There is a fully faithful monoidal embedding of the category of vector bundles on M in the category* $\mathrm{MOD}(\mathcal{C}^\infty(M))$ *of modules over the corresponding smooth algebra.*
2. *The $\mathcal{C}^\infty(M)$-module $\Omega^1(M)$ of ordinary differential forms on M is identified with the module of differential forms Ω^1_X in the sense of Definition* 2.5.4.

3. *The smooth algebra* $\mathcal{C}^\infty(J^n M)$ *of smooth functions on ordinary jet space for smooth functions on* M *is identified with the smooth algebra* $\mathrm{Jet}^n(\mathcal{C}^\infty(M))$ *in the sense of Definition* 2.5.4.
4. *If* $C \to M$ *is a smooth bundle, the* $\mathcal{C}^\infty(M)$*-algebra* $\mathcal{C}^\infty(\mathrm{Jet}^n(C/M))$ *is identified with the smooth algebra of jets in the category of smooth* $\mathcal{C}^\infty(M)$*-algebras in the sense of Definition* 2.5.4.

Proof If $E \to M$ is a vector bundle, with zero section $\underline{0} : M \to E$, we may describe the smooth algebra $\mathcal{C}^\infty(M)$ as the quotient of $\mathcal{C}^\infty(E)$ by the class $\mathcal{I}$ of functions $f : E \to \mathbb{R}$ such that $f \circ \underline{0}$ is the zero function. The functor

$$\begin{aligned}\big(\mathrm{VECT}(M), \times_M, \otimes\big) &\to \big(\mathrm{MOD}\big(\mathcal{C}^\infty(M)\big), \oplus, \otimes\big)\\ [E \to M] &\mapsto \big[\mathcal{C}^\infty(E)/\mathcal{I}^2 \to \mathcal{C}^\infty(M)\big]\end{aligned}$$

is fully faithful and bimonoidal. This means that it sends the fiber product $E \times_M E$ to the relative fiber product of smooth algebras, which corresponds to the sum of modules, and the tensor product of vector bundles to the pointed tensor product of smooth algebras. Differential forms in the sense of Definition 2.5.4 are given by the smooth algebra

$$\Omega^1_M := \mathcal{C}^\infty(M \times M)/\mathcal{I}^2_\Delta$$

over $\mathcal{C}^\infty(M)$ defined by the square of the diagonal ideal. The identification with usual differential forms may be achieved by choosing one of the projections to make this algebra a module, and by trivializing on a small open subset. The proof of the other results are of a similar nature and are left to the reader. □

Example 3.3.3 Let $M = \mathbb{R}$ and consider the trivial bundle $C = M \times \mathbb{R}$ on M. The k-th jet bundle for $C \to M$ is (non-canonically) identified with the space

$$\mathrm{Jet}^k(C/M) = M \times \mathbb{R}[T]/\big(T^{k+1}\big) \cong M \times \mathbb{R}^{k+1},$$

whose coordinates $(x, u_i)_{i=0,\dots,k}$ play the role of formal derivatives (i.e., coefficients of the Taylor series) of the "functional variable" u_0. The canonical section $C \to \mathrm{Jet}^k(C/M)$ corresponds geometrically to the map

$$\begin{aligned}j_k : \Gamma(M, C) &\to \Gamma\big(M, \mathrm{Jet}^k C\big)\\ \big[x \mapsto f(x)\big] &\mapsto \left[x \mapsto \sum_{i=0}^{k} \frac{\partial^i f}{\partial x^i}(x).\frac{(X-x)^i}{i!}\right],\end{aligned}$$

that sends a real valued function to its truncated Taylor series. The infinite jet space is identified with

$$\mathrm{Jet}(C/M) = M \times \mathbb{R}^{\mathbb{N}}.$$

Its functions are given by smooth functions that depend on finitely many coordinates. It is naturally a crystal of smooth varieties over M (in the sense of Definition 2.5.9).

Example 3.3.4 More generally, if $M = \mathbb{R}^n$ and $C = M \times \mathbb{R}^m$, one obtains the smooth algebra bundle

$$\mathrm{Jet}^k C = M \times \oplus_{i \leq k} \mathrm{Sym}^i\left(\mathbb{R}^n, \mathbb{R}^m\right),$$

with coordinates (x, u_α) for $x \in X$ and $u_\alpha \in \mathbb{R}^m$, α being a multi-index in $\mathbb{N}^n$ with $|\alpha| \leq k$. The jet map corresponds to the higher dimensional Taylor series

$$\begin{aligned} j_k : \Gamma(M, C) &\to \Gamma\left(M, \mathrm{Jet}^k C\right) \\ \left[x \mapsto f(x)\right] &\mapsto \left[x \mapsto \sum_{|\alpha| \leq k} \left(\partial_x^\alpha f\right)(x). \frac{(X - x)^\alpha}{\alpha!}\right], \end{aligned}$$

where multi-index notations are understood.

Note that one can prove more generally that categories of spaces given by convenient subcategories of the category of locally ringed spaces can also be treated with the methods of Sect. 2.5. The reader used to this setting can thus just use the differential invariants he is used to working with.

3.3.3 Super and Graded Geometry

We now explain how to use our formalism of categorical infinitesimal calculus in the super and graded setting, to define differential invariants of supermanifolds, which play a fundamental role in the description of mathematical models of field theory (e.g. in quantum electrodynamics, to cite the most famous example). We refer to [Nis00] and [Sac09] for other approaches to synthetic super-geometry.

Definition 3.3.8 The category of *smooth open superspaces* is the category $\mathrm{OPEN}^s_{C^\infty}$ whose objects are pairs $U^{n|m} = (U, \mathbb{R}^m)$, with $U \subset \mathbb{R}^n$ an open subset, and whose morphisms are given by the set

$$\mathrm{Hom}\left(U^{n|m}, V^{k|l}\right) = \mathrm{Hom}_{\mathrm{ALG}^s_{\mathbb{R}}}\left(C^\infty\left(V, \wedge^* \mathbb{R}^l\right), C^\infty\left(U, \wedge^* \mathbb{R}^m\right)\right)$$

of morphisms of real superalgebras. The category of *smooth affine superspaces* is the full subcategory $\mathrm{AFF}^s_{C^\infty} \subset \mathrm{OPEN}^s_{C^\infty}$ given by pairs $\mathbb{R}^{n|m} = (\mathbb{R}^n, \mathbb{R}^m)$.

Definition 3.3.9 A *smooth super-diffeology* is a sheaf on the site $(\mathrm{OPEN}^s_{C^\infty}, \tau)$ of smooth open superspaces. A variety relative to this site is called a *supermanifold*.

Proposition 3.3.4 *A morphism $f : U^{n|m} \to V^{k|l}$ of superspaces induces a smooth map $|f| : U \to V$ between the underlying open sets and a commutative diagram*

$$\begin{array}{ccc} U^{m|n} & \xrightarrow{\ f\ } & U^{k|l} \\ {\scriptstyle 0}\uparrow\downarrow{\scriptstyle p} & & {\scriptstyle 0}\uparrow\downarrow{\scriptstyle p} \\ U^{m|0} & \xrightarrow{\ |f|\ } & U^{k|0} \end{array}$$

The product of two smooth open superspaces $(U, \mathbb{R}^m)$ and $(V, \mathbb{R}^l)$ is given by $(U \times V, \mathbb{R}^{m+l})$.

Proof Note first that if $A = \mathcal{C}^\infty(U, \wedge^*\mathbb{R}^m)$, then the odd and even parts of A are

$$A^0 = \mathcal{C}^\infty\big(U, \wedge^{2*}\mathbb{R}^m\big) \quad \text{and} \quad A^1 = \mathcal{C}^\infty\big(U, \wedge^{2*+1}\mathbb{R}^m\big).$$

The subset $A^0 \subset A$ is a subalgebra and the subset A^1 is a sub-A^0-module. If we consider the ideal $(A^1) \subset A$ generated by A^1, we get a quotient algebra

$$|A| = A/\big(A^1\big),$$

that is isomorphic to $\mathcal{C}^\infty(U) = \mathcal{C}^\infty(U^{m|0})$. This construction is functorial in morphisms of open superspaces. This implies that a morphism $U^{n|m} \to V^{k|l}$ between open superspaces induces a diagram of superalgebras

$$\begin{array}{ccc} U^{m|n} & \xrightarrow{\ f\ } & U^{k|l} \\ {\scriptstyle 0}\uparrow & & {\scriptstyle 0}\uparrow \\ U^{m|0} & \xrightarrow{\ |f|\ } & U^{k|0} \end{array}$$

The product of usual open subsets $U \subset \mathbb{R}^n$ and $V \subset \mathbb{R}^k$ is $U \times V$, and we have

$$\mathcal{C}^\infty(U) \otimes \wedge^*\mathbb{R}^m \cong \mathcal{C}^\infty\big(U, \wedge^*\mathbb{R}^m\big)$$

so that $U^{m|0} \times \mathbb{R}^{0|m} \cong U^{n|m}$. We also have

$$\wedge^*\mathbb{R}^m \otimes \wedge^*\mathbb{R}^n \cong \wedge^*\mathbb{R}^{m+n}.$$

Combining these results, we get

$$U^{n|m} \times V^{k|l} \cong (U \times V)^{n+k|m+l}.$$

This implies in particular that there is a natural projection $U^{n|m} \to U^{n|0}$, given by the morphism $\mathcal{C}^\infty(U) \to \mathcal{C}^\infty(U, \wedge^*\mathbb{R}^m)$. This projection is functorial in $U^{n|m}$, meaning that there is a commutative diagram

$$\begin{array}{ccc} U^{m|n} & \xrightarrow{f} & U^{k|l} \\ \downarrow{\scriptstyle p} & & \downarrow{\scriptstyle p} \\ U^{m|0} & \xrightarrow{|f|} & U^{k|0} \end{array}$$

because $p \circ 0 = \mathrm{id}_{U^{m|0}}$ and $|f| = p \circ f \circ 0$. □

A diagram

$$\begin{array}{ccc} & & U^{m|n} \\ & & \downarrow \\ V^{k|l} & \longrightarrow & W^{r|s} \end{array}$$

is called transversal if the tangent superspace $TW^{r|s}$ is generated by $TU^{m|n} \cup TV^{k|l}$.

Definition 3.3.10 A *finitary smooth superalgebra* is a functor

$$A : \left(\mathrm{AFF}^s_{C^\infty}, \times\right) \to (\mathrm{SETS}, \times)$$

that preserves finite products. A *smooth superalgebra* is a functor

$$A : \left(\mathrm{OPEN}^s_{C^\infty}, - \underset{-}{\overset{t}{\times}} -\right) \to \left(\mathrm{SETS}, - \underset{-}{\times} -\right)$$

that sends finite transversal fiber products to fiber products. We denote by $\mathrm{ALG}^{s,fin}_{C^\infty}$ and $\mathrm{ALG}^s_{C^\infty}$ the categories of finitary and general smooth superalgebras. If A is a given superalgebra, we define $A^0 := A(\mathbb{R}^{1|0})$ and $A^1 := A(\mathbb{R}^{0|1})$.

Since $\mathbb{R}^{m,n} \cong (\mathbb{R}^{1|0})^m \times (\mathbb{R}^{0|1})^n$, if A is a given superalgebra, one has by definition

$$A\left(\mathbb{R}^{m|n}\right) \cong \left(A^0\right)^m \times \left(A^1\right)^n.$$

Definition 3.3.11 A *smooth superspace* is a sheaf on the site $(\mathrm{ALG}^{s,op}_{C^\infty}, \tau)$ of smooth superalgebras. A variety relative to this site is called a *super-variety* or *smooth super-scheme*.

Proposition 3.3.5 *The forgetful functor*

$$\begin{aligned} \mathrm{ALG}^s_{C^\infty} &\to \mathrm{ALG}^s_{\mathbb{R}} \\ A &\mapsto A\left(\mathbb{R}^{1|1}\right) \end{aligned}$$

has a right adjoint geom *called the geometrization functor. Moreover, there is a well defined functor*

$$\begin{aligned} \mathrm{ALG}^s_{\mathcal{C}^\infty} &\to \mathrm{ALG}^s_{\mathcal{C}^\infty} \\ A &\mapsto A^0 : \left[U^{m|n} \mapsto A\left(U^{m|0}\right)\right] \end{aligned}$$

and for each A*, functorial morphisms* $0 : A^0 \to A$ *and* $p : A \to A^0$ *of smooth superalgebras such that* $p \circ 0 = \mathrm{id}_{A^0}$.

Proof If A is a real superalgebra, one defines its geometrization by

$$\mathrm{geom}(A)\left(U^{n|m}\right) := \mathrm{Hom}_{\mathrm{ALG}^s_{\mathbb{R}}}\left(\mathcal{C}^\infty\left(U, \wedge^*\mathbb{R}^m\right), A\right).$$

This is a smooth superalgebra by construction. The map $A \mapsto A^0$ is functorial because of Proposition 3.3.4: a morphism $f : U^{m|n} \to V^{k|l}$ induces a natural map $|f| : U^{m|0} \to V^{k|0}$. The morphisms of functors $0 : A^0 \to A$ and $p : A \to A^0$ are induced by the maps $0 : U^{n|0} \to U^{n|m}$ and $p : U^{n|m} \to U^{0|n}$ (that are also functorial by loc. cit.). □

Proposition 3.3.6 *The maps* $\tilde{A} : U \mapsto A(U^{n|1})$ *and* $\tilde{A}^0 : U \mapsto A(U^{n|0})$ *induce a functor*

$$\begin{aligned} \mathrm{mod} : \mathrm{ALG}^s_{\mathcal{C}^\infty} &\to \mathrm{MOD}(\mathrm{ALG}_{\mathcal{C}^\infty}) \\ A &\mapsto \left[\tilde{A} \to \tilde{A}^0\right] \end{aligned}$$

with values in smooth modules. This functor has a left adjoint, denoted by $M \mapsto \mathrm{Sym}(M[1])$.

Proof Both functors $\tilde{A}$ and $\tilde{A}^0$ commute with transversal pullbacks and are thus smooth algebras. Note that the underlying algebra of $\tilde{A}$ is the set $A(\mathbb{R}^{1|1}) = A^0 \times A^1$, with multiplication induced by the map

$$\begin{aligned} \times : \mathbb{R}^{1|1} \times \mathbb{R}^{1|1} &\to \mathbb{R}^{1|1}, \\ (x_1, \theta_1), (x_2, \theta_2) &\mapsto (x_1x_2, x_1\theta_2 + x_2\theta_1). \end{aligned}$$

The zero morphism $0 : U^{n|0} \to U^{n|1}$, the projection $p : U^{n|1} \to U^{n|0}$, and the addition

$$+ : U^{n|1} \times_{U^{n|0}} U^{n|1} \cong U^{n|2} \longrightarrow U^{n|1}$$

given by sending (u, θ_1, θ_2) to $(u, \theta_1 + \theta_2)$ are all functorial in U, and make $U^{n|1}$ an abelian group object over U. This implies that $\tilde{A}$ is an abelian group object over $\tilde{A}^0$. In particular, its addition is given by

$$+ : (\tilde{A} \times_{\tilde{A}^0} \tilde{A})(U) \cong A\left(U^{n|1} \times_{U^{n|0}} U^{n|1}\right) \longrightarrow A\left(U^{n|0}\right) = \tilde{A}(U).$$

We have thus shown that mod(A) is indeed a smooth module. The left adjoint to mod must satisfy

$$\mathrm{Hom}_{\mathrm{ALG}^s_{C^\infty}}\big(\mathrm{Sym}\big(M[1]\big), B\big) \cong \mathrm{Hom}_{\mathrm{MOD}_{C^\infty}}\big(M, \mathrm{mod}(B)\big).$$

Concretely, if $M \to A$ is the structure map, it is given by

$$\mathrm{Sym}\big(M[1]\big)\big(U^{n|m}\big) := \Big(A \underset{\mathrm{geom}(A(\mathbb{R}))}{\otimes} \mathrm{geom}\big(\wedge^{2*}_{A(\mathbb{R})} M\big)\Big)(U) \times \big(\wedge^{2*+1}_{A(\mathbb{R})} M\big)^m,$$

where $\mathrm{geom}(A(\mathbb{R})) \to A$ is the natural adjunction map and M also denotes its underlying $A(\mathbb{R})$-module. By construction, $\mathrm{Sym}(M[1])$ commutes with transversal fiber products. The adjunction morphism $\mathrm{Sym}(\mathrm{mod}(B)[1]) \to B$ is defined by the following: first, recall that $\mathrm{mod}(B) = [M \to A]$ is given by $M = B^1$ and $A = B^0$, and the desired map is given on the odd part by the multiplication map

$$\wedge^{2*+1}_{A(\mathbb{R})} M \to M = B^1.$$

There is also a natural multiplication morphism of real algebras

$$\wedge^{2*}_{A(\mathbb{R})} M \to B^0(\mathbb{R})$$

that corresponds to a morphism of smooth algebras

$$\mathrm{geom}\big(\wedge^{2*+1}_{A(\mathbb{R})} M\big) \to B^0.$$

Since B^0 is an A-algebra via the identity map $A = B^0$, we get by extension of scalars a morphism of smooth algebras

$$A \underset{\mathrm{geom}(A(\mathbb{R}))}{\otimes} \mathrm{geom}\big(\wedge^{2*}_{A(\mathbb{R})} M\big) \to B^0.$$

The other adjunction morphism $M \to \mathrm{mod}(\mathrm{Sym}(M[1]))$ associated to a module $[M \to A]$ is given by the natural morphism $A \to \mathrm{Sym}(M[1])^0$ and the map $M \to \wedge^{2*+1}_{A(\mathbb{R})} M$, is given more precisely by the map of sets

$$M(U) = A(U) \times M \longrightarrow \Big(A \underset{\mathrm{geom}(A(\mathbb{R}))}{\otimes} \mathrm{geom}\big(\wedge^{2*}_{A(\mathbb{R})} M\big)\Big)(U) \times \big(\wedge^{2*+1}_{A(\mathbb{R})} M\big). \qquad \square$$

Proposition 3.3.7 *Let M be a smooth manifold and Ω^1_M be the cotangent bundle of M. Let $\Omega^*_M := \mathrm{geom}(\Gamma(M, \wedge^* \Omega^1_M))$ be the smooth superalgebra of differential forms on M. There is a natural isomorphism*

$$\underline{\mathrm{Hom}}\big(\mathbb{R}^{0|1}, M\big) \cong \underline{\mathrm{Spec}}\big(\Omega^*_M\big).$$

Proof Note that $\Omega^*_M = \mathrm{Sym}_{\mathcal{O}_M}(\Omega^1_M[1])$, where Ω^1_M is the smooth module of differential forms. Let A be a smooth superalgebra. One has

$$\begin{aligned}
\underline{\mathrm{Spec}}(\Omega_M^*)(A) &:= \mathrm{Hom}_{\mathrm{ALG}^s_{\mathcal{C}^\infty}}(\Omega_M^*, A)\\
&\cong \mathrm{Hom}_{\mathrm{ALG}^s_{\mathcal{C}^\infty}}\big(\mathrm{Sym}_{\mathcal{O}_M}(\Omega_M^1[1]), A\big)\\
&\cong \mathrm{Hom}_{\mathrm{MOD}_{\mathcal{C}^\infty}}\big(\Omega_M^1, \mathrm{mod}(A)\big)\\
&\cong \mathrm{Hom}_{\mathrm{ALG}_{\mathcal{C}^\infty}}(\mathcal{O}_M, \tilde{A}).
\end{aligned}$$

Now we may also compute the other side of the desired isomorphism. Note that $\tilde{A} \cong (\mathcal{C}^\infty(\mathbb{R}^{0|1}) \otimes A)^0$. Indeed, we have by definition

$$\big(\mathcal{C}^\infty(\mathbb{R}^{0|1}) \otimes A\big)^0\big(U^{n|m}\big) := \big(\mathcal{C}^\infty(\mathbb{R}^{0|1}) \otimes A\big)\big(U^{n|0}\big) \cong A\big(U^{n|1}\big) =: \tilde{A}(U),$$

so that

$$\begin{aligned}
\underline{\mathrm{Hom}}(\mathbb{R}^{0|1}, M)(A) &:= \mathrm{Hom}_{\mathrm{ALG}^s_{\mathcal{C}^\infty}}\big(\mathcal{O}_M, \mathcal{C}^\infty(\mathbb{R}^{0|1}) \otimes A\big),\\
&\cong \mathrm{Hom}_{\mathrm{ALG}_{\mathcal{C}^\infty}}\big(\mathcal{O}_M, \big(\mathcal{C}^\infty(\mathbb{R}^{0|1}) \otimes A\big)^0\big),\\
&\cong \mathrm{Hom}_{\mathrm{ALG}_{\mathcal{C}^\infty}}(\mathcal{O}_M, \tilde{A}),
\end{aligned}$$

which concludes the proof. □

It is easy to generalize superspaces to spaces equipped with an additional linear structure (for example $\mathbb{Z}$-graded spaces). This generalization, which we shall explain shortly, will be useful in our study of gauge theories.

Definition 3.3.12 Let (LEGOS, τ) be a site equipped with a class of transversal fiber products, and $\mathcal{C} = (\mathcal{C}, \mathcal{C}_0, \otimes)$ be a symmetric monoidal category $(\mathcal{C}, \otimes)$ equipped with a full subcategory $\mathcal{C}_0$ and a functor

$$\begin{aligned}
\mathcal{O}\big(-, \mathrm{Sym}(-)\big) : \mathrm{LEGOS} \times \mathcal{C}_0 &\to \mathrm{ALG}(\mathcal{C}, \otimes)\\
(U, E) &\mapsto \mathcal{O}\big(U, \mathrm{Sym}(E)\big),
\end{aligned}$$

such that $\mathcal{O}(-, \mathrm{Sym}(E))$ is a sheaf for all E. The category $\mathrm{LEGOS}^{\mathcal{C}}$ of $\mathcal{C}$-enriched legos is the category of pairs (U, E) composed of a lego U of an object E of $\mathcal{C}_0$, with morphisms $f : (U, E) \to (V, F)$ given by morphisms of $\mathcal{C}$-algebras

$$f^* : \mathcal{O}\big(V, \mathrm{Sym}(F)\big) \to \mathcal{O}\big(U, \mathrm{Sym}(E)\big).$$

We now assume given a notion of transversal fiber product on the category $\mathrm{LEGOS}^{\mathcal{C}}$.

Definition 3.3.13 A $\mathcal{C}$-enriched algebra modeled on LEGOS is a set-valued functor on $\mathrm{LEGOS}^{\mathcal{C}}$ that commutes with transversal fiber products. We denote by $\mathrm{ALG}_{\mathrm{LEGOS},\mathcal{C}}$ the category of $\mathcal{C}$-enriched algebras, equipped with the Grothendieck topology induced by τ.

All the sites LEGOS described in this chapter have a natural notion of transversal pullback. The above definition specializes to our definition of smooth superalgebras if we work with the monoidal category $(\text{VECT}^s_{\mathbb{R}}, \otimes)$ of real super vector spaces, the subcategory $\text{VECT}^{odd}_{\mathbb{R}}$ of odd vector spaces, and with the site of smooth open sets.

Definition 3.3.14 The category of smooth graded algebras is defined as the category

$$\text{ALG}^g_{C^\infty} := \text{ALG}_{\mathcal{C},\text{OPEN}_{C^\infty}}$$

of $\mathcal{C}$-enriched smooth algebras, where $\mathcal{C} = (\text{VECT}^g_{\mathbb{R}}, \text{VECT}^{\pm}, \otimes)$ is the monoidal category of graded real vector spaces, equipped with the subcategory $\text{VECT}^{\pm}$ of spaces with no degree zero part.

A fake function on a graded smooth space is a morphism

$$S : X \to \mathbb{A}^1$$

where $\mathbb{A}^1(A) = A$ is the space of all elements in the graded algebra A. A real valued function on a graded smooth space is a morphism

$$S : X \to \underline{\mathbb{R}}$$

where $\underline{\mathbb{R}}(A) := A^0$ is the space of degree zero elements in the graded algebra A.

Example 3.3.5 Let M be a variety and $\Omega^*(M)$ be its algebra of differential form, regarded as a smooth superalgebra. Then a fake smooth function

$$f : \underline{\text{Spec}^s}(\Omega^*(M)) \to \mathbb{A}^1$$

is simply a differential form ω in $\Omega^*(M)$, and a smooth function

$$f : \underline{\text{Spec}^s}(\Omega^*(M)) \to \underline{\mathbb{R}}$$

is a differential form ω in $\Omega^{2*}(M)$. If now $\Omega^*(M)$ is considered as a smooth graded algebra, fake smooth functions don't change but a smooth function

$$f : \underline{\text{Spec}^g}(\Omega^*(M)) \to \underline{\mathbb{R}}$$

is an ordinary function f on M (of degree 0).

We finish by giving the example that provides the main motivation for studying super-spaces via their functors of points.

Example 3.3.6 Let $p : S \to M$ be a vector bundle of a smooth manifold M. The superalgebra

$$\mathcal{O}_{\Pi S} := \Gamma(M, \wedge^* S)$$

has a natural smooth superalgebra structure. Its spectrum is the corresponding geometric odd bundle

$$\Pi p : \Pi S := \underline{\mathrm{Spec}}(\mathcal{O}_{\Pi S}) \to M.$$

This bundle has no non-trivial sections, i.e., $\underline{\Gamma}(M, \Pi S) = \{0\}$, because a $\mathcal{C}^\infty(M)$-algebra morphism

$$\mathcal{O}_{\Pi S} = \Gamma\big(M, \wedge^* S\big) \to \mathcal{C}^\infty(M)$$

is necessarily trivial on the generators of the exterior algebra. Physicists, and among them, DeWitt in [DeW03], solve this problem by using points of $\underline{\Gamma}(M, \Pi S)$ with values in the free odd algebra on a countable number of generators

$$A = \Lambda_\infty := \widehat{\wedge^* \mathbb{R}^{(\mathbb{N})}}$$

(of course, his formulation differs a little from ours, but the idea is the same). These are given by $\mathcal{C}^\infty(M)$-algebra morphisms

$$\mathcal{O}_{\Pi S} \to \mathcal{C}^\infty(M) \otimes A,$$

and since A also has odd components, one can find plenty of them. There is no reason to choose this coefficient algebra and not another one, so that the natural space of sections $\underline{\Gamma}(M, S)$ must really be thought of as the functor on superalgebras

$$A \mapsto \underline{\Gamma}(M, S)(A)$$

as we have done throughout this chapter.

References

[AGV73] Artin, M., Grothendieck, A., Verdier, J.-L.: Théorie des Topos et Cohomologie Étale des Schémas. Tome 3, Lecture Notes in Mathematics, vol. 305, p. 640. Springer, Berlin (1973). Séminaire de Géométrie Algébrique du Bois-Marie 1963–1964 (SGA 4), Dirigé par M. Artin, A. Grothendieck et J.L. Verdier. Avec la collaboration de P. Deligne et B. Saint-Donat

[BD86] Bunge, M., Dubuc, E.J.: Archimedian local C^∞-rings and models of synthetic differential geometry. Cah. Topol. Géom. Différ. Catég. **27**(3), 3–22 (1986)

[BD04] Beilinson, A., Drinfeld, V.: Chiral Algebras. American Mathematical Society Colloquium Publications, vol. 51, p. 375. Am. Math. Soc., Providence (2004). ISBN 0-8218-3528-9

[Ber90] Berkovich, V.G.: Spectral Theory and Analytic Geometry over Non-Archimedean Fields. Mathematical Surveys and Monographs, vol. 33, p. 169. Am. Math. Soc., Providence (1990). ISBN 0-8218-1534-2

[BH11] Baez, J.C., Hoffnung, A.E.: Convenient categories of smooth spaces. Trans. Am. Math. Soc. **363**(11), 5789–5825 (2011). doi:10.1090/S0002-9947-2011-05107-X

[BM88] Bierstone, E., Milman, P.D.: Semianalytic and subanalytic sets. Publ. Math. IHES **67**, 5–42 (1988)

[Che77] Chen, K.T.: Iterated path integrals. Bull. Am. Math. Soc. **83**(5), 831–879 (1977)

[DeW03] DeWitt, B.: The Global Approach to Quantum Field Theory. Vol. 1, 2. International Series of Monographs on Physics, vol. 114. Clarendon/Oxford University Press, Oxford/New York (2003). ISBN 0-19-851093-4

[DT83] Dubuc, E., Taubin, G.: Analytic rings. Cah. Topol. Géom. Différ. **24**(3), 225–265 (1983)

[Dub81] Dubuc, E.J.: C^{∞}-schemes. Am. J. Math. **103**(4), 683–690 (1981). doi:10.2307/2374046

[DZ94] Dubuc, E.J., Zilber, J.G.: On analytic models of synthetic differential geometry. Cah. Topol. Géom. Différ. Catég. **35**(1), 49–73 (1994)

[Gab96] Gabrielov, A.: Complements of subanalytic sets and existential formulas for analytic functions. Invent. Math. **125**(1), 1–12 (1996). doi:10.1007/s002220050066

[Gro60] Grothendieck, A.: Éléments de géométrie algébrique. I. Le langage des schémas. Publ. Math. IHÉS **1–4**(4), 228 (1960)

[Gro67] Grothendieck, A.: Éléments de géométrie algébrique. IV. Étude locale des schémas et des morphismes de schémas. Publ. Math. IHÉS **32**, 361 (1967)

[GTZ09] Ginot, G., Tradler, T., Zeinalian, M.: A Chen model for mapping spaces and the surface product. arXiv e-prints (2009). arXiv:0905.2231

[HK94] Huber, R., Knebusch, M.: On valuation spectra. In: Recent Advances in Real Algebraic Geometry and Quadratic Forms (Berkeley, CA, 1990/1991; San Francisco, CA, 1991). Contemp. Math., vol. 155, pp. 167–206. Am. Math. Soc., Providence (1994)

[Hub96] Huber, R.: Étale Cohomology of Rigid Analytic Varieties and Adic Spaces. Aspects of Mathematics, vol. 30, p. 450. Vieweg, Braunschweig (1996). ISBN 3-528-06794-2

[IZ99] Iglesias Zeimour, P.: Diffeology. Volume 1 of Web book (1999). http://math.huji.ac.il/~piz/Site/The Book/The Book.html

[Joy09] Joyce, D.: On manifolds with corners. arXiv e-prints (2009). arXiv:0910.3518

[Koc06] Kock, A.: Synthetic Differential Geometry, 2nd edn. London Mathematical Society Lecture Note Series, vol. 333, p. 233. Cambridge University Press, Cambridge (2006). doi:10.1017/CBO9780511550812. ISBN 978-0-521-68738-6; 0-521-68738-1

[Law79] Lawvere, F.W.: Categorical dynamics. In: Topos Theoretic Methods in Geometry. Various Publ. Ser., vol. 30, pp. 1–28. Aarhus Universitet, Aarhus (1979)

[Law04] Lawvere, F.W.: Functorial semantics of algebraic theories and some algebraic problems in the context of functorial semantics of algebraic theories. Theory Appl. Categ. **5**, 1–121 (2004). Reprinted from Proc. Natl. Acad. Sci. USA **50**, 869–872 (1963) [MR0158921] and Reports of the Midwest Category Seminar. II, pp. 41–61, Springer, Berlin (1968) [MR0231882]

[Łoj93] Łojasiewicz, S.: Sur la géométrie semi- et sous-analytique. Ann. Inst. Fourier (Grenoble) **43**(5), 1575–1595 (1993)

[Lur09] Lurie, J.: On the classification of topological field theories. arXiv e-prints (2009). arXiv:0905.0465

[Lyc93] Lychagin, V.: Quantizations of braided differential operators (1993, preprint)

[MR91] Moerdijk, I., Reyes, G.E.: Models for Smooth Infinitesimal Analysis, p. 399. Springer, New York (1991). ISBN 0-387-97489-X

[Nes03] Nestruev, J.: Smooth Manifolds and Observables. Graduate Texts in Mathematics, vol. 220, p. 222. Springer, New York (2003). Joint work of A.M. Astashov, A.B. Bocharov, S.V. Duzhin, A.B. Sossinsky, A.M. Vinogradov and M.M. Vinogradov. Translated from the 2000 Russian edition by A.B. Sossinsky, I.S. Krasil'schik and S.V. Duzhin. ISBN 0-387-95543-7

[Nis00] Nishimura, H.: Supersmooth topoi. Int. J. Theor. Phys. **39**(5), 1221–1231 (2000). doi:10.1023/A:1003630618417

[Sac09] Sachse, C.: Global analytic approach to super Teichmueller spaces. arXiv e-prints (2009). arXiv:0902.3289

[Smi66] Smith, J.W.: The de Rham theorem for general spaces. Tohoku Math. J. **18**, 115–137 (1966)

[Sou97] Souriau, J.-M.: Structure of Dynamical Systems. Progress in Mathematics, vol. 149, p. 406. Birkhäuser Boston, Boston (1997). A symplectic view of physics. Translated from the French by C.H. Cushman-de Vries. Translation edited and with a preface by R.H. Cushman and G.M. Tuynman. ISBN 0-8176-3695-1

[TV08] Toën, B., Vezzosi, G.: Homotopical Algebraic Geometry. II. Geometric Stacks and Applications. Mem. Am. Math. Soc. 193(902):x+224 (2008)

[TV09b] Toën, B., Vaquié, M.: Au-dessous de Spec $\mathbb{Z}$. J. K-Theory **3**(3), 437–500 (2009). doi:10.1017/is008004027jkt048

[Zil90] Zilber, J.C.: Local analytic rings. Cah. Topol. Géom. Différ. Catég. **31**(1), 21–52 (1990)

Chapter 4
Functorial Analysis

In this chapter, we translate the classical methods of the analysis of linear partial differential equations into the language used in this book, using functions with particular given domains of definition. The advantage of this viewpoint of analysis is that it generalizes directly to the super or graded setting, which is strictly necessary to treat the basic models of particle physics. It is also better suited to the combination of the homotopical geometric methods of Chap. 10 with the usual methods of classical analysis. This combination is useful to better understand the conceptual meaning of gauge theoretical computations of Chap. 13.

We will describe in Chaps. 11 and 12 a coordinate-free approach to partial differential equations, very much in the spirit of the functor of points approach to the resolution of equations. One needs both approaches to fully understand the models of mathematical physics.

The analysis of partial differential equations can be based on the notion of a well posed problem, introduced by Hadamard. Roughly speaking, a partial differential equation E with initial/boundary value f in a topological space of functions $\mathcal{F}$ is called well-posed if:

1. there exists a solution to E around the given initial value;
2. this solution is unique;
3. the solution depends continuously on the initial value f.

The most general result on this problem is the Cauchy-Kowalewskaya theorem (see Theorem 11.6.3 for the formulation of a generalization of this statement adapted to the study of systems of partial differential equations on manifolds) that states that any reasonable real-analytic non-linear partial differential equation has a unique solution depending analytically on the given analytic initial condition. In the case of evolution equations, one also often asks for a persistence condition (the solution remains in the same functional space as the initial condition).

The main method of the analysis of partial differential equations is to construct a large space $\mathcal{F}$ of generalized functions (for example Schwartz distributions) in which the existence of a weak solution for a given type of equation is assured, and to find a subspace $\mathcal{H}$ (for example a Sobolev space) in which the uniqueness is guar-

F. Paugam, *Towards the Mathematics of Quantum Field Theory*,
Ergebnisse der Mathematik und ihrer Grenzgebiete. 3. Folge / A Series of Modern
Surveys in Mathematics 59, DOI 10.1007/978-3-319-04564-1_4,

anteed. The existence and unicity can be proved by invoking a fixed point theorem. Existence can also be proved by a compacity criterion (by extracting a converging sub-sequence of a given sequence). Being in $\mathcal{H}$ is then often seen as a regularity condition on solutions, which can be proved by a priori estimates on the operator. In the case of non-linear equations, one can also use the local inversion theorem and perturbative methods, regarding the equation as a perturbation of a linear equation by a non-linear term multiplied by a small parameter. We refer to Evans' textbook for a general overview of the analysis of non-linear equations [Eva98], and to Kriegl-Michor [KM97] for an approach to general non-linear analysis using convenient locally convex topological vector spaces (including the implicit function theorem) that is conceptually close to ours.

Note that in this book, we haven't discussed at all the theory of qualitative results on partial differential equations, which began with Poincaré's study of the three body problem, and that is now called the theory of dynamical systems. The approach that is the closest to the geometric theory of partial differential equations we discussed is Thom's transversality theorem for jets, generalized by Mather for multijets. One may also quote Gromov's generalization of Smale's embedding theorem, called the h-principle, which says roughly, that in good cases, a given formal solution to a partial differential system can be deformed to a smooth one.

In this chapter, we will mostly be interested in the existence of solutions. For this purpose, we introduce in Sects. 4.1 and 4.2 abstract notions of functionals and distributions that are more flexible than the usual ones because they are also adapted to the non-linear setting. We will thus use them extensively.

4.1 Functionals and Functional Derivatives

To illustrate the general definitions of spaces we have given in Sect. 3.1, we will use them to define a flexible notion of functional and distribution on a function space. The idea of a distribution is already at the heart of Dirac's work on quantum mechanics [Dir82].

We first recall from Bourbaki [Bou70] the definition of partially defined functions between sets.

Definition 4.1.1 A *span* $f : X \to Y$ between two sets is an arbitrary map $\Gamma_f \to X \times Y$. A *correspondence* $f : X \to Y$ between two sets is a span given by a subset $\Gamma_f \subset X \times Y$. A *partially defined function* $f : X \to Y$ between two sets is a correspondence $\Gamma_f \subset X \times Y$ whose projection on X is injective, with image denoted by D_f and called the *domain of definition* of f. The set of correspondences (resp. partially defined functions) $f : X \to Y$ is denoted by $\mathrm{Hom}_{\mathrm{COR}}(X, Y)$ (resp. $\mathrm{Hom}_{\mathrm{PAR}}(X, Y)$). The category of spans $f : X \to Y$ is denoted by $\mathrm{Mor}^1_{\mathrm{SPAN}}(X, Y)$.

Proposition 4.1.1 *Sets with correspondences* (*resp. partially defined functions*) *between them form a category, denoted by* COR (*resp.* PAR). *Sets with spans between them form a 2-category, denoted by* SPAN.

Proof For $f : X \to Y$ and $g : Y \to Z$ two spans given by $\Gamma_f \to X \times Y$ and $\Gamma_g \to Y \times Z$, we denote by $g \circ f$ the span given by the fiber product $\Gamma_f \times_Y \Gamma_g \subset X \times Y \times Z \to X \times Z$. By construction, one has natural isomorphisms

$$\Gamma_{(h\circ g)\circ f} \xrightarrow{i_{f,g,h}} \Gamma_{h\circ(g\circ f)},$$

and the universal property of fiber products imply that these isomorphisms fulfill the natural coherence condition to make SPAN a 2-category. For correspondences, the inclusion $R_f \subset X \times Y$ allows one to define composition as the projection in $X \times Z$ of the span composition $\Gamma_{f\circ g}$ of $f : X \to Y$ and $g : Y \to Z$. This reduces the 2-categorical structure to a usual category structure. □

We now define a general notion of functional, which is very close to the notion of functional used by physicists in quantum field theory books: they usually write down a formula without taking too much care over its domain of definition.

Definition 4.1.2 Let (LEGOS, τ) be a site and X and Y in SH(LEGOS, τ) be two spaces. A *correspondence* (resp. *partially defined function*) between X and Y is given by a functorial family

$$f_U : X(U) \to Y(U)$$

of partially defined functions (resp. correspondences) indexed by $U \in$ LEGOS. The *domain of definition* of a correspondence f is the sheaf $\underline{D}_f$ associated to the presheaf $U \mapsto D_f(U)$. For $\mathcal{C} =$ COR, PAR, the sheaf $\underline{\mathrm{Hom}}_{\mathcal{C}}(X, Y)$ is the sheaf associated to the presheaf

$$U \mapsto \mathrm{Hom}_{\mathcal{C}}(X \times U, Y).$$

Sheaves with correspondences and partially defined functions between them form categories, denoted respectively by COR(LEGOS, τ) and PAR(LEGOS, τ).

We now define the notion of a functional, which plays an important role in physics.

Definition 4.1.3 Let (LEGOS, τ) be a site and X, Y and Z in SH(LEGOS, τ) be three spaces. A partially defined function (resp. correspondence)

$$f : \underline{\mathrm{Hom}}(X, Y) \to Z$$

is called a *functional* (resp. *multivalued functional*) on the function space $\underline{\mathrm{Hom}}(X, Y)$ with values in Z.

The general geometrical setting presented in Chap. 3 for differential calculus on spaces of functions can be applied directly to functionals.

Definition 4.1.4 Let $\mathbf{C} = (\text{LEGOS}, \tau, \ldots)$ be a differential geometric setting, with category of algebras ALG and category of spaces $\text{Sp} := \text{SH}(\text{ALG}^{op}, \tau)$. Suppose given a free algebra $A[\epsilon]$ in one variable, and denote by $\mathbb{D}_k$ the space

$$\mathbb{D}_k(A) := \underline{\text{Spec}}\big(A[\epsilon]/(\epsilon^{k+1})\big).$$

The *k-th jet bundle of a space X* is the space

$$\text{Jet}^k(X) := \underline{\text{Hom}}(\mathbb{D}_k, X) \to \text{Jet}^0(X) = X.$$

The *k-th derivative of a span* $f : X \to Y$ is the jet span

$$\text{Jet}^k(f) : \text{Jet}^k(X) \to \text{Jet}^k(Y)$$

with graph $\text{Jet}^k(\Gamma_f) \to \text{Jet}^k(X) \times \text{Jet}^k(Y)$. In particular, the k-th derivative of a partially defined function $f : X \to Y$ has domain of definition $\text{Jet}^k(\underline{\text{D}}_f)$.

The jet space of a mapping space may be computed more explicitly.

4.2 Smooth Functionals and Distributions

Definition 4.2.1 Consider the site $(\text{OPEN}_{\mathcal{C}^\infty}, \tau)$ whose objects are open subsets $U \subset \mathbb{R}^k$ for varying k, with morphisms given by smooth maps between them, equipped with its ordinary topology τ. Let $\Omega \subset \mathbb{R}^n$ be an open subset and consider the space $\underline{\mathcal{C}}^\infty(\Omega, \mathbb{R}^m)$ of smooth $\mathbb{R}^m$-valued functions on Ω, whose U-points are given by the set of parametrized smooth functions

$$\underline{\mathcal{C}}^\infty(\Omega, \mathbb{R}^m)(U) := \text{Hom}_{\mathcal{C}^\infty}(\Omega \times U, \mathbb{R}^m).$$

We define $\underline{\mathcal{C}}^\infty(\Omega) := \underline{\mathcal{C}}^\infty(\Omega, \mathbb{R})$. Let $\underline{\mathbb{R}}$ be the *smooth affine line*, defined by $\underline{\mathbb{R}}(U) := \text{Hom}_{\mathcal{C}^\infty}(U, \mathbb{R})$. A functional

$$f : \underline{\mathcal{C}}^\infty(\Omega, \mathbb{R}^m) \to \underline{\mathbb{R}}$$

is called a *smooth functional.*

Example 4.2.1 Let $\Omega \subset \mathbb{R}^n$ be an open subset. Let $L : \Omega \times \mathbb{R}[x_1, \ldots, x_n] \to \mathbb{R}$ be a measurable function (called the Lagrangian density) that depends on Ω but also on finitely many multi-indexed coordinates in Ω (formal partial derivatives of the dependent variable). The corresponding smooth (action) functional

$$\begin{aligned} S : \underline{\mathcal{C}}^\infty(\Omega) &\to \underline{\mathbb{R}} \\ \varphi(x, u) &\mapsto \int_\Omega L\big(x, \varphi(x, u), \partial_\alpha \varphi(x, u)\big) dx \end{aligned}$$

has a domain of definition given (for example) by Lebesgue's dominated derivation condition

$$\underline{\mathrm{D}}_S(U) = \big\{\varphi(x,u) \mid \text{locally on } U, \forall\beta, \text{there exists } g \in L^1(\Omega) \text{ such that}$$
$$\big|\partial_\beta^u L\big(\varphi(x,u), \partial_\alpha\varphi(x,u)\big)\big| \le g(x)\big\}.$$

The above example shows that the domain of definition of a functional is usually very deeply related to the formula that defines the functional and that it is generally a difficult analytical problem to find a *common* domain of definition for a large class of functionals. This problem can be overcome in linear analysis but becomes extremely difficult in non-linear analysis. General functionals provide a natural setting to work with formulas without taking too much care over their domain of definition at first sight, but the analyst's work needs to be done at some point to be sure that the domain of definition is, at least, non-empty.

We now show that functionals in our sense are stable under the usual limit operation.

Proposition 4.2.1 *Let $\underline{\mathcal{C}}^\infty(\Omega)$ be the function space on an open subset $\Omega \subset \mathbb{R}^k$ and*

$$\big(f_n : \underline{\mathcal{C}}^\infty(\Omega) \to \underline{\mathbb{R}}\big)_{n\ge 0}$$

be a family of smooth functionals. Then the limit

$$f = \lim_{n\to\infty} f_n : \underline{\mathcal{C}}^\infty(\Omega) \to \underline{\mathbb{R}}$$

is a well defined functional.

Proof For an open subset $U \subset \mathbb{R}^n$, we define

$$\underline{\mathrm{D}}_f(U) := \Big\{\varphi \in \underline{\mathcal{C}}^\infty(\Omega)(U) \mid \exists N,\ \forall n \ge N,\ \varphi \in \underline{\mathrm{D}}_{f_n}(U) \text{ and } \lim_{n\to\infty} f_n(\varphi) \text{ exists}\Big\}.$$

Then f is a well defined functional with (possibly empty) domain of definition $\underline{\mathrm{D}}_f$. □

Proposition 4.2.2 *Let $\Omega \subset \mathbb{R}^n$ be an open subset. A smooth functional*

$$f : \underline{\mathcal{C}}^\infty(\Omega) \to \underline{\mathbb{R}}$$

is uniquely defined by the datum $(f_{\{.\}}, \underline{\mathrm{D}}_f)$ of the corresponding set theoretic (partially defined) function

$$f_{\{.\}} : \mathcal{C}^\infty(\Omega) \to \mathbb{R}$$

and the domain of definition $\underline{\mathrm{D}}_f \subset \underline{\mathcal{C}}^\infty(\Omega)$.

Proof Indeed, for every $u \in U$, there is a unique morphism $\{.\} \to U$ that sends the point to u. The commutativity of the diagram

$$\begin{array}{ccc} \underline{\mathcal{C}}^{\infty}(\Omega)(U) & \xrightarrow{f_U} & \underline{\mathbb{R}}(U) \\ \downarrow & & \downarrow \\ \underline{\mathcal{C}}^{\infty}(\Omega)(\{.\}) & \xrightarrow{f_{\{.\}}} & \underline{\mathbb{R}}(\{.\}) \end{array}$$

implies that if $\varphi : \Omega \times U \to \mathbb{R}$ is a point in $\underline{\mathrm{D}}_f(U)$, then one has

$$f_U(\varphi)(u) = f_{\{.\}}\big(\varphi(-,u)\big)$$

in $\underline{\mathbb{R}}(U) := \mathcal{C}^{\infty}(U)$, which shows that the values of f are uniquely determined by the values of the ordinary function $f_{\{.\}} : \mathcal{C}^{\infty}(\Omega) \to \mathbb{R}$. The domain of definition is contained in the set

$$\tilde{\underline{\mathrm{D}}}_f(U) = \big\{\varphi(x,u) | \forall u, \varphi(-,u) \in D_{f_{\{.\}}} \text{ and } u \mapsto f_{\{.\}}\big(\varphi(x,u)\big) \in \mathcal{C}^{\infty}(U)\big\}. \quad \square$$

We now define a refined class of functionals, using the category of smooth algebras. This class is useful when describing differential calculus on functionals as a purely algebraic theory, with the help of the tools of categorical infinitesimal calculus, described in Sect. 2.5.

Definition 4.2.2 Consider the site $(\mathrm{ALG}_{\mathcal{C}^{\infty}}^{fp,op}, \tau)$ whose objects are spectra $\overline{\mathrm{Spec}(A)}$ of finitely presented smooth algebras, equipped with their usual topology τ. For $\Omega \subset \mathbb{R}^n$ an open subset, we denote by $\underline{\mathbf{C}}^{\infty}(\Omega, \mathbb{R}^m)$ the space of smooth $\mathbb{R}^m$-valued functions on Ω, whose A-points are given by the set of parametrized smooth functions

$$\underline{\mathbf{C}}^{\infty}(\Omega, \mathbb{R}^m)(A) := \big(\big(\mathcal{C}^{\infty}(\Omega) \otimes A\big)(\mathbb{R})\big)^n.$$

We define $\underline{\mathbf{C}}^{\infty}(\Omega) := \underline{\mathbf{C}}^{\infty}(\Omega, \mathbb{R})$. Let $\underline{\mathbb{R}}$ be the *refined smooth affine line*, defined by $\underline{\mathbb{R}}(A) := A(\mathbb{R})$. A functional

$$\mathbf{f} : \underline{\mathbf{C}}^{\infty}\big(\Omega, \mathbb{R}^m\big) \to \underline{\mathbb{R}}$$

is called a *refined smooth functional.*

Proposition 4.2.3 *Smooth functionals and refined smooth functionals are related by a natural map*

$$F : \mathrm{Hom}_{\mathrm{PAR}}\big(\underline{\mathbf{C}}^{\infty}\big(\Omega, \underline{\mathbb{R}}^m\big), \underline{\mathbb{R}}\big) \longrightarrow \mathrm{Hom}_{\mathrm{PAR}}\big(\underline{\mathcal{C}}^{\infty}\big(\Omega, \mathbb{R}^m\big), \underline{\mathbb{R}}\big).$$

Proof It is enough to treat the case $m = 1$. Suppose given a refined smooth functional

$$\mathbf{f} : \underline{\mathbf{C}}^{\infty}(\Omega, \mathbb{R}) \to \underline{\mathbb{R}}.$$

One can evaluate it on the smooth algebras $A = \mathcal{C}^\infty(U)$, for any $U \subset \mathbb{R}^k$. For such an algebra, the natural isomorphism

$$\mathcal{C}^\infty(\Omega) \otimes \mathcal{C}^\infty(U) \cong \mathcal{C}^\infty(\Omega \times U)$$

of smooth algebras implies that one has the equalities

$$\underline{\mathbf{C}}^\infty(\Omega, \mathbb{R})(A) = \underline{\mathcal{C}}^\infty(\Omega, \mathbb{R})(U) \quad \text{and} \quad \underline{\mathbb{R}}(A) = \underline{\mathbb{R}}(U).$$

This therefore gives a map $\mathbf{f} \mapsto f$ from refined smooth functionals to usual smooth functionals. □

Remark 4.2.1 The above result remains true, with essentially the same proof, if one works with finitely presented non-finitary smooth algebras, defined in Sect. 3.2, as covariant functors

$$A : \text{OPEN}_{\mathcal{C}^\infty} \to \text{SETS}$$

that respect transversal products and final objects. The finite presentation condition on such an algebra A means that there exists a $U \subset \mathbb{R}^k$ and a finitely generated ideal $I \subset \mathcal{C}^\infty(U)$ such that $A \cong \mathcal{C}^\infty(U)/I$ as a non-finitary smooth algebra.

Definition 4.2.3 Let $\Omega \subset \mathbb{R}^n$ be an open subset. A *distribution* is a smooth $\underline{\mathbb{R}}$-linear functional

$$f : \underline{\mathcal{C}}^\infty\left(\Omega, \mathbb{R}^m\right) \to \underline{\mathbb{R}}$$

such that the domain of definition of the underlying function $f_{\{.\}}$ contains $\mathcal{C}_c^\infty(\Omega, \mathbb{R}^m)$. It is called compactly supported if it has full domain of definition $\underline{\mathrm{D}}_{f_{\{.\}}} = \mathcal{C}^\infty(\Omega, \mathbb{R}^m)$. We denote by $\mathcal{C}^{-\infty}(\Omega, \mathbb{R}^m)$ the space of distributions and $\mathcal{C}_c^{-\infty}(\Omega, \mathbb{R}^m)$ the space of compactly supported distributions.

Example 4.2.2 Let $\Omega \subset \mathbb{R}^n$ be an open subset and $f : \Omega \to \mathbb{R}^m$ be a measurable function. Suppose given a non-degenerate symmetric pairing (e.g., the Euclidean or Lorentzian bilinear form)

$$\langle -, - \rangle : \mathbb{R}^m \otimes \mathbb{R}^m \to \mathbb{R}.$$

Then the functional

$$\begin{aligned} f : \underline{\mathcal{C}}^\infty\left(\Omega, \mathbb{R}^m\right) &\to \underline{\mathbb{R}} \\ \varphi(x, u) &\mapsto \int_\Omega \langle f(x), \varphi(x, u) \rangle dx \end{aligned}$$

is a distribution with domain of definition given (for example) by Lebesgue's dominated derivation condition

$$\begin{aligned} \underline{\mathrm{D}}_f(U) = \Big\{ \varphi(x, u) | &\text{locally on } U, \forall \beta, \text{ there exists } g(x) \in L^1(\Omega) \text{ such that} \\ &\text{for all } u, \ \left| \partial_\beta^u \langle f(x), \varphi(x, u) \rangle \right| \leq g(x) \Big\}. \end{aligned}$$

Example 4.2.3 Let $\Omega \subset \mathbb{R}^n$ be an open subset and $x_0 \in \Omega$ be a fixed point. Then the functional

$$\delta_{x_0} : \underline{C}^\infty(\Omega) \to \underline{\mathbb{R}}$$
$$\varphi(x,u) \mapsto \varphi(x_0,u)$$

is a compactly supported distribution with full domain of definition $\underline{C}^\infty(\Omega)$.

We now compare our approach to distributions and functionals to the usual approach that uses functional analytic methods and topological vector spaces. Our setting is much more flexible since we are completely free of choosing domains of definition for the functionals we use, but the comparison is however useful to relate our work to what is usually done in books on mathematical physics. We will also need these results to give our mathematical formalization of causal perturbative quantum field theories in Chap. 22.

Proposition 4.2.4 *A distribution $f : \underline{C}^\infty(\Omega, \mathbb{R}^m) \to \underline{\mathbb{R}}$ is an ordinary distribution if its domain of definition contains the subspace $\underline{C}_p^\infty(\Omega, \mathbb{R}^m) \subset \underline{C}_c^\infty(\Omega, \mathbb{R}^m)$ of properly supported functions, defined by*

$$\underline{C}_p^\infty(\Omega, \mathbb{R}^m)(U) := \{\varphi(x,u) \mid \textit{locally on } U, \textit{there exists a compact subset } K \subset \Omega$$
$$\textit{such that for all } u,\ \operatorname{supp}(\varphi(-,u)) \subset K\}.$$

Proof A distribution is an $\mathbb{R}$-linear continuous function on the topological vector space $C_c^\infty(\Omega)$ equipped with the topology given by the family of seminorms

$$N_{K,k}(\varphi) := \sum_{|\alpha| \le k} \sup_{x \in K} |\partial_\alpha \varphi(x)|$$

indexed by compact subsets $K \subset \Omega$ and integers $k \in \mathbb{N}$. A properly supported function $\varphi(x,u)$ defines a smooth function between open subsets of locally convex vector spaces

$$\varphi : U \to C_c^\infty(\Omega)$$

and its composition with a continuous linear (thus smooth) functional $f : C_c^\infty(\Omega) \to \mathbb{R}$ defines a smooth function $f \circ \varphi : U \to \mathbb{R}$. This shows that if f is an ordinary distribution, then f gives a well defined natural transformation $f : \underline{C}_p^\infty(\Omega, \mathbb{R}^m) \to \underline{\mathbb{R}}$. Conversely, if f is a distribution in our sense, it defines a linear function $f_{\{.\}} : C_c^\infty(\Omega) \to \mathbb{R}$ that is automatically continuous because it sends smooth functions $\varphi : U \to C_c^\infty(\Omega)$ to smooth functions $f(\varphi) : U \to \mathbb{R}$. Indeed, if $\varphi_n : \Omega \to \mathbb{R}$ is a sequence of smooth maps that converges in the topological vector space $C_c^\infty(\Omega)$, one can find a smooth family $\psi : \mathbb{R} \times \Omega \to \mathbb{R}$ in $\underline{C}_p^\infty(\Omega, \mathbb{R}^m)(\mathbb{R})$ and a family of points $u_n \in \mathbb{R}$ converging to 0 such that

$$\psi(-,u_n) = \varphi_n(-) \quad \text{and} \quad \psi(-,0) = \lim_n \varphi_n(-).$$

The functoriality of f in all inclusions $\{.\} \subset \mathbb{R}$ implies that

$$\lim_n f(\varphi_n) = \lim_n f\big(\psi(u_n)\big) = \lim_{t\to 0} f\big(\psi(t)\big) = f\big(\psi(0)\big) = f\Big(\lim_n \varphi_n\Big),$$

which completes the proof. □

Definition 4.2.4 Let $E \to M$ be a vector bundle over a smooth variety. Let $\underline{\Gamma}_p(M^n, E^n) \subset \underline{\Gamma}_c(M^n, E^n)$ be the subspace given by properly supported sections. The space of *distributional polynomials* is defined by

$$\mathcal{O}^{poly}_{\Gamma(M,E)} := \oplus_{n\geq 0}\mathrm{Hom}_{\underline{\mathbb{R}}}\big(\underline{\Gamma}_p\big(M^n, E^n\big), \mathbb{R}\big)^{S_n}.$$

The space of *distributional formal power series* is defined by

$$\mathcal{O}^{formal}_{\Gamma(M,E)} := \prod_{n\geq 0}\mathrm{Hom}_{\underline{\mathbb{R}}}\big(\underline{\Gamma}_p\big(M^n, E^n\big), \mathbb{R}\big)^{S_n}.$$

Proposition 4.2.5 *The natural map*

$$\mathcal{O}^{poly}_{\Gamma(M,E)} \to \mathcal{O}_{\Gamma(M,E)} := \mathrm{Hom}_{\mathrm{PAR}}\big(\underline{\Gamma}(M,E), \underline{\mathbb{R}}\big)$$
$$\sum_{n\geq 0} F_n \mapsto [s : M \to E] \mapsto \sum_{n\geq 0} F_n(s \times \cdots \times s)$$

is injective and compatible with the natural algebra structures on both sides. Let $\mathcal{O}_0 \subset \mathcal{O}_{\underline{\Gamma}(M,E)}$ *be the algebra of functionals on* $\underline{\Gamma}(M,E)$ *defined in* $0 \in \Gamma(M,E)$ *and* $\mathfrak{m}_0 \subset \mathcal{O}_0$ *be the ideal of functions that annihilate at* 0. *Denote by*

$$\widehat{\mathcal{O}}_0 := \lim \mathcal{O}_0/\mathfrak{m}_0^n$$

the corresponding completion. The natural map

$$\mathcal{O}^{formal}_{\Gamma(M,E)} \to \widehat{\mathcal{O}}_0$$

is injective and compatible with the algebra structures on both sides.

Proof The injectivity of the first map follows from the result of linear algebra that says that a multilinear function on a module over a ring of characteristic 0 is uniquely determined by the corresponding homogeneous functional (e.g., a bilinear form is uniquely determined by the corresponding quadratic form). To be more precise, we follow Douady [Dou66] by defining, for any function $h : X \to Y$ between two vector spaces,

$$\Delta_x(h)(y) := \frac{1}{2}\big(h(y+x) - h(y-x)\big).$$

If $f : X \to Y$ is a polynomial function of degree n associated to the multilinear map $\tilde{f} : X^{\otimes n} \to Y$, one has

$$\tilde{f}(x_1, \dots, x_n) = \frac{1}{n!} \Delta_{x_n} \circ \dots \circ \Delta_{x_1} f(0).$$

The algebra $\mathcal{O}^{formal}$ is obtained as a completion of $\mathcal{O}^{poly}$ with respect to the ideal $\mathfrak{n}_0 = \oplus_{n \geq 0} \underline{\mathrm{Hom}}_{\underline{\mathbb{R}}}(\underline{\mathrm{D}}_n^{dist}, \underline{\mathbb{R}})^{S_n}$ of multilinear functionals that annihilate at 0. One can thus check the injectivity at the finite level, in which case it follows from the definition of $\mathfrak{n}_0$ and $\mathfrak{m}_0$. □

We illustrate the above proposition by defining a notion of analytic function on a function space that is very similar to the one used by Douady to define Banach manifolds [Dou66].

Definition 4.2.5 Let $E \to M$ be a vector bundle. A functional $f : \underline{\Gamma}(M, E) \to \underline{\mathbb{R}}$ is called *distributionally analytic* if for every point $x \in \underline{\Gamma}(M, E)(U)$, there exists an open subspace $U_x \subset \underline{\Gamma}(M, E)$ that contains x and a formal functional

$$G = \sum_n G_n \in \mathcal{O}^{formal}_{\Gamma(U_x, E_{|U_x})}$$

such that f is the sum of the corresponding absolutely converging series:

$$f_{|U_x} = \sum_{n \geq 0} G_n.$$

To conclude this section, we hope that the reader is now convinced that our approach to functional calculus is compatible with the usual functional analytic one based on topological vector spaces. However we need more flexibility, in particular for fermionic and graded calculus, so that we shall continue to stick to our methods throughout.

4.3 Generalized Weak Solutions

Let $\Omega \subset \mathbb{R}^n$ be an open subset and let $X = \underline{\mathcal{C}}^\infty(\Omega, \mathbb{R})$. Recall that the space of functionals $\mathcal{O}_X$ is the space $\underline{\mathrm{Hom}}_{\mathrm{PAR}}(X, \underline{\mathbb{R}})$ of partially defined functions

$$f : X \to \underline{\mathbb{R}}.$$

In this setting, one can easily define a notion of a weak solution to a (not necessarily linear) problem.

Definition 4.3.1 The space

$$X^{**} := \underline{\mathrm{Hom}}_{\mathrm{PAR}}(\mathcal{O}_X, \underline{\mathbb{R}})$$

is called the *nonlinear bidual* of X. Let $P : X \to X$ be an operator (morphism of spaces). An element $F \in X^{**}$ is called a *weak solution* to the equation $\{P = 0\}$ if it satisfies the equation

$$F\big(f\big(P(\varphi)\big)\big) = 0.$$

Example 4.3.1 Let $P(D) = \sum_\alpha a_\alpha \partial_\alpha$ be a linear partial differential operator on $\mathcal{C}^\infty(\Omega)$ with smooth coefficients $a_\alpha \in X$. A distributional weak solution f to $P(D)f = 0$ is a distribution $f : X \to \underline{\mathbb{R}}$ (as defined in Sect. 4.2) such that for all test functions $\varphi \in X$, one has

$$\big\langle f, {}^t P(D)\varphi\big\rangle = 0,$$

where ${}^t P(D).\varphi := \sum_\alpha (-1)^\alpha \partial_\alpha (a_\alpha.\varphi)$ is the transpose operator for the standard integration pairing

$$\int : X \otimes_{\underline{\mathbb{R}}} X \to \underline{\mathbb{R}}$$

$$(\varphi, \psi) \mapsto \int_\Omega \varphi(x)\psi(x)dx.$$

One can interpret such a solution as a weak solution in our sense by using the integration pairing to define an embedding $\iota : X \hookrightarrow \mathcal{O}_X$ by

$$\big\langle \iota(\varphi), \psi\big\rangle := \int_\Omega \varphi(x)\psi(x)dx.$$

One can then identify ordinary distributions (i.e., linear functionals $f : X \to \underline{\mathbb{R}}$) with particular elements of X^{**} (i.e., functionals on $\mathcal{O}_X$) with domain of definition $\iota(X) \subset \mathcal{O}_X$. This gives an embedding

$$\iota : \mathcal{C}^{-\infty}(\Omega) \hookrightarrow X^{**}.$$

One then checks that a weak solution that is $\underline{\mathbb{R}}$-linear with domain of definition contained in $\iota(\mathcal{C}^{-\infty}(\Omega))$ is the same as a distributional weak solution. One can also use different bilinear pairings on X and this gives other notions of linear weak solutions. The above construction also works with $\mathbb{R}^m$-valued distributions.

We now define the Fourier transform, which is a construction specific to the affine space $\Omega = \mathbb{R}^n$.

Definition 4.3.2 Let $\langle -, - \rangle : \mathbb{R}^n \times \mathbb{R}^n \to \mathbb{R}$ be a given non-degenerate bilinear pairing. If $f : \mathbb{R}^n \to \mathbb{R}$ is an L^1-function, its *Fourier transform* is defined as the function

$$\hat{f}(\xi) := \int_{\mathbb{R}^n} e^{-2i\pi\langle x, \xi\rangle} f(x)dx.$$

The *Fourier transform of a distribution* $f : \underline{\mathcal{C}}^{\infty}(\Omega) \to \underline{\mathbb{R}}$ is the distribution $\hat{f}$ given by the formula

$$\langle \hat{f}, \varphi \rangle = \langle f, \hat{\varphi} \rangle,$$

with domain of definition given by

$$\underline{\mathrm{D}}_{\hat{f}}(U) = \left\{ \varphi(x,u),\ e^{-2i\pi\langle x,\xi\rangle}\varphi(x) \in \underline{\mathrm{D}}_f(U) \right\}.$$

The main property of Fourier transform is that it replaces differential operators by multiplication operators.

Proposition 4.3.1 *If $P(\xi) = \sum_{|\alpha|\leq k} a_\alpha x^\alpha$ is a polynomial on $\mathbb{R}^n$ with complex coefficients, we define a differential operator*

$$P(D) = \sum_{|\alpha|\leq k} a_\alpha D^\alpha$$

with $D^\alpha = D_1^{\alpha_1}\cdots D_n^{\alpha_n}$ and $D_i = \frac{\langle x_i, x_i\rangle}{2i\pi}\partial_{x_i}$. We then have

$$\left\langle \widehat{P(D)f}, \varphi \right\rangle = \left\langle P(\xi)\hat{f}, \varphi \right\rangle.$$

The space $C_c^\infty(\Omega)$ is not stable under the Fourier transform, meaning that the Fourier transform of a function with compact support need not have compact support (in quantum mechanical terms, this means that one cannot simultaneously measure the position and the momentum of a given particle). We now define Schwartz type distributions, whose domain of definition remains stable under the operation of derivation and Fourier transform.

Definition 4.3.3 The *standard Schwartz space* is the subspace $\underline{\mathcal{S}}(\mathbb{R}^n) \subset \underline{\mathcal{C}}^\infty(\mathbb{R}^n)$ defined by

$$\underline{\mathcal{S}}(\mathbb{R}^n)(U) := \Big\{ \varphi(x,u) \mid \text{locally on } U, \text{for all } \alpha, \beta, \text{there exists a } C_{\alpha,\beta,U} \\ \text{such that } \forall u \in U,\ \left\| x^\alpha \partial_\beta \varphi(-,u) \right\|_\infty < C_{\alpha,\beta,U} \Big\}.$$

A distribution whose domain of definition contains $\underline{\mathcal{S}}(\mathbb{R}^n)$ is called a *Schwartz distribution*. The space of Schwartz distributions is denoted by $\mathcal{S}'(\mathbb{R}^n)$.

The space of Schwartz distributions is the continuous dual $\mathcal{S}'(\mathbb{R}^n)$ of the topological vector space $\mathcal{S}(\mathbb{R}^n) := \underline{\mathcal{S}}(\mathbb{R}^n)(\{.\})$ with respect to the topology induced by the family of seminorms

$$N_k(\varphi) := \sum_{|\alpha|,|\beta|\leq k} \sup_{x\in\mathbb{R}^n} \left| x^\beta \partial_\alpha(x) \right|.$$

It is stable under the Fourier transform. There are natural inclusions

$$\mathcal{C}_p^{\infty}(\mathbb{R}^n) \subset \underline{\mathcal{S}}(\mathbb{R}^n) \subset \underline{\mathcal{C}}^{\infty}(\mathbb{R}^n) \quad \text{and} \quad \mathcal{C}_c^{-\infty}(\mathbb{R}^n) \subset \mathcal{S}'(\mathbb{R}^n) \subset \mathcal{C}^{-\infty}(\mathbb{R}^n).$$

We conclude this section with a general formulation of the Schwartz kernel theorem (see Schwartz's reference book [Sch66]). The properties of operators between functional spaces

$$P : \mathcal{C}_c^{\infty}(\mathbb{R}^n) \to \mathcal{C}^{\infty}(\mathbb{R}^n)$$

defined by a kernel function $K \in \mathcal{C}^{\infty}(\mathbb{R}^n \times \mathbb{R}^n)$, i.e., by a formula of the form

$$P(f) = \int_{\mathbb{R}^n} K(x, y) f(x) dx,$$

can be read from the properties of its kernel. Schwartz's kernel theorem shows that essentially any reasonable operator is of this kind, if one uses distributions as kernels.

Theorem 4.3.1 *Let $\Omega_1 \subset \mathbb{R}^n$ and $\Omega_2 \subset \mathbb{R}^m$ be two open subsets. Let $P : \mathcal{C}_c^{\infty}(\Omega_1) \to \mathcal{C}^{-\infty}(\Omega_2)$ be a continuous operator. Then there exists a distribution $K \in \mathcal{C}^{-\infty}(\Omega_1 \times \Omega_2)$ such that*

$$\langle Pf, \varphi \rangle = \langle K, f \otimes \varphi \rangle.$$

More generally, let $P : \mathcal{C}_c^{\infty}(\Omega_1, \mathbb{R}^l) \to \mathcal{C}^{-\infty}(\Omega_2, \mathbb{R}^k)$ be a continuous operator. Then there exists a distribution $K \in \mathcal{C}^{-\infty}(\Omega_1 \times \Omega_2, (\mathbb{R}^l)^* \boxtimes \mathbb{R}^k)$ such that

$$\langle Pf, \varphi \rangle = \langle K, f \otimes \varphi \rangle.$$

Note that there is a complex analytic approach to distributions due to Sato, called the theory of hyperfunctions. We refer to Chap. 11 for an introduction to hyperfunction theory and other methods of linear algebraic analysis.

4.4 Spectral Theory

Since the formalism of quantum mechanics (to be explained in Chap. 17) is mainly based on the spectral theory of operators on Hilbert spaces, we here recall its main features, referring to von Neumann's book [vN96] for more details. This theory is a kind of generalization to infinite dimension of the theory of diagonalization of hermitian matrices.

Definition 4.4.1 A *Hilbert space* is a pair $\mathcal{H} = (\mathcal{H}, \langle \cdot, \cdot \rangle)$ composed of a topological vector space $\mathcal{H}$ and a sesquilinear pairing $\langle \cdot, \cdot \rangle : \mathcal{H} \times \mathcal{H} \to \mathbb{C}$ such that $\mathcal{H}$ is complete with respect to the norm $\| f \| = \sqrt{\langle f, f \rangle}$.

The main objective of spectral theory is the study of the spectrum of an operator on a Hilbert space.

Definition 4.4.2 If $A : \mathcal{H} \to \mathcal{H}$ is an operator (i.e., a linear map), its *spectrum* $\mathrm{Sp}(A)$ is the subset of $\mathbb{C}$ defined by

$$\mathrm{Sp}(A) = \{\lambda \in \mathbb{C} | \, A - \lambda.\mathrm{Id} \text{ non invertible}\}.$$

Definition 4.4.3 If $A : \mathcal{H} \to \mathcal{H}$ is an operator, its *adjoint* A^* is defined by

$$\langle Af, g\rangle = \langle f, A^* g\rangle.$$

We now define various classes of operations on the Hilbert space $\mathcal{H}$.

Definition 4.4.4 Let $A : \mathcal{H} \to \mathcal{H}$ be a linear operator.

1. One says that A is *bounded* if there exists a constant $C \geq 0$ such that

$$\|Ax\| \leq C\|x\|,$$

and a minimal such constant C will be denoted by $\|A\|$.
2. One says that A is *positive* (bounded) if $\mathrm{Sp}(A) \subset [0, +\infty[\subset \mathbb{C}$.
3. One says that A is *compact* if the image of the unit ball $B(0, 1) = \{f \in \mathcal{H} \mid \|f\| \leq 1\}$ of $\mathcal{H}$ by A has compact closure, i.e.,

$$\overline{A\big(B(0,1)\big)} \quad \text{compact}.$$

4. One says that A is *self-adjoint* if $A^* = A$.
5. One says that A is *a projection* if A is bounded and

$$A = A^* = A^2.$$

We denote by $\mathcal{B}(\mathcal{H})$ (resp. $\mathcal{B}(\mathcal{H})^+$, resp. $\mathcal{K}(\mathcal{H})$, resp. $\mathrm{Proj}(\mathcal{B}(\mathcal{H}))$, resp. $\mathcal{B}^{sa}(\mathcal{H})$) the set of bounded (resp. positive, resp. compact, resp. projection, resp. bounded self-adjoint) operators on $\mathcal{H}$. If $A, B \in \mathcal{B}(\mathcal{H})$ are two bounded operators, one says that A *is smaller than* B if $B - A$ is positive, i.e.,

$$A \leq B \quad \Leftrightarrow \quad B - A \in \mathcal{B}(\mathcal{H})^+.$$

Theorem 4.4.1 *Let $A : \mathcal{H} \to \mathcal{H}$ be a bounded operator. One has an inclusion $\mathrm{Sp}(A) \subset \mathbb{R}$ if and only if A is self-adjoint. If $A \in \mathcal{K}(\mathcal{H})$ is compact then $\mathrm{Sp}(A) \subset \mathbb{C}$ is discrete.*

Proof See Reed-Simon [RS80], Theorem VI.8. □

Definition 4.4.5 We denote by $\mathrm{Borel}(\mathrm{Sp}(A))$ the Borel σ-algebra on $\mathrm{Sp}(A)$. A *spectral measure* for a bounded operator $A \in \mathcal{B}(\mathcal{H})$ is a multiplicative measure on the spectrum of A with values in the space of projectors. More precisely, this is a map

$$\begin{aligned} E : \mathrm{Borel}\big(\mathrm{Sp}(A)\big) &\to \mathrm{Proj}\big(\mathcal{B}(\mathcal{H})\big) \\ B &\mapsto E(B) \end{aligned}$$

such that

1. $E(\emptyset) = 0$;
2. $E(\mathrm{Sp}(A)) = 1$;
3. E is sigma-additive, i.e.,

$$E\left(\coprod_{i=1}^{\infty} B_i\right) = \sum_{i=1}^{\infty} E(B_i);$$

4. E is multiplicative, i.e., $E(B \cap C) = E(B) \cdot E(C)$.

The spectral theorem, which is a generalization to infinite dimensions of the diagonalization theorem of symmetric or hermitian endomorphisms, can be formulated as follows.

Theorem 4.4.2 *Let $A \in \mathcal{B}(\mathcal{H})$ be a bounded self-adjoint operator. There exists a unique spectral measure*

$$E_A : \mathrm{Borel}\big(\mathrm{Sp}(A)\big) \to \mathrm{Proj}\big(\mathcal{B}(\mathcal{H})\big)$$

such that

$$A = \int_{\mathrm{Sp}(A)} \lambda dE_A(\lambda).$$

Proof This theorem is also true for an unbounded self-adjoint operator (densely defined) and its proof can be found in Reed-Simon [RS80], Chaps. VII.2 and VIII.3. □

The main interest of this theorem is that it gives a measurable functional calculus, allowing the evaluation of a measurable positive function $f : \mathbb{R} \to \mathbb{R}_+$ on a given bounded self-adjoint operator A by

$$f(A) = \int_{\mathrm{Sp}(A)} f(\lambda) dE_A(\lambda).$$

This type of computation is very useful in quantum mechanics for the characteristic functions $\mathbb{1}_{[a,b]} : \mathbb{R} \to \mathbb{R}$, which allow the restriction to parts of the spectrum contained in an interval.

One also has the following theorem:

Theorem 4.4.3 *If two operators A and B of $\mathcal{B}(\mathcal{H})$ commute, there exists an operator $R \in \mathcal{B}(\mathcal{H})$ and two measurable functions $f, g : \mathbb{C} \to \mathbb{C}$ such that*

$$A = f(R) \quad \text{and} \quad B = g(R).$$

Stone's theorem allows one to translate the study of unbounded self-adjoint operators to the study of one parameter groups of unitary operators. A proof can be found in [RS80], Chap. VIII.4.

Theorem 4.4.4 (Stone's Theorem) *The datum of a strongly continuous one parameter family of unitary operators on $\mathcal{H}$, i.e., of a group morphism $U : \mathbb{R} \to \mathcal{U}(\mathcal{H})$ satisfying*

$$\text{if } \varphi \in \mathcal{H} \quad \text{and} \quad t \to t_0 \quad \text{in } \mathbb{R}, \quad \text{then } U(t)\varphi \to U(t_0)\varphi,$$

is equivalent to the datum of a self-adjoint operator A, via the map $A \mapsto e^{itA}$.

4.5 Green Functions and Parametrices

Definition 4.5.1 Let $P(D) = \sum_\alpha a_\alpha \partial_\alpha$ be a partial differential operator on $\Omega \subset \mathbb{R}^n$.

- A *Green function* (also called a *fundamental solution*) for the operator $P(D)$ is a distribution f on Ω^2 such that

$$P(D)f(x, y) = \delta_x(y)$$

  whenever $x, y \in \Omega$.
- A *parametrix* for the operator $P(D)$ is a distribution f on Ω^2 such that

$$P(D)f(x, y) = \delta_x(y) + \omega(x, y)$$

  where ω is a smooth function on Ω^2.

A Green function on $\Omega = \mathbb{R}^n$ is called a fundamental solution because of the following result: if g is a function that can be convoluted with the fundamental solution f (for example, if g has compact support), one can find a solution to the inhomogeneous equation

$$P(D)h = g$$

by simply defining $h = f * g$ to be the convolution of the fundamental solution and g.

Example 4.5.1 In Euclidean field theory on flat space, linear trajectories are related to the solutions and eigenvalues of the Laplacian $\Delta = \sum_{i=1}^{n} \partial_{x_i}^2$. This operator corresponds to the polynomial $\|x\|^2 = \sum_{i=1}^{n} x_i^2$. Using the formula

$$\int_{\mathbb{R}^n} f(x)dx = \int_{\mathbb{R}_+} \int_{S^{n-1}} f(x.\omega) r^{n-1} dr d\omega,$$

which is always true for a positive function, one gets that the function $\frac{1}{\|x\|^2}$ is locally integrable if $n > 2$ so that it defines a distribution D. One can check that it is tempered and that its inverse Fourier transform f is a fundamental solution of the Laplacian.

Example 4.5.2 The first model of electromagnetic waves is given by the wave operator

$$\frac{1}{c^2}\partial_t^2 - \sum_{i=2}^{n} \partial_{x_i}^2.$$

Its solutions are quite different to those of the Laplacian. One first observes that the corresponding polynomial $\frac{1}{c^2}t^2 - \sum_{i=2}^{n} x_i^2$ has real zeroes so that it cannot be naively inverted as a locally integrable function. The easiest way to solve the wave equation is to apply the space Fourier-transform (on the variable x) transforming it into the ordinary differential operator

$$\frac{1}{c^2}\partial_t^2 - \|x\|^2.$$

One can also use the complex family of distributions parametrized by a real nonzero constant ϵ given by the locally integrable function

$$\frac{\mathbb{1}_X}{-\frac{1}{c^2}t^2 + \sum_{i=2}^{n} x_i^2 + i\epsilon}$$

where X is the light cone (where the denominator is positive for $\epsilon = 0$). The Fourier transform of the limit of this distribution for $\epsilon \to 0$ gives a fundamental solution of the wave equation called the Feynman propagator. We refer to Chap. 11 for more details on this complex analytic approach to distributions, called algebraic analysis.

4.6 Complex Analytic Regularizations

The Sato-Bernstein polynomial gives an elegant solution to the following problem of Gelfand: if P is a real polynomial in n variables with positive coefficients, show that the distribution

$$\langle P^s, f \rangle := \int_{\mathbb{R}^n} P(x)^s f(x) dx,$$

defined and holomorphic in s for $\mathrm{Re}(s) > 0$, can be meromorphically continued to the whole complex plane.

One can try to answer this for $P(x) = x^2$. An integration by parts gives

$$(2s+1)(2s+2)\langle x^{2s}, f\rangle = \int_{\mathbb{R}} x^{2(s+1)} f''(x)dx =: \langle \partial_x^2 x^{2(s+1)}, f\rangle.$$

This functional equation gives the desired meromorphic continuation to the complex plane with poles in the set $\{-1/2, -1, -3/2, -2, \dots\}$.

The generalization of this method to any real polynomial with positive coefficients can be done using a similar functional equation, given by the following theorem due to Sato (in the analytic setting) and Bernstein (in the algebraic setting).

Theorem 4.6.1 (Sato-Bernstein) *If $f(x)$ is a non-zero polynomial in several variables with positive real coefficients, there is a non-zero polynomial $b(s)$ and a differential operator $P_x(s)$ with polynomial coefficients such that*

$$P_x(s) \cdot f(x)^{s+1} = b(s) f(x)^s.$$

Proof A full and elementary proof of this theorem can be found in the memoir of Chadozeau and Mistretta [CM00]. One may also refer to the original paper of Bernstein [Ber71] and to Kashiwara's book [Kas03], Chap. 6. □

As an example, if $f(x) = x_1^2 + \cdots + x_n^2$, one has

$$\sum_{i=1}^{n} \partial_i^2 f(x)^{s+1} = 4(s+1)\left(s + \frac{n}{2}\right) f(x)^s$$

so the Bernstein-Sato polynomial is

$$b(s) = (s+1)\left(s + \frac{n}{2}\right).$$

This gives a method to compute the fundamental solution of the Laplacian.

More generally, if $f(x) = \sum_{i=1}^{n} a_i x_i^2$ is a quadratic form on $\mathbb{R}^n$, and $f^{-1}(D) = \sum_{i=1}^{n} \frac{1}{a_i} \partial_i^2$ is the second order differential operator corresponding to the inverse of the given quadratic form, one has

$$f^{-1}(D).f(x)^{s+1} = 4(s+1)\left(s + \frac{n}{2}\right) f(x)^s.$$

This gives a method to compute the fundamental solution of the wave equation.

Theorem 4.6.2 (Malgrange-Ehrenpreis) *Let $P(D)$ be a differential operator in $\mathbb{R}^n$ with constant coefficients. There exists a distribution $f \in \mathcal{S}'(\mathbb{R}^n)$ such that*

$$P(D) f = \delta_0.$$

Proof By using the Fourier transform

$$\mathcal{F} : \mathcal{S}(\mathbb{R}^n) \to \mathcal{S}(\mathbb{R}^n),$$

we reduce the problem to finding a distributional inverse to the polynomial $P \in \mathbb{C}[x_1, \ldots, x_n]$ that defines the differential operator $P(D)$. Indeed, the Fourier transformed equation of $P(D)f = \delta_0$ is

$$P(x) \cdot \mathcal{F}(f) = 1.$$

If the given polynomial has no real zeroes, its inverse is continuous on $\mathbb{R}^n$ and thus locally integrable. The corresponding distribution solves the problem. Otherwise, we need more efforts. Replacing P by $P \cdot \bar{P}$ reduces the question to a polynomial P with positive coefficients. One can then apply the Malgrange-Ehrenpreis Theorem 4.6.2 to P to get a non-zero polynomial $b(s)$ and a differential operator $P_x(s)$ with polynomial coefficients such that

$$P_x(s) \cdot f(x)^{s+1} = b(s) f(x)^s.$$

This gives a meromorphic continuation of the Schwartz distribution $P(x)^s$. If we take the constant term of the Laurent series of this meromorphic function at -1, we get a distribution D that satisfies

$$P(x) \cdot D = 1,$$

so that the problem is solved. □

The Malgrange-Ehrenpreis theorem is very elegant but one can often find fundamental solutions more easily by other methods.

4.7 Heat Kernel Regularization

We refer to [BGV92] for a detailed study of heat kernels. Let (M, g) be a Riemannian manifold. The Laplace-Beltrami operator Δ_g is defined by

$$\Delta_g = \operatorname{div} \circ \overrightarrow{\operatorname{grad}},$$

where $\operatorname{div} : \Theta_M \to \mathcal{C}^\infty(M)$ is the unique operator associated to the volume form ω_g such that

$$\mathcal{L}_{\omega_g} X = \operatorname{div}(X)\omega_g,$$

and $\overrightarrow{\operatorname{grad}} : \mathcal{C}^\infty(M) \to \Theta_M$ is the unique operator such that

$$g\big(X, \overrightarrow{\operatorname{grad}}(f)\big) = df(X).$$

The heat kernel $K_\tau(x,y)$ on a Riemannian manifold (M,g) is the solution of the Heat equation

$$\partial_\tau K_\tau(x,y) = \Delta_g K_\tau(x,y),$$

where Δ_g is the Laplace-Beltrami operator, subject to the boundary condition

$$\lim_{\tau\to 0} K_\tau(x,y) = \delta_x(y).$$

Using the functional calculus presented in Sect. 4.4, one can also write the corresponding integral operator $\varphi(y) \mapsto (T_\tau\varphi)(y) := \int K_\tau(x,y)\varphi(x)dx$ as

$$T_\tau = e^{\tau\Delta}.$$

One can use this to write down the heat kernel presentation of the Green function P (also called the Euclidean propagator) for the Laplacian operator with mass $\Delta_g + m^2$ as the integral

$$P(x,y) = \int_0^\infty e^{\tau m^2} K_\tau(x,y)d\tau.$$

The heat kernel regularization is given simply by cutting off the above integral

$$P_{\epsilon,L} = \int_\epsilon^L e^{-\tau m^2 K_\tau} d\tau.$$

It is a smooth function that approximates the Euclidean propagator.

4.8 Wave Front Sets of Distributions

The wave front set is a refinement in the cotangent space of the singular support of a distribution (domain where a distribution is not smooth). It gives information on the propagation of singularities, and gives precise conditions for the existence of pull-backs (and in particular products) of distributions. Problems that occur when trying to form the product of distributions are central in the renormalization of quantum field theory.

We refer to Hormander [Hör03], Chap. VIII for a concise presentation of the smooth version of wave front sets. A more analytic approach, including a complete overview of algebraic microlocal analysis is given in Chap. 11.

The definition of the wave front set is based on the fact that one can analyze the regularity of a Schwartz function by studying how its Fourier transform increases.

Definition 4.8.1 Let M be a manifold and $u \in C^{-\infty}(M)$ be a distribution on M. One says that u is *microlocally smooth near a given point* $(x_0, \xi_0) \in T^*M$ if there exists a smooth cutoff function $\varphi \in C_c^\infty(M)$, with $\varphi(x_0) \neq 0$, such that the Fourier transform $\widehat{(\varphi u)}$ (in a given chart U of M containing the support of φ) is of rapid decrease at infinity in a small conic neighborhood of ξ_0, i.e, there exist $\epsilon > 0$ and C_ϵ such that

$$\forall N > 0,\ \forall \xi,\quad \left\| \frac{\xi}{\|\xi\|} - \frac{\xi_0}{\|\xi_0\|} \right\| < \epsilon \quad \Rightarrow \quad \|\xi\|^N \widehat{(\varphi u)}(\xi) \leq C_\epsilon.$$

The *wave front set* of u is the set $\mathrm{WF}(u)$ of points of T^*M at which f is not microlocally smooth.

The projection of $\mathrm{WF}(u)$ to M is the singular support of u (points around which u is not equal to a smooth function).

Theorem 4.8.1 *Let $f : N \to M$ be a smooth map between smooth manifolds and u be a distribution on M. Suppose that*

$$\mathrm{WF}(u) \cap T_N^* M = \emptyset,$$

i.e., no point of $\mathrm{WF}(u)$ *is normal to $df(TN)$. Then the pullback f^*u of u to a distribution on N can be defined such that* $\mathrm{WF}(f^*u) \subset f^*\mathrm{WF}(u)$. *This pullback is the unique way to continuously extend the pullback of functions to distributions subject to the above wavefront condition.*

Proof We only give a sketch of the proof, referring to Hormander [Hör03], Chap. VIII for more details. Let $u_i \to u$ be a sequence of smooth functions converging to u. One defines the pullback of u_i by $f^*u_i := u_i \circ f$. The wave front set condition ensures that f^*u_i converges to a well defined distribution f^*u on N. □

Corollary 4.8.1 *There is a unique way to continuously extend the product of smooth functions to a product $(u, v) \mapsto u \cdot v$ on distributions subject to the wave front condition*

$$\mathrm{WF}(u) \cap \mathrm{WF}(v)^o = \emptyset,$$

where $\mathrm{WF}(v)^o := \{(x, \xi) \in T^*M,\ \text{such that } (x, -\xi) \in \mathrm{WF}(v)\}$.

Proof The product is defined as the pullback of the tensor product $u \otimes v \in \mathcal{C}^{-\infty}(M \times M)$ along the diagonal. The result then follows from the estimation of the wave front set of the tensor product

$$\mathrm{WF}(u \otimes v) \cap \big(\mathrm{WF}(u) \times \mathrm{WF}(v)\big) \cup \big((\operatorname{supp} u \times \{0\}) \times \mathrm{WF}(v)\big)$$
$$\cup \big(\mathrm{WF}(u) \times (\operatorname{supp} v) \times \{0\}\big).$$

□

References

[Ber71] Bernšteĭn, I.N.: Modules over a ring of differential operators. An investigation of the fundamental solutions of equations with constant coefficients. Funkc. Anal. Prilozh. **5**(2), 1–16 (1971)

[BGV92] Berline, N., Getzler, E., Vergne, M.: Heat Kernels and Dirac Operators. Grundlehren der Mathematischen Wissenschaften [Fundamental Principles of Mathematical Sciences], vol. 298, p. 369. Springer, Berlin (1992). ISBN 3-540-53340-0

[Bou70] Bourbaki, N.: Éléments de Mathématique. Théorie des Ensembles, p. 349. Hermann, Paris (1970)

[CM00] Chadozeau, A., Mistretta, E.: Le polynôme de Bernstein-Sato (2000, preprint)

[Dir82] Dirac, P.A.M.: The Principles of Quantum Mechanics (International Series of Monographs on Physics). Oxford University Press, London (1982). ISBN 0198520115. http://www.amazon.ca/exec/obidos/redirect?tag=citeulike09-20&path=ASIN/0198520115

[Dou66] Douady, A.: Le problème des modules pour les sous-espaces analytiques compacts d'un espace analytique donné. Ann. Inst. Fourier (Grenoble) **16**(fasc. 1), 1–95 (1966)

[Eva98] Evans, L.C.: Partial Differential Equations. Graduate Studies in Mathematics, vol. 19, p. 662. Am. Math. Soc., Providence (1998). ISBN 0-8218-0772-2

[Hör03] Hörmander, L.: The Analysis of Linear Partial Differential Operators. i. Classics in Mathematics, p. 440. Springer, Berlin (2003). Distribution theory and Fourier analysis. Reprint of the second (1990) edition [Springer, Berlin; MR1065993 (91m:35001a)]. ISBN 3-540-00662-1

[Kas03] Kashiwara, M.: D-Modules and Microlocal Calculus. Translations of Mathematical Monographs, vol. 217, p. 254. Am. Math. Soc., Providence (2003). Translated from the 2000 Japanese original by Mutsumi Saito. Iwanami Series in Modern Mathematics. ISBN 0-8218-2766-9

[KM97] Kriegl, A., Michor, P.W.: The Convenient Setting of Global Analysis. Mathematical Surveys and Monographs, vol. 53, p. 618. Am. Math. Soc., Providence (1997). ISBN 0-8218-0780-3

[RS80] Reed, M., Simon, B.: Methods of Modern Mathematical Physics. I, 2nd edn., p. 400. Academic Press/Harcourt Brace Jovanovich, New York/San Diego (1980). Functional analysis. ISBN 0-12-585050-6

[Sch66] Schwartz, L.: Théorie des Distributions. Publications de L'Institut de Mathématique de L'Université de Strasbourg, p. 420. Hermann, Paris (1966). No. IX–X. Nouvelle édition, entiérement corrigée, refondue et augmentée

[vN96] von Neumann, J.: Mathematical Foundations of Quantum Mechanics. Princeton Landmarks in Mathematics, p. 445. Princeton University Press, Princeton (1996). Translated from the German and with a preface by Robert T. Beyer, Twelfth printing, Princeton Paperbacks. ISBN 0-691-02893-1

Chapter 5
Linear Groups

This chapter mostly contains very classical results, and is included for later reference in the description of the physical examples of Chaps. 14, 15 and 16, since these classical results are somewhat disseminated in the mathematical literature.

Symmetry considerations are nowadays an important tool in the mathematical description of most of the theories of modern physics.

We start here by introducing linear algebraic and Lie groups via their functor of points, following a path between SGA 3 [DG62] and Waterhouse's book [Wat79]. Here, our presentation of Lie groups gives a concrete illustration of the interest of the notion of a smooth algebra for basic differential geometric constructions. We also extensively use the book of involutions [KMRT98] because algebras with involutions give the smoothest approach to the structural theory of algebraic groups over non-algebraically closed fields (e.g., over $\mathbb{R}$).

5.1 Generalized Algebraic and Lie Groups

Definition 5.1.1 Let ALG and C be two categories and assume given a faithful functor

$$\mathrm{Forget} : C \to \mathrm{SETS}$$

which allows us to think of C as a category of sets with additional structures. A functor

$$F : \mathrm{ALG} \to C$$

is called a *C-structured functor* on ALG. If ALG is the category ALG_K of commutative algebras over a field K, we call F an algebraic C-object functor.

Example 5.1.1 Let $\mathrm{ALG} = \mathrm{ALG}_{\mathbb{R}}$ be the category of real algebras and $C = \mathrm{ASS}$ be the category of associative unital algebras. The functor

$$\begin{aligned} \mathrm{M}_n : \mathrm{ALG}_{\mathbb{R}} &\to \mathrm{ASS} \\ A &\mapsto \big(\mathrm{M}_n(A), \times, +, \underline{0}, \mathrm{id}\big) \end{aligned}$$

F. Paugam, *Towards the Mathematics of Quantum Field Theory*,
Ergebnisse der Mathematik und ihrer Grenzgebiete. 3. Folge / A Series of Modern Surveys in Mathematics 59, DOI 10.1007/978-3-319-04564-1_5,

makes M_n an algebraic associative algebra functor. The functor $\mathrm{GL}_n : \mathrm{ALG}_{\mathbb{R}} \to \mathrm{GRP}$ defined by

$$\mathrm{GL}_n(A) := \{(M, N) \in M_n(A) \times M_n(A) | MN = NM = I\}$$

makes GL_n an algebraic group functor. Both these functors are representable respectively by the algebras

$$A_{\mathrm{M}_n} := \mathbb{R}[\{M_{i,j}\}_{i,j=1,\dots,n}]$$

and

$$A_{\mathrm{GL}_n} := \mathbb{R}[\{M_{i,j}\}_{i,j=1,\dots,n}, \{N_{i,j}\}_{i,j=1,\dots,n}]/(MN = NM = I),$$

meaning that there are natural bijections

$$\mathrm{M}_n(A) \cong \mathrm{Hom}_{\mathrm{ALG}_{\mathbb{R}}}(A_{\mathrm{M}_n}, A) \quad \text{and} \quad \mathrm{GL}_n(A) \cong \mathrm{Hom}_{\mathrm{ALG}_{\mathbb{R}}}(A_{\mathrm{GL}_n}, A).$$

Example 5.1.2 Let $\mathrm{ALG} = \mathrm{OPEN}_{\mathcal{C}^\infty}^{op}$ be opposite to the category $\mathrm{OPEN}_{\mathcal{C}^\infty}$ of open subsets U of $\mathbb{R}^n$ for varying n with smooth maps between them. The functor

$$\begin{aligned} \mathcal{C}^\infty : \mathrm{OPEN}_{\mathcal{C}^\infty}^{op} &\to \mathrm{ALG}_{\mathbb{R}} \\ U &\mapsto \mathcal{C}^\infty(U) \end{aligned}$$

allows us to define

$$\mathrm{M}_n^{sm} : \mathrm{OPEN}_{\mathcal{C}^\infty}^{op} \to \mathrm{ASS} \quad \text{and} \quad \mathrm{GL}_n^{sm} : \mathrm{OPEN}_{\mathcal{C}^\infty}^{op} \to \mathrm{GRP}$$

called the smooth matrix algebra functor and the smooth general linear group functor (also called the general linear Lie group). One can also show that both these functors are representable respectively by $U_{\mathrm{M}_n} = \mathbb{R}^{n^2}$ and by the open subspace $U_{\mathrm{GL}_n} = \det^{-1}(\mathbb{R}^*) \subset U_{\mathrm{M}_n}$.

Example 5.1.3 The smooth special linear group $\mathrm{SL}_n^{sm} : \mathrm{OPEN}_{\mathcal{C}^\infty}^{op} \to \mathrm{GRP}$ defined by

$$\mathrm{SL}_n^{sm}(U) := \{g \in \mathrm{M}_n(U), \det(g) = 1\}$$

is not clearly representable because it corresponds to a closed subset of $\mathbb{R}^{n^2}$. A better setting in which one can work with it is given by the category of (finitary) smooth algebras defined in Example 3.2.1. In this setting, the functor

$$\mathrm{SL}_n^{sm} : \mathrm{ALG}_{\mathcal{C}^\infty}^{fin} \to \mathrm{GRP}$$

defined by

$$\mathrm{SL}_n^{sm}(A) := \{g \in \mathrm{M}_n(A), \deg(g) = 1\}$$

is representable by the smooth algebra

$$\mathcal{C}^\infty(\mathrm{SL}_n) = \mathcal{C}^\infty(\mathrm{M}_n)/\big(\det(M) = 1\big).$$

Similarly, the functor

$$\mathrm{GL}_n^{sm} : \mathrm{ALG}_{\mathcal{C}^\infty}^{fin} \to \mathrm{GRP}$$

is representable by the smooth algebra

$$\mathcal{C}^\infty(\mathrm{GL}_n) = \mathcal{C}^\infty\big(\mathrm{M}_n^2\big)/(MN = NM = I).$$

The setting of smooth algebras thus gives a better analogy between algebraic and Lie groups.

We now give two examples of group functors that are not representable. This kind of group is of common use in physics, where it is called a (generalized) gauge group.

Example 5.1.4 Let $\Omega \subset \mathbb{R}^d$ be an open subset and consider the group functor

$$\begin{aligned} \underline{\mathrm{Aut}}(\Omega) : \mathrm{OPEN}_{\mathcal{C}^\infty}^{op} &\to \mathrm{GRP} \\ U &\mapsto \big\{\varphi : U \times \Omega \to \Omega \text{ smooth such that } \varphi(u,-) : \Omega \to \Omega \\ &\qquad \text{diffeomorphism } \forall u \in U\big\} \end{aligned}$$

describing diffeomorphisms of Ω. It is not representable because of its infinite dimensionality. Such diffeomorphism groups play an important role as symmetry groups in generally covariant theories, such as Einstein's general relativity.

Example 5.1.5 Let $G = \mathrm{GL}_n^{sm}$ be the general linear Lie group and $\Omega \subset \mathbb{R}^p$ be an open subset. The functor

$$\begin{aligned} \underline{\mathrm{Hom}}(\Omega, G) : \mathrm{OPEN}_{\mathcal{C}^\infty}^{op} &\to \mathrm{SETS} \\ U &\mapsto \{\varphi : U \times \Omega \to U_{\mathrm{GL}_n} \text{ smooth}\} \end{aligned}$$

is naturally equipped with a group functor structure induced by that of GL_n^{sm}. These kinds of groups play an important role as symmetry groups in Yang-Mills gauge theories, such as Maxwell's electromagnetism.

To finish this section, we remark that the setting of smooth algebras allows us to give a completely algebraic treatment of linear Lie groups. We will thus only present the theory of algebraic matrix groups, since it is enough for our needs. However, we finish this section by defining the notion of an affine smooth group, which includes Lie groups but is a bit more flexible and is the perfect analog of affine group schemes in the smooth setting.

Definition 5.1.2 A *smooth affine group* is a representable functor

$$G : \text{ALG}_{\mathcal{C}^\infty} \to \text{GRP}$$

from the category of (finitary) smooth algebras (see Example 3.2.1) to the category of groups. More precisely, there exists a smooth algebra A_G and for any smooth algebra A, a natural isomorphism

$$G(A) \xrightarrow{\sim} \text{Hom}_{\text{ALG}_{\mathcal{C}^\infty}}(A_G, A).$$

A *Lie group* is a smooth affine group whose smooth algebra of functions is formally smooth and finitely presented.

Example 5.1.6 The universal covering $\widetilde{\text{SL}_2}$ of SL_2 is a smooth affine group that may not be described as a subgroup of the general linear group on a given vector space. In particular, it is an example of a non-algebraic smooth affine group, all of whose faithful linear representations are infinite dimensional.

5.2 Affine Group Schemes

The reader who wants to have a partial and simple summary can use Waterhouse's book [Wat79]. For many more examples and a full classification of linear algebraic groups on non-algebraically closed fields (like $\mathbb{R}$ for example, which is useful for physics), we refer to the very complete book of involutions [KMRT98].

Definition 5.2.1 A *linear algebraic group* over a field K is an algebraic group functor that is representable by a quotient of a polynomial ring, i.e., it is a functor

$$G : \text{ALG}_K \to \text{GRP}$$

and a function algebra A_G such that for any algebra A, there is a natural bijection

$$G(A) \xrightarrow{\sim} \text{Hom}_{\text{ALG}_K}(A_G, A).$$

One can show that any linear algebraic group G (with a finitely generated algebra of functions A_G) is a Zariski closed subgroup of the group GL_n for some n that we defined in the previous section by

$$\text{GL}_n(A) := \left\{(M, N) \in M_n(A) \times M_n(A) \cong A^{n^2} \times A^{n^2} \mid MN = NM = I\right\}.$$

This finite dimensional representation is actually obtained as a sub-representation of the regular representation

$$\rho_{reg} : G \to \text{GL}(A_G)$$

of G on its algebra of functions. For example, the additive group $\mathbb{G}_a$ whose points are given by

$$\mathbb{G}_a(A) = A$$

with its additive group structure is algebraic with function algebra $K[X]$ and it can be embedded in GL_2 by the morphism

$$\begin{aligned} r : \mathbb{G}_a &\to \mathrm{GL}_2 \\ a &\mapsto \begin{pmatrix} 1 & a \\ 0 & 1 \end{pmatrix}. \end{aligned}$$

Note that if V is a real vector space (without a particular choice of a basis), one can also define $\mathrm{GL}(V)$ by

$$\mathrm{GL}(V)(A) := \{(M, N) \in \mathrm{End}_A(V \otimes A)^2 \mid MN = NM = \mathrm{id}_V\}.$$

A representation of an algebraic group, also called a G-module, is a group morphism

$$G \to \mathrm{GL}(V)$$

where V is a finite dimensional vector space.

Definition 5.2.2 The *Lie algebra* of the group G is the space

$$\begin{aligned} \mathrm{Lie}(G) : \mathrm{ALG}_K &\to \mathrm{GRP} \\ B &\mapsto \{X \in \mathrm{Hom}_{\mathrm{ALG}_K}(A_G, B[\epsilon]/(\epsilon^2)) | X \mod \epsilon = I\}. \end{aligned}$$

Remark 5.2.1 In the setting of smooth Lie groups, the same definition works by replacing $K[\epsilon]/(\epsilon^2)$ by the smooth algebra $\mathbb{C}^\infty(\mathbb{R})/(\epsilon^2)$ where ϵ is the standard coordinate function on $\mathbb{R}$.

One can think of elements of the Lie algebra as points $I + \epsilon X \in G(B[\epsilon]/(\epsilon^2))$ where X plays the role of a tangent vector to G at the identity point. There is a natural action of G on its Lie algebra called the *adjoint action*

$$\rho_{ad} : G \to \mathrm{GL}(\mathrm{Lie}(G))$$

defined by

$$g \mapsto [X \in G(R[\epsilon]/(\epsilon^2)) \mapsto gXg^{-1} \in G(R[\epsilon]/(\epsilon^2))].$$

The derivative of this action at the identity gives a natural map

$$\begin{aligned} \mathrm{ad} : \mathrm{Lie}(G) &\to \mathrm{End}_K(\mathrm{Lie}(G)) \\ X &\mapsto [Y \mapsto \mathrm{ad}(X)(Y)] \end{aligned}$$

called the *adjoint action* of the Lie algebra on itself. This allows us to define the Lie bracket by

$$[\cdot,\cdot]:\mathrm{Lie}(G)\times\mathrm{Lie}(G)\to\mathrm{Lie}(G)$$
$$(X,Y)\mapsto[X,Y]:=\mathrm{ad}(X)(Y).$$

The symmetric bilinear form

$$\langle -,-\rangle:\mathrm{Lie}(G)\times\mathrm{Lie}(G)\to\mathrm{Lie}(G)$$
$$(X,Y)\mapsto\mathrm{Tr}\big(\mathrm{ad}(X)\mathrm{ad}(Y)\big)$$

is called the *Killing form*. The group G is called *semisimple* if this form is non-degenerate.

5.3 Algebras with Involutions and Algebraic Groups

The study of algebraic groups on non-algebraically closed fields (for example, unitary groups over $\mathbb{R}$) is easier to do in the setting of algebras with involutions because there one can perform a computation that works for all classical groups. For example, the computation of Lie algebras is carried out only once for all classical groups (orthogonal, symplectic, linear and unitary) in this section. We refer to the book of involutions [KMRT98] for a very complete account of this theory.

On non-algebraically closed fields, algebraic groups are defined using a generalization of matrix algebras called central simple algebras, and involutions on them.

Definition 5.3.1 Let K be a field. A *central simple algebra* A over K is a finite dimensional nonzero algebra A with center K which has no two-sided ideals except $\{0\}$ and A. An *involution* on (a product of) central simple algebra(s) A is a map $\sigma:A\to A$ such that for all $x,y\in A$:

1. $\sigma(x+y)=\sigma(x)+\sigma(y)$;
2. $\sigma(xy)=\sigma(y)\sigma(x)$;
3. $\sigma^2=\mathrm{id}$.

The center K of A is stable under σ and $\sigma_{|K}$ is either an automorphism of order 2 or the identity. In the first case, K is of degree 2 over the fixed subfield L.

Theorem 5.3.1 (Wedderburn) *Let A be a finite dimensional algebra over K. The following conditions are equivalent*:

1. *A is a central simple algebra over K.*
2. *There exists a finite extension K'/K on which A becomes a matrix algebra, i.e., such that*

$$A\otimes_K K'\cong\mathrm{M}_n(K).$$

Corollary 5.3.1 *Let A be a central simple algebra of dimension n^2 over K. Let*

$$i : A \to \mathrm{End}_K(A)$$

be the action of A on itself by left multiplication. The characteristic polynomial of $i(a)$ is the n-th power of a polynomial

$$P_{A,a} = X^n - \mathrm{Tr}_A(a)X^{n-1} + \cdots + (-1)^n \mathrm{Nm}_A(a)$$

of degree n called the reduced characteristic polynomial. The coefficients Tr_A and Nm_A are respectively called the reduced norm and reduced trace on A.

Before giving some examples of algebras with involutions, we define the corresponding algebraic groups.

Definition 5.3.2 Let (A, σ) be an algebra (that is a product of central simple algebras) with involution over (a product of fields) K.

1. The group $\mathrm{Isom}(A, \sigma)$ of *isometries* of (A, σ) is defined by

$$\mathrm{Isom}(A, \sigma)(R) = \{a \in A_R^\times | a.\sigma_R(a) = 1\}.$$

2. The group $\mathrm{Aut}(A, \sigma)$ of *automorphisms* of (A, σ) is defined by

$$\mathrm{Aut}(A, \sigma)(R) = \{\alpha \in \mathrm{Aut}_R(A_R) | \alpha \circ \sigma_R = \sigma_R \circ \alpha\}.$$

3. The group $\mathrm{Sim}(A, \sigma)$ of *similitudes* of (A, σ) is defined by

$$\mathrm{Sim}(A, \sigma)(R) = \{a \in A_R^\times | a.\sigma(a) \in K_R^\times\}.$$

The first example of an algebra with involution is the one naturally associated to a non-degenerate bilinear form $b : V \times V \to K$ on a K-vector space. One defines an algebra with involution (A, σ_b) by setting $A = \mathrm{End}_K(V)$ and the involution σ_b defined by

$$b\big(x, f(y)\big) = b\big(\sigma_b(f)(x), y\big)$$

for $f \in A$ and $x, y \in V$. If the bilinear form b is symmetric, the corresponding groups are the classical groups

$$\mathrm{Iso}(A, \sigma) = \mathrm{O}(V, b), \qquad \mathrm{Sim}(A, \sigma) = \mathrm{GO}(V, b), \qquad \mathrm{Aut}(A, \sigma) = \mathrm{PGO}(V, b).$$

If the bilinear form b is antisymmetric, the corresponding groups are

$$\mathrm{Iso}(A, \sigma) = \mathrm{Sp}(V, b), \qquad \mathrm{Sim}(A, \sigma) = \mathrm{GSp}(V, b), \qquad \mathrm{Aut}(A, \sigma) = \mathrm{PGSp}(V, b).$$

Another instructive example is given by the algebra

$$A = \mathrm{End}_K(V) \times \mathrm{End}_K\big(V^\vee\big)$$

where V is a finite dimensional vector space over K, equipped with the involutions sending (a, b) to $(b^\vee, a^\vee)$. It is not a central simple algebra but a product of such over $K \times K$ and $\sigma_{|K}$ is the exchange involution. One then obtains

$$\begin{aligned} \mathrm{GL}(V) &\xrightarrow{\sim} \mathrm{Iso}(A, \sigma) \\ m &\mapsto \left(m, \left(m^{-1}\right)^\vee\right) \end{aligned}$$

and

$$\mathrm{PGL}(V) \cong \mathrm{Aut}(A, \sigma).$$

If A is a central simple algebra over a field K with σ-invariant subfield $L \subset K$ of degree 2, one gets the unitary groups

$$\mathrm{Iso}(A, \sigma) = \mathrm{U}(V, b), \qquad \mathrm{Sim}(A, \sigma) = \mathrm{GU}(V, b), \qquad \mathrm{Aut}(A, \sigma) = \mathrm{PGU}(V, b).$$

It is then clear that by scalar extension to K, one obtains an isomorphism

$$U(V, b)_K \cong \mathrm{GL}(V_K)$$

by the above construction of the general linear group. One also defines $\mathrm{SU}(V, q)$ as the kernel of the reduced norm map

$$\mathrm{Nm} : \mathrm{U}(V, b) \to K^\times.$$

We now compute at once all the classical Lie algebras.

Proposition 5.3.1 *Let (A, σ) be a central simple algebra with involution. We have*

$$\mathrm{Lie}\big(\mathrm{Sim}(A, \sigma)\big) = \big\{m \in A,\ m + \sigma(m) = 0\big\}.$$

Proof If $1 + \epsilon m \in \mathrm{Sim}(A, \sigma)(K[\epsilon]/(\epsilon^2))$ is a generic element equal to 1 modulo ϵ, the equation

$$(1 + m\epsilon)\big(1 + \sigma(m)\epsilon\big) = 0$$

implies $a + \sigma(a) = 0$. □

The above proposition gives us the simultaneous computation of the Lie algebras of $\mathrm{U}(V, b)$, $\mathrm{O}(V, b)$ and $\mathrm{Sp}(V, b)$.

5.4 Clifford Algebras and Spinors

This section is treated in details in the book of involutions [KMRT98], in Deligne's notes on spinors [Del99] and in Chevalley's book [Che97]. The physical motivation for the study of the Clifford algebra is the necessity to define a square root of the Laplacian (or of the d'Alembertian) operator, as is nicely explained in Penrose's book [Pen05], Sect. 24.6.

5.4.1 Clifford Algebras

Let (V, q) be a symmetric bilinear space over a field K.

Definition 5.4.1 The *Clifford algebra* $\mathrm{Cliff}(V, q)$ on the bilinear space (V, q) is the universal associative and unitary algebra that contains V and in which q defines the multiplication of vectors in V. More precisely, it satisfies the universal property

$$\begin{aligned}&\mathrm{Hom}_{\mathrm{ALG}_K}\big(\mathrm{Cliff}(V,q), B\big)\\&\quad\cong \big\{ j \in \mathrm{Hom}_{\mathrm{VECT}_K}(V, B) | j(v)\cdot j(w) + j(w)\cdot j(v) = q(v,w)\cdot 1_B \big\}\end{aligned}$$

for every associative unitary K-algebra B.

One explicitely defines $\mathrm{Cliff}(V, q)$ as the quotient of the tensor algebra $T(V)$ by the bilateral ideal generated by expression of the form

$$m \otimes n + n \otimes m - q(m, n)\cdot 1 \quad \text{with } m, n \in V.$$

Note that the degree filtration on the tensor algebra $T(V)$ induces a natural filtration F on the Clifford algebra whose graded algebra

$$\mathrm{gr}^F\big(\mathrm{Cliff}(V, q)\big) = \wedge^* V$$

is canonically isomorphic to the exterior algebra. One can extend this isomorphism to a linear isomorphism

$$\wedge^* V \xrightarrow{\sim} \mathrm{Cliff}(V, q)$$

by showing that the linear maps $f_k : \wedge^k V \to \mathrm{Cliff}(V, q)$ defined by

$$f_k(v_1, \ldots, v_k) = \frac{1}{k!}\sum_{\sigma\in S_k} \mathrm{sgn}\,(\sigma) v_{\sigma(1)} \cdots v_{\sigma(k)}$$

induces the above isomorphism of graded algebras.

Consequently, one has $\dim_K(\mathrm{Cliff}(V, q)) = 2^{\dim_K V}$. The automorphism $\alpha : V \to V$ given by $m \mapsto -m$ induces an automorphism $\alpha : \mathrm{Cliff}(V) \to \mathrm{Cliff}(V)$ such that $\alpha^2 = \mathrm{id}$. One thus gets a decomposition

$$\mathrm{Cliff}(V, q) = \mathrm{Cliff}^0(V, q) \oplus \mathrm{Cliff}^1(V, q),$$

giving $\mathrm{Cliff}(V, q)$ a superalgebra structure.

Proposition 5.4.1 *Let (V, q) be an even dimensional quadratic space. The Clifford algebra $\mathrm{Cliff}(V, q)$ is a central simple algebra and $\mathrm{Cliff}^0(V, q)$ is either central simple over a quadratic extension of K or a product of two central simple algebras over K.*

Proof The general structure theorem for even Clifford algebras can be found in the book of involution [KMRT98], p. 88. Here we explicitly describe the case of hyperbolic quadratic spaces. The general case follows from the following two facts:

1. Every even dimensional quadratic space over an algebraically closed field is an orthogonal sum of a hyperbolic quadratic space and of a non-degenerate quadratic space.
2. If $(W, q) = (V_1, q_1) \perp (V_2, q_2)$ is an orthogonal sum, the natural morphism

$$C(V_1, q_2) \otimes_g C(V_2, q_2) \to C(W, q)$$

is an isomorphism of graded algebras (where the tensor product is the graded algebra tensor product).

A quadratic space (V, q) with V of dimension $2n$ is called hyperbolic if it is isomorphic to a hyperbolic quadratic space of the form $(\mathbb{H}(U), q_h)$, with U a subspace of dimension n, U^* its linear dual and $\mathbb{H}(U) = U \oplus U^*$ its hyperbolic quadratic space equipped with the bilinear form $q_h(v \oplus \omega) = \omega(v)$. Let

$$S := \wedge^* U.$$

For $u + \varphi \in \mathbb{H}(U)$, let ℓ_u be the left exterior multiplication by u on S and d_φ be the unique derivation on S extending φ, given by

$$d_\varphi(x_1 \wedge \cdots \wedge x_r) = \sum_{i=1}^{r} (-1)^{i+1} x_1 \wedge \cdots \wedge \hat{x_i} \wedge x_r \varphi(x_i).$$

One shows that the map

$$\mathbb{H}(U) \to \operatorname{End}(\wedge^* U)$$
$$\varphi + u \mapsto \ell_u + d_\varphi$$

extends to an isomorphism

$$\operatorname{Cliff}(V, q) \xrightarrow{\sim} \operatorname{End}(S).$$

This isomorphism is actually an isomorphism

$$\operatorname{Cliff}(V, q) \xrightarrow{\sim} \underline{\operatorname{End}}(S)$$

of superalgebras. Moreover, if we define the even and odd parts of S by

$$S^+ := \wedge^{2*} U \quad \text{and} \quad S^- := \wedge^{2*+1} U,$$

the restriction of this isomorphism to $\operatorname{Cliff}^0(V, q)$ induces an isomorphism

$$\operatorname{Cliff}^0(V, q) \xrightarrow{\sim} \operatorname{End}_K\left(S^+\right) \times \operatorname{End}_K\left(S^-\right)$$

which can be viewed as the isomorphism

$$\mathrm{Cliff}^0(V,q) \xrightarrow{\sim} \underline{\mathrm{End}}^0(S)$$

induced by the above isomorphism on even parts of the algebras in play. □

Remark 5.4.1 The fact that $\mathrm{gr}^F \mathrm{Cliff}(V,q) \cong \wedge^* V$ can be interpreted by saying that the Clifford algebra is the canonical quantization of the exterior algebra. Indeed, the Weyl algebra (of polynomial differential operators) also has a filtration whose graded algebra is the polynomial algebra, and this is interpreted by physicists as the fact that the Weyl algebra is the canonical quantization of the polynomial algebra. More precisely, if V is a vector space, and ω is a symplectic 2-form on V (for example, the fiber of a cotangent bundle with its canonical 2-form), one can identify the algebra of algebraic differential operators on V with the algebra that satisfies the universal property

$$\begin{aligned}&\mathrm{Hom}_{\mathrm{ALG}_{\mathbb{R}}}(\mathcal{D}_V, B)\\&\quad\cong \bigl\{ j \in \mathrm{Hom}_{\mathrm{VECT}_{\mathbb{R}}}(V, B) | j(v).j(w) - j(w).j(v) = \omega(v,w).1_B \bigr\}.\end{aligned}$$

In this case, the graded algebra

$$\mathrm{gr}^F \mathcal{D}_V \cong \mathrm{Sym}^*(V)$$

is identified with the polynomial algebra $\mathrm{Sym}^*(V)$. The above isomorphism can be extended to a linear isomorphism

$$\mathcal{D}_V \to \mathrm{Sym}^*(V)$$

by using the symmetrization formula

$$a_1 \cdots a_n \mapsto \frac{1}{n!} \sum_{\sigma \in S_n} a_{\sigma(1)} \cdots a_{\sigma(n)}.$$

The corresponding product on $\mathrm{Sym}^*(V)$ is called the Moyal product. A combination of these two results can be given in the setting of Clifford superalgebras with respect to complex valued super-quadratic forms. The introduction of Clifford algebras in the definition of the Dirac operator and in fermionic field theory is thus, in this sense, equivalent to the use of classical anti-commuting coordinates on fermionic bundles.

Remark 5.4.2 The above analogy between differential operators and elements of the Clifford algebra can be used in geometry, as is explained in the paper of Getzler [Get83] on the super proof of the Atiyah-Singer index theorem. Let M be a differential manifold. One can think of differential forms $\Omega^*(M)$ as functions on the super-bundle $T[1]M$, and the Clifford algebra of $TM \oplus T^*M$, being isomorphic to $\mathrm{End}(\wedge^* T^*M)$, gives operators on the state space $\Omega^*(M)$. This situation is analogous to the Weyl quantization of the symplectic manifold (T^*M, ω) by the action of differential operators $\mathcal{D}(M)$ on $L^2(M)$.

5.4.2 Spin Group and Spinorial Representations

Definition 5.4.2 The *Clifford group* of (V, q) is the group

$$\Gamma(V, q) := \{c \in \mathrm{Cliff}(V, q)^{\times} | cv\alpha(c)^{-1} \in v \text{ for all } v \in V\}.$$

One has by definition a natural action (by conjugation) of this group on V that induces a morphism

$$\Gamma(V, q) \to \mathrm{O}(V, q).$$

The special Clifford group of (V, q) is the subgroup $\Gamma^{+}(V, q)$ of even elements in $\Gamma(V, q)$ given by

$$\Gamma^{+}(V, q) = \Gamma(V, q) \cap \mathrm{Cliff}^{0}(V, q).$$

The tensor algebra $T(V)$ is naturally equipped with an anti-automorphism given by $m_1 \otimes \cdots \otimes m_k \mapsto m_k \otimes \cdots \otimes m_1$ on the homogeneous elements of degree k. Since the defining ideal for the Clifford algebra is stable under this anti-automorphism (because q is symmetric), one obtains an anti-automorphism of $\mathrm{Cliff}(V, q)$ called the transposition, and which we denoted by $c \mapsto {}^{t}c$.

Definition 5.4.3 The *spinor group* is the subgroup of $\Gamma^{+}(V, q)$ defined by

$$\mathrm{Spin}(V, q) = \{c \in \Gamma^{+}(V, q) | {}^{t}cc = 1\}.$$

One has a natural action (by multiplication) of $\mathrm{Spin}(V, q)$ on the real vector space $\mathrm{Cliff}(V, q)$ that commutes with the idempotent $\alpha : \mathrm{Cliff}(V, q) \to \mathrm{Cliff}(V, q)$ and thus decomposes into two representations $\mathrm{Cliff}^{0}(V, q)$ and $\mathrm{Cliff}^{1}(V, q)$.

Definition 5.4.4 Let (V, q) be an even dimensional quadratic space. A rational representation S of $\mathrm{Spin}(V, q)$ will be called:

1. a *rational spinorial representation* if it is a $\mathrm{Cliff}(V, q)$-irreducible submodule of the action of $\mathrm{Cliff}(V, q)$ on itself;
2. a *rational semi-spinorial representation* if it is a $\mathrm{Cliff}^{0}(V, q)$-irreducible submodule of the action of $\mathrm{Cliff}^{0}(V, q)$ on itself.

If K is algebraically closed and $(V, q) = (\mathbb{H}(U), q_h)$ is the hyperbolic quadratic space, one has $\mathrm{Cliff}(V, q) \cong \mathrm{End}(\wedge^{*} U)$, so that the spinorial representation is simply $\wedge^{*} U$ and the semi-spinorial representations are $S^{+} = \wedge^{2*} U$ and $S^{-} = \wedge^{2*+1} U$.

If (V, q) is the Lorentzian space, given by $q(t, x) = -c^2 t^2 + x_1^2 + x_2^2 + x_3^2$ on Minkowski space $V = \mathbb{R}^{3,1}$, the algebra $\mathrm{Cliff}^{0}(V, q)$ is isomorphic to the algebra $\mathrm{Res}_{\mathbb{C}/\mathbb{R}} M_{2,\mathbb{C}}$ of complex matrices viewed as a real algebra. The group $\mathrm{Spin}(3, 1) :=$ $\mathrm{Spin}(V, q)$ is then identified with the group $\mathrm{Res}_{\mathbb{C}/\mathbb{R}} \mathrm{SL}_{2,\mathbb{C}}$ of complex matrices of determinant 1 viewed as a real algebraic group. There is a natural representation of $\mathrm{Res}_{\mathbb{C}/\mathbb{R}} M_{2,\mathbb{C}}$ on the first column $\mathrm{Res}_{\mathbb{C}/\mathbb{R}} \mathbb{C}^2$ of $\mathrm{Cliff}^{0}(V, q) \cong \mathrm{Res}_{\mathbb{C}/\mathbb{R}} M_{2,\mathbb{C}}$ and this is the representation of Spin(3, 1) that we will call the real spinor representation. Its extension to $\mathbb{C}$ decomposes into two representations isomorphic to $\mathbb{C}^2$ that are exchanged by complex conjugation. These two representations are called the semi-spinorial representations and correspond in physics to the electron and the anti-electron.

For a more concrete and motivated approach to these matters, the reader is advised to read Sect. 24.6 of Penrose's book [Pen05].

5.4.3 Pairings of Spinorial Representations

Let (V,q) be an even dimensional quadratic space. Note that the natural action of the vector space $V \subset \mathrm{Cliff}(V,q)$ on a rational spinorial representation S of $\mathrm{Cliff}(V,q)$ induces a morphism of $\mathrm{Spin}(V,q)$-representations

$$V \otimes S \to S.$$

This pairing will be useful when describing the Dirac operator on the spinor bundle.

If (V,q) is a quadratic form over an algebraically closed field, one has two semi-spinorial representations S^+ and S^- of $\mathrm{Spin}(V,q)$ and the above morphism induces two morphisms of $\mathrm{Spin}(V,q)$ representations

$$V \otimes S^{\pm} \to S^{\mp}$$

which are also useful when defining Dirac operators between semi-spinorial bundles. More general semi-spinorial representations have to be studied on a case by case basis (see [Del99], part I, for a more complete description).

We now define the two natural pairings that are necessary for the description of the Dirac Lagrangian in Sect. 15.2.

Proposition 5.4.2 *Let (V,q) be an even dimensional quadratic space over an algebraically closed field. The spinorial representation S is naturally equipped with three pairings*

$$\epsilon : S \otimes S \to K,$$

and

$$\Gamma : S \otimes S \to V, \qquad \tilde{\Gamma} : S \otimes S \to V.$$

Proof The form $\epsilon : S \otimes S \to K$ is the non-degenerate bilinear form on S for which

$$\epsilon(vs,t) = \epsilon(s,vt)$$

for all $v \in V$. It is the bilinear form whose involution on

$$\underline{\mathrm{End}}(S) \cong \mathrm{Cliff}(V,q)$$

is the standard anti-involution of $\mathrm{Cliff}(V,q)$. The Clifford multiplication map

$$c : V \otimes S \to S$$

induces a morphism $c : S^\vee \otimes S \to V^\vee$, that gives a morphism

$$\Gamma := q^{-1} \circ c \circ \left(\epsilon^{-1} \otimes \mathrm{id}_S\right) : S \otimes S \to V.$$

One then defines $\tilde{\Gamma}$ by

$$\tilde{\Gamma} := \Gamma \circ (\epsilon \otimes \epsilon) : S^\vee \otimes S^\vee \to V.$$ □

We now give the real Minkowski version of the above proposition, which is useful when defining the super Poincaré group, and whose proof can be found in [Del99], Theorem 6.1.

Proposition 5.4.3 *Let (V,q) be a quadratic space over $\mathbb{R}$ of signature $(1, n-1)$. Let S be an irreducible real spinorial representation of* $\mathrm{Spin}(V,q)$. *The commutant Z of $S_\mathbb{R}$ is $\mathbb{R}$, $\mathbb{C}$ or $\mathbb{H}$.*

1. *Up to a real factor, there exists a unique symmetric morphism $\Gamma : S \otimes S \to V$. It is invariant under the group Z^1 of elements of norm 1 in Z.*
2. *For $v \in V$, if $Q(v) > 0$, the form $(s,t) \mapsto q(v, \Gamma(s,t))$ on S is positive or negative definite.*

5.5 General Structure of Linear Algebraic Groups

We refer to the Grothendieck-Demazure seminar [DG62] for the general theory of root systems in linear algebraic groups and for the proof of the classification theorem for reductive groups.

Definition 5.5.1 An algebraic group is called:

1. *reductive* if the category of its representations is semi-simple, i.e., every exact sequence
$$0 \to U \to V \to W \to 0$$
of G-modules has an equivariant splitting, i.e., induces a direct sum decomposition
$$V \cong W \oplus U;$$
2. *unipotent* if it is a successive extension of additive groups;
3. *a torus* if after extending it to a finite extension of the base field, it becomes isomorphic to GL_1^n (one says the torus splits).

Let T be a torus. Its character group is the functor

$$X^*(T) : \text{ALG}_{\mathbb{R}} \to \text{GRAB}$$
$$B \mapsto \text{Hom}(T_B, \text{GL}_{1,B})$$

with values in the category of GRAB of abelian groups and its cocharacter group is the functor

$$X_*(T) : \text{RINGS} \to \text{GRAB}$$
$$B \mapsto \text{Hom}(\text{GL}_{1,B}, T_B).$$

One has a perfect pairing

$$\langle \cdot, \cdot \rangle : X_*(T) \times X^*(T) \to \underline{\mathbb{Z}}$$

given by $(r, s) \mapsto r \circ s$ with $\underline{\mathbb{Z}}$ the functor that associates to the product $A = \prod A_i$ of simple algebras A_i the group $\mathbb{Z}^i$.

Lemma 5.5.1 *If $T = \text{GL}_1^n$, a representation $T \to \text{GL}(V)$ decomposes in a sum of characters: $V \cong \bigoplus \chi_i$, with $\chi_i \in X^*(T)$.*

Proposition 5.5.1 *Let G be a reductive group. The family of sub-tori $T \subset G$ has maximal elements that one calls maximal tori of G.*

The first structure theorem of linear algebraic groups gives a "dévissage" of a general group into unipotent and reductive groups.

Theorem 5.5.1 *Let P be a linear algebraic group. There exists a maximal normal unipotent subgroup $R_u P$ of P called the unipotent radical of P and the quotient $P/R_u P$ is reductive.*

To give a more detailed study of the structure of reductive groups, one introduces new invariants called the roots.

Definition 5.5.2 Let G be a reductive group and T be a maximal torus of G. The *roots* of the pair (G, T) are the (non-trivial) weights of T in the adjoint representation of G on its Lie algebra. More precisely, they form a subspace of the space $X^*(T)$ of characters whose points are

$$R^* = \left\{\chi \in X^*(T) | \rho_{ad}(t)(X) = \chi(t) \cdot X\right\}.$$

The *coroots* $R_* \subset X_*(T)$ are the cocharacters of T that are dual to the roots with respect to the given perfect pairing between characters and cocharacters. The quadruple $\Phi(G, T) = (X_*(T), R_*, X^*(T), R^*, \langle \cdot, \cdot \rangle)$ is called the *root system* of the pair (G, T).

Example 5.5.1 The adjoint representation of the maximal torus

$$T = \left\{ \begin{pmatrix} t & 0 \\ 0 & t^{-1} \end{pmatrix} \right\}$$

of SL_2 on $\mathrm{Lie}(\mathrm{SL}_2) = \{m \in \mathrm{M}_2,\ \mathrm{Tr}(m) = 0\}$ is given by

$$\begin{pmatrix} m & n \\ p & q \end{pmatrix} \mapsto \begin{pmatrix} m & t^2 n \\ t^{-2} p & q \end{pmatrix}.$$

The roots are therefore given by the morphisms

$$r : t \mapsto t^2 \quad \text{and} \quad -r : t \mapsto t^{-2}.$$

These morphisms correspond to the two morphisms $r : \mathbb{G}_a \to \mathrm{SL}_2$ given by

$$\begin{aligned} x_r : \mathbb{G}_a &\to \mathrm{SL}_2 \\ a &\mapsto \begin{pmatrix} 1 & a \\ 0 & 1 \end{pmatrix} \end{aligned}$$

and

$$\begin{aligned} x_{-r} : \mathbb{G}_a &\to \mathrm{SL}_2 \\ a &\mapsto \begin{pmatrix} 1 & 0 \\ a & 1 \end{pmatrix}. \end{aligned}$$

One may show that SL_2 is generated by the images of these two morphism and by the maximal torus T, with the relations

$$t x_r(a) t^{-1} = x_r\big(r(t)a\big).$$

The main theorem of the classification theory of reductive group is that the root system uniquely determines the group and that every root system comes from a reductive group (whose maximal torus splits). Actually, the root system gives a system of generators and relations for the points of the given algebraic group. More precisely, every root corresponds to a morphism

$$x_r : \mathbb{G}_a \to G$$

that is given by exponentiating the corresponding element of the Lie algebra, so that it satisfies

$$t x_r(a) t^{-1} = x_r\big(r(t)a\big)$$

and the group is generated by the images of these morphisms and by its maximal torus, the relations between them being given by the definition of a root. Actually, for any root $r \in R$, there is a homomorphism $\varphi_r : \mathrm{SL}_2 \to G$ such that

$$\varphi_r \begin{pmatrix} 1 & a \\ 0 & 1 \end{pmatrix} = x_r(a) \quad \text{and} \quad \varphi_r \begin{pmatrix} 1 & 0 \\ a & 1 \end{pmatrix} = x_{-r}(a),$$

the image of the diagonal matrices being given by the image of the dual coroot to the given root.

We don't want to define the abstract notion of a root system, which would be a necessary step to explain that every abstract root system is the root system of a given reductive group. Instead, we just state the unicity result.

Theorem 5.5.2 *The functor $(G,T)\mapsto \Phi(G,T)$ from reductive groups to root systems is conservative: if two groups have the same root system, they are isomorphic.*

Note that this theorem is often formalized in the setting of split reductive groups but the methods of SGA 3 [DG62] allow us to state it in general, if we are working with spaces of characters as sheaves for the étale topology on the base field, i.e., as Galois modules.

5.6 Representation Theory of Reductive Groups

Because representation theory is very important in physics, and to be complete, we also recall the classification of representations of reductive groups (in characteristic 0). This can be found in the book of Jantzen [Jan87], part II.

A Borel subgroup B of G is a maximal closed and connected solvable subgroup of G. For example, the standard Borel subgroup of GL_n is the group of upper triangular matrices. If B is a Borel subgroup that contains the given maximal torus $T\subset G$, the set of roots whose morphism $x_r:\mathbb{G}_a\to G$ have image in B are called positive roots and is denoted $R^+\subset R$. One defines an order on $X^*(T)$ by saying that

$$\lambda\leq\mu \quad\Leftrightarrow\quad \mu-\lambda\in\sum_{\alpha\in R^+}\mathbb{N}\alpha.$$

Any G-module V decomposes into weight spaces (representations of T, i.e., sums of characters) as

$$V=\oplus_{\lambda\in X^*(T)}V_\lambda.$$

For the given order on $X^*(T)$, every irreducible representation of G has a highest non-trivial weight called the highest weight of V. We define the set of dominant weights by

$$X^*(T)_{dom}:=\left\{\lambda\in X^*(T),\ \left\langle\lambda,\alpha^\vee\right\rangle\geq 0 \text{ for all } \alpha\in R^+\right\}.$$

If λ is a dominant weight, we define the G-module $L(\lambda)$ by

$$L(\lambda)=\mathrm{ind}_B^G(\lambda):=(\lambda\otimes A_G)^B$$

where A_G denotes the algebra of functions on G.

The main classification theorem of irreducible representations of reductive groups is the following.

Theorem 5.6.1 *Suppose that G is reductive and split, i.e., $T \cong \mathbb{G}_m^n$. The map*

$$\begin{aligned} L : X^*(T)_{dom} &\to \text{RepIrr}/\sim \\ \lambda &\mapsto L(\lambda) \end{aligned}$$

is a bijection between the set of dominant weight and the set of isomorphism classes of irreducible representations of G.

If G is not split, i.e., if T is a twisted torus, one has to work a bit more to get the analogous theorem, due to Tits, and which can be found in the book of involutions [KMRT98], in the section on Tits algebras. Note that many groups in physics are not split, and this is why their representation theory is sometimes tricky to handle.

5.7 Structure and Representations of SL_2

As explained above, SL_2 is the main building block for all other algebraic groups. It is defined by

$$\mathrm{SL}_2 = \{M \in \mathrm{GL}_2, \det(M) = 1\}.$$

Its maximal torus is

$$T = \left\{\begin{pmatrix} t & 0 \\ 0 & t^{-1} \end{pmatrix}\right\} \cong \mathrm{GL}_1.$$

The space of characters of T is identified with $\mathbb{Z}$ by $t \mapsto t^n$. The Lie algebra of SL_2 is given by

$$\mathrm{Lie}(\mathrm{SL}_2)(A) = \{I + \epsilon M \in \mathrm{SL}_2(A[\epsilon]/(\epsilon^2)),\ \det(I + \epsilon M) = 1\}.$$

If $M = \left(\begin{smallmatrix} a & b \\ c & d \end{smallmatrix}\right)$, the determinant condition means $(1+\epsilon a)\cdot(1+\epsilon d) = 1+\epsilon(a+d) = 1$, so that $\mathrm{Tr}(M) = 0$. We thus get

$$\mathrm{Lie}(\mathrm{SL}_2) = \{M \in M_2,\ \mathrm{Tr}(M) = 0\} = \left\{\begin{pmatrix} a & b \\ c & -a \end{pmatrix}\right\}.$$

The adjoint action of $t \in T$ on $\mathrm{Lie}(\mathrm{SL}_2)$ is given by

$$\left\{\begin{pmatrix} a & b \\ c & -a \end{pmatrix}\right\} \mapsto \left\{\begin{pmatrix} a & t^2 b \\ t^{-2} c & -a \end{pmatrix}\right\}$$

so that the roots are $r_2 : t \mapsto t^2$ and $r_{-2} : t \mapsto t^{-2}$. They correspond to the root morphisms

$$r_2, r_{-2} : \mathbb{G}_a \to \mathrm{SL}_2$$

given by

$$a \mapsto \begin{pmatrix} 1 & a \\ 0 & 1 \end{pmatrix} \quad \text{and} \quad a \mapsto \begin{pmatrix} 1 & 0 \\ a & 1 \end{pmatrix}.$$

Note that SL_2 is generated by T and the images of the root morphisms.

One shows that the dominant weights of SL_2 are given by $\mathbb{Z}_{>0} \subset \mathbb{Z} = X^*(T)$ and the corresponding irreducible representations are given by

$$V_n := \mathrm{Sym}^n(V)$$

where V is the standard representation of SL_2.

For GL_2, the classification of representations is similar, and gives a family of irreducible representations

$$V_{n,m} := \mathrm{Sym}^n(V) \otimes (\det)^{\otimes m}$$

where det is the determinant representation and (n,m) is a pair of integers with $n > 0$ and $m \in \mathbb{Z}$.

5.8 Structure and Representations of SU_2

Recall that SU_2 is defined by

$$SU_2 = \left\{M \in \mathrm{Res}_{\mathbb{C}/\mathbb{R}} M_{2,\mathbb{C}},{}^t\bar{M} \cdot M = M^t \bar{M} = 1\right\}.$$

Its Lie algebra can be computed by using

$$\mathfrak{su}_2 = \mathrm{Lie}(SU_2) = \left\{M \in SU_2\big(\mathbb{R}[\epsilon]/(\epsilon^2)\big), M = \mathrm{id} \mod \epsilon\right\}.$$

Indeed, if $I + \epsilon M$ is in $\mathfrak{su}_2$, it satisfies

$${}^t\bar{I}I + \epsilon\left[{}^t\bar{M}I +{}^t \bar{I}M\right] = 1$$

so that $\mathfrak{su}_2$ is given by

$$\mathfrak{su}_2 = \left\{M \in \mathrm{Res}_{\mathbb{C}/\mathbb{R}} M_{2,\mathbb{C}}, {}^t\bar{M} + M = 0\right\}.$$

Actually, if one defines the quaternion algebra by

$$\mathbb{H} := \left\{M \in \mathrm{Res}_{\mathbb{C}/\mathbb{R}} M_{2,\mathbb{C}}, \exists\lambda \in \mathbb{R}, {}^t\bar{M} + M = \lambda \cdot \mathrm{id}\right\},$$

one can show that $\mathbb{H}$ is a non-commutative field with group of invertibles

$$U_2 := \mathbb{H}^\times.$$

The Lie algebra of U_2 is $\mathbb{H}$ itself. There is a natural norm map $\mathrm{Nm}: \mathbb{H} \to \mathbb{R}$ sending M to $\mathrm{Nm}(M) :={}^t\bar{M} \cdot M$ and one can identify SU_2 with the multiplicative kernel of

$$\mathrm{Nm}: U_2 = \mathbb{H}^\times \to \mathbb{R}^\times.$$

One has $\mathbb{H}_\mathbb{C} := \mathbb{H} \otimes_\mathbb{R} \mathbb{C} \cong \mathrm{M}_{2,\mathbb{C}}$, so that the special unitary group is a twisted version of SL_2, meaning that

$$\mathrm{SU}_2 \otimes_\mathbb{R} \mathbb{C} \cong \mathrm{SL}_{2,\mathbb{C}}.$$

This implies that their structure and representation theory are essentially equivalent. The maximal torus of SU_2 is isomorphic to SU_1, that is to the kernel of the norm map

$$\mathrm{Res}_{\mathbb{C}/\mathbb{R}}\mathrm{GL}_1 \to \mathrm{GL}_{1,\mathbb{R}}$$

that sends z to $z\bar{z}$. It is also a twisted version of the maximal torus GL_1 of SL_2.

The representations of SU_2 are given by taking irreducible representations of $\mathrm{SL}_{2,\mathbb{C}}$, i.e., the representations $V_n = \mathrm{Sym}^n(V)$ for V the standard representation, making their scalar restriction to $\mathbb{R}$, for example $\mathrm{Res}_{\mathbb{C}/\mathbb{R}} V_n$ and then taking irreducible sub-representations of these for the natural action of $\mathrm{SU}(2)$.

For example, the representation $V_2 = \mathrm{Res}_{\mathbb{C}/\mathbb{R}}\mathbb{C}\langle X^2, XY, Y^2\rangle$ of real dimension 6 has a natural real sub-representation of dimension 3 that corresponds to the morphism

$$\mathrm{SU}_2 \to \mathrm{SO}_3.$$

5.9 Structure of $\mathrm{SO}(V, q)$

Let (V, q) be a quadratic space over $\mathbb{R}$. The special orthogonal group $\mathrm{SO}(V, q)$ is the subgroup of $\mathrm{GL}(V)$ defined by

$$\mathrm{SO}(V, q) = \{f \in \mathrm{GL}(V),\ q(f(v), f(w)) = q(v, w)\}.$$

Its Lie algebra is given by

$$\mathfrak{so}(V, q) = \{m \in \mathrm{End}(V),\ q(v, m(w)) + q(m(v), w) = 0\}.$$

Indeed, if $\mathrm{id} + \epsilon m \in \mathrm{SO}(V, q)(\mathbb{R}[\epsilon]/(\epsilon^2))$ is a generic element that reduces to id modulo ϵ, the equation

$$q((\mathrm{id} + \epsilon m)(v), (\mathrm{id} + \epsilon m)(w)) = q(v, w)$$

gives

$$\epsilon[q(v, m(w)) + q(m(v), w)] = 0.$$

There is a canonical identification

$$\begin{aligned} \wedge^2 V &\xrightarrow{\sim} \mathfrak{so}(V,q) \\ v \otimes w &\mapsto \big[x \mapsto q(w,x).v - q(v,x).w\big]. \end{aligned}$$

This identification can be better understood by using the natural isomorphism $\mathfrak{spin}(V,q) \xrightarrow{\sim} \mathfrak{so}(V,q)$, the natural embedding

$$\mathfrak{spin}(V,q) \to \mathrm{Cliff}(V,q)$$

and the linear isomorphism

$$\wedge^* V \xrightarrow{\sim} \mathrm{Cliff}(V,q)$$

given in degree 2 by

$$\begin{aligned} \wedge^2 V &\to \mathrm{Cliff}(V,q) \\ x \wedge y &\mapsto \frac{1}{2}(xy - yx). \end{aligned}$$

References

[Che97] Chevalley, C.: The Algebraic Theory of Spinors and Clifford Algebras, p. 214. Springer, Berlin (1997). Collected works, vol. 2. Edited and with a foreword by Pierre Cartier and Catherine Chevalley, with a postface by J.-P. Bourguignon. ISBN 3-540-57063-2

[Del99] Deligne, P.: Notes on spinors. In: Quantum Fields and Strings: A Course for Mathematicians, Vol. 1, 2 (Princeton, NJ, 1996/1997), pp. 99–135. Am. Math. Soc., Providence (1999)

[DG62] Demazure, M., Grothendieck, A.: SGA 3: Schémas en Groupes. Lecture Notes in Mathematics, vol. 152. Springer, Berlin (1962)

[Get83] Getzler, E.: Pseudodifferential operators on supermanifolds and the Atiyah-Singer index theorem. Commun. Math. Phys. **92**(2), 163–178 (1983)

[Jan87] Jantzen, J.C.: Representations of Algebraic Groups, p. 443. Academic Press, Boston (1987). ISBN 0-12-380245-8

[KMRT98] Knus, M.-A., Merkurjev, A., Rost, M., Tignol, J.-P.: The Book of Involutions, p. 593. Am. Math. Soc., Providence (1998). With a preface in French by J. Tits. ISBN 0-8218-0904-0

[Pen05] Penrose, R.: The Road to Reality, p. 1099. Knopf, New York (2005). A complete guide to the laws of the universe

[Wat79] Waterhouse, W.C.: Introduction to Affine Group Schemes, p. 164. Springer, New York (1979). ISBN 0-387-90421-2

Chapter 6
Hopf Algebras

We now make a small excursion into the setting of Hopf algebras, which are directly related to algebraic groups, but which will also be useful in forming a conceptual understanding of the combinatorial structures that appear in quantum field theory. Quantum group deformations are also interesting illustrations of the idea of categorical factorization quantization presented in Sect. 24.6. We refer to [KRT97, Kas95] and [Maj95] for an introduction to the modern theory of Hopf algebras and quantum groups.

6.1 Definitions and Examples

If G is an affine algebraic group over K, its algebra of functions is a commutative unital algebra, equipped with a coproduct $\Delta : A \to A \otimes A$, a counit $\epsilon : A \to K$ and an antipode $S : A \to A$ that are all algebra morphisms (given respectively by multiplication, unit and inverse in the corresponding algebraic group). The notion of a Hopf algebra may be regarded as a generalization of this setting, whose conceptual roots may (somewhat arbitrarily) be taken to be in deformation theory. Indeed, the category of representations $\rho : G \to \mathrm{GL}(V)$ of an algebraic group G with Hopf algebra A is a symmetric monoidal category

$$\big(\mathrm{R{\scriptstyle EP}}(A), \otimes\big) := \big(\mathrm{R{\scriptstyle EP}}(G), \otimes\big),$$

that we may view as a 3-tuply monoidal category (see Sect. 2.3). One may then deform this category over K to a 1-tuply or 2-tuply monoidal category over $K[\![\hbar]\!]$. This may be done by deforming A to a $K[\![\hbar]\!]$-bialgebra whose category of modules is monoidal or braided monoidal.

Definition 6.1.1 Let $(\mathcal{C}, \otimes)$ be a symmetric monoidal category. A *bialgebra* is an object A of $\mathcal{C}$, equipped with a monoid structure $(\mu : A \otimes A \to A, e : \mathbb{1} \to A)$ and a comonoid structure $(\Delta : A \to A \otimes A, \epsilon : A \to \mathbb{1})$ that are compatible, meaning that

F. Paugam, *Towards the Mathematics of Quantum Field Theory*,
Ergebnisse der Mathematik und ihrer Grenzgebiete. 3. Folge / A Series of Modern Surveys in Mathematics 59, DOI 10.1007/978-3-319-04564-1_6,

Δ and ϵ are monoid morphisms. A bialgebra is called *commutative* (resp. *cocommutative*) if the underlying monoid (resp. comonoid) is commutative (resp. cocommutative).

The interest of the notion of bialgebra is that one may form the tensor product of two modules V and W over a given bialgebra A by putting on the tensor product $V \otimes W$, that is a module over $A \otimes A$, the module structure over A given by the coproduct $\Delta : A \to A \otimes A$. This construction gives the following important result.

Lemma 6.1.1 *The monoidal structure (associativity condition) on* $(\mathcal{C}, \otimes)$ *induces a natural monoidal category structure on the category* $\mathrm{MOD}(A)$ *of modules over a bialgebra* A.

One may advantageously generalize the notion of bialgebra by weakening the above result.

Definition 6.1.2 A *quasi-bialgebra* is an object A of $\mathcal{C}$ equipped with a monoid structure $(\mu : A \otimes A \to A, e : \mathbb{1} \to 1)$ and a pair of monoid morphisms $(\Delta : A \to A \otimes A, \epsilon : A \to \mathbb{1})$, and with the additional datum of a monoidal structure (associativity, and left and right unit constraints) on the category $\mathrm{MOD}(A)$ with its usual tensor functor induced by that of $\mathcal{C}$ and by Δ and ϵ. A *braided quasi-bialgebra* is a quasi-bialgebra equipped with a braiding of the corresponding monoidal category $(\mathrm{MOD}(A), \otimes)$.

Definition 6.1.3 Let $(\mathcal{C}, \otimes)$ be a symmetric monoidal category with internal homomorphisms. Let (A, Δ, ϵ) be a comonoid and $(B, m, 1)$ be a monoid in $\mathcal{C}$. The internal homomorphism object $\underline{\mathrm{Hom}}(A, B)$ has a structure of a monoid, called the *convolution monoid*, and whose product

$$* : \underline{\mathrm{Hom}}(A, B) \otimes \underline{\mathrm{Hom}}(A, B) \to \underline{\mathrm{Hom}}(A, B)$$

may be defined (on the sets $\mathrm{Hom}(C, \underline{\mathrm{Hom}}(A, B))$, functorially in C) by the composition

$$A \xrightarrow{\Delta} A \otimes A \xrightarrow{f \otimes g} B \otimes B \xrightarrow{m} B.$$

Definition 6.1.4 A *Hopf algebra* is a bialgebra A whose identity endomorphism id_A has a two sided inverse S for convolution in $(\underline{\mathrm{End}}(A), *)$, called its *antipode*.

We now present two examples of Hopf algebras that play a fundamental role in the perturbative causal quantum field theory of Chap. 22.

Example 6.1.1

1. If V is a K-vector space, it corresponds to an algebraic additive group scheme $(\mathbb{V}, +)$, whose algebra of functions is the polynomial algebra $A = \mathrm{Sym}(V, \mathbb{R})$

given by

$$\mathrm{Sym}(V,\mathbb{R}) := \oplus_n \mathrm{Hom}_{\mathbb{R}}\big(V^{\otimes n},\mathbb{R}\big)^{S_n}.$$

The coproduct is then given by

$$\Delta\big(a(v_1,\ldots,v_n)\big) := a(v_1+u_1,\ldots,v_n+u_n).$$

If V is finite dimensional, we may also dualize and defined the coproduct on V^* by $\Delta(a) = 1\otimes a + a\otimes 1$ (extended to $S(V^*)$ by algebra morphism) and the counit by $\epsilon(a) = 0$.

2. Similarly, the above coproduct induced by addition on V extends to the associative algebra of multilinear functions $B = T(V,\mathbb{R})$ given by

$$T(V,\mathbb{R}) := \oplus_n \mathrm{Hom}_{\mathbb{R}}\big(V^{\otimes n},\mathbb{R}\big).$$

In the finite dimensional case, it may also be obtained as above.

Example 6.1.2 If (C,Δ,ϵ) is a coalgebra, the algebra $\mathrm{Sym}(C)$ may be equipped with a bialgebra structure Δ by setting $\Delta(c) = \Delta(c)$ for $c \in C$, and extending this formula by algebra morphism to $\mathrm{Sym}(C)$. Note that $\mathrm{Sym}(C)$ is a comodule algebra on the coalgebra C, meaning that the natural map

$$\mathrm{Sym}(C) \to C \otimes \mathrm{Sym}(C)$$

defined by extending comultiplication on C to an algebra morphism is compatible with comultiplication on C. For example, if we take $C = \mathrm{Sym}(V)$, we get a coaction

$$\mathrm{Sym}\big(\mathrm{Sym}(V)\big) \to \mathrm{Sym}(V) \otimes \mathrm{Sym}\big(\mathrm{Sym}(V)\big).$$

This coaction plays an important role in the Hopf algebraic interpretation of the deformation quantization of free fields, to be described in Sect. 22.4.

Example 6.1.3 The example that has stimulated a great deal of the recent work on Hopf algebras is given by the enveloping Hopf algebra $U(\mathfrak{g})$ of a (semisimple) Lie algebra, and the quasi-bialgebras that correspond to deformations of its category of representations as a monoidal (or braided monoidal) category, sexily named quantum groups by Drinfeld. Note that if $\mathfrak{g}$ is the Lie algebra of an algebraic group G with Hopf algebra of functions A, one may identify it with invariant differential operators on G, and this gives a duality

$$\begin{aligned} \langle -,-\rangle : U(\mathfrak{g}) \times A &\to K \\ (P,f) &\mapsto (Pf)(e) \end{aligned}$$

which also identifies $U(\mathfrak{g})$ with an algebra of distributions on G supported at the unit e. This duality becomes perfect only when one passes to the formal completion

$\hat{A}$ of the algebra A at e. Deforming the coproduct on $U(\mathfrak{g})$ is thus equivalent to deforming the product on the algebra $\hat{A}$ of formal functions on G around its unit. If $\mathfrak{g} = V$ is a vector space with its trivial commutative Lie algebra structure, then $U(\mathfrak{g})$ is the symmetric algebra $\mathrm{Sym}(V)$, and A is the Hopf algebra $\mathrm{Sym}(V, \mathbb{R})$ of polynomial functions on V. The completion $\hat{A}$ is given by

$$\hat{A} = \widehat{\mathrm{Sym}}(V, \mathbb{R}) := \prod_{n \geq 0} \mathrm{Hom}_{\mathbb{R}}\left(V^{\otimes n}, \mathbb{R}\right)^{S_n}.$$

Deforming the coproduct on $\mathrm{Sym}(V)$ is thus equivalent to deforming the product on $\widehat{\mathrm{Sym}}(V, \mathbb{R})$, which is what is done when one defines $*$-products in perturbative causal quantum field theory (see Chap. 22).

For reasons of completeness, we now give the definition of quantum groups.

Definition 6.1.5 Let K be a field. A *quantum enveloping algebra* for a Lie algebra $\mathfrak{g}$ is given by a noncommutative and noncocommutative braided quasi-bialgebra A over $K[\![\hbar]\!]$ with underlying module $U(\mathfrak{g})[\![\hbar]\!]$, and which gives back the enveloping bialgebra structure $U(\mathfrak{g})$ modulo $\hbar$.

An important result of the theory of quantum groups is given by the following existence result, whose proof may be found in Drinfeld's ICM address [Dri87].

Theorem 6.1.1 *If* $\mathfrak{g}$ *is a semisimple Lie algebra, then there exists a quantum enveloping algebra* $U_\hbar(\mathfrak{g})$ *for* $\mathfrak{g}$.

Definition 6.1.6 Let $(\mathcal{C}, \otimes)$ be a symmetric monoidal category that is additive. Let A be a bialgebra in $\mathcal{C}$. An element $a \in A$ is called:

1. *primitive* if

$$\Delta(a) = a \otimes 1 + 1 \otimes a,$$

i.e., if a is in the equalizer

$$\mathrm{Prim}(A) := \mathrm{Ker}\big((\Delta, \mathrm{id} \otimes e + e \otimes \mathrm{id}) : A \to A \otimes A\big);$$

2. *group like* if

$$\Delta(a) = a \otimes a,$$

i.e., if a is in the equalizer

$$\mathrm{Grp}(A) := \mathrm{Ker}\big((\Delta, (\mathrm{id} \otimes e) \cdot (e \otimes \mathrm{id})) : A \to A \otimes A\big).$$

6.2 Twists and Deformations

The deformation theory of Hopf algebras may be studied by using a cohomology theory for them, similar to the Hochschild cohomology of associative algebra. A geometrically intuitive way to look at this cohomology theory is to think of it as the tangent space to the moduli space of monoidal (resp. braided monoidal) dg-categories at the point corresponding to the category $(\mathrm{MOD}(A), \otimes)$ of modules over the bialgebra that one wants to deform. This situation is similar to the deformation theory of associative algebras, where the Hochschild cohomology complex may be interpreted as the tangent space to the moduli space of dg-categories (without monoidal structure) with Morita isomorphisms (bimodules) between them, at the point corresponding to the dg-category $\mathrm{MOD}(A)$ of modules over the given associative algebra to be deformed.

We now give a concrete description of the deformations of Hopf algebras that we will use to define $*$-products in quantum field theory. This may be found in [Maj95], Sect. 2.2. We adapt his definition to the general setting of symmetric monoidal categories.

Definition 6.2.1 Let $(\mathcal{C}, \otimes)$ be a symmetric monoidal category and A be a bialgebra in $\mathcal{C}$. For $\psi : A^{\otimes n} \to \mathbb{1}$ a linear functional with value in the unit object, we denote by $\cdot_i$ multiplication in the i-th place for $0 \leq i \leq n$, $\psi \circ \cdot_0 = \epsilon \otimes \psi$ and $\psi \circ \cdot_{n+1} = \psi \circ \epsilon$. An *n-cochain* on A is a linear functional $\psi : A^{\otimes n} \to \mathbb{1}$, invertible in the convolution algebra $\underline{\mathrm{Hom}}(A^{\otimes n}, \mathbb{1})$ (defined using the coalgebra structure on A, and the algebra structure on $\mathbb{1}$). The *coboundary operator* is defined by

$$\partial\psi := \prod_{i=0}^{n+1} \psi^{(-1)^i} \circ \cdot_i .$$

We know from Calaque and Etingof's survey [CE04] that:

- 2-cocycles classify monoidal structures on the forgetful functor

$$\big(\mathrm{COMOD}(A), \otimes\big) \to (\mathcal{C}, \otimes)$$

  from comodules on A to $\mathcal{C}$, which deform the trivial monoidal structure;
- 3-cocycles classify the associativity conditions that one may put on $\mathrm{COMOD}(A)$, without changing the tensor product, by deforming the usual associativity condition on $\otimes$.

We now use Sweedler's notations for the coproduct

$$\Delta(a) = \sum a_{(1)} \otimes a_{(2)}.$$

When using these notations in monoidal categories, one must carefully insert the necessary associativity and commutativity isomorphisms in all formulas. We will keep this important fact implicit in our presentation. This allows us to consistently interpret the classical formulas in the general setting of symmetric monoidal categories, and in particular for super, graded, or differential graded modules.

Definition 6.2.2 Let (A, m, Δ) be a bialgebra and $\psi : A \otimes A \to \mathbb{1}$ be a unital 2-cocycle on A. The twisted product $m_\psi : A \otimes A \to A$ is defined by the composition

$$m_\psi(a, b) = \sum \psi(a_{(1)} \otimes b_{(1)}) a_{(2)} b_{(2)} \psi^{-1}(a_{(3)} \otimes b_{(3)}).$$

Theorem 6.2.1 *A 2-cocycle ψ on A induces an isomorphism of monoidal categories*

$$F_\psi : \big(\mathrm{coMod}(A), \otimes\big) \to \big(\mathrm{coMod}(A_\psi), \otimes_\psi\big)$$

by the natural isomorphism

$$\begin{aligned} \sigma_\psi : F_\psi(V) \otimes_\psi F_\psi(W) &\to F_\psi(V \otimes W) \\ v \otimes_\psi w &\mapsto \psi(v_{(1)} \otimes w_{(1)}) v_{(2)} \otimes w_{(2)}. \end{aligned}$$

In particular, if (B, m) is an algebra in $(\mathrm{coMod}(A), \otimes)$, called a *comodule algebra*, the above equivalence sends it to a twisted comodule algebra (B_ψ, m_ψ), with multiplication given by

$$a \cdot_\psi b = \sum \psi(a_{(1)} \otimes b_{(1)}) a_{(2)} b_{(2)}.$$

More specifically, one may deform the algebra A as a comodule over the bialgebra A.

One may think of a comodule algebra as an object generalizing the notion of a homogeneous space X under an algebraic group G. Twisting such an object corresponds to a kind of deformation quantization for X, as a module over G. We will mainly be interested in the situation where $G = \mathbb{V}$ is the additive algebraic group of a module V, and $X = \mathbb{V}$ is the associated affine space. These deformations play an important role in the causal approach to perturbative quantum field theory, described in Chap. 22. We now describe it concretely.

Example 6.2.1 Let V be a real vector space and $\psi : V \otimes V \to \mathbb{R}$ be a bilinear pairing on V, which we regard as a bilinear pairing with values in the vector space $\hbar\mathbb{R} \cong \mathbb{R}$. Extend it to a unique 2-cocycle ψ on the symmetric bialgebra $(S(V)[\![\hbar]\!], \Delta, \mu)$ by the conditions

$$\begin{aligned} \psi(ab, c) &= \sum \psi(a, c_{(1)}) \psi(b, c_{(2)}), \\ \psi(a, bc) &= \sum \psi(a_{(1)}, b) \psi(a_{(2)}, c). \end{aligned}$$

It is known that all 2-cocycles on $S(V)$ are of this form, and that a cocycle is a boundary if and only if it is symmetric. Recall that the symmetric bialgebra is simply the Hopf algebra of functions on the algebraic group $(\mathbb{V}, +)$ given by the space V with its addition. The twisting by a bilinear pairing corresponds to defining a deformation of the affine space $\mathbb{V}$, viewed as a homogeneous space (comodule algebra) under the group $(\mathbb{V}, +)$. This gives an associative product on $S(V)[\![\hbar]\!]$ given by

$$a \overset{\hbar}{*} b = \sum \psi(a_{(1)}, b_{(1)}) a_{(2)} b_{(2)},$$

that we may call the *Moyal product*, if the bilinear form ψ is symplectic.

Example 6.2.2 One may refine Example 6.2.1 by the following. Let V be a vector space and $C = S(V)$ be the associated cofree coalgebra. By Example 6.1.2, one gets on $S(C)$ a comodule algebra structure under C, i.e., a morphism

$$\Delta : S(C) \to C \otimes S(C)$$

that is compatible with comultiplication on C. If $\psi : V \otimes V \to \mathbb{R}$ is a bilinear pairing on V, that one extends to a 2-cocycle on $C = S(V)$ as in loc. cit., one gets a twisted product via the formula

$$a \overset{\hbar}{*} b = \sum \psi(a_{(1)}, b_{(1)}) a_{(2)} b_{(2)},$$

where $\Delta(a) = \sum a_{(1)} \otimes a_{(2)}$ means that $a \in S(C)$, $a_{(1)} \in C$ and $a_{(2)}$ in $S(C)$.

References

[CE04] Calaque, D., Etingof, P.: Lectures on tensor categories. arXiv Mathematics e-prints (2004). arXiv:math/0401246

[Dri87] Drinfel'd, V.G.: Quantum groups. In: Proceedings of the International Congress of Mathematicians, Vol. 1, 2 (Berkeley, Calif., 1986), pp. 798–820. Am. Math. Soc., Providence (1987)

[Kas95] Kassel, C.: Quantum Groups. Graduate Texts in Mathematics, vol. 155, p. 531. Springer, New York (1995). ISBN 0-387-94370-6

[KRT97] Kassel, C., Rosso, M., Turaev, V.: Quantum Groups and Knot Invariants. Panoramas et Synthèses [Panoramas and Syntheses], vol. 5, p. 115. Société Mathématique de France, Paris (1997). ISBN 2-85629-055-8

[Maj95] Majid, S.: Foundations of Quantum Group Theory, p. 607. Cambridge University Press, Cambridge (1995). doi:10.1017/CBO9780511613104. ISBN 0-521-46032-8

Chapter 7
Connections and Curvature

The mathematical formulation of particle physics involves various notions of connections on bundles. The easiest way to relate them is to use the very general notion of a Grothendieck connection, which was described in categorical terms in Sect. 2.5. In each concrete case, the notion of connection is easier to handle, and we give for it a simpler description.

7.1 Koszul Connections

Definition 7.1.1 Let $F \to X$ be a vector bundle on a smooth manifold. Let $A = \mathcal{C}^\infty(X)$ and $\mathcal{F}$ be the A-module $\Gamma(X, F)$ of sections of F. A *Koszul connection* on F is an $\mathbb{R}$-linear map

$$\nabla : \mathcal{F} \to \mathcal{F} \otimes_A \Omega_A^1$$

that satisfies Leibniz's rule

$$\nabla(f \cdot s) = f \cdot \nabla(s) + df \otimes s.$$

Definition 7.1.2 Let X be a manifold with function algebra A. Let $F \to X$ be a vector bundle and let $\nabla : \mathcal{F} \to \mathcal{F} \otimes_A \Omega_A^1$ be a Koszul connection. The *curvature* of ∇ is the composition

$$C(\nabla) := \nabla_2 \circ \nabla \in \mathrm{End}_A(\mathcal{F}) \otimes_A \Omega_A^2,$$

where ∇_i is defined by

$$\nabla_i := \mathrm{id}_{\mathcal{F}} \otimes d + (-1)^i \nabla \wedge \mathrm{id}_{\Omega_A^1} : \mathcal{F} \otimes_A \Omega_A^i \to \mathcal{F} \otimes_A \Omega_A^{i+1}.$$

Note that the A-linearity of the curvature is not clear a priori.

F. Paugam, *Towards the Mathematics of Quantum Field Theory*,
Ergebnisse der Mathematik und ihrer Grenzgebiete. 3. Folge / A Series of Modern Surveys in Mathematics 59, DOI 10.1007/978-3-319-04564-1_7,

Proposition 7.1.1 *Let $F \to X$ be a vector bundle on a smooth manifold. The data of a linear Grothendieck connection and of a Koszul connection on F are equivalent.*

Proof We refer to [BO78] for more details on the link between connections and infinitesimal neighborhoods. Just recall that the exterior differential is a map $d : A \to \Omega^1_A \subset \mathrm{Jet}^1 A$. Denote by $\mathcal{F} = \Gamma(X, F)$ the space of sections of F. If $\tau : \mathcal{F} \otimes_A \mathrm{Jet}^1 A \to \mathcal{F} \otimes_A \mathrm{Jet}^1 A$ is a Grothendieck connection, and $d_{1,\mathcal{F}} := d \otimes \mathrm{id}_{\mathcal{F}} : \mathcal{F} \to \mathcal{F} \otimes \mathrm{Jet}^1 A$, then the morphism $\theta = \tau \circ d_{1,\mathcal{F}} : \mathcal{F} \to \mathcal{F} \otimes \mathrm{Jet}^1 A$ allows us to construct a Koszul connection by $\nabla(f) = \theta(f) - f \otimes 1$. Indeed, one has $\nabla(f) = \mathcal{F} \otimes \Omega^1$ because τ is the identity on X. The association $\tau \mapsto \nabla$ is grounded on various universal properties that makes it unique. □

7.2 Ehresmann Connections

We refer to the book [KMS93], 17.1, for a differential geometric approach to Ehresmann connections. We will mostly be interested in the more algebraic approach presented in the book of Krasilshchik and Verbovetsky [KV98], Sect. 5.2.

Definition 7.2.1 Let $p : B \to M$ be a surjective submersion of varieties. An *Ehresmann connection* on B is the datum of a section a of the natural projection

$$\mathrm{Jet}^1 B \to B.$$

Proposition 7.2.1 *Let $p : B \to X$ be a surjective submersion of varieties. Then the following data on B are equivalent:*

1. *An Ehresmann connection a.*
2. *A section $a \in \Gamma_B(TB, VB)$ of the canonical exact sequence*

$$0 \to VB \to TB \to \pi^* TX \to 0$$

   of bundles on B.
3. *A connection one form $v \in \Omega^1(B, TB) = \mathrm{End}(TB)$ such that $v^2 = v$ and the image of v is VB.*

Proof We first show that the two last data are equivalent. If $a : TB \to VB$ is a section of the natural map $VB \to TB$, the projection $v : TB \to TB$ on VB along the kernel of a gives a connection one form. Since its image is VB, the data of a and v are equivalent. Now we use the Krasilshchik-Verbovetsky approach to connections in [KV98], Sect. 5.2. Note that the section $a : TB \to VB$ induces a natural morphism of functors on the category of admissible $\mathcal{O}_B$-modules (chosen to be the category of all modules in the case of algebraic varieties and the category of geometric modules for smooth varieties)

$$\nabla_a : \mathrm{Der}(\mathcal{O}_X, -) \to \mathrm{Der}(\mathcal{O}_B, -).$$

This corresponds on the representing objects to a morphism

$$\nabla_a : \mathcal{O}_B \to \mathcal{O}_B \otimes_{\mathcal{O}_X} \Omega^1_{B/X},$$

which can be extended to

$$s_a^* := \mathrm{id} + \nabla_a : \mathcal{O}_B \to \mathcal{O}_B \oplus \mathcal{O}_B \otimes_{\mathcal{O}_X} \Omega^1_{B/X} =: \mathcal{O}_{\mathrm{Jet}^1 B},$$

and then to a Grothendieck connection

$$\epsilon : \mathcal{O}_{\mathrm{Jet}^1 B} \xrightarrow{\sim} \mathcal{O}_{\mathrm{Jet}^1 B}.$$

Composing with the canonical projection $\mathcal{O}_{\mathrm{Jet}^1 B} \to \mathcal{O}_B$ gives a section $s : B \to \mathrm{Jet}^1 B$. □

Definition 7.2.2 The *curvature* of an Ehresmann connection is given by

$$R = \frac{1}{2}[v, v]$$

where $[\cdot,\cdot]$ denotes the Frölicher-Nijenhuis bracket of $v \in \Omega^1(E, TE)$ with itself. Thus $R \in \Omega^2(E, TE)$ is defined by

$$R(X, Y) = v\big(\big[(\mathrm{id} - v)X, (\mathrm{id} - v)Y\big]\big).$$

Proposition 7.2.2 *Let $p : B \to X$ be a bundle given by a morphism of varieties. The following data are equivalent*:

1. *An Ehresmann connection on B.*
2. *A Grothendieck connection on B.*

Proof Let $\mathcal{A} = \mathcal{C}^\infty(X)$ and $\mathcal{B} = \mathcal{C}^\infty(B)$. Suppose given a Grothendieck connection

$$p_1^* B \xrightarrow{\sim} p_2^* B.$$

In this case, it is equivalent to a map

$$\mathcal{B} \otimes_{\mathcal{A}} \mathrm{Jet}^1 \mathcal{A} \xrightarrow{\sim} \mathcal{B} \otimes_{\mathcal{A}} \mathrm{Jet}^1 \mathcal{A}$$

that induces the identity on $\mathcal{A}$, and that we can compose with the projection $\mathcal{B} \otimes_{\mathcal{A}} \mathrm{Jet}^1 \mathcal{A} \to \mathcal{B}$ to get a map

$$\mathcal{B} \otimes_{\mathcal{A}} \mathrm{Jet}^1 \mathcal{A} \to \mathcal{B}$$

that is also a map

$$\mathrm{Jet}^1 \mathcal{B} \to \mathcal{B}$$

and gives a section of the natural projection $\mathrm{Jet}^1 B \to B$. These two data are actually equivalent because one can get back the Grothendieck connection by tensoring with $\mathcal{B}$ over $\mathcal{A}$. □

Corollary 7.2.1 *If $p : F \to X$ is a vector bundle on a smooth manifold, the datum of a linear Ehresmann connection on F and of a Koszul connection on F are equivalent.*

7.3 Principal Connections

We use here the definition of Giraud [Gir71], Chap. III, 1.4, since it is adapted to general spaces.

Definition 7.3.1 Let G be a Lie group and $P \to M$ be a space morphism equipped with an action $m : G \times_M P \to P$ of G. One says that (P, m) is a *principal homogeneous space* over M under G (also called a *torsor* under G over M) if:

1. $P \to M$ is an epimorphism, i.e., there exists a covering family $\{U_i \to M\}$ such that $P_M(U_i)$ are non-empty (i.e., the bundle has locally a section on a covering of M);
2. the natural morphism
$$\begin{aligned} G \times P &\to P \times P \\ (g, p) &\mapsto (p, gp) \end{aligned}$$
is an isomorphism (i.e., the action of G on P is simply transitive).

Definition 7.3.2 A *principal G-connection* on P is a G-equivariant connection on $P \to M$.

Proposition 7.3.1 *The following are equivalent:*

1. *A principal G-connection on $p : P \to M$.*
2. *An equivariant $\mathfrak{g}$-valued differential form A on P, i.e., $A \in \Omega^1(P, \mathfrak{g})^G$ such that the diagram*
$$\begin{array}{ccc} \mathfrak{g} = T_{1_G} G & \xrightarrow{\mathrm{id}_{\mathfrak{g}}} & \mathfrak{g} \\ {\scriptstyle T_{1_G} m} \downarrow & \nearrow {\scriptstyle A} & \\ \Theta_P & & \end{array}$$
commutes where $m : G \times P \to P$ is the action map.

Proof The equivalence between an equivariant Ehresmann connection $A \in \Omega^1(TP, TP)^G$ and a differential form in $\Omega^1(P, \mathfrak{g})$ follows from the fact that the derivative of the action map $m : G \times_M P \to P$ with respect to the G-variable at the identity $e \in G$ defines a bundle map

$$i = D_e m : \mathfrak{g}_P := \mathfrak{g} \times P \to VP,$$

called a parallelization, between the vertical tangent bundle to P defined by

$$0 \to VP \to TP \xrightarrow{Dp} p^*TM \to 0$$

and the trivial linear bundle with fiber the Lie algebra $\mathfrak{g}$ of G, which is an isomorphism. The fact that $i : \mathfrak{g}_P \to TP$ is valued in VP follows from the fact that G acts vertically on P. □

Proposition 7.3.2 *The curvature of a principal G-connection A is identified with the form*

$$F = dA + [A \wedge A]$$

in $\Omega^2(P, \mathfrak{g})$ *where the bracket exterior product is given by*

$$[\omega \otimes h \wedge \nu \otimes k] = \omega \wedge \nu \otimes [h, k].$$

7.4 Cartan Connections and Moving Frames

We refer to the excellent survey of Wise [Wis10] for a more complete description. Sharpe [Sha97] and Wise neatly explain the idea of Cartan geometry by a commutative diagram:

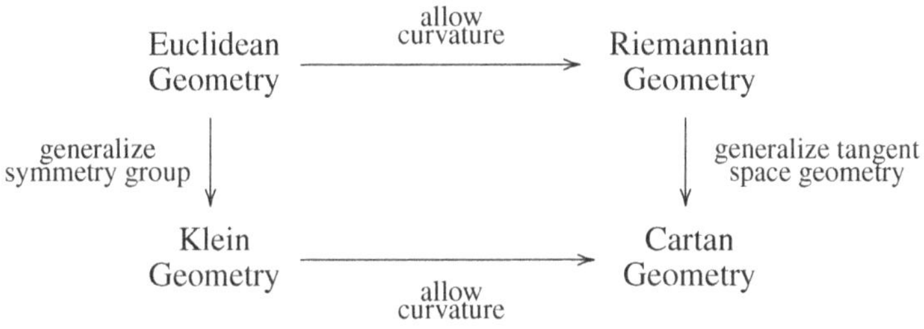

One can also say that a Cartan geometry is given by a space whose geometry is given by pasting infinitesimally some classical geometric spaces of the form G/H for $H \subset G$ two Lie groups. This can be nicely explained by the example of a sphere pasted infinitesimally on a space through tangent spaces, or, as in Wise's article [Wis10], by a Hamster ball moving on a given space as in Fig. 7.1.

Definition 7.4.1 Let M be a manifold and $H \subset G$ be two groups. A *Cartan connection* on M is the data of:

1. a principal G-bundle Q on M;
2. a principal G-connection A on Q;
3. a section $s : M \to E$ of the associated bundle $E = Q \times_G G/H$ with fibers G/H,

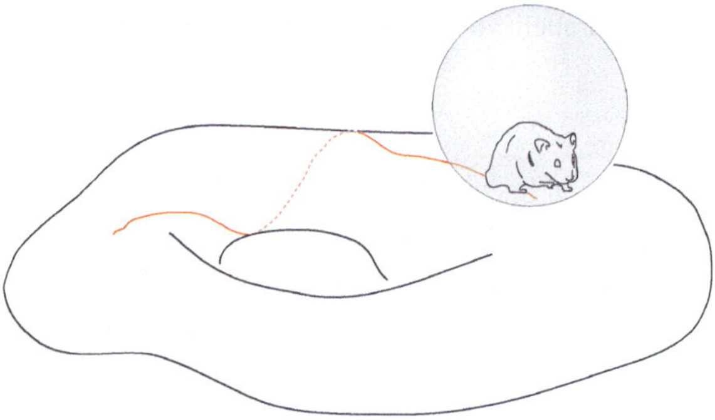

Fig. 7.1 Cartan geometry and the hamster ball

such that the pullback

$$e = s^*A \circ ds : TM \to VE,$$

called the moving frame (*vielbein*), for $A : TE \to VE$ the associated connection, is a linear isomorphism of bundles.

Here, the role of the section s is to "break the G/H symmetry". It is equivalent to the choice of a principal H-sub-bundle $P \subset Q$.

One may interpret the datum (s, Q, A) of a Cartan connection as a point of the stack (see Sect. 10.3 for a precise definition)

$$(s, Q, A) \in \underline{\mathrm{Hom}}\big(M, \big[G\backslash(G/H)\big]_{conn}\big)$$

satisfying the additional non-triviality condition.

In the cases of interest for this section, E is a linear bundle and the section s is simply the zero section that breaks the translation symmetry (action of $G/H = V$ on the sections of the vector bundle).

The first examples of Cartan connections are given by Klein geometries, i.e., by homogeneous spaces $E = G/H$ over the point space $M = \{.\}$. The corresponding Cartan connection is given by:

1. the trivial principal G-bundle $Q = G$ on M;
2. the trivial G-connection A on Q;
3. a section $s : M \to E$ of the associated bundle $E = Q \times_G G/H = G/H \to M$, i.e., a point $x = s(-) \in G/H$.

The pull-back $e = s^*A : TM = M \times \{0\} \to VE = M \times \{0\}$ is an isomorphism.

Three examples of Klein geometries that are useful in physics are given by $H = \mathrm{SO}(n-1, 1)$ and

$$G = \begin{cases} \mathrm{SO}(n,1) & \text{(de Sitter)} \\ \mathbb{R}^{n-1,1} \rtimes \mathrm{SO}(n-1,1) & \text{(Minkowski)} \\ \mathrm{SO}(n-1,2) & \text{(anti de Sitter)} \end{cases}$$

The de Sitter and anti de Sitter geometries are useful in the study of cosmological models with non-zero cosmological constant. Cartan geometries modeled on these Klein geometries are useful in general relativity. For example, the Minkowski Cartan geometry without torsion corresponds exactly to pseudo-Riemannian manifolds, which are the basic objects of general relativity.

Note that the G-connection A on the principal G-bundle Q is equivalent to an equivariant $\mathfrak{g}$-valued differential form

$$A : TQ \to \mathfrak{g},$$

and its restriction to $P \subset Q$ gives an H-equivariant differential form

$$A : TP \to \mathfrak{g}.$$

This is the original notion of a Cartan connection form.

Definition 7.4.2 The *curvature* of a Cartan connection is the restriction of the curvature of the corresponding Ehresmann connection on the principal G-bundle Q to the principal H-bundle P. It is given by the formula

$$F_A := dA + \frac{1}{2}[A \wedge A] \in \Omega^2(P, \mathfrak{g}).$$

The torsion of the Cartan connection is given by the composition of F_A with the projection $\mathfrak{g} \to \mathfrak{g}/\mathfrak{h}$.

Suppose that one can decompose $\mathfrak{g} = \mathfrak{h} \oplus \mathfrak{g}/\mathfrak{h}$ in an H-equivariant way (the Cartan geometry is called reductive). The Cartan connection form can thus be decomposed as

$$A = \omega + e$$

for $\omega \in \Omega^1(P, \mathfrak{h})$ and $e \in \Omega^1(P, \mathfrak{g}/\mathfrak{h})$. In this particular case, e can also be viewed as

$$e \in \Omega^1(M, \underline{\mathfrak{g}/\mathfrak{h}})$$

for $\underline{\mathfrak{g}/\mathfrak{h}}$ the H-bundle associated to $\mathfrak{g}/\mathfrak{h}$. This form is called the *vielbein* by Cartan. By definition of the Cartan connection, it gives an isomorphism

$$e : TM \to \underline{\mathfrak{g}/\mathfrak{h}}.$$

References

[BO78] Berthelot, P., Ogus, A.: Notes on Crystalline Cohomology, p. 243. Princeton University Press, Princeton (1978). ISBN 0-691-08218-9

[Gir71] Giraud, J.: Cohomologie Non Abélienne. Die Grundlehren der mathematischen Wissenschaften, vol. 179, p. 467. Springer, Berlin (1971)

[KMS93] Kolář, I., Michor, P.W., Slovák, J.: Natural Operations in Differential Geometry, p. 434. Springer, Berlin (1993). ISBN 3-540-56235-4

[KV98] Krasil'shchik, J., Verbovetsky, A.: Homological methods in equations of mathematical physics. arXiv (1998)

[Sha97] Sharpe, R.W.: Differential Geometry. Graduate Texts in Mathematics, vol. 166, p. 421. Springer, New York (1997). Cartan's generalization of Klein's Erlangen program, with a foreword by S.S. Chern. ISBN 0-387-94732-9

[Wis10] Wise, D.K.: MacDowell-Mansouri gravity and Cartan geometry. Class. Quantum Gravity **27**(15), 155010 (2010). 26. doi:10.1088/0264-9381/27/15/155010

Chapter 8
Lagrangian and Hamiltonian Systems

In this chapter, we describe Lagrange and Hamilton's geometric viewpoint of mechanics. We will mainly use Lagrangian methods in this book because their covariance is automatic in relativistic field theories and they allow a more conceptual treatment of symmetries. We thus only give a short account of the Hamiltonian formalism.

We will present in Chaps. 14, 15 and 16 a very large family of examples from classical, quantum and theoretical physics, formulated directly in the language of the following section.

8.1 Lagrangian Mechanics

The main interest of classical field theory is the notion of Lagrangian variational problem. We introduce here a very simple and general such notion, based on the parametrized approach to spaces, described in Chap. 3 (see also Chap. 10). We also use here the tools of functorial analysis from Chap. 4. We refer to Chap. 13 for a differential-algebraic treatment of symmetries in Lagrangian variational calculus, whose starting point is conceptually close to Lagrange's and Noether's original ideas on the subject.

Definition 8.1.1 A *Lagrangian variational problem* comprises the following data:

1. A space M called the parameter space for trajectories.
2. A space C called the configuration space for trajectories.
3. A morphism $\pi : C \to M$ (often assumed to be surjective).
4. A subspace $H \subset \Gamma(M, C)$ of the space of sections of π

   $$\Gamma(M, C) := \{x : M \to C,\ \pi \circ x = \mathrm{id}\},$$

   called the space of histories.

F. Paugam, *Towards the Mathematics of Quantum Field Theory*,
Ergebnisse der Mathematik und ihrer Grenzgebiete. 3. Folge / A Series of Modern
Surveys in Mathematics 59, DOI 10.1007/978-3-319-04564-1_8,

5. A functional (partial function with a domain of definition) $S : H \to R$ (where R is a space in rings that is often the real line $\mathbb{R}$, the complex line $\mathbb{C}$, or $\mathbb{R}[[\hbar]]$) called the *action functional*.

An *observable* is a functional $f : H \to R$.

The description of classical physical systems can often be based on the following.

Principle 1 (Least Action Principle) *The space T of physical trajectories of a given classical variational problem is the subspace T of the space of histories H given by critical points of the action S, i.e.,*

$$T = \{x \in H,\ d_x S = 0\}.$$

The advantage of the Lagrangian approach to classical mechanics is that it gives rise to a clear informal intuition on the notion of quantum trajectory. This viewpoint was introduced by Feynman in his thesis, and is at the basis of most of the modern developments of quantum field theory.

Principle 2 (Functional Integral Quantization) *The space of quantum trajectories is the space H of all histories. The probability of observing a given history $x : M \to C$ is proportional to $e^{\frac{i}{\hbar}S(x)}$. More precisely, if $f : H \to R$ is an observable, its probability value is given by the absolute value of the complex Gaussian mean value*[1]

$$\langle f \rangle_S := \frac{\int_H f(x) e^{\frac{i}{\hbar}S(x)}[dx]}{\int_H e^{\frac{i}{\hbar}S(x)}[dx]}.$$

Remark 8.1.1 The spaces of parameters and configurations that appear in the above definition may be manifolds, but one should also allow them to be more general types of spaces, such as manifolds with corners, stacks or derived stacks (see Chap. 10 for an introduction). In any case, one needs some finiteness or geometricity conditions (see Sect. 10.5 for the general notion of a geometric stack) on both of them (essentially, an analog of integration on M, and an analog of duality on C) to be able to define interesting variational problems.

One can apply this general approach to the case of so-called *local action functionals* (see Chaps. 12 and 13 for a detailed study of these objects), which are given by expressions of the form

$$S(x) = \int_M L\big(x(m), \partial_\alpha x(m)\big) dm$$

[1] The meaning of this expression has to be clearly specified mathematically, which is the central difficulty of this approach, which will be treated in Chaps. 20, 21 and 22.

for $L : \mathrm{Jet}(C) \to \mathbb{R}$ a function called the Lagrangian density, that depends on variables x_0 (functional variables) in the fiber of $\pi : C \to M$ and of finitely many formal additional variables x_α that play the role of formal derivatives of x_0. This will give a very nice description of the space T of physical trajectories for the classical system, as a solution space for a (usually nonlinear) partial differential equation called the Euler-Lagrange system.

Example 8.1.1 Consider the variational problem of classical Newtonian mechanics for a particle moving in a potential: one has $M = [0, 1]$, $C = [0, 1] \times \mathbb{R}^3$, $\Gamma(M, C) \cong \mathcal{C}^\infty([0, 1], \mathbb{R}^3)$ and for $(x_0, x_1, \mathbf{v}_0, \mathbf{v}_1)$ fixed in $(\mathbb{R}^3)^4$, the space of histories is given by

$$H := \{x \in \Gamma(M, C), x(0) = x_0, x(1) = x_1, \partial_t x(0) = \mathbf{v}_0, \partial_t x(1) = \mathbf{v}_1\}.$$

One equips $\mathbb{R}^3$ with its standard scalar product $\langle\cdot, \cdot\rangle$. The action functional is given by

$$S(x) = \int_M \frac{1}{2} m\langle\partial_t x(t), \partial_t x(t)\rangle - V(x(t))dt$$

where $V : \mathbb{R}^3 \to \mathbb{R}$ is a smooth function, called the potential. If $\mathbf{h} : \mathbb{R} \to \mathbb{R}^3$ is a function tangent to H at some point x, one shows that $h(0) = h(1) = 0$. Defining

$$d_x S(\mathbf{h}) := \lim_{\epsilon \to 0} \frac{S(x + \epsilon h) - S(x)}{\epsilon},$$

one gets

$$d_x S(\mathbf{h}) = \int_0^1 \langle m\partial_t x, \partial_t \mathbf{h}\rangle - \langle d_x V(x), \mathbf{h}\rangle dt$$

and by integrating by parts using that $h(0) = h(1) = 0$, finally,

$$d_x S(\mathbf{h}) = \int_0^1 \langle -m\partial_t^2 x - d_x V(x), \mathbf{h}\rangle dt.$$

The space of physical trajectories is thus the space of maps $x : \mathbb{R} \to \mathbb{R}^3$ such that

$$m \cdot \partial_t^2 x = -V'(x).$$

This is the standard law of Newtonian mechanics. For example, if $V = 0$, the physical trajectories are those with constant speed, which corresponds to Galilean inertial bodies.

Remark 8.1.2 The knowledge of the functional integral is not strictly necessary in order to carry out the functional integral quantization. One only needs to have a function

$$e^{\frac{i}{\hbar}S} : H \to S^1.$$

This viewpoint is used in the formalization of higher gauge theories, described in Sect. 16.5.

8.2 Symplectic and Poisson Manifolds

Definition 8.2.1 The algebra of *multi-vector fields* on a given manifold P is the antisymmetric algebra $\wedge^*\Theta_P$ on the $\mathcal{O}_P$-module of vector fields.

Proposition 8.2.1 *There exists a unique extension*

$$[\cdot,\cdot]_{NS}:\wedge^*\Theta_P\otimes\wedge^*\Theta_P\to\wedge^*\Theta_P$$

of the Lie bracket of vector fields, that is moreover a bigraded derivation. More precisely, this bracket, called the Schouten-Nijenhuis bracket, satisfies:

1. *The graded Jacobi identity*:

$$\big[X,[Y,Z]\big]=\big[[X,Y],Z\big]+(-1)^{(|X|-1)(|Y|-1)}\big[Y,[X,Z]\big].$$

2. *The graded anti-commutativity*:

$$[X,Y]=(-1)^{(|X|-1)(|Y|-1)}[Y,X].$$

3. *The bigraded derivation condition*:

$$[X,Y\wedge Z]=[X,Y]\wedge Z+(-1)^{(|X|-1)(|Y|-1)}.$$

Definition 8.2.2 A *symplectic manifold* is a pair (P,ω) comprising a manifold P and a non-degenerate closed 2-form $\omega\in\Omega^2(P)$:

$$d\omega=0,\qquad \omega^\flat:TP\xrightarrow{\sim}T^*P.$$

A *Poisson manifold* is a pair (P,π) comprising a manifold P and a bivector $\pi\in\wedge^2\Theta_P$ whose Schouten-Nijenhuis bracket with itself

$$[\pi,\pi]_{SN}=0$$

is zero. A *symplectomorphism*

$$f:(P_1,\omega_1)\to(P_2,\omega_2)$$

is a smooth map such that $f^*\omega_2=\omega_1$. A *Poisson morphism*

$$f:(P_1,\pi_1)\to(P_2,\pi_2)$$

is a smooth map such that $f_*\pi_1$ makes sense and is equal to π_2.

Proposition 8.2.2 *The datum of a symplectic manifold is equivalent to the datum of a Poisson structure that is non-degenerate, i.e., such that*

$$\pi^{\natural} : T^*P \to TP$$

is an isomorphism.

Proof The inverse $\pi^{\natural}$ of $\omega^{\flat} : TP \to T^*P$ gives the Poisson bivector. One checks that the nullity of the bracket is equivalent to the fact that ω is closed. □

Definition 8.2.3 If (P, π) is a Poisson manifold, we define its *Poisson bracket* by

$$\begin{aligned} \{.,.\}_{\pi} : \mathcal{O}_M \times \mathcal{O}_M &\to \mathcal{O}_M \\ (f, g) &\mapsto \langle \pi, df \wedge dg \rangle. \end{aligned}$$

The main example of a symplectic manifold is given by the cotangent bundle $P = T^*X$ of a given manifold. If we define the Legendre 1-form by

$$\theta(v) = \mathrm{ev}(Dp \circ v)$$

where

- $Dp : T(T^*X) \to p^*TX := TX \times_X T^*X$ is the differential of $p : T^*X \to X$;
- $\mathrm{ev} : TX \times_X T^*X \to \mathbb{R}_M$ is the natural evaluation duality; and
- $v \in \Theta_{T^*X} = \Gamma(T^*X, T(T^*X))$ is a vector field.

The symplectic 2-form on P is defined by

$$\omega = -d\theta.$$

On $P = T^*\mathbb{R}^n$, with coordinates (q, p), the Liouville form is given by

$$\theta = pdq = \sum_i p_i dq_i$$

and the symplectic form by

$$\omega = dq \wedge dp = \sum_i dq_i \wedge dp_i.$$

The Poisson bracket is then given by

$$\{f, g\} = \sum_i \frac{\partial f}{\partial q_i}\frac{\partial g}{\partial p_i} - \frac{\partial f}{\partial p_i}\frac{\partial g}{\partial q_i}.$$

8.3 Dynamics of a Hamiltonian System

Definition 8.3.1 A *Hamiltonian system* is a tuple $(P, \pi, I, \mathrm{Hist}, H)$ composed of:

1. a Poisson manifold (P, π) called the *phase space*; whose functions are called observables,
2. an interval $I \subset \mathbb{R}$ called the time *parameter space for trajectories*;
3. a subspace $\mathrm{Hist} \subset \mathrm{Hom}(I, P)$ called the *space of histories*;
4. a function $H : P \to \mathbb{R}$ called the *Hamiltonian.*

If $x : I \to P$ is a history in the phase space and $f \in \mathcal{O}_P$ is an observable, *Hamilton's equations* for x along f are given by

$$\frac{\partial(f \circ x)}{\partial t} = \{H, f\} \circ x.$$

The Hamiltonian function H defines a *Hamiltonian vector field*

$$X_H := \{H, .\} : \mathcal{O}_P \to \mathcal{O}_P.$$

On $T^*\mathbb{R}^n$, the Hamiltonian vector field is given by

$$X_H = \sum_i \frac{\partial H}{\partial p_i}\frac{\partial}{\partial q_i} - \sum_i \frac{\partial H}{\partial q_i}\frac{\partial}{\partial p_i}.$$

If the Poisson structure π is non-degenerate, i.e., if (P, π) comes from a symplectic manifold (P, ω), the Hamilton equation on a history $x : I \to P$ is given by

$$x^*(i_{X_H}\omega) = x^*(dH).$$

On $T^*\mathbb{R}^n$, this corresponds to

$$\begin{cases} \dfrac{\partial q_x}{\partial t} = \dfrac{\partial H}{\partial p}, \\ \dfrac{\partial p_x}{\partial t} = -\dfrac{\partial H}{\partial q}. \end{cases}$$

Theorem 8.3.1 (Symplectic Noether Theorem) *Given a Hamiltonian system (P, π, H), a function f on P is constant along the trajectories of the Hamiltonian vector field if and only if the Hamiltonian is constant under the Hamiltonian vector field induced by f.*

Proof This follows from the equality

$$X_f(H) = \{f, H\} = -\{H, f\} = -X_H(f).$$ □

8.4 Lagrangian Manifolds and Canonical Transformations

We now introduce the notion of a canonical transformation, with views towards applications to the gauge fixing procedure for gauge theories, in Sect. 13.5.

Definition 8.4.1 A *coisotropic submanifold* of a Poisson manifold (P, π) is a submanifold $L \subset P$ such that the ideal of functions on P that annihilate on L is stable under the Poisson bracket. An *isotropic submanifold* of a symplectic manifold (P, ω) is a submanifold L such that the symplectic form restricts to zero on L. A *Lagrangian submanifold* in a symplectic manifold is a submanifold that is both isotropic and coisotropic.

Example 8.4.1

1. Let $P = \mathbb{R}^{2n}$ be the flat symplectic space, with coordinates (x, x^*) and symplectic form $dx \wedge dx^*$, whose matrix with respect to the basis (x, x^*) is given by
$$dx \wedge dx^* := \begin{pmatrix} 0 & I \\ -I & 0 \end{pmatrix}.$$
Then the submanifolds L with coordinates only x or only x^* are Lagrangian. Submanifolds of dimension 1 are isotropic and submanifolds of codimension 1 are coisotropic.
2. The graph of a Poisson morphism $f : (P_1, \pi_1) \to (P_2, \pi_2)$ is a coisotropic submanifold in $(P_1 \times P_2, \pi_1 - \pi_2)$.
3. If $S : X \to \mathbb{R}$ is a smooth function, the image of its differential $dS : X \to T^*X$ is a Lagrangian submanifold.

Definition 8.4.2 Let (P, π) be a Poisson manifold. A *canonical transformation* is a Poisson isomorphism
$$f : (P, \pi) \to (P, \pi).$$
A canonical transformation of a symplectic manifold (P, ω) is a canonical transformation of the associated Poisson manifold. A canonical transformation f is called *Hamiltonian* if there exists a function $H : P \to \mathbb{R}$ such that f is the value at 1 of the flow of canonical transformations
$$\sigma_{X_H} : \mathbb{R} \times P \to P$$
associated to the Hamiltonian vector field $X_H := \{H, -\}$.

The following example will play a fundamental role in the general gauge fixing procedure for gauge theories, presented in Sect. 13.5.

Example 8.4.2 Let X be a smooth manifold and $\pi : P = T^*X \to X$ be the natural projection. Let $\psi : X \to \mathbb{R}$ be a smooth function, with differential $d\psi : X \to T^*X$.

Then the map

$$\mathrm{id}_P + d\psi \circ \pi : P \to P$$

(where $+ : P \times_X P \to P$ is the fiberwise addition map) is a Hamiltonian canonical transformation with Hamiltonian ψ, called a vertical Hamiltonian canonical transformation on T^*X. This function is also sometimes called the gauge fixing function.

8.5 Relation with Lagrangian Variational Problems

The main interest of the Hamilton formalism is that it translates the problem of solving the Euler-Lagrange equation of a Lagrangian variational problem, i.e., a partial differential equation, into the problem of solving an ordinary differential equation: the Hamilton equation. It is however hard to formalize without coordinates for a general Lagrangian variational problem. We refer to Vitagliano's paper [Vit09] for a jet space formulation of the relation between Lagrangian mechanics and multi-symplectic Hamiltonian mechanics, giving in this section only a coordinate dependent description of the Legendre transform.

Let $(\pi : C \to M, S)$ be a Lagrangian variational problem in which

$$\pi : C = I \times X \to I = M$$

with $I \subset \mathbb{R}$ an interval and X a manifold. We assume that the action functional $S : \underline{\Gamma}(M, C) \to \mathbb{R}$ is local and given by

$$S(x) = \int_I L(t, x, \partial_t x) dt,$$

with $L : \mathrm{Jet}^1 C = I \times TX \to \mathbb{R}$ a Lagrangian density. We also assume that we have a trivialization of TX and T^*X given respectively by the coordinates (x, x_1) and (q, p). The Legendre transformation is given by the map

$$\begin{aligned} \mathbb{F}L : TX &\to T^*X \\ (x, x_1) &\mapsto (q, p) := (x, \tfrac{\partial L}{\partial x_1}). \end{aligned}$$

If this map is an isomorphism, one defines the Hamiltonian by

$$H(q, p, t) = \langle p, x_1 \circ (\mathbb{F}L)^{-1}(q, p) \rangle - L\big(t, (\mathbb{F}L)^{-1}(q, p)\big).$$

In simplified notation, this gives

$$H(q, p, t) := p q_1(q, p) - L\big(t, q, q_1(q, p)\big).$$

The Euler-Lagrange equation

$$\frac{d}{dt}\left(\frac{\partial L}{\partial x_1^n}\right) - \frac{\partial L}{\partial x^n} = 0$$

can be written more explicitly as

$$(\partial_t^2 x^{n'}) \frac{\partial^2 L}{\partial x_1^{n'} \partial x_1^n}(t, x, \partial_t x) = -(\partial_t x^{n'}) \frac{\partial^2 L}{\partial x^{n'} \partial x_1^n}(t, x, \partial_t x) + \frac{\partial L}{\partial x^n}(t, x, \partial_t x).$$

If we assume the non-degeneracy condition that the matrix

$$\left(\frac{\partial^2 L}{\partial x_1^{n'} \partial x_1^n}(t, x, \partial_t x) \right)_{n,n'}$$

is always invertible, we can compute $\partial_t^2 x^{n'}$ from all the other variables. This then gives a translation of the Euler-Lagrange equation via the Legendre transform to the Hamilton equations

$$\begin{cases} \frac{\partial q}{\partial t} = \frac{\partial H}{\partial p}, \\ \frac{\partial p}{\partial t} = -\frac{\partial H}{\partial q}. \end{cases}$$

For example, if we start with the Lagrangian

$$L(t, x, x_1) = \frac{1}{2} m (x_1)^2 - V(x)$$

of Newtonian mechanics, we get the momentum variable $p = \frac{\partial L}{\partial x_1} = m x_1$, so that $x_1 = \frac{p}{m}$ and

$$H(t, q, p) = \sum p_i x_{i,1} - L(t, x, x_1) = \frac{p^2}{2m} + V(q).$$

The Hamilton equations are given by

$$\begin{cases} \frac{\partial q}{\partial t} = \frac{\partial H}{\partial p} = \frac{p}{m}, \\ \frac{\partial p}{\partial t} = -\frac{\partial H}{\partial q} = -V'(q) \end{cases}$$

and one recognizes in them the usual equations of Newtonian mechanics.

8.6 Hamilton-Jacobi Equations

The main interest of Hamilton-Jacobi equations is that they replace the problem of solving the second order Euler-Lagrange equations by a simpler, first order equation, whose formal solution space (in the sense of algebraic analysis, see Chap. 12) is a classical finite dimensional sub-variety of the Euler-Lagrange equation, which one can study with simple geometric tools. Vitagliano has expanded on this simple idea by defining a Hamilton-Jacobi problem for a general partial differential equation as a finite dimensional sub-equation (which gives a method to construct explicit solutions of a given complicated system).

Here, we first give a coordinate description, based on Arnold [Arn91], pp. 253–255. We then give a coordinate-free presentation, based on Vitagliano [Vit11] and Carinena et al. [CGM+06].

Definition 8.6.1 Let $\pi : C = I \times X \to I = M$ be a bundle of classical mechanics as in the last section and let $S : \underline{\Gamma}(M, C) \to \mathbb{R}$ be a local action functional of the form $S(x) = \int_I L(t, x, \partial_t x)dt$. Let $(t_0, x_0) \in C$ be a fixed point. The *Hamilton-Jacobi* action function

$$S_{hj} : C \to \mathbb{R}$$

is defined as the integral

$$S_{hj,(t_0,x_0)}(t, x) = \int_I L(t, x_{ext}, \partial_t x_{ext})dt$$

where $x_{ext} : M \to C$ is the extremal trajectory starting at (t_0, x_0) and ending at (x, t).

For this definition to make sense, one has to assume that the mapping $(x_{0,1}, t) \mapsto (x, t)$, given by solving the equations of motion with initial condition $x(0) = x_0$ and $\partial_t x(0) = x_{0,1}$, is non-degenerate. This can be shown to be possible in a small neighborhood of the initial condition. We thus stick to this case.

Theorem 8.6.1 *The differential of the Hamilton-Jacobi action function (for a fixed initial point) is equal to*

$$dS_{hj} = pdq - Hdt$$

where $p = \frac{\partial L}{\partial q_1}$ *and* $H = pq_1 - L$ *are defined with the help of the terminal velocity* $q_1 := \partial_t x_{ext}(t, x)$ *of the extremal trajectory.*

Corollary 8.6.1 *The action function satisfies the Hamilton-Jacobi equation*

$$\frac{\partial S}{\partial t} + H\left(\frac{\partial S}{\partial x}, x, t\right) = 0.$$

We now give a coordinate-free presentation of the Hamilton-Jacobi formalism, basing ourselves on Vitagliano [Vit11] and Carinena et al. [CGM+06]. As before, we denote by $\pi : C = I \times X \to I = M$ the bundle of classical mechanics and $\mathrm{Jet}^1 C = TX \times I$ the corresponding jet space. Let $\mathrm{Jet}^\dagger C := T^*X \times I$ be its dual bundle. One has a natural isomorphism

$$\mathrm{Jet}^1 C \times_C \mathrm{Jet}^\dagger C \cong I \times TX \times_X T^*X.$$

Definition 8.6.2 Let $L : \mathrm{Jet}^1 C \to \mathbb{R}$ be a Lagrangian density. Its Legendre correspondence

$$\mathrm{Leg}_L : \mathrm{Jet}^1 C \to \mathrm{Jet}^\dagger C,$$

is defined as the subspace $\Gamma_{\mathrm{Leg}_L} \subset \mathrm{Jet}^1 C \times_C \mathrm{Jet}^\dagger C$ given by

$$\Gamma_{\mathrm{Leg}_L} := \left\{ \left(t, x, x_1, \frac{\partial L}{\partial x_1} \right) \right\} \subset \mathrm{Jet}^1 C \times_C \mathrm{Jet}^\dagger C.$$

One says that a section $s \in \Gamma(C, \mathrm{Jet}^1 C \times_C \mathrm{Jet}^\dagger C)$, given by a pair of sections (X, W) of the form

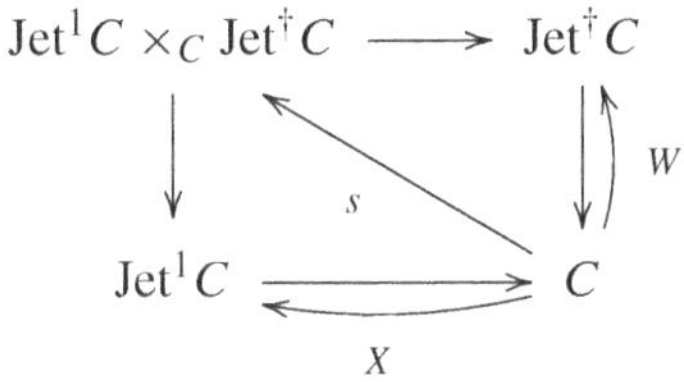

is a solution to the *generalized Hamilton-Jacobi equation* if

$$i_X s^* \omega = s^* dE$$

and $\mathrm{im}(s) \subset \Gamma_{\mathrm{Leg}}$, where $E(t, x, x_1, p) = p x_1 - L(t, x, x_1)$ is the Hamiltonian function and ω is the canonical symplectic form on $T^* X$.

The following theorem was proved by Carinena et al. [CGM+06].

Theorem 8.6.2 *The two following conditions are equivalent:*

1. $s = (X, W)$ *is a solution to the generalized Hamilton-Jacobi equation;*
2. *every integral curve of* X *is a solution of the Euler-Lagrange equations.*

If, further, the Legendre transform is invertible, then $X = \mathrm{Leg}^{-1} \circ W$ *and the combination of the generalized Hamilton-Jacobi equation with the equation*

$$s^* \omega = 0$$

is locally equivalent to the condition that $W = dS$ *and* S *is a solution to the classical Hamilton-Jacobi equation.*

This formulation of the Hamilton-Jacobi formalism has been extended to higher field theories by Vitagliano (see [Vit10] and [Vit11]).

8.7 Poisson Reduction

We refer to the book [Ber01] for an introduction and to [GS90] for a more complete account of the theory in the symplectic setting. For the Poisson setting, we refer to [But07], Part A, 1, 6.1. We use also the article of Kostant-Sternberg [KS87] and Marsden-Weinstein [MW74].

Definition 8.7.1 Let G be a Lie group. A *Hamiltonian G-space* is a tuple (M, π, ϕ, δ) composed of:

1. a Poisson manifold (M, π);
2. a G-action $\phi : G \times M \to M$ that is canonical, i.e., respects the Poisson bracket;
3. a linear map $\delta : \mathfrak{g} \to \mathcal{O}_M$ from the Lie algebra $\mathfrak{g} = \mathrm{Lie}(G)$ to the Poisson algebra of functions on M such that

$$\{\delta(\xi), \cdot\} = \xi_M(-)$$

   where $\mathfrak{g} \to \Theta_M, \xi \mapsto \xi_M$ is the infinitesimal action.

The *moment map* of a Hamiltonian G-space is defined as the G-equivariant map

$$\begin{aligned} J : M &\to \mathfrak{g}^* \\ m &\mapsto \left[\xi \mapsto \delta(\xi)(m)\right]. \end{aligned}$$

Following Marsden and Weinstein, one says that $\mu \in \mathfrak{g}^*$ is a weakly regular value of J if:

1. $J^{-1}(\mu)$ is a sub-manifold of M; and
2. for every $m \in J^{-1}(\mu)$, the inclusion $T_m(J^{-1}(\mu)) \subset \mathrm{Ker}(D_m J)$ is an equality.

The main theorem of Marsden-Weinstein's symplectic reduction is the following.

Theorem 8.7.1 *Let (M, π, ϕ, δ) be a Hamiltonian G-space (which is assumed to be symplectic). Suppose that $\mu \in \mathfrak{g}^*$ is a regular value for the moment map J and that the isotropy group G_μ of μ acts freely and properly on the constraint surface $C_\mu := J^{-1}(\mu)$. Then the manifold*

$$C_\mu / G_\mu$$

is equipped with a natural symplectic form.

An algebraic Poisson version of this construction can be given by the following.

Proposition 8.7.1 *Let (M, π, ϕ, δ) be a Hamiltonian G-space. Let $\mu \in \mathfrak{g}^*$ be an element and suppose that the ideal $\mathcal{I}_C$ of the constraint subspace $J^{-1}(\mu) \subset M$ is closed under the Poisson bracket (one talks of first class constraints). Then the algebra*

$$(\mathcal{O}_M / \mathcal{I}_C)^{\mathcal{I}_C}$$

of $\mathcal{I}_C$-invariant functions (such that $\{\mathcal{I}_C, f\} = 0$) in the quotient algebra is a Poisson algebra.

Note that the above Poisson reduction coincides with the Marsden-Weinstein reduction of the above theorem with its hypothesis.

8.8 The Finite Dimensional BRST Formalism

The presentation of this section is based on Kostant-Sternberg [KS87]. The BRST formalism gives a simplified Hamiltonian analog of the Batalin-Vilkovisky homotopical Poisson reduction of gauge theories, which is treated in Chap. 13.

Let (M, π, ϕ, δ) be a Hamiltonian G-space and suppose that $0 \in \mathfrak{g}^*$ is a regular value of the moment map $J : M \to \mathfrak{g}^*$. This implies that the ideal $\mathcal{I}$ of the subspace $C := J^{-1}(0)$ is generated by the image of δ.

We consider δ as a dg-module

$$\mathcal{C} := \left[\underset{-1}{\mathfrak{g} \otimes \mathcal{O}_M} \xrightarrow{\delta} \underset{0}{\mathcal{O}_M}\right]$$

and define the Koszul resolution as the symmetric algebra

$$\mathcal{K} := \mathrm{Sym}_{\mathcal{O}_M - dg}(\mathcal{C}).$$

As a graded algebra, we have $\mathcal{K} \cong \oplus \wedge^* \mathfrak{g} \otimes \mathcal{O}_M$, and the differential is given by defining $\delta(1 \otimes f) = 0$ and $\delta(\xi \otimes 1) = 1 \otimes \delta(\xi)$. One then has that

$$H^0(\mathcal{K}, \delta) = \mathcal{O}_M / \mathcal{I}_C = \mathcal{O}_C.$$

We now consider $\mathcal{K}$ as a module over $\mathfrak{g}$ and define

$$d : \mathcal{K} \to \mathfrak{g}^* \otimes \mathcal{K} = \mathrm{Hom}(\mathfrak{g}, \mathcal{K})$$

by

$$(dk)(\xi) = \xi k, \quad \xi \in \mathfrak{g},\ k \in \mathcal{K}.$$

It is a morphism of modules over the dg-algebra $\mathcal{K}$. One extends this differential to the algebra

$$\mathcal{A}^{\bullet,\bullet} := \mathrm{Sym}_{\mathcal{K}-dg}\left(\left[\underset{0}{K} \xrightarrow{d} \underset{1}{\mathfrak{g}^* \otimes K}\right]\right).$$

This bidifferential bigraded algebra is called the BRST algebra. One gets as zero cohomology

$$H^0(\mathcal{A}^{\bullet,\bullet}, d) = \mathcal{K}^{\mathfrak{g}}$$

for the differential d the space of $\mathfrak{g}$-invariants in $\mathcal{K}$ and the cohomology

$$H^0\big(H^0(\mathcal{A}^{\bullet,\bullet},\delta),d\big) = \mathcal{O}_C^{\mathfrak{g}}$$

is the space of functions on the Poisson reduction on the given Hamiltonian G-space.

The total complex $(\mathrm{Tot}(\mathcal{A}^{\bullet,\bullet}), D)$ is called the BRST complex. One can make the hypothesis that its zero cohomology is equal to the above computed space $\mathcal{O}_C^{\mathfrak{g}}$ of functions on the Poisson reduction.

Now observe that one has a canonical isomorphism

$$\wedge^*\mathfrak{g} \otimes \wedge^*\mathfrak{g}^* \cong \wedge^*(\mathfrak{g} \oplus \mathfrak{g}^*)$$

that induces a split scalar product on $\wedge^*(\mathfrak{g} \oplus \mathfrak{g}^*)$ given by the evaluation of linear forms. Let $C(\mathfrak{g} \oplus \mathfrak{g}^*)$ be the corresponding Clifford superalgebra.

Lemma 8.8.1 *The super-commutator in $C(\mathfrak{g} \oplus \mathfrak{g}^*)$ induces a super-Poisson structure on $\wedge^*(\mathfrak{g} \oplus \mathfrak{g}^*)$.*

Proof Let c_i and c_j be representative elements in the Clifford algebra for some elements in the exterior algebra of respective degrees i and j. The class of the super-commutator $[c_i, c_j]$ in the degree $i + j$ exterior power depends only on the classes of c_i and c_j. This operation satisfies the axioms of a super Poisson algebra. □

The main theorem of homological perturbation theory is the following.

Theorem 8.8.1 *Under all the hypotheses given above, there exists an odd element $\Theta \in \mathcal{A}^{\bullet,\bullet}$, called the BRST generator, such that the Poisson bracket by Θ is precisely the BRST differential D, i.e.,*

$$\{\Theta, \cdot\} = D$$

on the total complex $\mathrm{Tot}(\mathcal{A}^{\bullet,\bullet})$. *Since $D^2 = 0$, this element satisfies the so-called classical master equation*

$$\{\Theta, \Theta\} = 0.$$

Note that in the case of a group action on a point $X = \{\bullet\}$, the BRST generator $\Theta \in \wedge^3(\mathfrak{g} \oplus \mathfrak{g}^*)$ is simply given by the Lie bracket on $\mathfrak{g}$

$$\Theta = [\cdot, \cdot] \in \mathfrak{g}^* \wedge \mathfrak{g}^* \wedge \mathfrak{g}$$

and the BRST cohomology identifies with the Lie algebra cohomology.

References

[Arn91] Arnol'd, V.I.: Mathematical Methods of Classical Mechanics. Graduate Texts in Mathematics, vol. 60, p. 516. Springer, New York (1991). Translated from the 1974 Russian original by K. Vogtmann and A. Weinstein. Corrected reprint of the second (1989) edition. ISBN 0-387-96890-3

[Ber01] Berndt, R.: An Introduction to Symplectic Geometry. Graduate Studies in Mathematics, vol. 26, p. 195. Am. Math. Soc., Providence (2001). Translated from the 1998 German original by Michael Klucznik. ISBN 0-8218-2056-7

[But07] Butterfield, J.: Philosophy of Physics, pp. 1–1369. Elsevier/North-Holland, Amsterdam (2007)

[CGM+06] Carinena, J.F., Gracia, X., Marmo, G., Martinez, E., Munoz-Lecanda, M., Roman-Roy, N.: Geometric Hamilton-Jacobi theory. arXiv Mathematical Physics e-prints (2006). arXiv:math-ph/0604063

[GS90] Guillemin, V., Sternberg, S.: Symplectic Techniques in Physics, 2nd edn., p. 468. Cambridge University Press, Cambridge (1990). ISBN 0-521-38990-9

[KS87] Kostant, B., Sternberg, S.: Symplectic reduction, BRS cohomology, and infinite-dimensional Clifford algebras. Ann. Phys. **176**(1), 49–113 (1987)

[MW74] Marsden, J., Weinstein, A.: Reduction of symplectic manifolds with symmetry. Rep. Math. Phys. **5**(1), 121–130 (1974)

[Vit09] Vitagliano, L.: The Lagrangian-Hamiltonian formalism for higher order field theories. arXiv e-prints (2009). arXiv:0905.4580

[Vit10] Vitagliano, L.: The Hamilton-Jacobi formalism for higher order field theories. arXiv e-prints (2010). arXiv:1003.5236

[Vit11] Vitagliano, L.: Hamilton-Jacobi diffieties. arXiv e-prints (2011). arXiv:1104.0162

Chapter 9
Homotopical Algebra

One may think of homotopical algebra as a tool for computing and systematically studying obstructions to the resolution of (not necessarily linear) problems. Since most of the problems that occur in physics and mathematics carry obstructions (see the introduction of this book, and Sect. 10.1 for an intuitive account of the main examples that we will encounter), one needs tools to study these and give an elegant presentation of the physicists' ideas (who often invented some of these techniques for their own safety, as one can see in the example of the BRST formalism of Sect. 8.8).

In this book, we chose to base our presentation on the setting of doctrines and of higher categorical logic (see Sect. 2.1), generalizing Lawvere and Ehresmann's categorical approach to mathematics. This setting carries in itself the methods of obstruction theory, because all the obstructions we want to study may be defined, at some point, as some kinds of (higher) Kan extensions of models of theories in some given doctrine.

In this chapter, we want to present the main tools that allow the explicit computation of these abstractly defined obstructions, by giving a concrete flavor of the idea of general obstruction theory. Along the way, we will also explain the various approaches that allow us to base higher category theory on usual category theory, described in a set theoretical fashion. These results are useful to have a precise definition of what higher categories are.

In the next chapter, we will also explain how this concrete approach to obstruction theory may be used to give a geometrical presentation of part of it, called homotopical geometry. The use of these geometrical methods gives fruitful intuitions to better understand the physicists' constructions, particularly for gauge theories.

An exhaustive treatment of homotopical algebraic methods would require a book with a number of five hundred page volumes strictly bigger than six (see Hovey's "model categories" [Hov99], Toen-Vezzosi's "Homotopical algebraic geometry I and II" [TV02, TV08] and Lurie's "Higher topos theory" [Lur09d], "Higher algebra" [Lur09c] and "Derived algebraic geometry" [Lur09a] for an idea of the scope). We refer to the above literature (and other introductory references along the way) for the interested reader.

F. Paugam, *Towards the Mathematics of Quantum Field Theory*,
Ergebnisse der Mathematik und ihrer Grenzgebiete. 3. Folge / A Series of Modern Surveys in Mathematics 59, DOI 10.1007/978-3-319-04564-1_9,

9.1 Localizations and Derived Functors

The main objects of homotopical algebra are localizations of categories. Given a category $\mathcal{C}$ and a multiplicative class of morphisms, the localization $\mathcal{C}[W^{-1}]$ is the universal category in which all morphisms of W become isomorphisms. This universal property should be taken in a 2-categorical sense to be sufficiently flexible. Recall that the symmetric monoidal category

$$(\text{CAT}, \times)$$

of categories with their cartesian product have an internal homomorphism object given for $\mathcal{C}$, $\mathcal{D}$ two categories by the category $\underline{\text{Mor}}(\mathcal{C}, \mathcal{D})$ of functors $F : \mathcal{C} \to \mathcal{D}$, with morphisms given by natural transformations.

We first give a definition of the localization of categories adapted to n-categories and ${}^{\infty}n$-categories. In applications, we will often use ${}^{\infty}1$-localizations of usual categories, so that this level of generality is necessary.

Definition 9.1.1 Let $\mathcal{C}$ be an n-category and W be a multiplicative class of morphisms in $\mathcal{C}$. If $\mathcal{D}$ is an n-category, we denote by $\underline{\text{Mor}}_{W^{-1}}(\mathcal{C}, \mathcal{D})$ the n-category of functors $F : \mathcal{C} \to \mathcal{D}$ that send morphisms in W to isomorphisms. A *localization* of $\mathcal{C}$ with respect to W is the datum of a pair $(\mathcal{C}[W^{-1}], L)$ composed of an n-category $\mathcal{C}[W^{-1}]$ and a functor $L : \mathcal{C} \to \mathcal{C}[W^{-1}]$ such that for every n-category $\mathcal{D}$, the functor

$$\begin{aligned} \underline{\text{Mor}}\big(\mathcal{C}\big[W^{-1}\big], \mathcal{D}\big) &\to \underline{\text{Mor}}_{W^{-1}}(\mathcal{C}, \mathcal{D}) \\ G &\mapsto G \circ L \end{aligned}$$

is an equivalence of n-categories.

We now describe the existence result in the case of usual categories. Higher categories will be treated later.

Theorem 9.1.1 *Localizations of categories exist and are unique up to a unique equivalence of categories.*

The proof of the above theorem, which may be found in [GZ67], is "easy" and constructive: objects of the localized category are given by sequences

$$X_1 \xleftarrow{\sim} X_2 \longrightarrow \cdots \xleftarrow{\sim} X_{n-1} \longrightarrow X_n$$

where left arrows are in the class W. Two such strings are equivalent if, after forcing them to the same length, they can be connected by a commutative diagram with

vertical arrows in W of the form

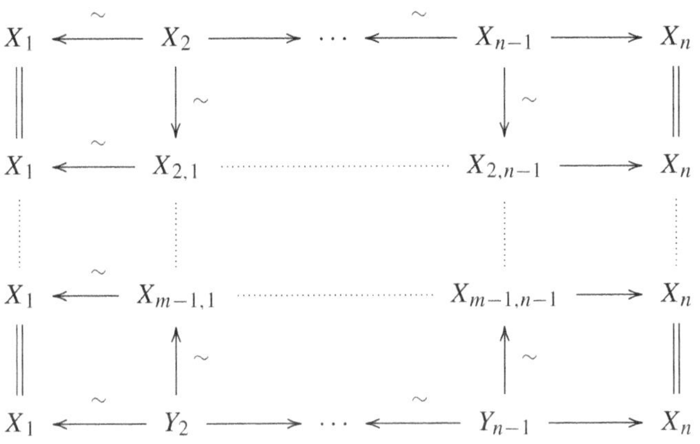

From the construction, it is not clear that the class $\mathrm{Hom}_{\mathcal{C}[W^{-1}]}(X, Y)$ of morphisms between two given objects of the localized category form a set. This is not so problematic because we allowed ourselves to work with categories where homomorphisms form classes. However, a more critical problem with the above construction is that one can't compute anything with it because its morphisms are not given in a concrete and explicit way.

We now define left and right derived functors, which are particular examples of left and right Kan extensions (see Definition 2.1.10). We refer to [KM08] for a complete treatment of derived functors using 2-categorical methods.

Definition 9.1.2 Let $\mathcal{C}$ be a category, W be a multiplicative set of morphisms in $\mathcal{C}$ and $L : \mathcal{C} \to \mathcal{C}[W^{-1}]$ be the corresponding localization. Let $F : \mathcal{C} \to \mathcal{D}$ be a functor. A *right derived functor for* F is a pair $(\mathbb{R}F, a)$ composed of a functor $\mathbb{R}F : \mathcal{C}[W^{-1}] \to \mathcal{D}$ and a natural transformation $a : \mathbb{R}F \circ L \Rightarrow F$ giving a diagram

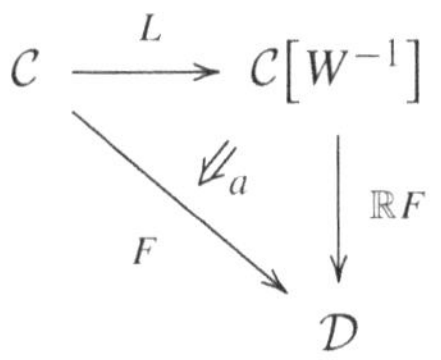

such that for every pair (G, b) composed of a functor $G : \mathcal{C}[W^{-1}] \to \mathcal{D}$ and of a natural transformation $b : G \circ L \Rightarrow F$, there exists a unique natural transformation

$c : G \Rightarrow \mathbb{R}F$ such that the equality $b = a \circ c$, illustrated by the following diagrams

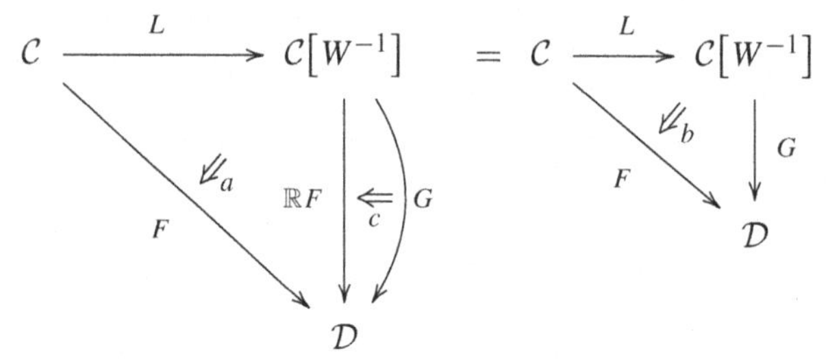

is true. A *total right derived functor* is a pair $(\mathbb{R}F, a)$ such that for all functors $H : N \to N'$, the composed pair $(H \circ \mathbb{R}F, H * a)$ is a right derived functor for $H \circ F$. One defines a *left derived functor for* F by the same property with the reversed direction for 2-arrows.

Note that Definition 2.1.10 also works without change in the ${}^{\infty}2$-category of a ${}^{\infty}1$-categories, giving a notion of ${}^{\infty}1$-derived functor. It may also be formulated in the ${}^{\infty}1$-category of ${}^{\infty}1$-categories, using that fact that functors between them also form an ${}^{\infty}1$-category (see Note 2.1.1 and Corollary 9.6.1).

As said above, one can't really compute with localized categories because their morphism sets are combinatorially very big (and sometimes not even sets). The model category setting gives a method of replacing objects of the localization with simpler objects called fibrant or cofibrant resolutions, so that morphisms in the localization are given by length two zigzags, and derived functors become explicitly computable. It is usually a tedious work to show that a given category satisfies the model category axioms, but in the examples of interest for physics, this work has already been done by topologists and algebraists, so that we can use them as black boxes to make explicit computations with physical theories.

9.2 Model Categories

The setting of model categories was first defined by Quillen [Qui67] to make homotopy theory functorial and symmetric, and to treat with unified methods:

- classical homotopy theory, which involves a model category structure on the category of topological spaces; and
- homological algebra (derived categories), which involves a model category structure on the category of complexes of modules over a ring.

For short presentations, we refer to Toen's various introductory articles on his webpage, and in particular the course [Toe02] and to Keller's notes [Kel06], 4.1. For more details, we use Hovey [Hov99] and Dwyer-Spalinski [DS95].

The aim of the theory of model categories is to give a workable notion of localization of a category C with respect to a given multiplicative class W of morphisms, whose elements will be denoted by arrows with tilde $\overset{\sim}{\longrightarrow}$.

The general definition of model categories can be found in [Hov99], 1.1.3 (see also Quillen's original book [Qui67], and [Cis10] for weakened axioms).

Definition 9.2.1 A *model category* is a complete and cocomplete category C (i.e., a category that has all small limits and colimits) together with the following data:

- three distinguished classes $(W, \mathrm{Fib}, \mathrm{Cof})$ of maps in C, (whose elements are respectively called *weak equivalences*, *fibrations* and *cofibrations*, and respectively denoted by $\overset{\sim}{\longrightarrow}$, $\twoheadrightarrow$ and $\rightarrowtail$);
- two functorial factorizations $f \mapsto (p, i')$, $f \mapsto (p', i)$ (i.e., $f = p \circ i'$ and $f = p' \circ i$, functorially in f),

subject to the following axioms:

1. (2 out of 3) If (f, g) are composable arrows (i.e. $f \circ g$ exists), then f, g and $f \circ g$ are in W if any two of them are.
2. (Left and right lifting properties) Calling maps in $W \cap \mathrm{Fib}$ (resp. $W \cap \mathrm{Cof}$) *trivial fibrations* (resp. *trivial cofibrations*), in each commutative square solid diagram of the following two types

   the dotted arrow exists so as to make the two triangles commutative.
3. (Retracts) If a morphism f is a retract of a morphism g, meaning that there is a commutative diagram

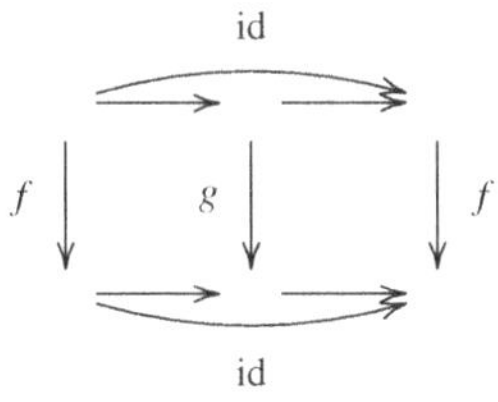

   and f is a weak equivalence (resp. a fibration, resp. a cofibration), then g is a weak equivalence (resp. a fibration, resp. a cofibration).
4. (Factorizations) For the functorial factorizations $f \mapsto (p, i')$, $f \mapsto (p', i)$, p is a fibration, i' a trivial cofibration, p' is a trivial fibration and i a cofibration.

The *homotopy category* of a model category C is defined as the localization

$$h(C) := C[W^{-1}].$$

Note that these axioms are self-dual, meaning that the opposite category is also a model category (with exchanged fibrations and cofibrations), and that the two classes Fib and Cof are determined by the datum of one of them and the class W of weak equivalences (by the lifting axiom). We now define the notion of homotopy of maps.

Definition 9.2.2 Let $X \in \mathcal{C}$. A *cylinder* for X is a factorization

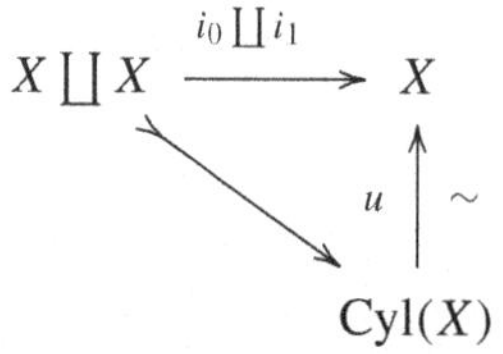

of the canonical map $X \coprod X \to X$ into a cofibration followed by a weak equivalence. A cylinder for X in the opposite category is called a *path object* or a *cocylinder*. If $f, g : X \to Y$ are maps, a *left homotopy* is a map $h : \mathrm{Cyl}(X) \to Y$ such that the following diagram commutes

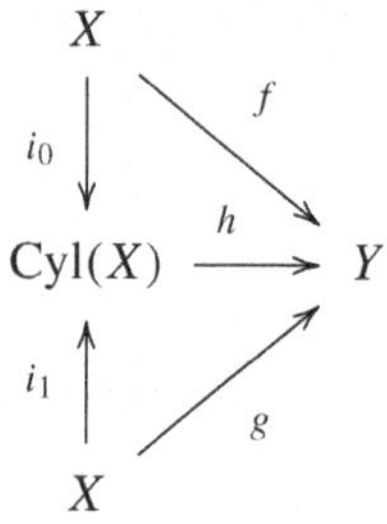

A *right homotopy* is a left homotopy in the opposite category. Left and right homotopies are denoted by $h : f \sim_l g$ and $h : f \sim_r g$.

The factorization axiom of model categories ensures the existence of functorial cylinders and cocylinders.

Example 9.2.1 In the category Top of (Hausdorff and compactly generated) topological spaces, one defines the classical cylinder of X by $\mathrm{Cyl}_c(X) = X \times [0, 1]$. It is then clear that the injections $i_0, i_1 : X \to \mathrm{Cyl}_c(X)$ are closed and the projection $u : \mathrm{Cyl}_c(X) \to X$ is a homotopy equivalence because $[0, 1]$ can be contracted to a point. One can use this classical cylinder to define in an ad-hoc way the notion of (left) homotopy of maps between topological spaces. Two maps $f, g : X \to Y$ are then called homotopic if there is a map $h : X \times [0, 1] \to Y$ such that $h(0) = f$ and $h(1) = g$. This is the usual notion of homotopy. One then defines the homotopy groups of a pointed topological space $(X, *)$ as the homotopy classes of pointed continuous maps

$$\pi_i(X) := \mathrm{Hom}_{\mathrm{TOP}}\big((S^i, *), (X, *)\big)/\sim_l$$

from the pointed spheres to $(X, *)$. These can be equipped with a group structure by identifying S^n with the quotient of the hypercube $[0, 1]^n$ by the relation that identifies its boundary with a point and setting the composition to be the map induced by

$$(f * g)(x_1, \dots, x_n)$$
$$= \begin{cases} f(2x_1, x_2, \dots, x_n) & \text{for } x_1 \in [0, 1/2] \text{ and} \\ (f * g)(x_1, x_2, \dots, x_n) = g(2x_1 - 1, x_2, \dots, x_n) & \text{for } x_1 \in [1/2, 1] \end{cases}$$

on the quotient space. The main example of a model category is given by the category Top of topological spaces, where:

- weak equivalences are given by weak homotopy equivalences (i.e., maps inducing isomorphisms on all the π_i's, for any choice of base-point);
- fibrations (also called Serre fibrations) $p : X \to Y$ have the right lifting property with respect to the maps $D^n \times \{0\} \to D^n \times I$ for all discs D^n:

$$\begin{array}{ccc} D^n & \longrightarrow & X \\ {\scriptstyle i}\Big\downarrow{\scriptstyle \sim} & \nearrow & \Big\downarrow{\scriptstyle p} \\ D^n \times I & \longrightarrow & Y \end{array}$$

- cofibrations $i : X \to Y$ have the left lifting property with respect to fibrations (these can also be described as retracts of transfinite compositions of pushouts of the generating cofibrations $S^{n-1} \to D^n$).

In this model category, the classical cylinder $\mathrm{Cyl}_c(X)$ is a cylinder for X if X is cofibrant. The mapping space $\mathrm{Cocyl}(X) := \underline{\mathrm{Hom}}([0, 1], X)$, equipped with the topology generated by the subsets $V(K, U)$ indexed by opens U in X and compacts K in $[0, 1]$, and defined by

$$V(K, U) = \big\{f : [0, 1] \to X,\ f(K) \subset U\big\},$$

is a path (or cocylinder) for X if X is fibrant.

Recall that for $\mathcal{C}$ a model category, the associated homotopy category is the localization $h(\mathcal{C}) = \mathcal{C}[W^{-1}]$. It thus doesn't depend on fibrations and cofibrations. Since $\mathcal{C}$ is complete and cocomplete, it contains an initial object $\emptyset$ and a final object $*$. An object x is called fibrant (resp. cofibrant) if the natural map $x \to *$ (resp. $\emptyset \to x$) is a fibration (resp. a cofibration). The existence of functorial factorizations applied to these maps imply the existence of two functors $R, Q : \mathcal{C} \to \mathcal{C}$ together with natural transformations $\mathrm{id} \to R, Q$ that are object-wise natural equivalences. The functor R is called the fibrant replacement functor and Q the cofibrant replacement. They are valued respectively in the subcategory $\mathcal{C}_f$ and $\mathcal{C}_c$ of fibrant and cofibrant objects. Denote by $\mathcal{C}_{cf}$ the subcategory of objects that are fibrant and cofibrant.

The main structure theorem of model category theory is the following (see [Hov99], Theorem 1.2.10).

Theorem 9.2.1 *Let $\mathcal{C}$ be a model category.*

1. (*Simple description of objects*) *The natural inclusion $\mathcal{C}_{cf} \to \mathcal{C}$ induces an equivalence of categories*
$$h(\mathcal{C}_{cf}) \to h(\mathcal{C})$$
whose quasi-inverse is induced by RQ.
2. (*Simple description of morphisms*) *The homotopy category $h(\mathcal{C}_{cf})$ is the quotient category of $\mathcal{C}_{cf}$ by the (left or right) homotopy relation $\sim$, i.e.,*
$$\mathrm{Hom}_{h(\mathcal{C}_{cf})}(x, y) \cong \mathrm{Hom}_{\mathcal{C}_{cf}}(x, y)/\sim.$$
3. *There is a natural isomorphism*
$$\mathrm{Hom}_{h(\mathcal{C})}(x, y) \cong \mathrm{Hom}_{\mathcal{C}}(QX, RY)/\sim.$$

This gives a nice description of maps in the homotopy category $h(\mathcal{C})$ as homotopy classes of maps in $\mathcal{C}$:

$$\begin{aligned}[x, y] := \mathrm{Hom}_{h(\mathcal{C})}(x, y) &\cong [RQx, RQy] \cong \mathrm{Hom}_{\mathcal{C}}(RQx, RQy)/\sim \\ &\cong \mathrm{Hom}_{\mathcal{C}}(Qx, Ry)/\sim.\end{aligned}$$

We now introduce the notion of homotopy limits (see Toen's lecture notes [Toe02] for an introduction and Cisinski [Cis10] for an optimal formalization). These give tools to describe limits (defined in Sect. 2.1) in ∞-categories, by restriction to the case of higher groupoids through Yoneda's lemma.

Definition 9.2.3 Let I be a small category and $\mathcal{C}$ be a category, equipped with a class W of morphisms. Let $\mathcal{C}^I := \underline{\mathrm{Mor}}_{\mathrm{CAT}}(I, \mathcal{C})$ be the category of I-diagrams in $\mathcal{C}$ and W_I be the class of diagrams that are objectwise in W. Let $L : \mathcal{C} \to \mathcal{C}[W^{-1}]$ and $L_I : \mathcal{C}^I \to \mathcal{C}^I[W_I^{-1}]$ be the localizations and $c : (\mathcal{C}, W) \to (\mathcal{C}^I, W_I)$ be the constant diagram functor. A *homotopy limit functor* (resp. *homotopy colimit functor*) is a right (resp. left) adjoint to the localized constant functor
$$\bar{c} : \mathcal{C}[W^{-1}] \to \mathcal{C}^I[W_I^{-1}].$$

The main theorem on classical homotopy limits can be found in the book [DHKS04] or in [Cis10].

Theorem 9.2.2 *If $(\mathcal{C}, W)$ is a model category, then small homotopy limits exist in $\mathcal{C}$.*

9.3 Quillen Functors and Derived Functors

One then wants to study functors between homotopy categories induced by morphisms of model categories. These are called Quillen functors. We refer to Dwyer-Spalinski [DS95] for the following approach to derived functors via their 2-universal property, which is an improvement on Hovey's presentation in [Hov99].

Definition 9.3.1 Let $\mathcal{C}$ and $\mathcal{D}$ be two model categories, and

$$F : \mathcal{C} \leftrightarrows \mathcal{D} : G$$

be a pair of adjoint functors.

1. If F preserves cofibrations and G preserves fibrations, one calls the pair (F, G) a *Quillen adjunction*, F a *left Quillen functor* and G a *right Quillen functor*. If (F, G) is a Quillen adjunction, the natural functor

$$\mathcal{C}_c \xrightarrow{F} \mathcal{D} \xrightarrow{L_{\mathcal{D}}} h(\mathcal{D})$$

$$\left(\text{resp. } \mathcal{D}_f \xrightarrow{G} \mathcal{C} \xrightarrow{L_{\mathcal{C}}} h(\mathcal{C})\right)$$

sends local equivalences in $\mathcal{C}_c$ (resp. local equivalences in $\mathcal{D}_f$) to isomorphisms, and defines an adjoint pair

$$\mathbb{L}F : h(\mathcal{C}) \cong h(\mathcal{C}_c) \leftrightarrows h(\mathcal{D}_f) \cong h(\mathcal{D}) : \mathbb{R}G.$$

This adjoint pair between $h(\mathcal{C})$ and $h(\mathcal{D})$ can be explicitly computed by setting

$$\mathbb{L}F = L_{\mathcal{D}} \circ F \circ Q \quad \text{and} \quad \mathbb{L}G = L_{\mathcal{C}} \circ G \circ R.$$

2. Suppose in addition that for each cofibrant object A of $\mathcal{C}$ and fibrant object X of $\mathcal{D}$, a map $f : A \to G(X)$ is a weak equivalence in $\mathcal{C}$ if and only if its adjoint $f^\sharp : F(A) \to X$ is a weak equivalence in $\mathcal{D}$, then the pair (F, G) is called a *Quillen equivalence* and $\mathbb{L}F$ and $\mathbb{R}G$ are inverse equivalences of categories.

The main property of derived functors is that they are compatible with composition and identity. This follows easily from the above definition and the 2-universal property of localizations.

Theorem 9.3.1 *For every model category $\mathcal{C}$, there is a natural isomorphism $\epsilon : \mathbb{L}(\mathrm{id}_{\mathcal{C}}) \to \mathrm{id}_{h(\mathcal{C})}$. For every pair of left Quillen functors $F : \mathcal{C} \to \mathcal{D}$ and $G : \mathcal{D} \to \mathcal{E}$ between model categories, there is a natural isomorphism $m_{GF} : \mathbb{L}G \circ \mathbb{L}F \to \mathbb{L}(G \circ F)$.*

The above compatibility isomorphism moreover satisfies additional associativity and coherence conditions that can be found in [Hov99], Theorem 1.3.7. One obtains the right derived version of this theorem by passing to the opposite category.

9.4 Bousfield's Homotopical Localizations

An important operation in homotopical sheaf theory is the notion of the Bousfield localization of a model category, which solves the following problem: given a model category $\mathcal{C}$ and a subset S of maps in $\mathcal{C}$, find another model category $L_S\mathcal{C}$ in which the maps in S are weak equivalences. The localization functor must be a left Quillen functor

$$f : \mathcal{C} \to L_S\mathcal{C}$$

that is universal with respect to left Quillen functors $f : \mathcal{C} \to \mathcal{D}$ that send images of morphisms in S in $h(\mathcal{C})$ to isomorphisms in $h(\mathcal{D})$.

Let $\mathcal{C}[W^{-1}]$ be the ∞-localization of $\mathcal{C}$ (also called the Hammock localization, to be described in Theorem 9.6.2). This is an ${}^{\infty}1$-category, and in particular, its morphism spaces are ∞-groupoids (which we will model on simplicial sets in Sect. 9.5).

An object x of $\mathcal{C}$ is called S-local if it sees maps in S as weak equivalences, i.e., if it is fibrant and for any $y \to y'$ in S, the induced map

$$\underline{\mathrm{Mor}}_{\mathcal{C}[W^{-1}]}(y', x) \to \underline{\mathrm{Mor}}_{\mathcal{C}[W^{-1}]}(y, x)$$

is an equivalence of ∞-groupoids. A map $f : x \to x'$ is called an S-local equivalence if it is seen as an equivalence by any S-local map, i.e., if for any S-local object y in $\mathcal{C}$, the induced map

$$\underline{\mathrm{Mor}}_{\mathcal{C}}(x', y) \to \underline{\mathrm{Mor}}_{\mathcal{C}}(x, y)$$

is an equivalence of ∞-groupoids.

Theorem 9.4.1 *Let $\mathcal{C}$ be a standard model category. Then the following classes of maps in $\mathcal{C}$*

- $W_S :=$ *S-local equivalence*;
- $\mathrm{Cof}_S :=$ *cofibrations in $\mathcal{C}$*,

are part of a model structure on $\mathcal{C}$, denoted by $L_S\mathcal{C}$ and the identity functor induces a left Quillen functor

$$\mathcal{C} \to L_S\mathcal{C}$$

that is universal among left Quillen functors whose derived functor send morphisms in S to isomorphisms.

9.5 Simplicial Sets and Higher Groupoids

We refer to the book [GJ99] for an introduction to simplicial sets.

The notion of a simplicial set allows us to give a purely combinatorial description of the homotopy category of the category TOP of topological spaces. It is also a fundamental tool in higher category theory, because it is the universally accepted model for the theory of ∞-groupoids.

Let Δ be the category whose objects are finite ordered sets $[n] = [0, \ldots, n-1]$ and whose morphisms are nondecreasing maps.

Definition 9.5.1 A *simplicial set* is a contravariant functor

$$X : \Delta^{op} \to \text{SETS}.$$

We denote by SSETS the category of simplicial sets.

To give examples of simplicial sets, one can use the following description of the category Δ by generators and relations.

Proposition 9.5.1 *The morphisms*

$$\begin{aligned} d^i &: [n-1] \to [n] \quad 0 \leq i \leq n \text{ (cofaces)} \\ s^j &: [n+1] \to [n] \quad 0 \leq j \leq n \text{ (codegeneracies)} \end{aligned}$$

given by

$$d^i([0, \ldots, n-1]) = [0, \ldots, i-1, i+1, \ldots, n]$$

and

$$s^j([0, \ldots, n+1]) = [0, \ldots, j, j, \ldots, n]$$

satisfy the so-called cosimplicial identities

$$\begin{cases} d^j d^i = d^i d^{j-1} & \text{if } i < j \\ s^j d^i = d^i s^{j-1} & \text{if } i < j \\ s^j d^j = 1 = s^j d^{j+1} & \\ s^j d^i = d^{i-1} s^j & \text{if } i > j+1 \\ s^j s^i = s^i s^{j+1} & \text{if } i \leq j \end{cases}$$

and give a set of generators and relations for the category Δ.

Define the n simplex to be $\Delta^n := \mathrm{Hom}(-, [n])$. Yoneda's lemma implies that for every simplicial set X, one has

$$\mathrm{Hom}(\Delta^n, X) \cong X_n.$$

The boundary $\partial\Delta^n$ is defined by

$$(\partial\Delta^n)_m := \{f : [m] \to [n], \mathrm{im}(f) \neq [n]\}$$

and the k-th horn $\Lambda_k^n \subset \partial\Delta^n$ by

$$(\Lambda_k^n)_m := \{f : [m] \to [n],\ k \notin \mathrm{im}(f)\}.$$

The category SSETS has all limits and colimits (defined component-wise) and also internal homomorphisms defined by

$$\underline{\mathrm{Hom}}(X, Y) : [n] \mapsto \mathrm{Hom}_{\Delta_n}(X \times \Delta_n, Y \times \Delta_n) = \mathrm{Hom}(X \times \Delta^n, Y).$$

The geometric realization $|\Delta^n|$ is defined to be

$$|\Delta^n| := \left\{(t_0, \dots, t_n) \in [0, 1]^{n+1}, \sum t_i = 1\right\}.$$

It defines a covariant functor $|\cdot| : \Delta \to$ TOP with maps $\theta_* : |\Delta^n| \to |\Delta^m|$ for $\theta : [n] \to [m]$ given by

$$\theta_*(t_0, \dots, t_m) = (s_0, \dots, s_n)$$

where

$$s_i = \begin{cases} 0 & \text{if } \theta^{-1}(i) = \emptyset \\ \sum_{j \in \theta^{-1}(i)} t_j & \text{if } \theta^{-1}(i) \neq \emptyset. \end{cases}$$

The geometric realization of a general simplicial set X is the colimit

$$|X| = \operatorname*{colim}_{\Delta^n \to X} |\Delta^n|$$

indexed by the category of maps $\Delta^n \to X$ for varying n. The ∞-groupoid (also called the singular simplex) of a given topological space Y is the simplicial set

$$\Pi_\infty(Y) : [n] \mapsto \mathrm{Hom}_{\mathrm{TOP}}(|\Delta^n|, Y).$$

The geometric realization and ∞-groupoid functors are adjoint, meaning that

$$\mathrm{Hom}_{\mathrm{TOP}}(|X|, Y) \cong \mathrm{Hom}_{\mathrm{SSETS}}(X, \Pi_\infty(Y)).$$

Definition 9.5.2 The *simplicial cylinder* of a given simplicial set is defined as $\mathrm{Cyl}(X) := X \times \Delta^1$. Let $f, g : X \to Y$ be morphisms of simplicial sets. A *homo-*

topy between f and g is a factorization

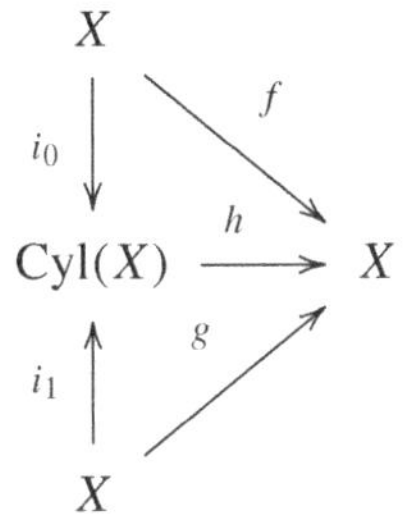

Fibrations of simplicial sets are defined as maps that have the right lifting property with respect to all the standard inclusions $\Lambda^n_k \subset \Delta^n$, $n > 0$.

Definition 9.5.3 A morphism $p : X \to Y$ of simplicial sets is called a *fibration* (or a *Kan fibration*) if in every commutative square

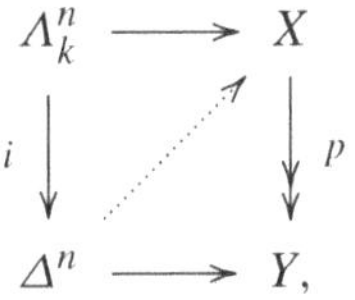

the dotted arrow exists so as to make the two triangles commutative. A simplicial set X is called *fibrant* (or a *Kan complex*) if the projection map $X \to \{*\}$ is a fibration.

Let X be a simplicial set. Its set of connected components is defined as the quotient

$$\pi_0(X) := X_0/\sim$$

by the equivalence relation $\sim$ generated by the relation $\sim_1$, called "being connected by a path", i.e., $x_0 \sim_1 x_1$ if there exists a $\gamma : \Delta^1 \to X$ such that $\gamma(0) = x_0$ and $\gamma(1) = x_1$.

Lemma 9.5.1 *The functor* $\pi_0 : \text{SSETS} \to \text{SETS}$ *commutes with direct products.*

Proof This follows from the fact that $\pi_0(X)$ identifies naturally with the set of connected components $\pi_0(|X|)$ of the geometric realization, and geometric realization commutes with products. □

The above lemma implies that simplicial sets may be equipped with a new category structure SSETS^{π_0}, whose morphisms are given by elements in $\pi_0(\underline{\text{Hom}}(X, Y))$, and whose composition is induced by the internal composition map

$$\underline{\text{Hom}}(X, Y) \times \underline{\text{Hom}}(Y, Z) \to \underline{\text{Hom}}(X, Z).$$

Definition 9.5.4 A morphism $f : X \to Y$ of simplicial sets is called a *weak equivalence* if for all fibrant objects Z, the natural map

$$\pi_0\big(\underline{\mathrm{Hom}}(f, Z)\big) : \pi_0\big(\underline{\mathrm{Hom}}(Y, Z)\big) \to \pi_0\big(\underline{\mathrm{Hom}}(X, Z)\big)$$

is bijective.

The following theorem is due to Quillen [Qui67].

Theorem 9.5.1 *The category of simplicial sets is equipped with a model category structure such that*:

1. *weak equivalences are weak equivalences of simplicial sets*;
2. *fibrations are Kan fibrations*;
3. *cofibrations are injections.*

Definition 9.5.5 The above defined model structure is called the model structure of higher groupoids, and is denoted by ${}^{\infty}\mathrm{GRPD}$.

Let X be a fibrant simplicial set and $x : \Delta^0 \to X$ be a base point of X. For $n \geq 1$, define the homotopy group $\pi_n(X, x)$ as the set of homotopy classes of maps $f : \Delta^n \to X$ that fit into a diagram

$$\begin{array}{ccc} \Delta^n & \xrightarrow{f} & X \\ \uparrow & & \uparrow \\ \partial\Delta^n & \longrightarrow & \Delta^0 \end{array}$$

The geometric realization of such maps correspond to maps

$$f : S^n \to |X|.$$

Theorem 9.5.2 *Weak equivalences of simplicial sets may be defined as morphisms that induce isomorphisms on homotopy groups, for any choice of base point* $x : \Delta^0 \to X$.

Theorem 9.5.3 *The geometric realization and the* ∞*-groupoid functors induce a Quillen adjunction*

$$|-| : {}^{\infty}\mathrm{GRPD} \leftrightarrows \mathrm{TOP} : \Pi_\infty$$

whose derived version is an equivalence of homotopy categories

$$\mathbb{L}|-| : h\big({}^{\infty}\mathrm{GRPD}\big) \overset{\sim}{\leftrightarrows} h(\mathrm{TOP}) : \mathbb{R}\Pi_\infty.$$

9.6 ∞-Categories

In this section, we will often use the basic notions of higher category theory described in Sect. 2.1. In particular, we suppose known the abstract notions of Kan extensions and limits given in Definitions 2.1.10 and 2.1.11, with the adaptation given in Remark 2.1.1.

The simplicial approach to higher category theory finds its roots and expands on the Dwyer-Kan simplicial localization methods [DK80]. Recall from Sect. 9.5 that there is a natural model structure $^{\infty}\mathrm{GRPD}$ on the category of simplicial sets whose equivalences are weak homotopy equivalences. We now describe another model structure $^{\infty}\mathrm{CAT}$ whose equivalences are categorical equivalences. This leads to Joyal's quasi-categories. This gives a nice model for a theory of $^{\infty}1$-categories (see Sect. 9.10), that we will simply call ∞-categories. We refer to [Joy11, Lur09d] and [Lur09c] for nice accounts of these structures and of their use in geometry and algebra. These are the simplest generalizations of nerves of categories that also allow the encoding of higher groupoids.

In this section, we give an elementary presentation of ∞-categories, which is easier to work with than the abstract one, whose definition will be sketched in Sect. 9.10. It is useful to present both approaches because the less elementary method of loc. cit. makes the generalization of usual categorical concepts more simple and natural than the present one.

Definition 9.6.1 Let $\mathcal{C}$ be a category. The *nerve* of $\mathcal{C}$ is the simplicial set $N(\mathcal{C})$ whose n-simplices are

$$N(\mathcal{C})_n := \mathrm{Hom}_{\mathrm{CAT}}\big([n], \mathcal{C}\big),$$

where $[n]$ is the linearly ordered set $\{0, \dots, n\}$. More concretely:

1. $N(\mathcal{C})_0$ is the set of objects of $\mathcal{C}$;
2. $N(\mathcal{C})_1$ is the set of morphisms in $\mathcal{C}$; and
3. $N(\mathcal{C})_n$ is the set of families

$$x_0 \xrightarrow{f_1} x_1 \xrightarrow{f_2} \cdots \xrightarrow{f_n} x_n$$

of composable arrows.

To define the simplicial structure of $N(\mathcal{C})$, one only needs to define maps

$$d_i : N(\mathcal{C})_n \to N(\mathcal{C})_{n-1}, \quad \text{for } 0 \leq i \leq n \text{ (faces)}$$
$$s_j : N(\mathcal{C})_n \to N(\mathcal{C})_{n+1}, \quad \text{for } 0 \leq j \leq n \text{ (degeneracies)}$$

satisfying the simplicial identities, described in Sect. 9.5. The faces d_i are given by composition

$$x_0 \xrightarrow{f_1} \cdots \xrightarrow{f_{i-1}} x_{i-1} \xrightarrow{f_{i+1}\circ f_i} x_{i+1} \xrightarrow{f_{i+2}} \cdots \xrightarrow{f_n} x_n$$

of two consecutive arrows and the degeneracies s_i by insertion of identities

$$x_0 \xrightarrow{f_1} \cdots \xrightarrow{f_i} x_i \xrightarrow{\mathrm{id}_{x_i}} x_i \xrightarrow{f_{i+1}} \cdots \xrightarrow{f_n} x_n.$$

Let $a_i : [1] \to [k]$ be the morphisms in the simplicial category Δ given by $a^i(0) = i$ and $a^i(1) = i + 1$ for $0 \leq i \leq k - 1$ and let a_i be the corresponding maps in Δ^{op}. Denote by X the simplicial set $N(\mathcal{C})(a_i)$. The map

$$X(a_i) = X_k \to X_1$$

simply sends a family of k composable arrows to the i-th one. Their fiber products give the so-called *Segal maps*

$$\varphi_k : X_k \to X_1 \underset{X_0}{\times} \cdots \underset{X_0}{\times} X_1,$$

and by definition of X_k, these maps are bijections.

Proposition 9.6.1 *Let* CAT *be the category of small categories. The nerve functor*

$$\begin{aligned} N : \mathrm{CAT} &\to \mathrm{SSETS} \\ \mathcal{C} &\mapsto [n] \mapsto \mathrm{Hom}_{\mathrm{CAT}}\big([n], \mathcal{C}\big) \end{aligned}$$

is a fully faithful embedding, with essential image the subcategory of simplicial sets whose Segal maps are bijections. Moreover, it has a left adjoint τ_1, called the fundamental category *functor.*

Proof The full faithfulness is clear from the above description of the nerve. The existence of an adjoint to the nerve functor follows formally from the completeness of the categories in play (see [GZ67]). We only describe it explicitly. Let X be a simplicial set. Two elements f and g of X_1 are called composable if there exists an $h \in X_2$ such that $h_{[0,1]} = f$ and $h_{[1,2]} = g$. Their composition is then given by $h_{[0,2]}$. We put on X_1 the smallest equivalence relation generated under this composition that is stable by composition. We define $\tau_1(X)$ as the category whose objects are elements in X_0, whose morphisms are families $(f_1, \ldots, f_n)$ of elements in X_1 that are composable, quotiented by the composition equivalence relation. □

Definition 9.6.2 A *quasi-fibration* of simplicial sets is a morphism $p : X \to Y$ such that every commutative square

$$\begin{array}{ccc} \Lambda^n_k & \longrightarrow & X \\ {\scriptstyle i}\downarrow & \nearrow & \downarrow{\scriptstyle p} \\ \Delta^n & \longrightarrow & Y \end{array}$$

with $0 < k < n$ can be completed by a dotted arrow so as to make the two triangles commutative. A *categorical equivalence* of simplicial sets is a morphism $f : X \to Y$ that induces an equivalence of categories

$$\tau_1(f) : \tau_1(X) \to \tau_1(Y).$$

A *quasi-category* is a simplicial set X such that the morphism $X \to \{*\}$ is a quasi-fibration. More precisely, it is a simplicial set such that for $0 < k < n$, there exists a dotted arrow rendering the diagram

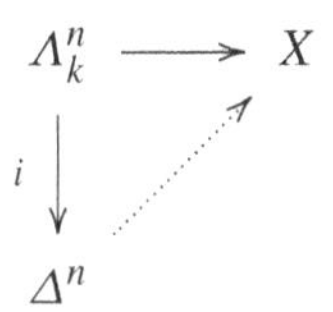

commutative. An ∞-*groupoid* is a quasi-category such that the dotted arrow in the above diagram exists for $0 \leq k \leq n$.

Note that an ∞-groupoid is exactly the same as a fibrant simplicial set for the model structure $^{\infty}$GRPD of Sect. 9.6 (i.e., a homotopy type), and a quasi-category such that the dotted arrow rendering the diagram

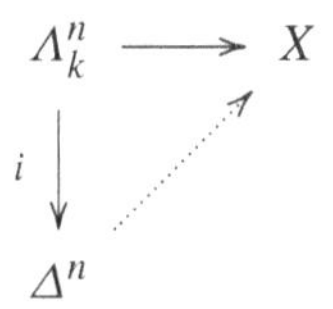

commutative is unique is simply a nerve.

We state a useful characterization of quasi-categories, whose proof can be found in [Joy11].

Lemma 9.6.1 *A simplicial set $\mathcal{C}$ is a quasi-category if and only if the projection*

$$\underline{\mathrm{Hom}}_{\mathrm{SSETS}}(\Delta^2, \mathcal{C}) \longrightarrow \underline{\mathrm{Hom}}_{\mathrm{SSETS}}(\Lambda^2_1, \mathcal{C})$$

is a trivial fibration.

We now give an intrinsic formulation of the fact that the collection of ∞-categories form an ∞-category.

Theorem 9.6.1 *There exists a model structure on* SSETS, *denoted by* $^{\infty}$CAT, *whose fibrations are quasi-fibrations, whose cofibrations are monomorphisms, and whose weak equivalences are categorical equivalences. Its fibrant objects are quasi-categories.*

Proof We refer to [Joy11], and [Lur09d] for a complete proof. □

Theorem 9.6.1 may be improved by saying that the collection of ∞-categories form an ${}^{\infty}2$-category (with non-invertible 2-morphisms). We will explain this shortly in Sect. 9.10. This is mandatory to our treatment of Kan extensions, limits, and colimits in Sect. 2.1, exactly as in the classical categorical setting, because these definitions use ${}^{\infty}2$-categorical universal properties. However, one may avoid passing to this 2-dimensional setting by using only the above results. This viewpoint was systematically treated by Joyal [Joy11] and Lurie [Lur09d].

Definition 9.6.3 An ∞-category is an object of the homotopy category of ${}^{\infty}\mathrm{CAT}$.

The following result shows that quasi-categories themselves form a quasi-category.

Proposition 9.6.2 *If X is a simplicial set and $\mathcal{C}$ is a quasi-category, the simplicial set*

$$\underline{\mathrm{Mor}}_{{}^{\infty}\mathrm{CAT}}(X, \mathcal{C}) := \underline{\mathrm{Hom}}_{\mathrm{SSETS}}(X, \mathcal{C})$$

is a quasi-category. Moreover, if $F : X \to Y$ or $G : \mathcal{C} \to \mathcal{D}$ are categorical equivalences, then

$$\underline{\mathrm{Mor}}(F, \mathcal{C}) : \underline{\mathrm{Mor}}_{{}^{\infty}\mathrm{CAT}}(Y, \mathcal{C}) \to \underline{\mathrm{Mor}}_{{}^{\infty}\mathrm{CAT}}(X, \mathcal{C})$$

and

$$\underline{\mathrm{Mor}}(X, G) : \underline{\mathrm{Mor}}_{{}^{\infty}\mathrm{CAT}}(X, \mathcal{C}) \to \underline{\mathrm{Mor}}_{{}^{\infty}\mathrm{CAT}}(X, \mathcal{D})$$

are categorical equivalences.

Corollary 9.6.1 *The model category ${}^{\infty}\mathrm{CAT}$ is cartesian closed. The collection of ∞-categories form an ∞-category ${}^{\infty}\mathrm{CAT}$.*

Proof This is proved in [Joy11] and [Lur09d]. The quasi-category ${}^{\infty}\mathrm{CAT}$ is naturally associated, via a simplicial nerve construction, to the simplicial category whose simplicial sets of morphisms are given by invertible ∞-functors, i.e., by a maximal Kan complex in the quasi-category $\underline{\mathrm{Mor}}_{{}^{\infty}\mathrm{CAT}}(\mathcal{C}, \mathcal{D})$. □

Definition 9.6.4 If $\mathcal{C}$ is a quasi-category representing an ∞-category, one calls

1. elements in $\mathcal{C}_0$ its *objects*,
2. elements in $\mathcal{C}_1$ its *morphisms*,
3. elements in $\mathcal{C}_2$ its *compositions*.

If $\mathcal{C}$ is a quasi-category, one may interpret elements in $\mathcal{C}_2$ as triangular 2-morphisms

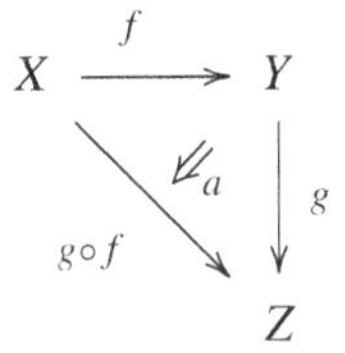

in the corresponding ∞-category, encoding compositions. Note that the arrow $g \circ f$ is only unique up to equivalence, and that the 2-morphism a is invertible only up to a 3-morphism, which may be drawn as a tetrahedron.

The following theorem is completely central and at the origin of the simplicial approach to higher category theory. It is essentially due to Dwyer and Kan [DK80].

Theorem 9.6.2 *Let $\mathcal{C}$ be an ∞-category and W be a class of morphisms in $\mathcal{C}$. There exists a localized ∞-category, i.e., a functor $L : \mathcal{C} \to \mathcal{C}[W^{-1}]$ such that for every ∞-category $\mathcal{D}$, the functor*

$$\underline{\mathrm{Mor}}(\mathcal{C}[W^{-1}], \mathcal{D}) \to \underline{\mathrm{Mor}}_{W^{-1}}(\mathcal{C}, \mathcal{D})$$
$$G \mapsto G \circ L$$

is an equivalence of ∞-categories.

Theorem 9.6.3 *If $F : (\mathcal{C}, W_{\mathcal{C}}) \leftrightarrows (\mathcal{D}, W_{\mathcal{D}}) : G$ is a Quillen adjunction between model categories, then there is an ∞-adjunction (in the sense of Definition 2.1.9)*

$$\mathbb{L}F : \mathcal{C}[W_{\mathcal{C}}^{-1}] \leftrightarrows \mathcal{D}[W_{\mathcal{D}}^{-1}] : \mathbb{R}G$$

between the associated ∞-localizations.

Proposition 9.6.3 *The identity functors*

$$^{\infty}\mathrm{GRPD} \leftrightarrows {}^{\infty}\mathrm{CAT}$$

induce a Quillen adjunction between the two given model categories on SSETS. *The corresponding adjunction on homotopy categories is also denoted by*

$$i : h(^{\infty}\mathrm{GRPD}) \leftrightarrows h(^{\infty}\mathrm{CAT}) : \tau_0,$$

and the functor i is a fully faithful embedding.

Definition 9.6.5 Let $\mathcal{C}$ be a quasi-category, and x and y be two objects of $\mathcal{C}$. The simplicial set of morphisms from x to y is defined by the fiber product

$$\begin{array}{ccc} \underline{\mathrm{Mor}}_{\mathcal{C}}(x,y) & \longrightarrow & \underline{\mathrm{Hom}}_{\mathrm{SSETS}}(\Delta^1,\mathcal{C}) \\ \downarrow & & \downarrow{\scriptstyle (s,t)} \\ \Delta_0 & \xrightarrow{(x,y)} & \mathcal{C}\times\mathcal{C} \end{array}$$

We now show that ∞-categories have the nice property of being weakly enriched over ∞-groupoids, in the sense of the following proposition. This implies that they are a good candidate for a theory of ${}^{\infty}1$-categories.

Proposition 9.6.4 *The simplicial set* $\underline{\mathrm{Mor}}_{\mathcal{C}}(x,y)$ *is a Kan complex. There exists a morphism of simplicial sets*

$$\circ : \underline{\mathrm{Mor}}_{\mathcal{C}}(x,y)\times\underline{\mathrm{Mor}}_{\mathcal{C}}(y,z)\to\underline{\mathrm{Mor}}_{\mathcal{C}}(x,z)$$

that gives an associative composition law in the category ${}^{\infty}\mathrm{GRPD}$, *unique up to isomorphism.*

Proof From Lemma 9.6.1, we know that the natural map

$$\underline{\mathrm{Hom}}_{\mathrm{SSETS}}(\Delta^2,\mathcal{C})\longrightarrow\underline{\mathrm{Hom}}_{\mathrm{SSETS}}(\Lambda^2_1,\mathcal{C})$$

is a trivial fibration, so that it has a section s. One has a natural cartesian square

$$\begin{array}{ccc} \underline{\mathrm{Hom}}_{\mathrm{SSETS}}(\Lambda^2_1,\mathcal{C}) & \longrightarrow & \underline{\mathrm{Hom}}_{\mathrm{SSETS}}(\Delta^1,\mathcal{C}) \\ \downarrow & & \downarrow{\scriptstyle t} \\ \underline{\mathrm{Hom}}_{\mathrm{SSETS}}(\Delta^1,\mathcal{C}) & \xrightarrow{s} & \underline{\mathrm{Hom}}_{\mathrm{SSETS}}(\Delta^0,\mathcal{C})=\mathcal{C}, \end{array}$$

so that the map $\underline{\mathrm{Hom}}(d_1,\mathcal{C}):\underline{\mathrm{Hom}}(\Delta^2,\mathcal{C})\to\underline{\mathrm{Hom}}(\Delta^1,\mathcal{C})$ induces, by composition with s, a composition law

$$\circ:\underline{\mathrm{Hom}}_{\mathrm{SSETS}}(\Delta^1,\mathcal{C})\underset{\mathcal{C}}{\times}\underline{\mathrm{Hom}}_{\mathrm{SSETS}}(\Delta^1,\mathcal{C})\to\underline{\mathrm{Hom}}_{\mathrm{SSETS}}(\Delta^1,\mathcal{C}).$$

This map reduces to a well defined composition map

$$\circ:\underline{\mathrm{Mor}}_{\mathcal{C}}(x,y)\times\underline{\mathrm{Mor}}_{\mathcal{C}}(y,z)\to\underline{\mathrm{Mor}}_{\mathcal{C}}(x,z).$$

By construction, this law is unique up to isomorphism and associative in ${}^{\infty}\mathrm{GRPD}$. □

We now define the notion of limits in ∞-categories. One may give a more explicit presentation in the setting of quasi-categories, but this would take us too far (see [Lur09d] and [Joy11]).

Definition 9.6.6 Let $\mathcal{C}$ and I be two ∞-categories. A limit functor (resp. colimit functor) is a right (resp. left) adjoint to the constant ∞-functor

$$\mathcal{C} \cong \underline{\mathrm{Mor}}_{\infty\mathrm{CAT}}(*, \mathcal{C}) \to \underline{\mathrm{Mor}}_{\infty\mathrm{CAT}}(I, \mathcal{C}).$$

The only limits that we really need to compute explicitly are given by limits of higher groupoids, because we will always use Yoneda's lemma to reduce computations of limits in ∞-categories to computations of limits for their morphism spaces.

We now use the nerve construction to give an explicit description of homotopy limits and homotopy colimits of simplicial sets (see [GJ99, BK72] and [Lur09d], Appendix A). The fact that every simplicial set is cofibrant implies that homotopy colimits have a simpler description than homotopy limits.

Definition 9.6.7 Let $f : I \to \mathrm{SSETS}$ be a diagram of simplicial sets indexed by a small category. For each n, denote by Ef_n the so-called *translation category*, with objects (i, x), with i an object of I and $x \in f(i)_n$, and morphisms $\alpha : (i, x) \to (j, y)$ given by morphisms $\alpha : i \to j$ of I such that $f(\alpha)_n(x) = y$. The *homotopy colimit* of f is the diagonal simplicial set of the bisimplicial set

$$\mathrm{hocolim} f = \left[n \mapsto N(Ef_n)\right]$$

obtained by composing with the diagonal functor $\Delta^{op} \to \Delta^{op} \times \Delta^{op}$. The *homotopy limit* of a diagram $f : I^{op} \to \mathrm{SSETS}$ *whose values are Kan complexes* is defined dually to colimits, as diagonals of nerves of translation categories of the diagram $f : I \to \mathrm{SSETS}^{op}$.

Concretely, one has

$$(\mathrm{hocolim} f)_{n,m} = \coprod_{i_0 \to i_1 \to \cdots \to i_m} f(i_0)_n.$$

We refer to loc. cit. for a proof of the following proposition.

Proposition 9.6.5 *The above defined homotopy colimits and limits of simplicial sets are indeed the homotopy limits and colimits for the model structure* ${}^{\infty}\mathrm{GRPD}$, *and also the ∞-limits and ∞-colimits in the ∞-category* ${}^{\infty}\mathrm{GRPD}$.

We now give various examples of concrete computations of homotopy limits in the ∞-category

$${}^{\infty}\mathrm{GRPD} = \mathrm{SSETS}\left[W_{eq}^{-1}\right]$$

obtained by localizing simplicial sets by weak homotopy equivalences (note that we use the same notation for the quasi-category and its homotopy category; its meaning will be clear from the context). These examples will play an important role in applications.

Example 9.6.1

1. Let G be a group, regarded as a category with one object and invertible arrows given by elements of G, and let BG be the associated nerve. Then one has a homotopy cartesian diagram

$$\begin{array}{ccc} G & \longrightarrow & \{*\} \\ \downarrow & & \downarrow \\ \{*\} & \longrightarrow & BG \end{array}$$

2. Let $(X, *)$ be a pointed topological space and $\Omega X := \underline{\mathrm{Hom}}((S^1, *), (X, *))$ be the topological space of continuous pointed loops in X. Then one has the homotopy cartesian diagram

$$\begin{array}{ccc} \Omega X & \longrightarrow & \{*\} \\ \downarrow & & \downarrow \\ \{*\} & \longrightarrow & X \end{array}$$

3. Let X be a topological space and $LX := \underline{\mathrm{Hom}}(S^1, X)$ be the topological space of (unpointed) loops in X. Then one has the homotopy cartesian diagram

$$\begin{array}{ccc} LX & \longrightarrow & X \\ \downarrow & & \downarrow{\scriptstyle \Delta} \\ X & \xrightarrow{\Delta} & X \times X \end{array}$$

   in $\mathrm{T{\small OP}}$ (meaning that applying $\Pi_\infty : \mathrm{T{\small OP}} \to {}^\infty\mathrm{G{\small RPD}}$ to it gives a homotopy cartesian diagram).
4. More generally, if $Y, Z \subset X$ are two closed subset of a topological space X, and

$$L_{Y \to Z} X \subset \underline{\mathrm{Hom}}\big([0, 1], X\big)$$

   is the subspace of paths starting on Y and ending on Z, then one has the homotopy cartesian diagram

$$\begin{array}{ccc} L_{Y \to Z} X & \xrightarrow{\mathrm{ev}_0} & Z \\ \downarrow{\scriptstyle \mathrm{ev}_1} & & \downarrow \\ Y & \longrightarrow & X \end{array}$$

5. Even more generally, if $f : Y \to X$ and $g : Z \to X$ are continuous maps of topological spaces, then one has a decomposition $f = p \circ i$ of the form

$$Y \xrightarrow{i} Y \times_X \underline{\mathrm{Hom}}(\Delta_1, X) \xrightarrow{p} X,$$

with p a fibration and i a trivial cofibration. The homotopy fiber product of f and g is then given by the fiber product

$$Y \times_X^h Z = \big(Y \times_X \underline{\mathrm{Hom}}(\Delta_1, X)\big) \times_X Z,$$

that is the relative space of paths $L_{f \to g}X$, given by triples (y, γ, z) with $y \in Y$ and $z \in Z$, and γ a path in X from $f(y)$ to $g(z)$. This construction also applies directly to a cartesian diagram of fibrant simplicial sets, i.e., of ∞-groupoids.

9.7 Stable ∞-Categories

We refer to Lurie [Lur09c], Chap. 1, for the content of this section. The reader may also refer to Quillen [Qui67] for a more classical treatment of similar matters, in terms of pointed model categories.

Definition 9.7.1 Let $\mathcal{C}$ be an ∞-category. A *zero object* in $\mathcal{C}$ is an object that is both initial and final. An ∞-category is called *pointed* if it has a zero object.

The existence of a zero object implies the existence of a zero morphism

$$0 : X \to Y$$

between every two objects in the homotopy category $h(\mathcal{C}) := \tau_1(\mathcal{C})$.

Definition 9.7.2 Let $\mathcal{C}$ be a pointed ∞-category. A *triangle* in $\mathcal{C}$ is a commutative diagram

$$\begin{array}{ccc} X & \xrightarrow{f} & Y \\ \downarrow & & \downarrow g \\ 0 & \longrightarrow & Z \end{array}$$

where 0 is a zero object of $\mathcal{C}$. A triangle in $\mathcal{C}$ is called a *fiber sequence* if it is a pullback square, and a *cofiber sequence* if it is a pushout square. If it is a fiber sequence, it is called a *fiber* for the morphisms $g : Y \to Z$. If it is a cofiber sequence, it is called a *cofiber* for the morphism f.

Definition 9.7.3 Let $\mathcal{C}$ be a pointed ∞-category. A *loop space object* is a fiber ΩX for the morphism $0 \to X$ and a *suspension object* ΣX is a cofiber for the morphism $0 \to X$.

Definition 9.7.4 An ∞-category $\mathcal{C}$ is called *stable* if it satisfies the following conditions:

1. There exists a zero object $0 \in \mathcal{C}$.
2. Every morphism in $\mathcal{C}$ admits a fiber and a cofiber.
3. A triangle in $\mathcal{C}$ is a fiber sequence if and only if it is a cofiber sequence.

In a stable ∞-category, the loop space and suspension are autoequivalences

$$\Omega : \mathcal{C} \to \mathcal{C} \quad \text{and} \quad \Sigma : \mathcal{C} \to \mathcal{C}$$

which are inverses of each other. Moreover, the homotopy category $h(\mathcal{C}) := \tau_1(\mathcal{C})$ is additive.

Theorem 9.7.1 *Let $\mathcal{C}$ be a stable ∞-category, let*

$$\begin{array}{ccc} X & \xrightarrow{f} & Y \\ \downarrow & & \downarrow g \\ 0 & \longrightarrow & Z \end{array}$$

be a fibration sequence, and let A be any object in $h(\mathcal{C})$. Then there is a long exact homotopy sequence

$$\cdots \to \operatorname{Hom}(A, \Omega^{n+1} Z) \to \operatorname{Hom}(A, \Omega^n X) \to \operatorname{Hom}(A, \Omega^n Y) \to \operatorname{Hom}(A, \Omega^n Z) \to \cdots$$

Definition 9.7.5 Let $\mathcal{C}$ be an ∞-category with finite limits. The *stabilization* of $\mathcal{C}$ is the limit

$$\operatorname{Stab}(\mathcal{C}) := \lim(\cdots \mathcal{C} \xrightarrow{\Omega} \mathcal{C})$$

in the ∞-category of ∞-categories.

Definition 9.7.6 The stable ∞-category Sp of *spectra* is the stabilization of the ∞-category of pointed ∞-groupoids (or equivalently, pointed topological spaces).

Note that there is a natural ∞-monoidal structure $\wedge$ on pointed spaces, and that this monoidal structure extends to spectra giving a symmetric monoidal ∞-category

$$(\mathrm{Sp}, \wedge).$$

Definition 9.7.7 A *pointed model category* is a model category whose initial and final object coincide.

If (M, W) is a pointed model category, its ∞-localization $\mathcal{C} = M[W^{-1}]$ is a pointed ∞-category.

Example 9.7.1 The category TOP_* of pointed (Hausdorff and compactly generated) topological spaces is an example of a pointed model category. In this setting, the suspension is given by the derived functor of the functor

$$X \mapsto \Sigma(X) := X \wedge S^1$$

of smash product with S^1 (the product in the category TOP_* of pointed spaces) and the loop is given by the pointed mapping space

$$X \mapsto \Omega(X) := \underline{\text{Hom}}_{\text{TOP}_*}(S^1, X).$$

Example 9.7.2 The category $\text{MOD}_{dg}(A)$ of dg-modules on a given ring (see Sect. 9.8) is another example of a pointed model category. In this setting, the suspension is given by the shift functor

$$X \mapsto X[1]$$

and the loop space is given by the functor

$$X \mapsto X[-1].$$

The cofiber of a map $f : M \to N$ of complexes is given by the graded module

$$C(f) = M[1] \oplus N$$

equipped with the differential

$$d_{C(f)} = \begin{pmatrix} d_{A[1]} & 0 \\ f[1] & d_B \end{pmatrix}$$

acting on matrix vectors and the inclusion $N \to C(f)$ and projection $C(f) \to M[1]$ are just the natural ones. If $f : M_0 \to N_0$ is concentrated in degree 0, then $C(f)$ is just given by f itself with M_0 in degree -1, so that $H^0(C(f))$ is the cokernel of f and $H^{-1}(C(f))$ is its kernel. If $f : M_0 \to N_0$ is injective, i.e., a cofibration for the injective model structure, one gets an exact sequence

$$0 \to M_0 \xrightarrow{f} N_0 \to H^0(C_f) \to 0.$$

9.8 Derived Categories and Derived Functors

Let A be an associative unital ring. Let $\text{MOD}_{dg}(A)$ be the category of graded (left) A-modules, equipped with a linear map $d : C \to C$ of degree -1 such that $d^2 = 0$, with graded morphisms that commute with d.

Note that this category is pointed, meaning that its initial and final objects are identified (the zero dg-module). It is moreover additive, meaning that morphisms are abelian groups, and even an abelian category. We would like to localize this category with respect to the class W of quasi-isomorphisms, (i.e., morphisms that induce isomorphisms of cohomology space $H^* := \mathrm{Ker}(d)/\mathrm{Im}(d)$). This is usually done in textbooks via of the formalism of derived and triangulated categories. Here we directly provide the homotopical description of this construction.

The projective model structure on the category $\mathrm{MOD}_{dg}(A)$ is defined by saying that:

- weak equivalences are quasi-isomorphisms;
- fibrations are degree-wise surjections; and
- cofibrations are maps with the left lifting property with respect to trivial fibrations.

The injective model structure on the category $\mathrm{MOD}_{dg}(A)$ is defined by saying that:

- weak equivalences are quasi-isomorphisms;
- cofibrations are degree-wise injections; and
- fibrations are maps with the left lifting property with respect to trivial cofibrations.

Note that it is hard to prove that these really define model category structures on $\mathrm{MOD}_{dg}(A)$ (see [Hov99]). Recall that a module P (resp. I) over A is called projective (resp. injective) if for all surjective (resp. injective) module morphisms $M \to N$, the morphism

$$\mathrm{Hom}_{\mathrm{MOD}(A)}(P, M) \to \mathrm{Hom}_{\mathrm{MOD}(A)}(P, N)$$
$$\big(\text{resp. } \mathrm{Hom}_{\mathrm{MOD}(A)}(M, I) \to \mathrm{Hom}_{\mathrm{MOD}(A)}(N, I)\big)$$

is surjective.

One can show that any cofibrant object in the projective model structure has projective components, and that having this property is enough to be cofibrant for bounded below dg-modules. Using this one can show that two morphisms $f, g : P \to Q$ between cofibrant dg-modules are homotopic if and only if there is an A-linear morphism $h : X \to Y$ homogeneous of degree -1 (called a chain homotopy) such that

$$f - g = d \circ h + h \circ d.$$

This can be seen by observing that given a complex X, a path object for X is given by the chain complex

$$\mathrm{Cocyl}(X)_n = X_n \oplus X_n \oplus X_{n+1}$$

with the differential

$$\partial(x, y, z) = (\partial x, \partial y, -\partial z + x - y).$$

One also defines the cylinder of X as the complex defined by

$$\mathrm{Cyl}(X)_n := X_n \oplus X_{n-1} \oplus X_n$$

equipped with the differential

$$\partial(x, y, z) := (\partial x - y, -\partial x, x + \partial y).$$

This can be used to show that the left homotopy in the injective model structure is given by a chain homotopy.

Definition 9.8.1 The homotopy category $h(\mathrm{MOD}_{dg}(A))$ is called the *derived category* of A and is denoted by $D(A)$.

Note that in this case, a dg-module is cofibrant and fibrant if and only if it is fibrant. Applying Theorem 9.2.1, one can compute $D(A)$ by taking the quotient category of the category of cofibrant dg-modules by the chain homotopy equivalence relation $\sim_{ch}$, i.e., the natural functor

$$\mathrm{MOD}_{dg,c}(A)/\sim_{ch} \longrightarrow h\big(\mathrm{MOD}_{dg}(A)\big)$$

is an equivalence.

By definition of derived functors, if

$$F : \mathrm{MOD}(A) \leftrightarrows \mathrm{MOD}(B) : G$$

is an adjunction of additive functors such that F is left exact (respects injections) and G is right exact (respect surjections), one can extend it to a Quillen adjunction

$$F : \mathrm{MOD}_{dg,inj}(A) \leftrightarrows \mathrm{MOD}_{dg,proj}(B) : G$$

and this defines a derived adjunction

$$\mathbb{L}F : D(A) \leftrightarrows D(B) : \mathbb{L}G.$$

One can compute this adjunction by setting $\mathbb{L}F(M) := F(Q(M))$ and $\mathbb{L}G(M) := G(R(M))$ where R and Q are respective fibrant and cofibrant replacements of M. This gives back the usual derived functors of homological algebra, defined in [Har70].

For example, let A and B be two rings and M be an (A, B) bimodule. The pair

$$-\otimes_A M : \mathrm{MOD}_{dg,proj}(A) \leftrightarrows \mathrm{MOD}_{dg,inj}(B) : \mathrm{Hom}_B(M, -)$$

is a Quillen adjunction that corresponds to a pair of adjoint derived functors

$$-\overset{\mathbb{L}}{\otimes}_A M : D(A) \leftrightarrows D(B) : \mathbb{R}\mathrm{Hom}_B(M, -).$$

Similarly, the pair (we carefully inform the reader that one works with the opposite category on the right hand side, to get covariant functors)

$$\mathrm{Hom}_A(-,M):\mathrm{MOD}_{dg,proj}(A)\leftrightarrows\big(\mathrm{MOD}_{dg,proj}(B)\big)^{op}:\mathrm{Hom}_B(-,M)$$

is a Quillen adjunction that corresponds to a pair of adjoint derived functors

$$\mathbb{R}\mathrm{Hom}_A(-,M):D(A)\leftrightarrows D(B)^{op}:\mathbb{R}\mathrm{Hom}_B(-,M).$$

The two constructions above give back all the usual constructions of derived functors on bounded derived categories.

If $\mathrm{MOD}_{dg,+}$ denotes the subcategory of bounded below dg-modules, one can define a (different from the standard) cofibrant replacement functor $Q:\mathrm{MOD}_{dg,+}(A)\to\mathrm{MOD}_{dg}(A)$ by

$$Q(M)=\mathrm{Tot}\big(L(M)\big),$$

where $L(M)$ is the free resolution of M given by setting $L(M)_0=A^{(M)}$, $f_0: L(M)_0\to M$ the canonical morphism and

$$f_{i+1}:L(M)_{i+1}:=A^{(\mathrm{Ker}(f_i))}\to L(M)_i,$$

and $\mathrm{Tot}(L(M))_k=\sum_{i+j=k}L(M)_{i,j}$ is the total complex associated to the bicomplex $L(M)$ (with one of the differentials given by M and the other by the free resolution degree).

Note that the category $\mathrm{MOD}_{dg,+}(A)$ of positively graded dg-modules can be equipped with a model structure (see [DS95], Theorem 7.2) by defining

- weak equivalences to be quasi-isomorphisms;
- cofibrations to be monomorphisms with component-wise projective cokernel;
- fibrations to be epimorphisms component-wise in strictly positive degree.

The dual result is that the category $\mathrm{MOD}_{dg,-}(A)$ of negatively graded dg-modules can be equipped with a model structure by defining:

- weak equivalences to be quasi-isomorphisms;
- fibrations to be epimorphisms with component-wise injective kernel;
- cofibrations to be component-wise monomorphisms in strictly negative degree.

Of course, one has Quillen adjunctions

$$i:\mathrm{MOD}_{dg,+}(A)\leftrightarrows\mathrm{MOD}_{dg}(A):\tau_{\geq 0}$$

and

$$i:\mathrm{MOD}_{dg,-}(A)\leftrightarrows\mathrm{MOD}_{dg}(A):\tau_{\leq 0}$$

given by the inclusion and truncation functors. These allow us to define derived functors on bounded complexes, and to compute them by the usual method of projective and injective resolutions.

Bounded differential graded modules may also be studied with simplicial methods. Simplicial A-modules are defined as contravariant functors $M : \Delta^{op} \to \textsc{Mod}(A)$.

Proposition 9.8.1 *There is an equivalence of categories*

$$N : \textsc{Mod}_s(A) \leftrightarrows \textsc{Mod}_{dg,+}(A) : \Gamma$$

between the category $\textsc{Mod}_s(A)$ *of simplicial modules (which is equipped with the model structure induced by the one given on the underlying simplicial sets) and the category of positively graded complexes, which preserves the model structures.*

Proof The dg-module $N(M)$ associated to a simplicial module M has graded parts

$$N(M)_n = \cap_{i=1}^{n} \mathrm{Ker}\big(\partial_n^i\big)$$

and is equipped with the differential

$$d_n(a) = \partial_n^0$$

where ∂_n^i is induced by the non-decreasing map $[0, \ldots, \hat{i}, \ldots, n+1] \to [0, \ldots, n]$. The simplicial set associated to a given dg-module C is the family of sets

$$\Gamma(C)_n := \oplus_{n \twoheadrightarrow k} C_k$$

with naturally defined simplicial structure, for which we refer the reader to [GJ99], Chap. III. □

This gives a way of introducing simplicial methods in homological algebra.

Definition 9.8.2 Let K be a commutative unital ring. A differential graded algebra is an algebra in the symmetric monoidal category $\textsc{Mod}_{dg}(K)$, i.e., a pair (A, d) composed of a graded algebra A and a derivation $d : A \to A$, i.e., a morphism of K-module $A \to A$ of degree -1 and such that

$$d(ab) = d(a) \cdot b + (-1)^{\deg(a)} a \cdot d(b)$$

and $d \circ d = 0$.

Theorem 9.8.1 *The model category structure on* $\textsc{Mod}_{dg}(K)$ *induces a model category structure on the category* $\textsc{Alg}_{dg,K}$ *of differential graded algebras.*

9.9 Derived Operations on Sheaves

We refer to Hartshorne's book [Har77], Chaps. II.1 and III.1, for a short review of the basic notions regarding abelian sheaves and to Kashiwara and Schapira's book [KS06] for a more complete account. The extension of the above methods to derived categories of Grothendieck abelian categories (for example, categories of abelian sheaves on spaces) can be found in [CD09].

Recall that a sheaf on a category $X = (\text{LEGOS}, \tau)$ with Grothendieck topology is a contravariant functor

$$X : \text{LEGOS}^{op} \to \text{SETS}$$

that satisfies the sheaf condition given in Definition 2.4.1. More precisely, the sheaf condition is that for every covering family $\{f_i : U_i \to U\}$, the sequence

$$X(U) \longrightarrow \prod_i X(U_i) \rightrightarrows \prod_{i,j} X(U_i \times_U U_j)$$

is exact.

We first translate the sheaf condition in terms of nerves of coverings.

Definition 9.9.1 Let $\mathcal{U} := \{f_i : U_i \to U\}$ be a covering family for a Grothendieck topology. The *nerve* $N(\mathcal{U})$ is the simplicial object of LEGOS with n-vertices

$$N(\mathcal{U})_n := \coprod_{i_1,\dots,i_n} U_{i_1} \times_U \cdots \times_U U_{i_n},$$

and faces and degenerations given by restrictions and inclusions.

Proposition 9.9.1 *A contravariant functor* $X : \text{LEGOS}^{op} \to \text{SETS}$ *is a sheaf if and only if it commutes to colimits along nerves of coverings, i.e., one has*

$$\lim_{[n]\in\Delta} X\big(N(\mathcal{U})_n\big) \cong X\Big(\operatorname*{colim}_{[n]\in\Delta} N(\mathcal{U})_n\Big) \cong X(U)$$

for every covering $\mathcal{U}$ *of* U.

An abelian sheaf on X is a sheaf with values in the category GRAB of abelian groups. We denote by $\text{SHAB}(X)$ the category of abelian sheaves on X.

Theorem 9.9.1 (Beke) *The category of complexes of abelian sheaves has a so-called injective model category structure defined by setting:*

- *weak equivalences to be quasi-isomorphisms; and*
- *cofibrations to be monomorphisms (injective in each degree).*

The projective model structure is harder to define because it involves a cohomological descent condition. We first define it in a simple but non-explicit way (explained to the author by D.-C. Cisinski) and then briefly provide the description of fibrations. We refer to [CD09] for more details on this result.

Theorem 9.9.2 *The category of complexes of abelian sheaves has a so-called projective model category structure defined by setting:*

- *weak equivalences to be quasi-isomorphisms;*
- *cofibrations to have the left lifting property with respect to trivial fibrations, which are defined ad-hoc as morphisms $f : K \to L$ that are object-wise surjective in each degree (i.e., $f_{n,X} : K(X)_n \to L(X)_n$ is surjective for every n and X) and such that* $\mathrm{Ker}(f_X)$ *is acyclic for every X; and*
- *fibrations to have the right lifting property with respect to trivial cofibrations.*

We now give, just for culture, the description of fibrations in the projective model structure on complexes of abelian sheaves.

Definition 9.9.2 Let $X \in$ LEGOS. A *hyper-cover* of X is a simplicial lego $U : \Delta^{op} \to$ LEGOS equipped with a morphism $p : U \to X$ of simplicial legos such that each U_n is a coproduct of representables, and $U \to X$ is a local acyclic fibration (locally for the given topology on LEGOS, it is an object-wise acyclic fibration of simplicial sets).

Theorem 9.9.3 *The fibrations in the projective model category structure on the category of complexes of abelian sheaves are given by epimorphisms of complexes (surjective in each degree), whose kernel K has the cohomological descent property, meaning that for every object X in* LEGOS *and every hyper-covering $U \to X$, the natural morphism*

$$K(X) \to \mathrm{Tot}\big(N\big(K(U)\big)\big)$$

is a quasi-isomorphism.

Another way to define the left bounded derived category of sheaves for the projective model structure is given by making a Bousfield homotopical localization (see Sect. 9.4) of the category of left bounded complexes of abelian presheaves (or equivalently, simplicial presheaves), with its standard model structure, by the morphisms of complexes of presheaves that induce isomorphisms on the cohomology sheaves.

Definition 9.9.3 Let $f : X \to Y$ be a morphism of topological spaces. The *direct image* of an abelian sheaf $\mathcal{F}$ on X is defined by

$$f_*\mathcal{F}(U) := \mathcal{F}\big(f^{-1}(U)\big)$$

and the *inverse image* is defined as the sheaf associated to the presheaf

$$f^{-1}\mathcal{F}(U) := \lim_{f(U)\subset V} \mathcal{F}(V).$$

Theorem 9.9.4 *The direct and inverse image functors form a Quillen adjunction between categories of complexes of abelian sheaves*

$$f^* : \mathrm{CSHAB}(X)_{proj} \leftrightarrows \mathrm{CSHAB}(Y)_{inj} : f_*$$

that induce an adjoint pair of derived functors

$$\mathbb{L}f^* : D(\textsc{ShAb}(X)) \leftrightarrows D(\textsc{ShAb}(Y)) : \mathbb{R}f_*.$$

The above constructions can be extended to sheaves of modules over ringed spaces. A ringed space is a pair $(X, \mathcal{O}_X)$ comprising a space and a sheaf of commutative rings, and a morphism of ringed spaces

$$f : (X, \mathcal{O}_X) \to (Y, \mathcal{O}_Y)$$

is a pair comprising a continuous map $f : X \to Y$ and a morphism of sheaves of rings $f^{-1}\mathcal{O}_Y \to \mathcal{O}_X$.

Definition 9.9.4 The *inverse image functor* $f^* : \textsc{Mod}(\mathcal{O}_Y) \to \textsc{Mod}(\mathcal{O}_X)$ is defined by

$$f^*\mathcal{F} := f^{-1}\mathcal{F} \otimes_{f^{-1}\mathcal{O}_Y} \mathcal{O}_X.$$

Theorem 9.9.5 *The model category structures on the category of complexes of abelian sheaves induce model category structures on the category of complexes of $\mathcal{O}_X$-modules on a given ringed space $(X, \mathcal{O}_X)$. The direct and inverse image functors form a Quillen adjunction*

$$f^* : C\textsc{Mod}(\mathcal{O}_X)_{proj} \leftrightarrows C\textsc{Mod}(\mathcal{O}_Y)_{inj} : f^*.$$

9.10 Higher Categories

We present here one model for the definition of the notion of higher category, which was essentially developed by Rezk [Rez09]. We base our presentation on the work of Ara (see [Ara10] and [Ara12], where this theory is compared to higher quasi-categories, and presented in the language of model categories; we had to slightly adapt these references to our doctrinal viewpoint, based on ${}^{\infty}1$-categories).

The approach to higher categories using an inductive construction and cartesian closed model category structures originates in the work of Simpson and Tamsamani (see [Tam95], and [Sim10] for historical background and more references) and of Joyal [Joy07]. There are various other models for higher categories. Comparison results may be found in Bergner's survey [Ber06] and in Toen's article [Toe04] for the one-dimensional case, and in [BR12] and [Ara12] in higher dimensions. An axiomatization characterizing the higher dimensional case may be found in Barwick and Schommer-Pries' paper [BS11].

We refer to Definition 2.1.3 for an informal definition of the general notion of higher category.

Definition 9.10.1 The *globular category* $\mathbb{G}$ is the category generated by the graph

$$D_0 \underset{s}{\overset{t}{\leftleftarrows}} D_1 \underset{s}{\overset{t}{\leftleftarrows}} D_2 \underset{s}{\overset{t}{\leftleftarrows}} D_3 \cdots,$$

with relations given by

$$ss = st, \qquad ts = tt.$$

Definition 9.10.2 A *dimension array* T is an array of natural integers $(i_k, j_k) \in \mathbb{N}_{>0}^{n+1} \times \mathbb{N}^n$ of the form

$$T = \begin{pmatrix} i_1 & & i_2 & & i_3 & \cdots & i_{n+1} \\ & j_1 & & j_2 & \cdots & j_n & \end{pmatrix},$$

with $i_k > j_k$ and $i_{k+1} > j_k$ for all $0 \leq k \leq n$. The dimension of T is the maximal integer $\dim(T) = \max_k(i_k)$ that appears in T.

To each dimension array T, one associates a diagram in $\mathbb{G}$ given by iteratively composing the source and target maps

$$[T] = \begin{array}{ccccccccc} D_{i_1} & & D_{i_2} & & D_{i_3} & & \cdots & & D_{i_{n+1}} \\ & {\scriptstyle s}\searrow \quad \swarrow{\scriptstyle t} & & {\scriptstyle s}\searrow \quad \swarrow{\scriptstyle t} & & & & \swarrow{\scriptstyle t} & \\ & D_{j_1} & & D_{j_2} & & \cdots & D_{j_n} & & \end{array}.$$

Here we choose a convention which differs from the usual one, denoting by Θ_0 the opposite to Joyal's category [Joy07].

Definition 9.10.3 The category Θ_0 is the universal category under $\mathbb{G}$ with all limits corresponding to the dimension arrays, i.e., the sketch associated to the category $\mathbb{G}$ and the family of limit diagrams $[T]$.

Definition 9.10.4 A*globular set* is a covariant functor

$$X : \mathbb{G} \to \text{SETS}.$$

The category of globular sets is denoted by $\widehat{\mathbb{G}}$.

The natural fully faithful embedding

$$\mathbb{G} \hookrightarrow \widehat{\mathbb{G}}^{op}$$

extends to a fully faithful embedding

$$\Theta_0 \hookrightarrow \widehat{\mathbb{G}}^{op}.$$

To make a clear difference between the notions of:

- higher categories with invertible higher morphisms of arbitrary degrees, which we called ∞-categories; and
- higher categories with non-invertible higher morphisms of arbitrary degrees,

we will call the latter *ω-categories*.

Definition 9.10.5 A *strict* 0*-category* is a set. For $n \geq 1$, a *strict n-category* is a category enriched (in the sense of Definition 2.2.7) over the cartesian monoidal category of strict $(n-1)$-categories. The category of *strict ω-categories* is the limit

$$^{\omega}\mathrm{CAT}_{str} := \lim_{n} n\mathrm{CAT}_{str}$$

of the categories of strict n-categories.

There is a forgetful contravariant functor

$$^{\omega}\mathrm{CAT}_{str} \longrightarrow \widehat{\mathbb{G}}^{op}$$

that sends C to the globular set $\{\mathrm{Mor}^i(C)\}_{i\geq 0}$ of morphisms in C.

Proposition 9.10.1 *The above forgetful functor commutes with arbitrary limits and thus has a right adjoint, denoted by F_∞ and called the free strict ∞-category functor.*

We now define (the opposite category to) Joyal's category.

Definition 9.10.6 *Joyal's category* Θ is the full subcategory of $(\mathrm{CAT}^{\infty}_{str})^{op}$ whose objects are dimension arrays and whose morphisms are given by

$$\mathrm{Hom}_{\Theta}(T_1, T_2) := \mathrm{Hom}_{\mathrm{CAT}^{\infty}_{str}}\big(F_\infty([T_2]), F_\infty([T_1])\big).$$

The subcategory with objects given by arrays of dimension smaller than n is denoted by Θ_n.

There is a sequence of covariant embeddings

$$\mathbb{G} \to \Theta_0 \to \Theta.$$

Each of these functors are obtained by extending the given categories: the first one by addition of limits to dimension array diagrams, and the second by addition of composition and unit morphisms between diagrams, such as for example

$$m : D_1 \times_{D_0} D_1 \to D_1 \quad \text{or} \quad 1 : D_i \to D_{i+1}.$$

The basic idea is that a strict ω-category is a functor

$$C : \Theta \to \mathrm{SETS}$$

that respects the colimits associated to array diagrams, so that $(\Theta, \{[T]\})$ is the limit sketch of higher categories. One may naively extend this by considering ${}^{\infty}\omega$-categories, which may be regarded as ∞-functors

$$\mathcal{C} : \Theta \to {}^{\infty}\text{GRPD}$$

that send array diagrams to limits. This has internal homomorphisms in the ${}^{\infty}1$-categorical sense, meaning that if $\mathcal{C}_1$ and $\mathcal{C}_2$ is an ${}^{\infty}\omega$-category, there is a naturally associated ${}^{\infty}\omega$-category $\underline{\mathrm{Mor}}(\mathcal{C}_1, \mathcal{C}_2)$. We now give a more precise definition.

For T an array, regarded as the limit of a diagram $[T]$ in Θ, we denote by I_T the limit of $[T]$ in $\widehat{\Theta}^{op}$. There is a natural *higher Segal morphism*

$$T \to I_T.$$

For example, in $\Theta_0 = \Delta^{op}$, the array of length k corresponds to the objects Δ_k and the above morphism is the Segal morphism

$$\Delta_k \to I_k := \Delta_1 \underset{\Delta_0}{\times} \Delta_1 \cdots \underset{\Delta_0}{\times} \Delta_1.$$

Asking that a functor $\mathcal{C} : \Theta \to \text{SETS}$ commutes with limits of array diagrams is equivalent to asking that its extension to presheaves sends the Segal morphisms $T \to I_T$ to bijections.

Let J_1 be the free groupoid on two objects 0 and 1, depicted (we omit the identity morphisms) as

$$J_1 := \quad 0 \rightleftarrows 1$$

There is a natural functor

$$J_1 \to \Delta_0 = \{*\}$$

that corresponds to a morphism $\Delta_0 \to J_1$ in $\widehat{\Theta}^{op}$. For $i \geq 1$, we denote by J_{i+1} the strict $(i+1)$-category with objects 0 and 1, $\mathrm{Mor}(1, 0) = \emptyset$ and only non-trivial morphism i-category $\mathrm{Mor}(0, 1) := J_i$. Explicitly, one has

$J_2 := \quad 0$ 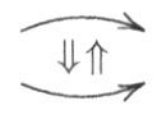 1

and

$J_3 := \quad 0$ 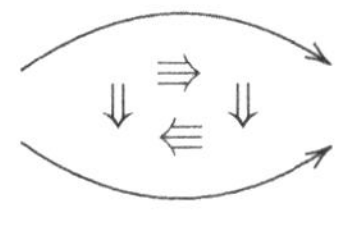 $1.$

There is a natural morphism $J_{i+1} \to D_i$ that sends the only two non-trivial i-arrows to the only non-trivial i-arrow. It corresponds to a morphism $D_i \to J_{i+1}$ in $\widehat{\Theta}^{op}$.

We now define ${}^{\infty}\omega$-categories as ∞-functors that send the above morphisms to equivalences. This corresponds to sending the corresponding limit diagrams to ∞-limit diagrams.

Definition 9.10.7 An ${}^{\infty}\omega$*-category* is an ∞-functor

$$\mathcal{C} : \Theta \to {}^{\infty}\text{GRPD}$$

that sends the Segal morphisms $T \to I_T$ and the morphisms $D_i \to J_{i+1}$ for $i \geq 0$ to equivalences. An ${}^{\infty}n$*-category* is an ∞-functor

$$\mathcal{C} : \Theta_n \to {}^{\infty}\text{GRPD}$$

satisfying the same condition for all arrays of dimension smaller than n and for $i \geq n$.

The main theorem of Rezk [Rez09] (which is similar in spirit to Simpson's main result [Sim10]) can be formulated in the ∞-categorical language by the following.

Theorem 9.10.1 *The* ∞*-category of* ${}^{\infty}n$*-categories is cartesian closed.*

9.11 Theories up to Homotopy and the Doctrine Machine

We now define a general notion of homotopical doctrine and homotopical theory, basing ourselves on the theory of homotopical higher categories discussed in Sect. 9.10. The main interest of this notion is that it gives a wide generalization of the theory of operadic structures up to homotopy, which is necessary to describe all the examples presented in this book, in terms of their universal properties. The notion of categorification was given an important impetus in Baez and Dolan's paper [BD98]. Our approach through sketches is new, even if it is already widely applied in particular examples.

The main motivation for studying the homotopy theory of theories is to understand how algebraic or geometric structures get transfered along homotopies. These homotopy transfer operations are often used in mathematical physics, particularly in the context of deformation theory. They give a conceptual understanding of some aspects of perturbative expansions (see for example Mnev-Merkulov's work in [Mer10] on the quantum BV formalism, and Tamarkin's formulation [Tam98] of the deformation quantization of Poisson manifolds).

Here is a standard example: consider the monoidal model category $(\text{MOD}_{dg}(A), \otimes, W_{qis})$, and suppose given a chain homotopy

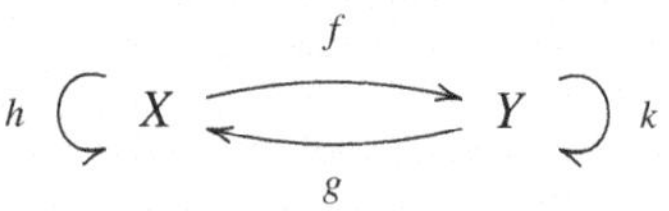

between two dg-modules X and Y over a ring A. One has

$$fg - \mathrm{id}_Y = \partial(k) := d_Y k + k d_Y \quad \text{and} \quad gf - \mathrm{id}_X = \partial(h) := d_X h + h d_X,$$

where ∂ denotes the differential on graded internal endomorphisms of X and Y induced by the graded morphisms d_X and d_Y. Suppose given on Y an associative (differential graded) multiplication

$$m : Y \otimes Y \to Y.$$

Then

$$m_2 := g \circ m \circ (f \times f) : X \otimes X \to X$$

is an operation on X, but it is no longer associative in general. However, the obstruction for it being associative (the associator) is homotopic to zero, so that there is a ternary operation (the homotopy)

$$m_3 : X^{\otimes 3} \to X$$

such that

$$m_2\big(a, m_2(b,c)\big) - m_2\big(m_2(a,b), c\big) = \partial(m_3)(a,b,c),$$

where ∂ is the differential induced on the internal homomorphisms $\underline{\mathrm{Hom}}(X^{\otimes 3}, X)$ by the formula

$$\partial(f) = d_X \circ f - (-1)^{\deg(f)} f \circ d_{X^{\otimes 3}}.$$

One can combine m_2 and m_3 to get new relations that are also true up to a higher homotopy $m_4 : X^{\otimes 4} \to X$, and so on... The combinatorics that appear here are a bit intricate, and the homotopy theory of operads is here to help us to deal with that problem.

We refer to Loday and Vallette's book [LV12], and Fresse's book [Fre09], for the standard algebraic approach to operads, which gives a way of treating linear and monoidal theories, and the corresponding algebra structures, up-to-homotopy. An ∞-categorical presentation of the theory of operads up to homotopy may also be found in [Lur09c]. The main results in the model category approach may also be found in Berger and Moerdijk [BM03]. An important aspect of this theory, which we will not describe here, is its explicit computational aspect, based on Koszul duality results, that give small resolutions of the theories in play (see [LV12], Chap. 11).

We now define a setting, based on higher categories, that allows us to define very general types of higher theories up to homotopy, and work with them essentially as with the usual theories, presented in Sect. 2.1. These have a built-in abstract notion of homotopy transfer.

Definition 9.11.1 An ∞-doctrine is an ${}^{\infty}(n+1)$-category $\mathcal{D}$ of ${}^{\infty}n$-categories with additional structures and properties. A theory of type $\mathcal{D}$ is an object $\mathcal{C}$ of $\mathcal{D}$. A model for a theory of type $\mathcal{D}$ in another theory of the same type is an object

$$M : \mathcal{C}_1 \to \mathcal{C}_2$$

of the ${}^{\infty}n$-category $\underline{\mathrm{Mor}}(\mathcal{C}_1, \mathcal{C}_2)$.

Of course, every doctrine $\mathcal{D}$ gives an ${}^{\infty}(n+1)$-doctrine. For example, the doctrine of symmetric monoidal categories is contained in the doctrine of symmetric monoidal ∞-categories. As an example, a dg-PROP (symmetric monoidal category enriched in dg-modules) can be viewed as a particular kind of symmetric monoidal ∞-category, and a cofibrant replacement of it for the natural model structure will encode the corresponding structure up to homotopy.

Important examples of theories are given by classical Lawvere (finite product) or Ehresmann (sketches) theories, embedded in the ${}^{\infty}1$ setting. They give notions of algebraic and other objects up to homotopy, such as ∞-Lie algebras, or ∞-associative algebras.

We now sum up the general idea of categorification of theories, by defining a natural categorification technique, called the *doctrine machine*, that allows us to define homotopical versions of any of the theories we might be interested in, provided it is described by an Ehresmann sketch in the sense of Definition 2.1.16. This very general categorification method was already used to define ${}^{\infty}\omega$-categories, and will also be used in Chap. 10 to build homotopical analogs of geometrical spaces. One may easily extend this method to other kinds of universal structures on categories, such as inner homomorphisms, to be able to treat cartesian closed categories. It is also possible to apply the following construction process inductively, as was already done for example in Sects. 2.2 and 2.3 for monoidal theories. These generalizations are all contained in the general notion of doctrine, so we shall stick to the setting of sketches.

Definition 9.11.2 (The Doctrine Machine) Let $(\mathfrak{L}, \mathfrak{C}) = (\{I\}, \{J\})$ be two classes of categories, called categories of indices, and let $\mathcal{T} = (\mathcal{T}, \mathcal{L}, \mathcal{C})$ be an $(\mathfrak{L}, \mathfrak{C})$-sketch, in the sense of Definition 2.1.16, i.e., a category $\mathcal{T}$ with given families of I and J indexed diagrams in $\mathcal{T}$ for all $I \in \mathfrak{L}$ and $J \in \mathfrak{C}$. The *${}^{\infty}n$-categorification* of this theory is simply the theory $\mathcal{T}$, regarded as an ${}^{\infty}n$-categorical sketch. The ${}^{\infty}(n+1)$-category of its models

$$A : (\mathcal{T}, \mathcal{L}, \mathcal{C}) \longrightarrow \left({}^{\infty}n\mathrm{CAT}, \lim_{\mathfrak{L}}, \mathrm{colim}_{\mathfrak{C}}\right)$$

with values in the ${}^{\infty}(n+1)$-$(\mathfrak{L}, \mathfrak{C})$-sketch of ${}^{\infty}n$-categories is denoted by $\mathrm{ALG}_{\infty_n}(\mathcal{T})$ and is called the ${}^{\infty}(n+1)$-category of ${}^{\infty}n$-$\mathcal{T}$-algebras.

One can of course apply the same idea to more general types of structures, and this complication sometimes simplifies the study of the corresponding theories. For example, Lurie [Lur09] has shown that the theory of topological quantum field theories up to homotopy, given by the symmetric monoidal ${}^{\infty}n$-category of cobordisms, is the free symmetric monoidal ${}^{\infty}n$-category with strong duality on one object, i.e., the free n-monoidal theory with strong duality.

We now define ∞-Lie algebras more concretely, since they will play an important role in the quantum part of this book.

Definition 9.11.3 Let K be a base field. Let $(\mathcal{T}_{\mathrm{LIE}}, \otimes)$ be the linear theory of Lie algebras whose underlying monoidal category is opposite to that of free finitely generated Lie algebras. We view this theory as a linear symmetric monoidal theory. If $(\mathcal{C}, \otimes)$ is a symmetric monoidal linear ∞-category, a Lie algebra in $(\mathcal{C}, \otimes)$ is a monoidal ∞-functor

$$L : (\mathcal{T}_{\mathrm{LIE}}, \otimes) \longrightarrow (\mathcal{C}, \otimes).$$

One may describe Lie algebras in the ∞-category of differential graded modules by a more concrete "tree style" linear theory, called the L_∞ operad. Here we only define the notion of an L_∞-algebra.

Definition 9.11.4 A *curved L_∞-algebra* $\mathfrak{g}$ over a ring A of characteristic zero is the datum of:

1. a positively graded flat module $\mathfrak{g}$;
2. a square zero differential $D : \vee^*\mathfrak{g} \to \vee^*\mathfrak{g}$ of degree -1 on the free graded co-commutative coalgebra on $\mathfrak{g}$.

The graded parts $\{l_i\}_{i \geq 0}$ of the associated canonical (cofreeness) map

$$l := \sum l_i : \vee^* \to \mathfrak{g}$$

are called the *higher brackets* of the L_∞-algebra. If $l_0 = 0$, one calls $\mathfrak{g}$ an *L_∞-algebra.*

References

[Ara10] Ara, D.: The groupoidal analogue Theta˜ to Joyal's category Theta is a test category. arXiv e-prints (2010). arXiv:1012.4319

[Ara12] Ara, D.: Higher quasi-categories vs higher Rezk spaces. arXiv e-prints (2012). arXiv: 1206.4354

[BD98] Baez, J.C., Dolan, J.: Categorification. arXiv Mathematics e-prints (1998). arXiv: math/9802029

[Ber06] Bergner, J.E.: A survey of $(\infty, 1)$-categories. arXiv Mathematics e-prints (2006). arXiv:math/0610239

[BK72] Bousfield, A.K., Kan, D.M.: Homotopy Limits, Completions and Localizations. Lecture Notes in Mathematics, vol. 304, p. 348. Springer, Berlin (1972)

[BM03] Berger, C., Moerdijk, I.: Axiomatic homotopy theory for operads. Comment. Math. Helv. **78**(4), 805–831 (2003). doi:10.1007/s00014-003-0772-y

[BR12] Bergner, J.E., Rezk, C.: Comparison of models for (∞, n)-categories, I. arXiv e-prints (2012). arXiv:1204.2013

[BS11] Barwick, C., Schommer-Pries, C.: On the unicity of the homotopy theory of higher categories. arXiv e-prints (2011). arXiv:1112.0040

[CD09] Cisinski, D.-C., Déglise, F.: Local and stable homological algebra in Grothendieck Abelian categories. Homol. Homotopy Appl. **11**(1), 219–260 (2009)

[Cis10] Cisinski, D.-C.: Catégories dérivables. Bull. Soc. Math. Fr. **138**(3), 317–393 (2010)

[DHKS04] Dwyer, W.G., Hirschhorn, P.S., Kan, D.M., Smith, J.H.: Homotopy Limit Functors on Model Categories and Homotopical Categories. Mathematical Surveys and Monographs, vol. 113, p. 181. Am. Math. Soc., Providence (2004). ISBN 0-8218-3703-6

[DK80] Dwyer, W.G., Kan, D.M.: Simplicial localizations of categories. J. Pure Appl. Algebra **17**(3), 267–284 (1980). doi:10.1016/0022-4049(80)90049-3

[DS95] Dwyer, W.G., Spaliński, J.: Homotopy theories and model categories. In: Handbook of Algebraic Topology, pp. 73–126. North-Holland, Amsterdam (1995). doi:10.1016/B978-044481779-2/50003-1

[Fre09] Fresse, B.: Modules over Operads and Functors. Lecture Notes in Mathematics, vol. 1967, p. 308. Springer, Berlin (2009). doi:10.1007/978-3-540-89056-0. ISBN 978-3-540-89055-3

[GJ99] Goerss, P.G., Jardine, J.F.: Simplicial Homotopy Theory. Progress in Mathematics, vol. 174, p. 510. Birkhäuser, Basel (1999). ISBN 3-7643-6064-X

[GZ67] Gabriel, P., Zisman, M.: Calculus of Fractions and Homotopy Theory. Ergebnisse der Mathematik und ihrer Grenzgebiete, vol. 35, p. 168. Springer, New York (1967)

[Har70] Hartshorne, R.: Residues and Duality. Springer, Berlin (1970)

[Har77] Hartshorne, R.: Algebraic Geometry. Graduate Texts in Mathematics, vol. 52, p. 496. Springer, New York (1977). ISBN 0-387-90244-9

[Hov99] Hovey, M.: Model Categories. Mathematical Surveys and Monographs, vol. 63, p. 209. Am. Math. Soc., Providence (1999). ISBN 0-8218-1359-5

[Joy07] Joyal, A.: Discs, duality and theta-categories (2007, preprint)

[Joy11] Joyal, A.: Notes on quasi-categories (2011, preprint)

[Kel06] Keller, B.: A_∞-algebras, modules and functor categories. In: Trends in Representation Theory of Algebras and Related Topics. Contemp. Math., vol. 406, pp. 67–93. Am. Math. Soc., Providence (2006)

[KM08] Kahn, B., Maltsiniotis, G.: Structures de dérivabilité. Adv. Math. **218**(4), 1286–1318 (2008)

[KS06] Kashiwara, M., Schapira, P.: Categories and Sheaves. Grundlehren der Mathematischen Wissenschaften [Fundamental Principles of Mathematical Sciences], vol. 332, p. 497. Springer, Berlin (2006). ISBN 978-3-540-27949-5; 3-540-27949-0

[Lur09] Lurie, J.: On the classification of topological field theories. arXiv e-prints (2009). arXiv:0905.0465

[Lur09a] Lurie, J.: Derived algebraic geometry. Doctoral thesis (2009, preprint)

[Lur09c] Lurie, J.: Higher algebra (2009, preprint)

[Lur09d] Lurie, J.: Higher Topos Theory. Annals of Mathematics Studies, vol. 170, p. 925. Princeton University Press, Princeton (2009). ISBN 978-0-691-14049-0; 0-691-14049-9

[LV12] Loday, J.-L., Vallette, B.: Algebraic Operads. Grundlehren der Mathematischen Wissenschaften [Fundamental Principles of Mathematical Sciences], vol. 346. Springer, Heidelberg (2012). ISBN 978-3-642-30361-6. xxiv+634 pp.

[Mer10] Merkulov, S.A.: Wheeled pro(p)file of Batalin-Vilkovisky formalism. Commun. Math. Phys. **295**, 585–638 (2010). doi:10.1007/s00220-010-0987-x. arXiv:0804.2481

[Qui67] Quillen, D.G.: Homotopical Algebra. Lecture Notes in Mathematics, vol. 43, p. 156. Springer, Berlin (1967)

[Rez09] Rezk, C.: A cartesian presentation of weak n-categories. arXiv e-prints (2009). arXiv:0901.3602

[Sim10] Simpson, C.T.: Homotopy theory of higher categories. arXiv e-prints (2010). arXiv:1001.4071

[Tam95] Tamsamani, Z.: On non-strict notions of n-category and n-groupoid via multisimplicial sets (1995). arXiv:math.AG/9512006

[Tam98] Tamarkin, D.E.: Another proof of M. Kontsevich formality theorem. arXiv Mathematics e-prints (1998). arXiv:math/9803025

[Toe02] Toen, B.: From homotopical algebra to homotopical algebraic geometry: lectures in Essen (2002). http://www.math.univ-toulouse.fr/~toen/essen.pdf

[Toe04] Toen, B.: Towards an axiomatization of the theory of higher categories. arXiv Mathematics e-prints (2004). arXiv:math/0409598

[TV02] Toen, B., Vezzosi, G.: Homotopical algebraic geometry I: topos theory. arXiv Mathematics e-prints (2002). arXiv:math/0207028

[TV08] Toën, B., Vezzosi, G.: Homotopical Algebraic Geometry. II. Geometric Stacks and Applications. Mem. Am. Math. Soc. 193(902):x+224 (2008)

Chapter 10
A Glimpse at Homotopical Geometry

Homotopical geometry is a mathematical device designed to study and formulate obstruction theoretic problems in geometry from a geometric viewpoint. We may briefly say that it is an enhancement of the parametrized and functional geometry approach presented in Chap. 3, obtained by replacing there everywhere the (classical) category SETS of classical sets by the (homotopical) ∞-category ${}^{\infty}$GRPD of ∞-groupoids.

Homotopical geometry started with the work of Serre [Ser70] (on local derived intersection multiplicities), Illusie [Ill71] (on the deformation theory of singular spaces), and Quillen [Qui70] (on the deformation theory of algebras) to cite some of the many grounders. It has found a new impetus with Grothendieck's long manuscript [Gro13], Simpson-Hirschowitz' paper [SH98], and the work of the homotopy theory community (see e.g., Jardine's work [Jar87] on simplicial presheaves).

The algebraic version of this theory is now fully developed, with the foundational works given by Toen-Vezzosi's "Homotopical algebraic geometry I and II" [TV02, TV08], and Lurie's "Higher topos theory" [Lur09d], "Higher algebra" [Lur09c] and "Derived algebraic geometry" [Lur09a].

The non-algebraic setting is currently in development (see Spivak's thesis [Spi08] and Lurie's work [Lur09a]), with inspirations coming from Lawvere's thesis [Law04], the synthetic differential geometry community (see e.g., Moerdijk-Reyes' book [MR91]), and Dubuc and Zilber's work [DZ94], that we have also used in the classical setting, to make the full presentation of homotopical geometric spaces clear and consistent.

We give here, basing ourselves on the doctrinal approach to categorical logic, described in Sect. 2.1, a systematic construction method for homotopical spaces, which is somewhat more flexible than the methods present in the literature, and that is robust and general enough to be adapted. This presentation is also fully compatible with the pedagogical spirit of our approach based on the higher Yoneda lemma.

F. Paugam, *Towards the Mathematics of Quantum Field Theory*,
Ergebnisse der Mathematik und ihrer Grenzgebiete. 3. Folge / A Series of Modern Surveys in Mathematics 59, DOI 10.1007/978-3-319-04564-1_10,

10.1 Motivations

In physics, for example in gauge theory, one often has to study spaces defined by (say Euler-Lagrange) equations that are not smooth, because they are given by non-transverse intersections (in gauge theory, this is related to the so-called Noether relations, to be treated in Chap. 13). This non-smoothness means that one does not have a system of local coordinates on the solution space. For example, consider a functional $S : X \to \mathbb{R}$ on a given space. Its space of critical points may be described by the intersection

$$T = \mathrm{im}(dS) \cap T_0^* X \subset T^* X$$

of the image of the differential and of the zero section in the cotangent bundle. For gauge theories, this intersection is not transverse, and one has to compute the *derived* (or *homotopical*) intersection

$$T^h = \mathrm{im}(dS) \cap^h T_0^* X$$

to get a better space, with nice local coordinates. To illustrate this general problem of transversality on a simple example, consider two lines D_1 and D_2 in the algebraic affine space $\mathbb{R}^2$, defined by the equations $x = 0$ and $y = 0$. Their intersection is given by the origin $(0,0) \in \mathbb{R}^2$ whose coordinate ring may be computed as the tensor product

$$\mathcal{O}(D_1 \cap D_2) = \mathcal{O}(D_1 \times_{\mathbb{R}^2} D_2) = \mathcal{O}(D_1) \otimes_{\mathcal{O}(\mathbb{R}^2)} \mathcal{O}(D_2)$$

that gives explicitly

$$\mathcal{O}(D_1 \cap D_2) = A/(x) \otimes_A A/(y) \cong A/(x,y) \cong \mathbb{R},$$

where $A = \mathbb{R}[x, y]$. This space is of dimension zero and the intersection is transverse. Now consider the intersection of D_1 with itself. It is a non-transverse intersection whose usual points are given by D_1, that is of dimension 1. We would like to force the transversality of this intersection to get a space of dimension 0. This is easily done by defining the derived intersection to be the space $D_1 \cap^h D_1$ with function algebra

$$\mathcal{O}\big(D^1 \cap^h D_1\big) := \mathcal{O}(D_1) \otimes^{\mathbb{L}}_{\mathcal{O}(\mathbb{R}^2)} \mathcal{O}(D_1)$$

given by the derived tensor product. This may be computed, in this case of a linear equation, by taking the projective resolution of $A/(x)$ as an A-module

$$0 \to A \xrightarrow{.x} A \to A/(x) \to 0.$$

The derived tensor product (of A-modules) is then equal to the complex of modules over $A/(x)$ given by

$$[A \xrightarrow{.x} A] \otimes_A A/(x) = \big[A/(x) \xrightarrow{0} A/(x)\big].$$

It is thus the zero map with cohomology spaces $H^0 = A/(x)$ and $H^1 = A/(x)$. If we define the derived dimension of the derived intersection as the Euler characteristic of this complex, we get

$$\dim\bigl(D_1 \cap^h D_1\bigr) = \dim\bigl(H^0\bigr) - \dim\bigl(H^1\bigr) = 1 - 1 = 0,$$

so that the derived intersection is transverse, at least in terms of dimension. This idea of improving on intersection theory by using cohomological methods was already present in Serre's work [Ser70] and in Illusie [Ill71] and Quillen [Qui70]. The case of spaces of trajectories is computed in a similar way, but one has to pass to the setting of $\mathcal{D}$-geometry. This will be described in Chap. 13, where we describe general gauge theories.

The above example is also related to the problem of defining a nice deformation theory and differential calculus on a non-smooth space. To illustrate this, take a smooth variety (for example the affine plane $\mathbb{A}^2 := \mathrm{Spec}(\mathbb{R}[x, y])$) and an equation in it whose solution space M is not smooth (for example $xy = 0$). Then one cannot do a reasonable differential calculus directly on M. However, one can replace the quotient ring $A = \mathbb{R}[x, y]/(xy)$ by a dg-algebra A' that is a cofibrant resolution of A as an $\mathbb{R}[x, y]$-algebra, and do differential calculus on this new dg-algebra. Morally, we may think of $\mathrm{Spec}(A')$ in $\mathbb{A}^2$ as a kind of homotopical infinitesimal tubular neighborhood of M in $\mathbb{A}^2$. This will give a much better behaved non-smooth calculus. For example, the space of differential forms on A' gives the full cotangent complex of A.

To formalize this, one has to define a differential calculus "up to homotopy". This can be done in the general setting of *model* categories of algebras (see Toen-Vezzosi's seminal work [TV08] and Hovey's book [Hov99]), which also encompasses a good notion of higher stack, necessary to correctly formalize quotients and moduli spaces in covariant gauge theory. Here we will only give a sketch of this theory, whose application to physics will certainly be very important, particularly in gauge theory, where physicists independently discovered similar mathematical structures in the BRST-BV formalism.

We may also motivate the introduction of homotopical methods in physics by the fact that many of the physicists' spaces of fields are equivariant objects. For example, in Yang-Mills theory, one works with a variable principal G-bundle P over a given spacetime M, which one may formalize properly in equivariant geometry (the theory of ∞-stacks) as a map

$$P : M \to BG$$

where BG is the smooth classifying space for principal G-bundles. It is very convenient to think of principal bundles in this way because this gives improved geometric tools to work on them. The proper setting for differential calculus on spaces like BG is given by derived geometry, which has been alluded to above.

Another good reason for using methods of homotopical geometry is the fact that the quantization problem itself is in some sense a deformation problem: one wants to pass from a commutative algebra, say over $\mathbb{R}$, to an associative (or more generally a factorization) algebra, say over $\mathbb{R}[[\hbar]]$, by an infinitesimal deformation process (see Costello and Gwilliam's work [CG10]). The modern methods of deformation theory are deeply rooted in homotopical geometry, as one can see in the expository work of Lurie [Lur10] of his formalization of the derived deformation theory program of Deligne, Drinfeld and Kontsevich. The conceptual understanding of the mathematical problem solved by the quantization and renormalization process thus uses homotopical methods.

10.2 Homotopical Spaces

The main idea of homotopical geometry is to generalize the functorial approach to spaces by the following systematic process, which is an almost direct ∞-categorical generalization of the functorial approach to geometry presented in Chap. 3. This categorification, which is very much in the doctrinal spirit of Sect. 2.1, is obtained by replacing the category SETS of usual sets (i.e., 0-categories) by the ∞-category ${}^{\infty}$GRPD of higher groupoids (i.e., ${}^{\infty}$0-categories).

1. Define an ∞-category ALG_h of homotopical algebras (functors of functions) as the ∞-category of functors

$$A : (\text{LEGOS}, \mathcal{C}) \to \left({}^{\infty}\text{GRPD}, \{\text{all limits}\}\right)$$

   commuting with particular limits (e.g., finite products or transversal pullbacks), where ${}^{\infty}$GRPD is the ∞-category of ∞-groupoids, and LEGOS is a category of building blocks with a Grothendieck topology τ.
2. Extend the Grothendieck topology τ to a homotopical analog of the Grothendieck topology τ_h on $\text{LEGOS}_h := \text{ALG}_h^{op}$, by defining the class of morphisms

$$Y \to X$$

   of simplicial presheaves on LEGOS^h that are hypercoverings (see Definition 10.2.3) as those whose pullbacks along all points $U \to X$ of X with values in $U \in$ LEGOS are hypercoverings for the original topology τ. Denote by τ_h the corresponding class of cocones in LEGOS_h.
3. Define homotopical functors of points (derived spaces) as ∞-functors

$$X : \left(\text{LEGOS}_h^{op}, \tau_h\right) \to \left({}^{\infty}\text{GRPD}, \{\text{all limits}\}\right)$$

   sending hypercovering cocones to limits.

The analogous construction with SETS-valued functors describes the category of sheaves of sets on LEGOS as a localization of the category of presheaves by local equivalences. One may find a refined way of defining general derived stacks, based on Lurie's notion of geometry, in [Lur09b].

We now give a concrete presentation of the above general methods in the setting of Quillen's model categories. This gives, for example, a finer take on the computational aspects of mapping spaces between stacks. The reader only interested in formal properties of higher and derived stacks may skip the rest of this section.

The homotopical analog of a Grothendieck topology τ is the following.

Definition 10.2.1 Let (LEGOS, W) be a model category with homotopy fiber products. A *model topology* τ on LEGOS is the data, for every lego U, of covering families $\{f_i : U_i \to U\}_{i\in I}$ of morphisms in $h(\text{LEGOS})$, satisfying:

1. (Homotopy base change) For every morphism $f : V \to U$ in $h(\text{LEGOS})$ and every covering family $\{f_i : U_i \to U\}$ of U, $f \times_U^h f_i : V \times_U^h V_i \to V$ is a covering family.
2. (Local character) If $\{f_i : U_i \to U\}$ is a covering family and $\{f_{i,j} : U_{i,j} \to U_i\}$ are covering families, then $\{f_i \circ f_{i,j} : U_{i,j} \to U\}$ is a covering family.
3. (Isomorphisms) If $f : U \to V$ is an isomorphism in $h(\text{LEGOS})$, it is a covering family.

A triple (LEGOS, W, τ) composed of a model category (LEGOS, W) and a model topology τ is called a *model site*.

We now define the notion of a hypercovering, which replaces the classical notion of (nerve of a) covering in the setting of homotopical geometry.

Definition 10.2.2 Let $\Delta_{\leq n} \subset \Delta^{op}$ be the full subcategory on the objects [1], [2], ..., [n]. The truncation functor

$$\tau_n : \text{SSETS} = \underline{\text{Mor}}_{\text{CAT}}(\Delta^{op}, \text{SETS}) \to \underline{\text{Mor}}_{\text{CAT}}(\Delta_{\leq n}, \text{SETS}) =: \text{SSETS}_{\leq n}$$

has a left adjoint

$$\text{sk}_n : \text{SSETS}_{\leq n} \to \text{SSETS},$$

called the *n-skeleton*, and a right adjoint

$$\text{cosk}_n : \text{SSETS}_{\leq n} \to \text{SSETS}$$

called the *n-coskeleton*.

Definition 10.2.3 Let (LEGOS, τ) be a site. A morphism $X \to Y$ of simplicial presheaves $X, Y \in \text{PR}_{\text{SSETS}}(\text{LEGOS})$ is called an *hypercovering for the topology* τ if for all n, the canonical morphism

$$Y_n \to (\text{cosk}_{n-1} Y)_n \underset{(\text{cosk}_{n-1} X)_n}{\times} X_n$$

is a local epimorphism for τ.

We can now give a sketch of the definition of the category of general homotopical spaces. We refer to Toen-Vezzosi [TV08] for more precise definitions.

Definition 10.2.4 Let (LEGOS, τ, W) be a model category equipped with a model topology. Let (C, W_C) be a model category. For $x \in \text{LEGOS}$ and $A \in C$, one defines a functor

$$\begin{aligned} A \otimes h_x : \text{LEGOS}^{op} &\to C \\ y &\mapsto \coprod_{\text{Hom}(y,x)} A. \end{aligned}$$

- The model category

$$\left(\text{PR}_{(C,W_C)}(\text{LEGOS}, W), W_{global}\right)$$

  of C-valued *prestacks* on LEGOS is the category of functors

$$X : \text{LEGOS}^{op} \to C,$$

  equipped with the model structure obtained by left Bousfield localization of the object-wise model structure (induced by the model structure W_C on C) by the weak equivalences $A \otimes h_f$ for $f \in W$ and A in a family of objects that generates C by homotopical colimits.
- The model category

$$\left(\text{SH}_{(C,W_C)}(\text{LEGOS}, W, \tau), W_{local}\right)$$

  of C-valued *stacks* on LEGOS is the left Bousfield localization of

$$\text{PR}_{(C,W_C)}(\text{LEGOS}, W, \tau)$$

  by the class of homotopy τ-hypercoverings.

Definition 10.2.5 Let (LEGOS, τ) be a site, equipped with the trivial model structure whose weak equivalences are the isomorphisms, and $(C, W_C) = (\text{SSETS}, W_{eq})$ be the model category of simplicial sets. The ∞-category of *higher stacks* is the ∞-localization

$$\text{STACKS}_\infty(\text{LEGOS}, \tau) := \text{SH}_{(\text{SSETS}, W_{eq})}(\text{LEGOS}, W_{triv}, \tau)\left[W_{local}^{-1}\right].$$

Definition 10.2.6 Let (LEGOS, τ) be a site, equipped with a family $\mathcal{C}$ of limit cones (e.g. transversal pullbacks), and ALG_h be the model category of covariant functors

$$A : \text{LEGOS} \to \text{SSETS}$$

that commute homotopically with limits in $\mathcal{C}$, and that are quotients of algebras of the form $\mathcal{O}(U)$. Equip ALG_h with the Grothendieck topology τ_h induced by τ, and assume that $(\mathrm{ALG}_h^{op}, \tau_h)$ is a model site. Let $(C, W_C) = (\mathrm{SSETS}, W_{eq})$ be the model category of simplicial sets. The ∞-category of *derived higher stacks* is the ∞-localization

$$\mathrm{DSTACKS}_\infty(\mathrm{LEGOS}, W, \tau) := \mathrm{SH}_{(\mathrm{SSETS}, W_{eq})}(\mathrm{LEGOS}_h, W, \tau_h)\big[W_{local}^{-1}\big].$$

If LEGOS is the category of affine schemes over $\mathbb{Q}$ (i.e., opposite to $\mathbb{Q}$-algebras), one can also replace simplicial algebras by differential graded algebras in the above construction, since both model categories are Quillen equivalent.

One can also work with unbounded differential graded algebras (or with ring spectra, in the simplicial language). This also gives a notion of (unbounded) derived stacks.

10.3 Non-abelian Cohomology

The main interest of derived and homotopical geometry is that it gives a setting, called non-abelian cohomology, that allows one to treat the problems of obstruction theory geometrically. These ideas of using homotopy theory in geometry were developed by many people, starting with Quillen [Qui70] and Illusie [Ill71]. The idea of non-abelian cohomology was already present in Grothendieck's long letter to Quillen [Gro13]. It was also developed by Simpson and Hirschowitz [SH98]. A systematic treatment is now available in the work of Toen-Vezzosi (see [TV02] and [TV08]), and Lurie (see [Lur09d] and [Lur09a]). We refer to Sati-Schreiber-Stasheff [SSS09] and to Schreiber's book [Sch11] for nice applications of this formalism in the description of higher gauge theories.

Let (LEGOS, τ) be a site with a family $\mathcal{C}$ of limit cones. In this section, we will denote by STACKS_∞ the ∞-category of ∞-stacks on the site $(\mathrm{LEGOS}, \mathcal{C})$, and by $\mathrm{DSTACKS}_\infty$ the ∞-category of derived ∞-stacks (given by simplicial sheaves on the model site of simplicial generalized algebras $A : \mathrm{LEGOS} \to \mathrm{SSETS}$). Objects in STACKS_∞ will simply be called stacks, and objects in $\mathrm{DSTACKS}_\infty$ will be called derived stacks.

Definition 10.3.1 Let X and G be two stacks or derived stacks. One defines:

1. *the non-abelian G-cohomology space on X* as the ∞-groupoid $\underline{\mathrm{Mor}}(X, G)$ of morphisms from X to G;
2. *non-abelian G-valued cocycles on X* as objects of $\underline{\mathrm{Mor}}(X, G)$;
3. *non-abelian coboundaries* between these cocycles as morphisms in $\underline{\mathrm{Mor}}(X, G)$;
4. *non-abelian cohomology of X with coefficients in G* as the set

$$H(X, G) := \pi_0\big(\underline{\mathrm{Mor}}(X, G)\big)$$

of connected components of $\underline{\mathrm{Mor}}(X, G)$.

Example 10.3.1 If A is a sheaf of abelian groups, and $A[-n]$ is the corresponding complex concentrated in degree n, the Dold-Kan correspondence (see Proposition 9.8.1) gives a stack $K(A,n)$ such that

$$H^n(X,A) = H\big(X, K(A,n)\big).$$

This shows that non-abelian cohomology generalizes the usual cohomology of abelian sheaves.

Definition 10.3.2 The *pointed loop space* ΩA of a pointed object $\{*\} \to A$ is defined as the homotopy pullback

$$\begin{array}{ccc} \Omega A & \longrightarrow & \{*\} \\ \downarrow & & \downarrow \\ \{*\} & \longrightarrow & A \end{array}$$

A *delooping* $\{*\} \to BA$ of $\{*\} \to A$ is a pointed object such that

$$A \cong \Omega BA.$$

A pointed object that has a delooping is called an ∞*-group*. If A is an ∞-group and $X \to BA$ is a non-abelian BA-valued cocycle, the homotopy fiber product

$$\begin{array}{ccc} P & \longrightarrow & \{*\} \\ \downarrow & & \downarrow \\ X & \longrightarrow & BA \end{array}$$

is called the corresponding *principal* ∞*-bundle*.

Note that a section $s : X \to P$ of a principal ∞-bundle is equivalent to a morphism $s : X \to \{*\}$ over BA, i.e., a homotopy commutative diagram

$$\begin{array}{ccc} & & \{*\} \\ & \overset{s}{\nearrow} & \downarrow \\ X & \longrightarrow & BA \end{array}$$

Example 10.3.2 Abelian sheaves of groups have the particular property of being deloopable arbitrarily many times. If A is a sheaf of abelian groups, one has

$$K(A,n) \cong B^n A.$$

The principal ∞-bundles associated to cocycles $X \to K(A, n)$ are called n-gerbes. Their equivalence classes correspond exactly to the cohomology set $H^n(X, A)$.

Example 10.3.3 Suppose we are working in the algebraic, differential geometric, or analytic setting. A semi-simple group G is only once deloopable, with delooping the moduli space BG of principal G-bundles. In general, if G is an ∞-group and A is an arbitrarily deloopable ∞-stack, one thinks of

$$H^n(G, A) := \pi_0\big(\underline{\mathrm{Mor}}(BG, B^n A)\big)$$

as the group cohomology of G with values in A.

Example 10.3.4 As before, suppose we are working in the algebraic, differential geometric, or analytic setting. Let G be a semisimple group and $G \to \mathrm{GL}(V)$ be a representation of G. The push-forward

$$\begin{array}{ccc} V \times G & \longrightarrow & V \\ \downarrow & & \downarrow \\ \{*\} & \longrightarrow & [V/G] \end{array}$$

defines a fiber bundle $[V/G] \to BG$ with fiber V over the point. This bundle has an n-delooping

$$K_{BG}(V, n) \to BG$$

relative to BG for all n, with fiber $K(V, n)$ over the point. If $P : X \to BG$ is a principal G-bundle on X, the bundle $V^P := P \times_G [V/G]$ associated to the representation V is given by the homotopy pullback

$$\begin{array}{ccc} V^P & \longrightarrow & [V/G] \\ \downarrow & & \downarrow \\ X & \longrightarrow & BG \end{array}$$

A morphism $s : X \to [V/G]$ of stacks is the same thing as a section $s : X \to V^P$ of the bundle associated to the principal bundle $P : X \xrightarrow{s} [V/G] \to BG$. More generally, the bundle $K(V^P, n)$ is defined as the pullback

$$\begin{array}{ccc} K(V^P, n) & \longrightarrow & K_{BG}(V, n) \\ \downarrow & & \downarrow \\ X & \longrightarrow & BG \end{array}$$

A morphism $s : X \to K_{BG}(V, n)$ is the same thing as a cohomology class $s \in H^n(X, V^P)$.

Example 10.3.5 Let X be a smooth variety, regarded as a (non-derived) stack modeled on algebras in the given geometrical setting. The infinitesimal ∞-groupoid of X is the stack associated to the prestack whose degree n component is

$$\Pi_{\infty}^{inf}(X)_n(T) := \left\{x \in X^n(T),\ \exists U \to T^n \text{ thickening },\ \forall i, j,\ x_{i|U} = x_{j|U}\right\}.$$

It is equipped with its natural simplicial structure induced by its inclusion in the simplicial space $n \mapsto X^n$. Let G be an ∞-group. The stack of principal G-bundles with flat connection on X is given by the stack

$$\text{BUNConn}_G(X) := \underline{\text{Hom}}\left(\Pi_{\infty}^{inf}(X), BG\right).$$

This space plays an important role in higher gauge theory. If G is a semisimple group (in the algebraic, smooth or analytic context), one can give a simplified description of the above space as

$$\text{BUNConn}_G(X) \cong \underline{\text{Hom}}\left(\Pi_1^{DR} X, BG\right),$$

where $\Pi_1^{DR} X := \tau_1(\Pi_{\infty}^{inf}(X))$ is the groupoid associated to the infinitesimal groupoid. This is a subgroupoid of the groupoid $\text{Pairs}(X)$ of pairs of points in X, whose maps to BG identify with flat G-equivariant Grothendieck connections on principal G-bundles.

Example 10.3.6 If (X, ω) is a smooth symplectic manifold and L_1 and L_2 are two Lagrangian submanifolds, their derived intersection gives the stack whose points with values in a simplicial smooth algebra A are

$$\left(L_1 \cap^h L_2\right)(A) = \left\{f : [0, 1] \to \left|X(A)\right|,\ f(0) \in \left|L_1(A)\right|,\ f(1) \in \left|L_2(A)\right|\right\},$$

where $|-| : \text{SSETS} \to \text{TOP}$ is the geometric realization functor. This derived stack is actually representable by a simplicial smooth algebra, i.e., it may be described as the derived spectrum

$$L_1 \cap^d L_2 = \mathbb{R}\underline{\text{Spec}}\left(\mathcal{C}^{\infty}(L_1) \otimes^{\mathbb{L}}_{\mathcal{C}^{\infty}(X)} \mathcal{C}^{\infty}(L_2)\right).$$

The above applies in particular to the case of a cotangent space $X = T^*M$ and the Lagrangian submanifolds are given by the zero section $L_1 = T_0^*M$ and the image $L_2 = \text{im}(dS)$ of the differential of a given function $S : M \to \mathbb{R}$. The corresponding intersections are the usual, homotopical and derived critical points of S. One usually uses the stack of derived critical points to get a well behaved differential calculus.

Definition 10.3.3 The *free loop space* LX of a derived stack X is the homotopy pullback

$$\begin{array}{ccc} LX & \longrightarrow & X \\ \downarrow & & \downarrow \Delta \\ X & \xrightarrow[\Delta]{} & X \times X \end{array}$$

The simplicial algebra of functions on LX is called the *Hochschild homology* of X.

Note that this definition really gives the right Hochschild homology in algebraic, smooth and analytic geometry, on a uniform ground.

10.4 Differential Cohomology

Differential cohomology is a differential version of non-abelian cohomology which is used to formalize gauge theories on higher bundles with connections. We will give examples of such theories in Sect. 16.5. We refer to Sati-Schreiber-Stasheff [SSS09] for a short introduction to non-abelian cohomology and its application in physics. We base our presentation on Schreiber's book [Sch11], which gives a systematic and elegant treatment of higher gauge theories in the setting of cohesive homotopical geometry.

The general setting is given by an ∞-topos (i.e., ∞-category of sheaves on a homotopical site, see [Lur09d] for a thorough study and various characterizations)

$$\mathrm{H} = \mathrm{SH}_{\infty\mathrm{GRPD}}(C, \tau)$$

whose global section adjunction

$$\mathrm{Disc} : {}^{\infty}\mathrm{GRPD} \leftrightarrows \mathrm{H} : \Gamma$$

is equipped with additional adjunctions

$$\Pi : \mathrm{H} \leftrightarrows {}^{\infty}\mathrm{GRPD} : \mathrm{Disc} \quad \text{and} \quad \Gamma : \mathrm{H} \leftrightarrows {}^{\infty}\mathrm{GRPD} : \mathrm{coDisc},$$

called the cohesive ∞-groupoid and the codiscrete functor, satisfying additional conditions. We will work with the cohesive ∞-topos

$$\mathrm{H} = \mathrm{STACKS}_{\infty}(\mathrm{OPEN}_{\mathcal{C}^{\infty}}, \tau)$$

of smooth ∞-stacks, but all constructions of this section generalize to other cohesive situations. As usual, we denote by $\underline{\mathrm{Mor}}(X, Y)$ the ∞-groupoid of morphisms between two smooth ∞-stacks.

Definition 10.4.1 Let $[c] \in H(B, C) = \pi_0(\underline{\mathrm{Mor}}(B, C))$ be a non-abelian cohomology class, called a *characteristic class*, and denote by A its homotopy fiber

$$\begin{array}{ccc} A & \longrightarrow & \{*\} \\ \downarrow & & \downarrow \\ B & \xrightarrow{c} & C \end{array}$$

The *c-twisted A-valued non-abelian cohomology space of X* is defined as the homotopy pullback

$$\begin{array}{ccc} \underline{\mathrm{Mor}}_{c-tw}(X, A) & \xrightarrow{tw} & H(X, C) \\ \downarrow & & \downarrow s \\ \underline{\mathrm{Mor}}(X, B) & \xrightarrow{\theta_*} & \underline{\mathrm{Mor}}(X, C) \end{array}$$

where $s : H(X, C) := \pi_0(\underline{\mathrm{Mor}}(X, C)) \to \underline{\mathrm{Mor}}(X, C)$ is a given choice of section.

We now define the notion of higher flat connection and flat differential cohomology.

Definition 10.4.2 Let X be an object of H. The *smooth ∞-groupoid of X* is the smooth singular space

$$\Pi(X) := \mathrm{LConst}\big(\Pi_\infty(|X|)\big)$$

where $\Pi_\infty(|X|)$ is the singular ∞-groupoid of the geometric realization $|X|$ of X and LConst is the locally constant ∞-groupoid stack associated to a given ∞-groupoid.

We refer to Schreiber [Sch11] for the proof of the following result.

Theorem 10.4.1 *The smooth ∞-groupoid functor $\Pi : \mathrm{H} \to \mathrm{H}$ has a right adjoint denoted by*

$$\flat : \mathrm{H} \to \mathrm{H}.$$

For A an object of H, we define $\flat_{dR} A := \{*\} \times^h_A \flat A$. If G is an ∞-group, the defining fibration sequence of $\flat_{dR} G$ extends to

$$\flat_{dR} G \to G \to \flat G \xrightarrow{curv} \flat_{dR} BG,$$

and induces in particular a morphism

$$\theta : G \to \flat_{dR} BG$$

called the Mauer-Cartan form *on G.*

Definition 10.4.3 The *flat cohomology* of X with coefficients in A is defined by

$$H_{flat}(X, A) := H\big(\Pi(X), A\big) = H(X, \flat A).$$

The *de Rham cohomology* of X with coefficients in A is defined by

$$H_{dR}(X, A) := H(X, \flat_{dR} A).$$

The corresponding spaces are denoted by $\underline{\mathrm{Mor}}_{flat}(X, A)$ and $\underline{\mathrm{Mor}}_{dR}(X, A)$. The *differential cohomology space* of X with coefficients in an ∞-group A is the θ-twisted cohomology, where $\theta : A \to \flat_{dR} BA$ is the Mauer-Cartan form. More concretely, it is the homotopy fiber product

$$\begin{array}{ccc} \underline{\mathrm{Mor}}_{diff}(X, A) & \longrightarrow & H_{dR}(X, BA) \\ \downarrow & & \downarrow \\ \underline{\mathrm{Mor}}(X, A) & \xrightarrow{\theta_*} & \underline{\mathrm{Mor}}_{dR}(X, BA) \end{array}$$

where the right arrow is given by a choice of representative for each class in π_0. The *differential cohomology* of X with coefficients in an n-deloopable ∞-group A is

$$H^n_{diff}(X, A) := \pi_0\big(\underline{\mathrm{Mor}}_{diff}(X, B^n A)\big).$$

Example 10.4.1 If A is a Lie group, the cohomology

$$H^1(X, A) := H(X, BA)$$

is the set of isomorphism classes of smooth principal A-bundles on X. The flat cohomology set

$$H^1_{flat}(X, A) := H_{flat}(X, BA)$$

is the set of isomorphism classes of smooth principal A-bundles on X equipped with a flat connection. If $A = U(1)$, the differential cohomology set

$$H^n_{diff}(X, A) := H_{diff}\big(X, B^n A\big)$$

is the Deligne cohomology of X, usually defined as the homotopy pullback of integral cohomology and the complex of differential forms along real cohomology.

Definition 10.4.4 For $X \in \mathrm{H}$, the space $\underline{\mathrm{Mor}}_{conn}(X, BG)$ of *∞-connections* is defined as the homotopy pullback

$$\begin{array}{ccc} \underline{\mathrm{Mor}}_{conn}(X, BG) & \longrightarrow & \prod_{c_i \in H(BG, B^{n_i} A), i \geq 1} \underline{\mathrm{Mor}}_{diff}(X, B^{n_i} A) \\ \downarrow \eta & & \downarrow \\ \underline{\mathrm{Mor}}(X, BG) & \longrightarrow & \prod_{c_i \in H(BG, B^{n_i} A), i \geq 1} \underline{\mathrm{Mor}}(X, B^{n_i} A) \end{array}$$

One defines:

- an ∞-connection as a cocycle $\nabla \in \mathrm{H}_{conn}(X, BG) := \pi_0(\underline{\mathrm{Mor}}_{conn}(X, BG))$;
- the underlying ∞-bundle as $P := \eta(\nabla)$;
- a gauge transformation as a morphism in $\underline{\mathrm{Mor}}_{conn}(X, BG)$;
- the refined Chern-Weil homomorphism as the homomorphism

$$c : H_{conn}(BG, A) \to H^n_{diff}(X, A)$$

associated to $c \in H^n(BG, A)$.

We will see explicit examples of higher gauge field theories, defined by applying the present formalism of non-abelian cohomology, in Chap. 16.

Another way to formulate the above Chern-Weil theory is given by the following procedure, which may be found in [Sch11], Sect. 2.9.

Define

$$\Omega^1_{cl}(-, A) := \flat_{dR} BA,$$

and for each $n \in \mathbb{N}$, recursively choose a morphism

$$\Omega^{n+1}_{cl}(-, A) \to B\Omega^n_{cl}(-, A)$$

out of a 0-truncated abelian group object (i.e., a sheaf of abelian groups).

Definition 10.4.5 For any $n \in \mathbb{N}$, the *differential refinement $B^n A_{conn}$ of the classifying space of n-bundles* is the ∞-pullback

$$\begin{array}{ccc} B^n A_{conn} & \longrightarrow & \Omega^{n+1}_{cl}(-, A) \\ \downarrow & & \downarrow \\ B^n A & \xrightarrow{curv} & \flat_{dR} B^{n+1} A \end{array}$$

For X a manifold, define

$$H^n_{conn}(X, A) := \pi_0\big(\underline{\mathrm{Mor}}\big(X, B^n A_{conn}\big)\big).$$

Definition 10.4.6 Let $c : BG \to B^n A$ be a smooth characteristic map and $B^n A_{conn}$ be a differential refinement. A differential refinement of BG along c is an object BG_{conn} that fits into a factorization

$$\begin{array}{ccc} \flat BG & \xrightarrow{\flat c} & \flat B^n A \\ \downarrow & & \downarrow \\ BG_{conn} & \xrightarrow{\hat{c}} & B^n A_{conn} \\ \downarrow & & \downarrow \\ BG & \xrightarrow{c} & B^n A \end{array}$$

of the above natural vertical rectangle diagram.

If M is a manifold and $\hat{c} : BG_{conn} \to B^n A_{conn}$ is a differentially refined characteristic class, one may define the differentially refined Chern-Weil homomorphism

$$\hat{c} : \underline{\mathrm{Hom}}(M, BG_{conn}) \longrightarrow \underline{\mathrm{Hom}}\big(M, B^n A_{conn}\big)$$

from the moduli space of G-bundles with connection to the moduli space of A-n-gerbes with connection on M.

10.5 Geometric Stacks

We now give a short account of the notion of geometric higher stacks, originally due to Simpson and Hirschowitz [SH98]. We refer to [TV08], Sect. 1.3 for a complete treatment of geometric stacks, and to [LMB00] for the classical theory of algebraic stacks (i.e., geometric 1-stacks). We also refer to [Pri11] for a concrete description of geometric stacks in terms of simplicial diagrams of building blocks (which are affine schemes, in his case, but may be any reasonable kinds of generalized algebras in a geometric context). We prefer the approach through general stacks because it is also adapted to infinite dimensional spaces of fields (i.e., spaces of functions). The relation between geometricity and our considerations is that this is the most general finiteness hypothesis that one may impose on both the space M of parameters and the configuration space C of a classical physical system.

Definition 10.5.1 Let STACKS be the ∞-category of stacks (resp. derived stacks) on a site (LEGOS, τ) (resp. homotopical site ($\overline{\text{LEGOS}}$, τ)). Let $\mathbb{P}$ be a class of morphisms in the given site that is local with respect to the given topology. We define geometric stacks with respect to the class $\mathbb{P}$ by the following inductive process for $n \geq 0$:

1. A stack is (-1)-geometric if it is representable.
2. A morphism $f : X \to Y$ of stacks is (-1)-representable if for any representable stack Z and any morphism $Z \to Y$, the fiber product $Y \times_Z X$ is geometric.
3. A morphism $f : X \to Y$ is in (-1)-$\mathbb{P}$ if it is (-1)-representable and the pullback of f along a point $Z \to Y$ with values in a representable stack is a morphism in $\mathbb{P}$.
4. An n-atlas on a stack X is a family of morphisms $\{U_i \to X\}_{i \in I}$ such that:

   (a) each U_i is representable;
   (b) each morphism $U_i \to X$ is in $(n-1)$-$\mathbb{P}$;
   (c) the total morphism

   $$\coprod_{i \in I} U_i \to X$$

   is an epimorphism.
5. A stack X is n-geometric if its diagonal morphism $X \to X \times X$ is $(n-1)$-representable and X admits an n-atlas.
6. A morphism of stacks $X \to Y$ is n-representable if for any representable stack Z and any morphism $Z \to Y$, the pullback $X \times_Y Z$ is n-geometric.
7. A morphism of stacks $X \to Y$ is in n-$\mathbb{P}$ if it is n-representable and for any morphism $Z \to Y$, there exists an n-atlas $\{U_i\}$ of $X \times_Y Z$ such that each composite morphism $U_i \to X$ is in $\mathbb{P}$.

Definition 10.5.2 In algebraic, differential or analytic geometry, a stack that is geometric with respect to the class of smooth morphisms is called an Artin stack, and a stack that is geometric with respect to the class of étale morphisms is called a Deligne-Mumford stack.

Example 10.5.1 Most of the stacks used in non-abelian cohomology and gauge theory are geometric. At least, we will always suppose that the parameter stack M and the configuration stack C for trajectories are both geometric in a convenient sense.

1. If G is a linear Lie group acting on a manifold M, the stacky quotient $[M/G]$ is a geometric Artin stack. In particular, $BG := [*/G]$ is a geometric stack.
2. More generally, if A is a commutative Lie group, then the stack $B^n A$ is also geometric.
3. Lurie's version of Artin's representability criterion (see [TV08], Appendix and [Lur09a]) implies that if M is a proper smooth scheme and G is a linear algebraic group, then the mapping stack

   $$\underline{\mathrm{Hom}}(M, BG)$$

   is also geometric. This may not remain true in the smooth or analytic setting.

10.6 Homotopical Infinitesimal Calculus

We now briefly describe differential calculus relative to a given ∞-category $\mathcal{C}$, explaining the necessary modifications that need to be carried out to adapt the methods of Sect. 2.5 to the ∞-categorical setting. This presentation of differential invariants is originally due to Quillen [Qui70] (see also Illusie [Ill71]). We refer to Lurie [Lur09c] and Toen-Vezzosi [TV08] for thorough treatments of the cotangent complex.

Definition 10.6.1 Let $\mathcal{C}$ be an ∞-category with finite limits. The tangent fibration $T\mathcal{C} \to [I, \mathcal{C}]$ to an ∞-category is the fiberwise stabilization of its codomain fibration

$$\text{codom} : [I, \mathcal{C}] \to \left[\{0\}, \mathcal{C}\right] \cong \mathcal{C}.$$

More precisely, it is a categorical fibration $T\mathcal{C} \to \mathcal{C}$ (in the sense of, say, the model category ${}^{\infty}\text{CAT}$) universal with the property that for each $A \in \mathcal{C}$, its fiber $T\mathcal{C}_A$ over A is the stabilization of the overcategory $\mathcal{C}/A$. If $A \in \mathcal{C}$, we define $\text{MOD}(A) := T\mathcal{C}_A$.

Definition 10.6.2 A *cotangent complex functor* for an ∞-category $\mathcal{C}$ is an adjoint

$$\text{dom} : T\mathcal{C} \to \left[\{0\}, \mathcal{C}\right] \cong \mathcal{C} : \Omega^1$$

to the domain map on the tangent ∞-category.

Definition 10.6.3 A *square zero extension* in $\mathcal{C}$ is a morphism $A' \to A$ that is a torsor under a module $M \in T\mathcal{C}_A$. More precisely, if BM is a *delooping* of M, satisfying $\Omega BM \cong M$, this torsor structure is given by a morphism

$$A' \to BM$$

in the overcategory $\mathcal{C}/A$.

Using the above definition of square zero extension, one may extend what was done in Sect. 2.5 to the ∞-categorical setting. This gives, in particular, a notion of thickenings and jets functors, and of crystals of module and spaces. We carefully inform the reader that working with non-flat connections in the derived setting is a bit involved, since the curvature must be regarded as an additional structure on a given connection, not as a property.

We finish this section by an explicit definition of the de Rham space of a stack, generalized directly from the algebraic one studied by [GR11], and adapted to the various geometrical settings we will use in this book.

Definition 10.6.4 Let $X : \text{ALG}_h^{op} \to {}^{\infty}\text{GRPD}$ be a derived higher stack. The *de Rham stack* of X is defined by

$$X_{DR}(A) := X\big(\text{disc}\big(\pi_0(A)_{red}\big)\big),$$

where

- the adjunction

$$\pi_0 : \text{ALG}_h \leftrightarrows \text{ALG} : \text{disc}$$

  between the ∞-category ALG_h of homotopical algebras and the category ALG of usual algebras is induced by the usual adjunction

$$\pi_0 : {}^{\infty}\text{GRPD} \leftrightarrows \text{SETS} : \text{disc};$$

- the functor $\text{red} : \text{ALG} \to \text{ALG}$ sends an algebra A to the final object (which may be regarded as a functor, if it doesn't exist) B in the category of nilpotent morphisms $A \to B$ (see Definition 2.5.4), which correspond to classical infinitesimal thickenings $\underline{\text{Spec}}(B) = U \hookrightarrow T = \underline{\text{Spec}}(A)$.

10.7 Derived Symplectic Structures

We now give a short sketch of the notion of a closed form on a derived stack, and of a derived symplectic form, which can be found in [PTVV11] (see also [BN10] and [TV09a]). This gives a direct generalization of the usual notions that is quite well adapted to a global and canonical study of gauge theories. A striking difference between the classical and the derived setting is the following: the fact that a differential form is closed is not a property but an additional structure, in general.

From now on, we work in the ∞-category $\text{DSTACKS}_\infty(\text{LEGOS}, \tau)$ of derived stacks on one of the classical sites (algebraic, smooth or analytic geometry). We have already seen in Proposition 3.3.7 that one may view differential forms on a manifold M as functions on the superspace of maps $\underline{\text{Hom}}(\mathbb{R}^{0|1}, M)$. We will now present a refined version of this, in the setting of derived geometry, replacing $\mathbb{R}^{0|1}$ by the circle stack S^1.

Definition 10.7.1 The *circle stack* S^1 is the delooping $B\mathbb{Z}$ of the constant group stack $\mathbb{Z}$. The *derived loop space* is the stack

$$LX := \mathbb{R}\underline{\text{Hom}}\big(S^1, X\big).$$

The *unipotent loop space* is the stack

$$L^u X := \mathbb{R}\underline{\text{Hom}}(B\mathbb{G}_a, X).$$

The *formal loop space* $\hat{L}X$ is the formal completion of LX at the constant loop $X \subset LX$.

The natural morphism $S^1 = B\mathbb{Z} \to B\mathbb{G}_a$ induced by the inclusion $\mathbb{Z} \to \mathbb{G}_a$ induces a morphism

$$L^u X := \mathbb{R}\underline{\text{Hom}}(B\mathbb{G}_a, X) \to LX.$$

In the affine case, i.e., $X = \mathrm{Spec}(A)$, the above map is an equivalence. This means that the group stack $\underline{\mathrm{Aut}}(B\mathbb{G}_a)$ acts on LX. The more general case may be treated by using the following theorem of Ben-Zvi and Nadler [BN10].

Theorem 10.7.1 *Formal loops are unipotent, i.e., the natural map $\hat{L}X \to LX$ factors as*

$$\hat{L}X \to L^u X \to LX.$$

In particular, $\hat{L}X$ inherits an action of $\underline{\mathrm{Aut}}(B\mathbb{G}_a)$.

We can form the quotient stack

$$p : \big[\hat{L}X/\underline{\mathrm{Aut}}(B\mathbb{G}_a)\big] \to B\underline{\mathrm{Aut}}(B\mathbb{G}_a).$$

Definition 10.7.2 The *de Rham complex* $\mathrm{DR}(X)$ of X is the module

$$p_*\mathcal{O}_{[\hat{L}X/\underline{\mathrm{Aut}}(B\mathbb{G}_a)]}$$

over $B\underline{\mathrm{Aut}}(B\mathbb{G}_a)$.

Concretely, the group stack $\underline{\mathrm{Aut}}(B\mathbb{G}_a)$ is isomorphic to the semi-direct product $\mathbb{G}_m \rtimes B\mathbb{G}_a$, and the de Rham algebra is the module over its functions with underlying complex the derived exterior product

$$\mathrm{DR}(X) = \oplus_p (\wedge^p \Omega^1_X)[p],$$

and action of the functions on $\underline{\mathrm{Aut}}(B\mathbb{G}_a)$ given by the de Rham differential, for the $B\mathbb{G}_a$ part, and the grading, for the $\mathbb{G}_m$ part. This looks like a complicated formulation, but it is necessary to treat the derived case canonically.

More generally, one may also formalize analogs of $\mathcal{D}$-modules, and quasi-coherent sheaves with non-flat connections on derived Artin stacks using the loop space formalism, as in [BN10]. Here we only give a short sketch of these notions. We denote by Aff Toen's affinization functor (see [Toe00]), which sends a derived stack X to the universal morphism $X \to \mathrm{Aff}(X)$ from X to an affine stack. We have $\mathrm{Aff}(S^1) = \mathrm{Aff}(B\mathbb{Z}) = B\mathbb{G}_a$. Let $S^1 \to S^3 \to S^2$ be the Hopf fibration and consider its rotation

$$\begin{array}{ccc} \Omega S^3 & \longrightarrow & \Omega S^2 \\ \downarrow & & \downarrow \\ \{e\} & \longrightarrow & S^1 \end{array}$$

Definition 10.7.3 Let X be a derived Artin stack. A *graded de Rham module over* X is a quasi-coherent sheaf on the derived stack $[\hat{L}X/\mathrm{Aff}(S^1) \rtimes \mathbb{G}_m]$. A *curved graded de Rham module over* X is a quasi-coherent sheaf on the derived stack

$[\hat{L}X/\mathrm{Aff}(\Omega S^2) \rtimes \mathbb{G}_m]$. The *curvature* of a curved graded de Rham module $\mathcal{F}$ is the associated module over the derived stack $[\hat{L}X/\mathrm{Aff}(\Omega S^3) \rtimes \mathbb{G}_m]$.

We refer to [BN10] for a proof of the following theorem and of other related results.

Theorem 10.7.2 *Let X be a usual scheme. There are natural functors from $\mathcal{D}$-modules to graded de Rham modules over X, and from quasi-coherent sheaves with connections to curved graded de Rham modules.*

We now define the cyclic complex as the complex of $B\mathbb{G}_a$-equivariant (i.e., S^1-equivariant) de Rham forms on X.

Definition 10.7.4 The *negative cyclic complex* of X is the direct image

$$NC(X) := q_*\big(\mathrm{DR}(X)\big),$$

where $q : B\underline{\mathrm{Aut}}(B\mathbb{G}_a) \to B\mathbb{G}_m$ is the natural projection.

There is a natural morphism

$$NC(X) \to \mathrm{DR}(X)$$

that induces a morphism

$$NC(X)(p) \longrightarrow \wedge^p \Omega_X^1[p]$$

for all $p \geq 0$.

For E a complex of modules, we denote by $|E|$ the simplicial set associated to the truncation $\tau_{\leq 0}E$.

Definition 10.7.5 The simplicial set

$$\mathcal{A}^p(X, n) := \big|\wedge^p \Omega_X^1[n]\big|$$

is called the *space of p-forms of degree n* on the derived stack X. The simplicial set

$$\mathcal{A}^{p,cl}(X, n) := \big|NC(X)[n-p](p)\big|$$

is called the *space of closed p-forms of degree n* on the derived stack X. The homotopy fiber of the natural morphism

$$\mathcal{A}^{p,cl}(X, n) \to \mathcal{A}^p(X, n)$$

at a given form ω is called the *space of keys* of ω.

Note that keys of ω are precisely the structures needed to close the form ω. From now on we denote $\mathbb{L}_X$ the cotangent complex Ω_X^1.

Definition 10.7.6 A degree n derived symplectic form on a derived stack X is a closed 2-form $\omega \in \mathcal{A}^{2,cl}(X,n)$ of degree n such that the associated morphism

$$\omega : \mathbb{T}_X \longrightarrow \mathbb{L}_X[n]$$

is an isomorphism.

We now quote the important structure theorems of [PTVV11], which is valid in the algebraic setting.

Theorem 10.7.3 *Let X be a derived Artin stack endowed with an $\mathcal{O}$-orientation of dimension d, and (F,ω) be a derived Artin stack with an n-shifted symplectic structure ω. Then the derived mapping stack $\underline{\mathrm{Hom}}(X,F)$ carries a natural $(n-d)$-shifted symplectic structure.*

We now present an example that gives a clear relation between Yang-Mills theory and differential calculus on derived stacks. Let G be a smooth affine group with Lie algebra $\mathfrak{g}$. An invariant pairing on $\mathfrak{g}$ induces a closed 2-form on BG. The above theorem allows in particular the construction of a shifted $2(1-d)$ symplectic form on the derived stack

$$\mathrm{B\scriptstyle{UN}}\mathrm{Conn}^G_{flat}(M) := \underline{\mathrm{Hom}}(M_{DR}, BG)$$

of principal G-bundles with flat connection, from a fundamental class $[M] \in H^{2d}_{DR}(M,\mathcal{O})$.

We now discuss derived isotropic and Lagrangian structures.

Definition 10.7.7 Let (F,ω) be an Artin stack equipped with an n-shifted symplectic form ω. An *isotropic structure* on a morphism $f : X \to F$ of derived Artin stacks is a path h between 0 and the closed 2-form $f^*\omega$ in the space $\mathcal{A}^{2,cl}(X,n)$. An isotropic structure is called a *Lagrangian structure* if the associated morphism

$$\Theta_h : \mathbb{T}_{X/F} \to \mathbb{L}_X[n-1]$$

is a quasi-isomorphism of perfect complexes.

Theorem 10.7.4 *Derived Lagrangian intersections in an n-shifted derived Artin stack are equipped with an $(n-1)$-shifted symplectic form.*

10.8 Deformation Theory and Formal Geometry

The basis for this section is called the derived deformation theory program, and is essentially due to Deligne, Drinfeld and Kontsevich. We base our presentation on Lurie's ICM talk [Lur10] (see also [Lur11]), on Hinich's [Hin01] and [Hin99], and on Quillen's article [Qui69]. The theory given here is fully developed over a field, and is in active development over other bases by Toen-Vezzosi and Gaitsgory-Rozenblyum. There is no reasonable doubt about the fact that a version of it will pass through to this more general setting.

Let k be a field. A spectral k-algebra is an algebra in the stable monoidal ∞-category $\mathrm{Sp}(k)$ of spectra over k (see Definition 4.4.5). We could equivalently work with differential graded algebras, for k of characteristic zero. Let $\mathrm{Art}_k = \mathrm{ALG}_k^{sm}$ be the category of spectral Artinian algebras over k, also called small augmented spectral k-algebras, given by pairs $(A, \epsilon : A \to k)$ of a nilpotent small spectral k-algebra A and a morphism ϵ such that if $\mathfrak{m}_A = \mathrm{Ker}(\epsilon)$ is the corresponding prime ideal, one has that $k = \pi_0(A/\mathfrak{m}_A)$. Recall that a k-module spectra is small if its homotopy groups are all finite dimensional and if they vanish in negative degrees and in big enough degrees. This category is in the jet category of the category of spectral k-algebras, in the sense of Sect. 2.5.

We also describe the coalgebra approach to deformation theory (see [Hin01]). Denote by COALG_k^{sm} the category of small spectral coalgebras as the category of unital spectral coalgebras $(A, \epsilon : k \to A)$ obtained as linear duals of small spectral k-algebras.

By definition, duality gives an equivalence

$$\mathrm{Hom}_k(-,k) : \mathrm{ALG}_k^{sm} \leftrightarrows \mathrm{COALG}_k^{sm} : \mathrm{Hom}_k(-,k),$$

however, in infinite dimensional settings, it is preferable to work directly with small coalgebras. This will be the case when we pass to the chiral setting.

Note that given a derived moduli functor

$$X : \mathrm{ALG}(\mathrm{Sp})_k \to \mathrm{SSETS}$$

from spectral k-algebras to simplicial sets, and a point $x \in X(k)$, one defines its formal completion $\hat{X}_x$ by sending a small spectral algebra $A \in \mathrm{Art}_k$ to the fiber of the map $X(A) \to X(k)$. The properties of this functor are axiomatized by the following definition.

Definition 10.8.1 A formal deformation problem is a functor

$$X : \mathrm{Art}_k \to \mathrm{SSETS}$$

that satisfies a kind of homotopical dévissage condition, which guarantees its pro-representability:

1. The space $X(k)$ is contractible.
2. If $A \to B$ and $A' \to B$ are maps between small spectral algebras with induced surjections on π_0, then X commutes homotopically with the fiber product $A \times_B A'$.

If X is a deformation problem, one may associate to it its tangent spectrum TX, given by the suspension compatible family of spaces $\{X(k \oplus k[n])\}_{k \geq 0}$.

The basic idea of deformation theory is that Artinian algebras break into pieces of the form $k \oplus k[n]$, and that the dévissage axiom allows us to retrieve a formal deformation problem from its values on these simpler pieces, which are encoded by the tangent spectrum, so that the datum of the latter is essentially equivalent to the knowledge of the deformation problem.

Derived Koszul duality is a natural ∞-adjunction

$$\mathrm{Prim}[-1] : \mathrm{COALG} \leftrightarrows \mathrm{LIE} : \mathrm{CE}$$

between commutative unital coalgebras and Lie algebras given by taking the primitive elements and the Chevalley-Eilenberg complex. If A is a commutative counital coalgebra, we denote by $\bar{A} := \mathrm{Ker}(\epsilon)$ the kernel of its counit.

Definition 10.8.2 Let $\mathfrak{g}$ be a spectral Lie algebra and A be a commutative coalgebra. The ∞-groupoid of Mauer-Cartan elements for the pair is defined as

$$\mathrm{MC}(A, \mathfrak{g}) := \mathrm{Mor}_{\mathrm{COALG}}\big(A, \mathrm{CE}(\mathfrak{g})\big) \cong \mathrm{Mor}_{\mathrm{Lie}}\big(\mathrm{Prim}[-1](A), \mathfrak{g}\big).$$

To every commutative coalgebra B, one associates a formal deformation problem by

$$F_B : A \mapsto \mathrm{Mor}\big(A^*, B\big),$$

where the linear dual $A^* = \mathrm{Hom}_k(A, k)$ of an Artinian algebra is equipped with its canonical coalgebra structure. One can characterize deformation problems that are corepresentable.

Theorem 10.8.1 *The functor*

$$T : \mathrm{LIE}_k \to \mathrm{DEF}_k$$

from Lie algebras to deformation problems that sends a Lie algebra $\mathfrak{g}$ to

$$F_{\mathfrak{g}}(A) := \mathrm{MC}\big(A^*, \mathfrak{g}\big)$$

is fully faithful. It has a quasi-inverse on its image given by the tangent map functor

$$F \mapsto TF[-1] := \big\{F\big(k \oplus k[i+1]\big)\big\}_{i \geq 0},$$

where the spectral Lie algebra structure comes from the canonical identification $TF[-1] \cong T\Omega F$ and the infinitesimal composition of loops in F.

References

[BN10] Ben-Zvi, D., Nadler, D.: Loop spaces and connections. arXiv e-prints (2010). arXiv: 1002.3636

[CG10] Costello, K., Gwilliam, O.: Factorization algebras in perturbative quantum field theory (2010, preprint)
[DZ94] Dubuc, E.J., Zilber, J.G.: On analytic models of synthetic differential geometry. Cah. Topol. Géom. Différ. Catég. **35**(1), 49–73 (1994)
[GR11] Gaitsgory, D., Rozenblyum, N.: Crystals and D-modules. arXiv e-prints (2011). arXiv:1111.2087
[Gro13] Grothendieck, A.: A la Poursuite des Champs (1983–2013). SMF, Maltsiniotis edition
[Hin99] Hinich, V.: Deformations of homotopy algebras. arXiv Mathematics e-prints (1999). arXiv:math/9904145
[Hin01] Hinich, V.: DG coalgebras as formal stacks. J. Pure Appl. Algebra **162**(2–3), 209–250 (2001). doi:10.1016/S0022-4049(00)00121-3
[Hov99] Hovey, M.: Model Categories. Mathematical Surveys and Monographs, vol. 63, p. 209. Am. Math. Soc., Providence (1999). ISBN 0-8218-1359-5
[Ill71] Illusie, L.: Complexe Cotangent et Déformations. I. Lecture Notes in Mathematics, vol. 239, p. 355. Springer, Berlin (1971)
[Jar87] Jardine, J.F.: Simplicial presheaves. J. Pure Appl. Algebra **47**(1), 35–87 (1987). doi:10.1016/0022-4049(87)90100-9
[Law04] Lawvere, F.W.: Functorial semantics of algebraic theories and some algebraic problems in the context of functorial semantics of algebraic theories. Theory Appl. Categ. **5**, 1–121 (2004). Reprinted from Proc. Natl. Acad. Sci. USA **50**, 869–872 (1963) [MR0158921] and Reports of the Midwest Category Seminar. II, pp. 41–61, Springer, Berlin (1968) [MR0231882]
[LMB00] Laumon, G., Moret-Bailly, L.: Champs Algébriques. Ergebnisse der Mathematik und ihrer Grenzgebiete. 3. Folge. A Series of Modern Surveys in Mathematics [Results in Mathematics and Related Areas. 3rd Series. A Series of Modern Surveys in Mathematics], vol. 39, p. 208. Springer, Berlin (2000). ISBN 3-540-65761-4
[Lur09a] Lurie, J.: Derived algebraic geometry. Doctoral thesis (2009, preprint)
[Lur09b] Lurie, J.: Derived algebraic geometry V: structured spaces. arXiv e-prints (2009). arXiv:0905.0459
[Lur09c] Lurie, J.: Higher algebra (2009, preprint)
[Lur09d] Lurie, J.: Higher Topos Theory. Annals of Mathematics Studies, vol. 170, p. 925. Princeton University Press, Princeton (2009). ISBN 978-0-691-14049-0; 0-691-14049-9
[Lur10] Lurie, J.: Moduli problems for ring spectra. In: Proceedings of the International Congress of Mathematicians. Volume II, pp. 1099–1125. Hindustan Book Agency, New Delhi (2010)
[Lur11] Lurie, J.: Derived algebraic geometry X: formal moduli problems (2011, preprint)
[MR91] Moerdijk, I., Reyes, G.E.: Models for Smooth Infinitesimal Analysis, p. 399. Springer, New York (1991). ISBN 0-387-97489-X
[Pri11] Pridham, J.P.: Notes characterising higher and derived stacks concretely. arXiv e-prints (2011). arXiv:1105.4853
[PTVV11] Pantev, T., Toen, B., Vaquie, M., Vezzosi, G.: Quantization and derived moduli spaces I: shifted symplectic structures. arXiv e-prints (2011). arXiv:1111.3209
[Qui69] Quillen, D.: Rational homotopy theory. Ann. Math. (2) **90**, 205–295 (1969)
[Qui70] Quillen, D.: On the (co-) homology of commutative rings. In: Applications of Categorical Algebra (Proc. Sympos. Pure Math., Vol. XVII, New York, 1968), pp. 65–87. Am. Math. Soc., Providence (1970)
[Sch11] Schreiber, U.: Differential cohomology in a cohesive ∞-topos. Nlab website (2011)
[Ser70] Serre, J.P.: Algèbre Locale—Multiplicités. Springer, Berlin (1970)
[SH98] Simpson, C., Hirschowitz, A.: Descente pour les n-champs (descent for n-stacks) (1998). arXiv:math.AG/9807049
[Spi08] Spivak, D.I.: Derived smooth manifolds. arXiv e-prints (2008). arXiv:0810.5174
[SSS09] Sati, H., Schreiber, U., Stasheff, J.: Differential twisted string and Fivebrane structures. arXiv e-prints (2009). arXiv:0910.4001

[Toe00] Toen, B.: Affine stacks (champs affines). arXiv Mathematics e-prints (2000). arXiv: math/0012219

[TV02] Toen, B., Vezzosi, G.: Homotopical algebraic geometry I: topos theory. arXiv Mathematics e-prints (2002). arXiv:math/0207028

[TV08] Toën, B., Vezzosi, G.: Homotopical Algebraic Geometry. II. Geometric Stacks and Applications. Mem. Am. Math. Soc. 193(902):x+224 (2008)

[TV09a] Toen, B., Vezzosi, G.: S^1-Equivariant simplicial algebras and de Rham theory. arXiv e-prints (2009). arXiv:0904.3256

Chapter 11
Algebraic Analysis of Linear Partial Differential Equations

In this chapter, we present the natural coordinate-free approach to systems of linear partial differential equations on manifolds, in the setting of $\mathcal{D}$-modules. These tools will be used in Chaps. 12 and 13 for the algebraic study of nonlinear partial differential equations and gauge theories, and in Chap. 24 for the study of the quantization of factorization algebras.

Linear algebraic analysis was founded by Mikio Sato in the sixties, with his groundbreaking work on hyperfunctions and microfunctions. The theory of $\mathcal{D}$-modules was then developed by Kashiwara and others. We refer to [Sch07, And07] and [Sch09a] for historical background on these methods. In this chapter, we describe Kashiwara and Schapira's version of linear algebraic analysis, using subanalytic sheaves, that gives a convenient generalization of Sato's ideas, allowing:

- a short presentation of microlocalization techniques, that provide a conceptual and general solution to the problem of multiplication (or more generally pullback) of distributional solutions of partial differential systems on manifolds; and
- the treatment of functions with growth conditions, that provides a precise link between the methods of algebraic analysis introduced by Sato and the classical methods of the analysis of partial differential equations, as they are presented by Hörmander in his reference book [Hör03].

We remark that algebraic analysis is usually used in the setting of real analytic geometry, meaning that one works with differential operators with real analytic coefficients, even if the solutions considered may have quite wild singularities. However, most of the tools and results can be adapted to the smooth setting, as was done by Hörmander and others (see [Hör07]) for microlocal analysis. The main result that does not pass through is the Cauchy-Kowalewskaya-Kashiwara theorem on the Cauchy problem. Many of the physical models we will describe do indeed have an underlying real analytic flavor.

This chapter is based on the articles [Sch10, Sch12, KS99] and [Pre07]. We refer to [KS90, KSIW06] and [KS01] for a complete account of this theory and to [KKK86] and [SKK73] for a more classical presentation and many examples of explicit applications.

F. Paugam, *Towards the Mathematics of Quantum Field Theory*,
Ergebnisse der Mathematik und ihrer Grenzgebiete. 3. Folge / A Series of Modern Surveys in Mathematics 59, DOI 10.1007/978-3-319-04564-1_11,

11.1 $\mathcal{D}$-Modules and Linear Partial Differential Equations

We refer to Schneiders' review [Sch94a] and Kashiwara's book [Kas03] for an introduction to $\mathcal{D}$-modules.

Let M be a smooth (real or complex) analytic manifold of dimension n and $\mathcal{D}$ be the algebra of differential operators on M. Recall that locally on M, one can write an operator $P \in \mathcal{D}$ as a finite sum

$$P = \sum_{\alpha} a_\alpha \partial^\alpha$$

with $a_\alpha \in \mathcal{O}_M$,

$$\partial = (\partial_1, \ldots, \partial_n) : \mathcal{O}_M \to \mathcal{O}_M^n$$

the universal derivation and α some multi-indices.

To write down the equation $Pf = 0$, with f in an $\mathcal{O}_M$-module $\mathcal{S}$, one needs to define the universal derivation $\partial : \mathcal{S} \to \mathcal{S}^n$. This is equivalent to giving $\mathcal{S}$ the structure of a $\mathcal{D}$-module. The solution space of the equation with values in $\mathcal{S}$ is then given by

$$\mathcal{S}ol_P(\mathcal{S}) := \{f \in \mathcal{S},\ Pf = 0\}.$$

Note that

$$\mathcal{S}ol_P : \text{MOD}(\mathcal{D}) \to \text{VECT}_{\mathbb{R}_M}$$

is a functor that one can think of as representing the space of solutions of P. Denote by $\mathcal{M}_P$ the cokernel of the $\mathcal{D}$-linear map

$$\mathcal{D} \xrightarrow{.P} \mathcal{D}$$

given by right multiplication by P. Applying the functor $\mathcal{H}om_{\text{MOD}(\mathcal{D})}(-, \mathcal{S})$ to the exact sequence

$$\mathcal{D} \xrightarrow{.P} \mathcal{D} \longrightarrow \mathcal{M}_P \to 0,$$

we get the exact sequence

$$0 \to \mathcal{H}om_{\text{MOD}(\mathcal{D})}(\mathcal{M}_P, \mathcal{S}) \to \mathcal{S} \xrightarrow{P.} \mathcal{S},$$

which gives a natural isomorphism

$$\mathcal{S}ol_P(\mathcal{S}) \cong \mathcal{H}om_{\text{MOD}(\mathcal{D})}(\mathcal{M}_P, \mathcal{S}).$$

This means that the $\mathcal{D}$-module $\mathcal{M}_P$ represents the solution space of P, so that the category of $\mathcal{D}$-modules is a convenient setting for the functor of points approach to linear partial differential equations.

Note that it is even better to consider the derived solution space

$$\mathbb{R}\mathcal{S}ol_P(\mathcal{S}) := \mathbb{R}\mathcal{H}om_{\text{MOD}(\mathcal{D})}(\mathcal{M}_P, \mathcal{S})$$

because it also encodes information on the inhomogeneous equation

$$Pf = g.$$

Indeed, applying $\mathcal{H}om_{\mathcal{D}}(-, \mathcal{S})$ to the exact sequences

$$0 \to \mathcal{I}_P \to \mathcal{D} \to \mathcal{M}_P \to 0$$
$$0 \to \mathcal{N}_P \to \mathcal{D} \to \mathcal{I}_P \to 0$$

where $\mathcal{I}_P$ is the image of P and $\mathcal{N}_P$ is its kernel, one gets the exact sequences

$$0 \to \mathcal{H}om_{\mathcal{D}}(\mathcal{M}_P, \mathcal{S}) \to \mathcal{S} \to \mathcal{H}om_{\mathcal{D}}(\mathcal{I}_P, \mathcal{S}) \to \mathcal{E}xt^1_{\mathcal{D}}(\mathcal{M}_P, \mathcal{S}) \to 0$$
$$0 \to \mathcal{H}om_{\mathcal{D}}(\mathcal{I}_P, \mathcal{S}) \to \mathcal{S} \to \mathcal{H}om_{\mathcal{D}}(\mathcal{N}_P, \mathcal{S}) \to \mathcal{E}xt^1_{\mathcal{D}}(\mathcal{I}_P, \mathcal{S}) \to 0$$

If $Pf = g$, then $QPf = 0$ for $Q \in \mathcal{D}$ implies $Qg = 0$. The second exact sequence implies that this system, called the algebraic compatibility condition for the inhomogeneous equation $Pf = g$, is represented by the $\mathcal{D}$-module $\mathcal{I}_P$, because

$$\mathcal{H}om_{\mathcal{D}}(\mathcal{I}_P, \mathcal{S}) = \{g \in \mathcal{S},\ Q.g = 0,\ \forall Q \in \mathcal{N}_P\}.$$

The first exact sequence shows that $\mathcal{E}xt^1_{\mathcal{D}}(\mathcal{M}_P, \mathcal{S})$ are classes of vectors $f \in \mathcal{S}$ satisfying the algebraic compatibility conditions modulo those for which the system is truly compatible. Moreover, for $k \geq 1$, one has

$$\mathcal{E}xt^k_{\mathcal{D}}(\mathcal{I}_P, \mathcal{S}) \cong \mathcal{E}xt^{k+1}_{\mathcal{D}}(\mathcal{M}_P, \mathcal{S})$$

so that all the $\mathcal{E}xt^k_{\mathcal{D}}(\mathcal{M}_P, \mathcal{S})$ give interesting information about the differential operator P.

Recall that the sub-algebra $\mathcal{D}$ of $\text{End}_{\mathbb{R}}(\mathcal{O})$ is generated by the left multiplication by functions in $\mathcal{O}_M$ and by the derivation induced by vector fields in Θ_M. There is a natural right action of $\mathcal{D}$ on the $\mathcal{O}$-module Ω^n_M by

$$\omega \cdot \partial = -L_\partial \omega$$

where L_∂ is the Lie derivative.

There is a tensor product in the category $\text{MOD}(\mathcal{D})$ given by

$$\mathcal{M} \otimes \mathcal{N} := \mathcal{M} \otimes_{\mathcal{O}} \mathcal{N}.$$

The $\mathcal{D}$-module structure on the tensor product is given on vector fields $\partial \in \Theta_M$ by Leibniz's rule

$$\partial(m \otimes n) = (\partial m) \otimes n + m \otimes (\partial n).$$

There is also an internal homomorphism object $\mathcal{H}om(\mathcal{M},\mathcal{N})$ given by the $\mathcal{O}$-module $\mathcal{H}om_{\mathcal{O}}(\mathcal{M},\mathcal{N})$ equipped with the action of derivations $\partial \in \Theta_M$ by

$$\partial(f)(m) = \partial\big(f(m)\big) - f(\partial m).$$

An important system is given by the $\mathcal{D}$-module of functions $\mathcal{O}$, which can be presented by the de Rham complex

$$\mathcal{D} \otimes \Theta_M \to \mathcal{D} \to \mathcal{O} \to 0,$$

meaning that $\mathcal{O}$, as a $\mathcal{D}$-module, is the quotient of $\mathcal{D}$ by the sub-$\mathcal{D}$-module generated by vector fields. The family of generators ∂_i of the kernel of $\mathcal{D} \to \mathcal{O}$ form a regular sequence, i.e., for every $k = 1, \dots, n$, ∂_k is not a zero divisor in $\mathcal{D}/(\partial_1, \dots, \partial_{k-1})$ (where $\partial_{-1} = 0$ by convention). This implies (see Lang [Lan93], XXI §4 for more details on Koszul resolutions) the following:

Proposition 11.1.1 *The natural map*

$$\mathrm{Sym}_{(\mathrm{MOD}_{dg}(\mathcal{D}),\otimes)}\big([\mathcal{D} \otimes \Theta_M \to \mathcal{D}]\big) \longrightarrow \mathcal{O}$$

is a quasi-isomorphism of dg-$\mathcal{D}$-modules. The left hand side gives a free resolution of $\mathcal{O}$ as a $\mathcal{D}$-module called the universal Spencer complex.

Proof This is the classical Koszul resolution of the regular ideal Θ_M in $\mathcal{D}$. For more details, see [Kas03], Proposition 1.6. □

Proposition 11.1.2 *The functor*

$$\mathcal{M} \mapsto \Omega^n_M \otimes_{\mathcal{O}} \mathcal{M}$$

induces an equivalence of categories between the categories $\mathrm{MOD}(\mathcal{D})$ *and* $\mathrm{MOD}(\mathcal{D}^{op})$ *of left and right $\mathcal{D}$-modules whose quasi-inverse is*

$$\mathcal{N} \mapsto \mathcal{H}om_{\mathcal{O}_M}\big(\Omega^n_M, \mathcal{N}\big).$$

The monoidal structure induced on $\mathrm{MOD}(\mathcal{D}^{op})$ *by this equivalence is denoted by* $\otimes^!$.

Proof The two functors are quasi-inverses of each other because Ω^n_M is a locally free $\mathcal{O}$-module of rank 1. A short computation shows that this equivalence is compatible with the $\mathcal{D}$-module structures. □

Definition 11.1.1 Let $\mathcal{S}$ be a right $\mathcal{D}$-module. The *de Rham functor* with values in $\mathcal{S}$ is the functor

$$\mathrm{DR}_{\mathcal{S}} : \mathrm{MOD}(\mathcal{D}) \to \mathrm{VECT}_{\mathbb{R}_M}$$

that sends a left $\mathcal{D}$-module to

$$\mathrm{DR}_{\mathcal{S}}(\mathcal{M}) := \mathcal{S} \otimes^{\mathbb{L}}_{\mathcal{D}} \mathcal{M}.$$

The *de Rham functor* with values in $\mathcal{S} = \Omega^n_M$ is denoted by DR and simply called the de Rham functor. One also writes $\mathrm{DR}^r_{\mathcal{S}}(\mathcal{M}) = \mathcal{M} \otimes^{\mathbb{L}}_{\mathcal{D}} \mathcal{S}$ if $\mathcal{S}$ is a fixed left $\mathcal{D}$-module and $\mathcal{M}$ is a varying right $\mathcal{D}$-module, and $\mathrm{DR}^r := \mathrm{DR}^r_{\mathcal{O}}$.

Proposition 11.1.3 *The natural map*

$$\begin{aligned} \Omega^n_M \otimes_{\mathcal{O}} \mathcal{D} &\to \Omega^n_M \\ \omega \otimes Q &\mapsto \omega(Q) \end{aligned}$$

extends to a $\mathcal{D}^{op}$-linear quasi-isomorphism

$$\Omega^*_M \otimes_{\mathcal{O}} \mathcal{D}[n] \xrightarrow{\sim} \Omega^n_M.$$

Proof This follows from the fact that the above map is induced by tensoring the Spencer complex by Ω^n_M, and by the internal product isomorphism

$$\begin{aligned} \mathrm{Sym}_{\mathrm{MOD}_{dg}(\mathcal{D})}\big(\big[\underset{-1}{\mathcal{D} \otimes \Theta_M} \to \underset{0}{\mathcal{D} \otimes \mathcal{O}}\big]\big) \otimes \Omega^n_M &\longrightarrow \big(\Omega^*_M \otimes \mathcal{D}[n], d\big) \\ X \otimes \omega &\longmapsto i_X \omega. \end{aligned}$$ □

We will see that in the super setting, this proposition can be taken as a definition of the right $\mathcal{D}$-modules of volume forms, called Berezinians.

The $\mathcal{D}$-modules we will use are usually not $\mathcal{O}$-coherent but only $\mathcal{D}$-coherent. The right duality to be used in the monoidal category $(\mathrm{MOD}(\mathcal{D}), \otimes)$ to get a biduality statement for coherent $\mathcal{D}$-modules is thus not the internal duality $\underline{\mathrm{Hom}}_{\mathcal{O}}(\mathcal{M}, \mathcal{O})$ but the derived dual $\mathcal{D}^{op}$-module

$$\mathbb{D}(\mathcal{M}) := \mathbb{R}\mathcal{H}om_{\mathcal{D}}(\mathcal{M}, \mathcal{D}).$$

The non-derived dual works well for projective $\mathcal{D}$-modules, but most of the $\mathcal{D}$-modules used in field theory are only coherent, so that one often uses the derived duality operation. We now describe the relation (based on biduality) between the de Rham and duality functors.

Proposition 11.1.4 *Let $\mathcal{S}$ be a coherent $\mathcal{D}^{op}$-module and $\mathcal{M}$ be a coherent $\mathcal{D}$-module. There is a natural quasi-isomorphism*

$$\mathbb{R}\mathrm{Sol}_{\mathbb{D}(\mathcal{M})}(\mathcal{S}) := \mathbb{R}\mathcal{H}om_{\mathcal{D}^{op}}\big(\mathbb{D}(\mathcal{M}), \mathcal{S}\big) \cong \mathrm{DR}_{\mathcal{S}}(\mathcal{M}),$$

where $\mathbb{D}(\mathcal{M}) := \mathbb{R}\mathrm{Hom}_{\mathcal{D}}(\mathcal{M}, \mathcal{D})$ is the $\mathcal{D}^{op}$-module dual of $\mathcal{M}$.

Proof This follows from the biduality isomorphism

$$\mathcal{M} \cong \mathbb{R}\mathcal{H}om_{\mathcal{D}^{op}}\big(\mathbb{D}(\mathcal{M}), \mathcal{D}\big).$$ □

The use of $\mathcal{D}$-duality will be problematic in the study of covariant operations (such as Lie bracket on local vector fields). We will come back to this in Sect. 12.4.

Definition 11.1.2 Let $\mathcal{M}$ be a coherent $\mathcal{D}$-module. A *coherent filtration* of $\mathcal{M}$ is a sequence $\{F^m(\mathcal{M})\}_{m\geq 0}$ of subsheaves of $\mathcal{M}$ such that:

1. $\mathcal{M} = \bigcup_m F^m(\mathcal{M})$;
2. $F^m(\mathcal{D})F^l(\mathcal{M}) \subset F^{m+l}(\mathcal{M})$;
3. $\oplus_m F^m(\mathcal{M})$ is locally finitely generated as a $\oplus_m F^m(\mathcal{D})$-module.

The *characteristic ideal* of a filtered $\mathcal{D}$-module $(\mathcal{M}, F^\bullet\mathcal{M})$ is the annihilator $\mathcal{I}_{F^\bullet\mathcal{M}}$ of the $\mathcal{O}_{T^*M}$-module $\mathcal{O}_{T^*M} \otimes_{\pi^{-1}\mathrm{gr}(\mathcal{D})} \mathrm{gr}^F(\mathcal{M})$.

Proposition 11.1.5 *Let $\mathcal{M}$ be a coherent $\mathcal{D}$-module. There exists locally a coherent filtration $F^\bullet\mathcal{M}$ on $\mathcal{D}$. Moreover, the radical ideal*

$$\mathcal{I}_{\mathcal{M}} = \mathrm{rad}(\mathcal{I}_{F^\bullet\mathcal{M}})$$

is independent of the local choice of a coherent filtration.

Proof This follows from the local finiteness condition on $\oplus_m F^m(\mathcal{D})$. See [Kas03], Sect. 2.2 for more details. □

Definition 11.1.3 Let $\mathcal{M}$ be a coherent $\mathcal{D}$-module. The ideal $\mathcal{I}_{\mathcal{M}}$ defined in Proposition 11.1.5 is called the characteristic ideal of $\mathcal{M}$ and the corresponding variety char($\mathcal{M}$) is called the characteristic variety of $\mathcal{M}$.

11.2 Berezinians and Operations on Super-$\mathcal{D}$-Modules

We refer to Penkov's article [Pen83] for a complete study of the Berezinian in the $\mathcal{D}$-module setting and to Deligne-Morgan's lectures [DM99] and Manin's book [Man97] for more details on supermanifolds.

Let M be a supermanifold of dimension $n|m$ and denote by Ω^1_M the $\mathcal{O}_M$-module of differential forms on M and Ω^*_M the super-$\mathcal{O}_M$-module of higher differential forms on M, defined as the exterior (i.e., odd symmetric) power

$$\Omega^*_M := \wedge^*\Omega^1_M := \mathrm{Sym}_{\textsc{Mod}(\mathcal{O}_M)}\Omega^1_M[1].$$

Note that Ω^*_M is strictly speaking a $\mathbb{Z}/2$-bigraded $\mathbb{R}$-module, but we can view it as a $\mathbb{Z}/2$-graded module because its diagonal $\mathbb{Z}/2$-grading identifies with $\mathrm{Sym}_{\textsc{Mod}(\mathcal{O}_M)}T\Omega^1_M$, where $T : \textsc{Mod}(\mathcal{O}_M) \to \textsc{Mod}(\mathcal{O}_M)$ is the grading exchange. Thus from now on, we consider Ω^*_M as a mere $\mathbb{Z}/2$-graded module.

The super version of Proposition 11.1.3 can be taken as a definition of the Berezinian, as a complex of $\mathcal{D}$-modules, up to quasi-isomorphism.

Definition 11.2.1 The *Berezinian* of M is defined in the derived category of $\mathcal{D}_M$-modules by the formula

$$\mathrm{Ber}_M := \Omega^*_M \otimes_{\mathcal{O}} \mathcal{D}[n].$$

The *complex of integral forms* $I_{*,M}$ is defined by

$$I_{*,M} := \mathbb{R}\mathrm{Hom}_{\mathcal{D}}(\mathrm{Ber}_M, \mathrm{Ber}_M).$$

The following proposition (see [Pen83], 1.6.3) gives a description of the Berezinian as a $\mathcal{D}$-module.

Proposition 11.2.1 *The Berezinian complex is concentrated in degree* 0, *and is equal there to*

$$\mathrm{Ber}_M := \mathcal{E}xt^n_{\mathcal{D}}(\mathcal{O}, \mathcal{D}).$$

It is moreover projective of rank 1 *over* $\mathcal{O}$.

Proof This follows from the fact that

$$\wedge^* \Theta_M \otimes_{\mathcal{O}} \mathcal{D}[-n] \to \mathcal{O}$$

is a projective resolution such that

$$\mathrm{Ber}_M := \Omega^*_M \otimes_{\mathcal{O}} \mathcal{D}[n] = \mathbb{R}\mathcal{H}om_{\mathcal{D}}\big(\wedge^* \Theta_M \otimes_{\mathcal{O}} \mathcal{D}[-n], \mathcal{D}\big)$$

and this de Rham complex is exact (Koszul resolution of a regular module) except in degree zero where it is equal to

$$\mathcal{E}xt^n_{\mathcal{D}}(\mathcal{O}, \mathcal{D}).$$ □

Proposition 11.2.2 *Suppose M is a supermanifold of dimension $m|n$. The Berezinian is a locally free $\mathcal{O}$-module of rank* 1 *on M with generator denoted by $D(dx_1, \ldots, dx_m, d\theta_1, \ldots, d\theta_n)$. If $f : M \to M$ is an isomorphism of super-varieties (change of coordinate) with local tangent map $D_x f$ described by the even matrix*

$$D_x f = \begin{pmatrix} A & B \\ C & D \end{pmatrix}$$

acting on the real vector space

$$T_x M = (T_x M)^0 \oplus (T_x M)^1,$$

the action of $D_x f$ on $D(dx_1, \ldots, dx_m)$ is given by the Berezin determinant

$$\mathrm{Ber}(D_x f) := \det\big(A - BD^{-1}C\big)\det(D)^{-1}.$$

Proof This is a classical result (see [DM99] or [Man97]). □

In the super setting, the equivalence of left and right $\mathcal{D}$-modules is given by the functor

$$\mathcal{M} \mapsto \mathcal{M} \otimes_{\mathcal{O}} \mathrm{Ber}_M$$

that twists by the Berezinian right $\mathcal{D}$-module, which can be computed by using the definition

$$\mathrm{Ber}_M := \Omega^*_M \otimes_{\mathcal{O}} \mathcal{D}[n]$$

and passing to degree 0 cohomology.

A more explicit description of the complex of integral forms (up to quasi-isomorphism) is given by

$$I_{*,M} := \mathbb{R}\mathcal{H}om_{\mathcal{D}}(\mathrm{Ber}_M, \mathrm{Ber}_M) \cong \mathcal{H}om_{\mathcal{D}}\big(\Omega^*_M \otimes_{\mathcal{O}} \mathcal{D}[n], \mathrm{Ber}_M\big)$$

so that we get

$$I_{*,M} \cong \mathcal{H}om_{\mathcal{O}}\big(\Omega^*_M[n], \mathrm{Ber}_M\big) \cong \mathcal{H}om_{\mathcal{O}}\big(\Omega^*_M[n], \mathcal{O}\big) \otimes_{\mathcal{O}} \mathrm{Ber}_M$$

and in particular $I_{n,M} \cong \mathrm{Ber}_M$.

Note that Proposition 11.1.3 shows that if M is a non-super manifold, then Ber_M is quasi-isomorphic with Ω^n_M, and this implies that

$$I_{*,M} \cong \mathcal{H}om_{\mathcal{O}}\big(\Omega^*_M[n], \mathcal{O}\big) \otimes_{\mathcal{O}} \mathrm{Ber}_M \cong \wedge^* \Theta_M \otimes_{\mathcal{O}} \Omega^n_M[-n] \xrightarrow{i} \Omega^*_M,$$

where i is the internal product homomorphism. This implies the isomorphism

$$I_{*,M} \cong \Omega^*_M,$$

so that in the purely even case, integral forms essentially identify with ordinary differential forms.

The main use of the module of Berezinians is given by its usefulness in the definition of integration on supermanifolds. We refer to Manin [Man97], Chap. 4 for the following proposition.

Proposition 11.2.3 *Let M be a supermanifold, with underlying manifold $|M|$ and orientation sheaf $\mathrm{or}_{|M|}$. There is a natural integration map*

$$\int_M \big[dt^1 \cdots dt^n d\theta^1 \cdots d\theta^q\big] : \Gamma_c(M, \mathrm{Ber}_M \otimes \mathrm{or}_{|M|}) \to \mathbb{R}$$

given in a local chart (i.e., an open subset $U \subset \mathbb{R}^{n|m}$) for $g = \sum_I g_I \theta^I \in \mathcal{O}$ by

$$\int_U \big[dt^1 \cdots dt^n d\theta^1 \cdots d\theta^q\big] g := \int_{|U|} g_{1,\dots,1}(t) d^n t.$$

We finish by describing the inverse and direct image functors in the supergeometric setting, following the presentation of Penkov in [Pen83].

Let $g : X \to Y$ be a morphism of supermanifolds. Recall that for $\mathcal{F}$ a sheaf of $\mathcal{O}_Y$-modules on Y, we denote by $g^{-1}\mathcal{F}$ the sheaf on X defined by

$$g^{-1}\mathcal{F}(U) := \lim_{g(U) \subset V} \mathcal{F}(V).$$

The $(\mathcal{D}_X, g^{-1}\mathcal{D}_Y)$ module of relative inverse differential operators is defined as

$$\mathcal{D}_{X\to Y} := \mathcal{O}_X \otimes_{g^{-1}\mathcal{O}_Y} g^{-1}\mathcal{D}_Y.$$

The $(g^{-1}\mathcal{D}_Y, \mathcal{D}_X)$ module of relative direct differential operators is defined as

$$\mathcal{D}_{X\leftarrow Y} := \mathrm{Ber}_X \otimes_{\mathcal{O}_X} \mathcal{D}_{X\to Y} \otimes_{g^{-1}\mathcal{O}_Y} \left(\mathrm{Ber}_Y^*\right).$$

Definition 11.2.2 The inverse image functor of $\mathcal{D}$-modules is defined by

$$g_{\mathcal{D}}^*(-) := \mathcal{D}_{X\to Y} \otimes^{\mathbb{L}}_{g^{-1}\mathcal{O}_Y} g^{-1}(-) : D(\mathcal{D}_Y) \to D(\mathcal{D}_X).$$

If $g : X \hookrightarrow Y$ is a locally closed embedding, the direct image functor is defined by

$$g_*^{\mathcal{D}}(-) := \mathcal{D}_{Y\leftarrow X} \otimes^{\mathbb{L}}_{\mathcal{D}_X} (-) : D(\mathcal{D}_X) \to D(\mathcal{D}_Y).$$

More generally, for any morphism $g : X \to Y$, one defines the right adjoint to $g_{\mathcal{D}}^*$ by the formula

$$g_*^{\mathcal{D}}(-) := \mathbb{R}f_*\left(- \otimes^{\mathbb{L}}_{\mathcal{D}} \mathcal{D}_{X\to Y}\right).$$

It may also be convenient to approach operations on $\mathcal{D}$-modules using spans (generalized correspondences) between varieties, as in [FG11].

11.3 Subanalytic Sheaves

We now introduce the main technical tool of the modern presentation of linear algebraic analysis, which allows a smooth treatment of microlocalization and of generalized functions with growth conditions. We refer to [KS01] and [Pre07] for a detailed treatment of these matters.

Recall from Example 3.2.3 the basics of semianalytic geometry.

Definition 11.3.1 A *rational domain* in the affine space $\mathbb{C}^n$ is a finite intersection of domains of the form

$$U_{f,g} := \left\{x \in \mathbb{C}^n, \left|f(x)\right| < \left|g(x)\right|\right\}$$

for $f, g \in \mathcal{O}(\mathbb{C}^n)$. We denote by LEGOS_{san} the category of relatively compact rational domains in $\mathbb{C}^n$ with holomorphic maps between them with its locally finite Grothendieck topology. A *semianalytic algebra* is a functor

$$A : \mathrm{LEGOS}_{san} \to \mathrm{SETS}$$

that commutes with transversal fiber products. It is called finitely generated if it is a quotient $\mathcal{O}(U)/(f_1, \dots, f_n)$ where $\mathcal{O}(U) := \mathrm{Hom}(U, -)$ is the semianalytic algebra associated to a given lego U and $f_i \in \mathcal{O}(U)(\mathbb{C})$. We denote by $(\overline{\mathrm{LEGOS}_{san}}, \tau)$ the site of semianalytic affine schemes. A *semianalytic variety* is a sheaf on this site that is locally representable.

We now recall the basic results on subanalytic subsets. We give here a purely analytic presentation of subanalytic subsets, using only inequalities on norms of functions.

Definition 11.3.2 Let X be a complex semianalytic variety. A subset $V \subset X$ is called subanalytic if it is locally given by a projection along

$$\mathbb{A}^n_X = X \times \mathbb{C}^n \to X$$

of a relatively compact subset in the algebra generated by rational domains by union, intersection and complement. If M is a real semianalytic variety, a subset $V \subset M$ is called subanalytic if it is locally given by the real trace of a subanalytic subset in the complexification $M_{\mathbb{C}}$.

We refer to Gabrielov for the proof of the following theorem [Gab96]. The main difficulty is to pass to the complement.

Theorem 11.3.1 *Subanalytic subsets are stable by union, intersection, complement and projection along proper maps.*

Definition 11.3.3 Let X be a real or complex analytic space. Let X_{sa} be the site with objects the open subanalytic subsets of X and coverings given by the usual coverings $U = \cup_{V \in S} V$, subject to the additional local finiteness hypothesis that for every compact subset K of X, there exists a finite subset $S_0 \subset S$ such that $K \cap U = K \cap (\cup_{V \in S_0} V)$.

There is a natural map of sites

$$\rho : X \to X_{sa},$$

and associated to this map, there are two pairs (ρ^{-1}, ρ_*) and $(\rho_!, \rho^{-1})$ of adjoint functors

$$\mathrm{MOD}(k_X) \begin{array}{c} \xrightarrow{\rho_*} \\ \xleftarrow{\rho^{-1}} \\ \xrightarrow[\rho_!]{} \end{array} \mathrm{MOD}(k_{X_{sa}}).$$

We are now able to define constructible sheaves (see [KS90], Chap. VIII for a detailed presentation).

Definition 11.3.4 Let k be a field and X be a complex analytic space. An object in $D^b(k_{X_{sa}})$ is called constructible if it has locally constant cohomology on a subanalytic stratification and its fibers are perfect complexes (i.e., quasi-isomorphic to bounded complexes of finitely generated modules). An object $\mathcal{F}$ in $D^b(k_X)$ is called $\mathbb{R}$-constructible if its direct image $\rho_*\mathcal{F}$ in $D^b(k_{X_{sa}})$ is constructible.

If Z is a locally closed subset of X, we define

$$\begin{aligned}\Gamma_Z : \text{Mod}(k_{X_{sa}}) &\to \text{Mod}(k_{X_{sa}})\\ \mathcal{F} &\mapsto \mathcal{H}om(\rho_* k_Z, \mathcal{F}),\end{aligned}$$

and

$$\begin{aligned}(-)_Z : \text{Mod}(k_{X_{sa}}) &\to \text{Mod}(k_{X_{sa}})\\ \mathcal{F} &\mapsto \mathcal{F} \otimes \rho_* k_Z.\end{aligned}$$

Let X and Y be two real analytic manifolds and $f : X \to Y$ be a real analytic map. There is a natural adjoint pair

$$f_* : \text{Mod}(k_{X_{sa}}) \leftrightarrows \text{Mod}(k_{Y_{sa}}) : f^{-1}$$

and there is also a proper direct image functor defined by

$$\begin{aligned}f_{!!} : \text{Mod}(k_{X_{sa}}) &\to \text{Mod}(k_{Y_{sa}})\\ \mathcal{F} &\mapsto \operatorname*{colim}_U f_* \mathcal{F}_U,\end{aligned}$$

where U ranges through the family of relatively compact open subanalytic subsets of X.

Let $\mathcal{R}$ be a sheaf of $k_{X_{sa}}$-algebras.

Theorem 11.3.2 *If $f : X \to Y$ is a real analytic map between real analytic manifolds, the functors f^{-1}, f_* and $f_{!!}$ extend to functors*

$$\begin{aligned}f^{-1} &: \text{Mod}(\mathcal{R}) \to \text{Mod}(f^{-1}\mathcal{R}),\\ f_* &: \text{Mod}(f^{-1}\mathcal{R}) \to \text{Mod}(\mathcal{R}),\\ f_{!!} &: \text{Mod}(f^{-1}\mathcal{R}) \to \text{Mod}(\mathcal{R}).\end{aligned}$$

Moreover, the derived functor $\mathbb{R}f_{!!} : D^+(f^{-1}\mathcal{R}) \to D^+(\mathcal{R})$ admits a right adjoint denoted by

$$f^! : D^+(\mathcal{R}) \to D^+(f^{-1}\mathcal{R}).$$

We now describe the main examples of subanalytic sheaves, given by functions with growth conditions. Let X be a real analytic manifold, $k_X = \mathbb{C}_X$, $\mathcal{C}_X^\infty$ be the sheaf of complex valued smooth functions and $\mathcal{D}_X$ be the ring of linear differential operators on X.

Definition 11.3.5 One says that a function $f \in \mathcal{C}^\infty(U)$ has *polynomial growth* at a point p of X if, for a local coordinate system $(x_1, \dots, x_n)$ around p, there exists a compact neighborhood K of p and a positive integer N such that

$$\sup_{x \in K \cap U} \big(\operatorname{dist}(x, X \backslash U)\big)^N \big|f(x)\big| < \infty.$$

One says that f is *tempered at* p if all of its derivatives are of polynomial growth at p. One says that f is *tempered* if it is tempered at every point p of U. We denote by $\mathcal{C}^{\infty,t}(U)$ the vector space of tempered functions.

We refer to [KS01] for a proof of the following theorem.

Theorem 11.3.3 *The mapping* $U \mapsto \mathcal{C}^{\infty,t}(U)$ *is a sheaf on* X_{sa}.

We now define Dolbeault complexes with growth conditions. These will be the main tools for the definition of generalized functions.

Definition 11.3.6 Let M be a real analytic manifold with complexification X and denote by $X^{\mathbb{R}} = \mathrm{Res}_{\mathbb{C}/\mathbb{R}} X$ the underlying real manifold, identified with the diagonal of $X^{\mathbb{R}} \otimes_{\mathbb{R}} \mathbb{C} \cong X \times \overline{X}$. The subanalytic sheaf of *tempered holomorphic functions* is the Dolbeault complex of X with coefficients in $\mathcal{C}_X^{\infty,t}$, defined by

$$\mathcal{O}_X^t := \mathbb{R}\mathcal{H}om_{\mathcal{D}_{\overline{X}}}\big(\mathcal{O}_{\overline{X}}, \mathcal{C}_X^{\infty,t}\big).$$

11.4 Sheaf Theoretic Generalized Functions

We refer to Schapira's surveys [Sch10] and [Sch12] for this section. Let M be a real analytic manifold and X be a complexification of M. We denote by $D^b(\mathbb{C}_X)$ the bounded derived category of sheaves of $\mathbb{C}$-vector spaces and $D'_X(-) := \mathbb{R}\mathcal{H}om_{\mathbb{C}_X}(-, \mathbb{C}_X)$ its duality functor. The constant sheaf on X with support on M is denoted by $\mathbb{C}_M$.

The intuitive idea behind Sato's approach to generalized functions is to define them as sums of boundary values on M of holomorphic functions on $X \backslash M$. More precisely:

Definition 11.4.1 The sheaf of *hyperfunctions* on M is defined as the cohomology with support

$$\mathcal{B}_M := \mathbb{R}\mathcal{H}om_{\mathbb{C}_X}\big(D'_X(\mathbb{C}_M), \mathcal{O}_X\big).$$

Note first that the sheaf $\mathcal{B}_M$ on X is a naturally equipped with a right $\mathcal{D}_X$-module structure. Using the Poincaré-Verdier duality isomorphism (see [KS90] for a proof of this statement) $D'_X(\mathbb{C}_M) \cong \mathrm{or}_M[-n]$, one shows that $\mathcal{B}_M$ is concentrated in degree zero and

$$\mathcal{B}_M \cong \mathcal{H}_M^n(X) \otimes \mathrm{or}_M$$

where or_M is the orientation sheaf on M.

Now consider a subanalytic open subset Ω of X and denote by $\overline{\Omega}$ its closure. Assume that $M \subset \overline{\Omega}$ and $D'_X(\mathbb{C}_{\overline{\Omega}}) \cong \mathbb{C}_\Omega$ (this is true if $\Omega \subset X$ is locally homeomorphic to the inclusion of an open convex subset into $\mathbb{C}^n$). The morphism $\mathbb{C}_{\overline{\Omega}} \to \mathbb{C}_M$

defines by duality the morphism $D'_X(\mathbb{C}_M) \to D'_X(\mathbb{C}_{\overline{\Omega}}) \cong \mathbb{C}_\Omega$. Applying the functor $\mathbb{R}\mathrm{Hom}_{\mathbb{C}_X}(-, \mathcal{O}_X)$, one gets the boundary value morphism

$$b : \mathcal{O}(\Omega) \to \mathcal{B}(M).$$

The source and the target of this map can both be written as

$$\mathbb{R}\mathrm{Hom}_{\mathbb{C}_X}(F, \mathcal{O}_X)$$

for F an $\mathbb{R}$-constructible sheaf on X (see Definition 11.3.4). We thus think of sheaves of this form as spaces of generalized analytic functions.

To illustrate the above general procedure, consider the case $M = \mathbb{R} \subset \mathbb{C} = X$ with $\Omega = \mathbb{C} \backslash \mathbb{R}$. One then has $\mathcal{B}_M \cong \mathcal{O}(\mathbb{C}\backslash\mathbb{R})/\mathcal{O}(\mathbb{C})$, the natural inclusion $\mathcal{O}(\mathbb{C}) \to \mathcal{B}(M)$ is given by $f \mapsto (f, 0)$ and the boundary value map is simply the projection

$$b : \mathcal{O}(\mathbb{C}\backslash\mathbb{R}) \to \mathcal{O}(\mathbb{C}\backslash\mathbb{R})/\mathcal{O}(\mathbb{C}).$$

For example, Dirac's δ function and Heaviside's step function are defined as the boundary values

$$\delta := b\left(\frac{1}{2\pi i z}, \frac{1}{2\pi i z}\right) \quad \text{and} \quad H := b\left(\frac{1}{2\pi i}\log z, \frac{1}{2\pi i}\log z - 1\right)$$

where $\log(1) = 0$. Contrary to the usual approach of analysis, boundary values are not computed as limits, but as equivalence classes. The δ function may also be written more intuitively as

$$\delta(x) = \frac{1}{2i\pi}\left(\frac{1}{x - i0} - \frac{1}{x + i0}\right).$$

The main drawback of the above sheaf theoretic construction of spaces of generalized functions is that it prevents us from using functions with growth conditions, like tempered functions, because they don't form sheaves. One can overcome this problem by working with subanalytic sheaves.

Definition 11.4.2 The subanalytic sheaf of *hyperfunctions* is defined by

$$\mathcal{B}_M = \mathbb{R}\mathcal{H}om_{\mathbb{C}_{X_{sa}}}\left(D'\mathbb{C}_{M_{sa}}, \mathcal{O}_X\right).$$

The subanalytic sheaf of *tempered distributions* is defined by the formula

$$\mathcal{D}b^t_M := \mathbb{R}\mathcal{H}om_{\mathbb{C}_{X_{sa}}}\left(D'\mathbb{C}_{M_{sa}}, \mathcal{O}^t_X\right).$$

The subanalytic sheaf of *distributions* is defined by

$$\mathcal{D}b_M := \rho_*\rho^{-1}\mathcal{D}b^t_M.$$

We will now use similar constructions of cohomology with support, to define differential operators. Recall first that the integral kernel of the identity map on functions on $X = \mathbb{C}$ is given by

$$K(\mathrm{id}) = \delta(x, y) := \frac{1}{2i\pi} \frac{dy}{(x-y)}.$$

Indeed, Cauchy's formula tells us that

$$\mathrm{id}\big(f(x)\big) = \int_{\mathbb{C}} \delta(x, y) f(y) = \frac{1}{2i\pi} \int_{\mathbb{C}} \frac{f(y)}{(x-y)} dy.$$

One gets the kernel of the differential operator ∂_x by differentiating the above equality, to yield

$$K(\partial_x) = \frac{1}{2i\pi} \frac{dy}{(x-y)^2}.$$

We now give a definition of infinite order differential operators, by viewing differential operators as integral operators for some kernels on X^2 supported on the diagonal X. This definition can be directly microlocalized, to obtain the notion of microlocal operators (see Definition 11.5.4).

Definition 11.4.3 The ring of *differential operators of infinite order* is defined as

$$\mathcal{D}_X^{\mathbb{R}} := \mathbb{R}\mathcal{H}om\big(\mathbb{C}_\Delta, \mathcal{O}_{X\times X}^{(0,d_X)}\big),$$

where $\mathcal{O}_{X\times X}^{(0,d_X)} := \mathcal{O}_{X\times X} \otimes^{\mathbb{L}}_{p_2^{-1}\mathcal{O}_X} p_2^{-1}\Omega_X$. The ring of *differential operators* is defined as

$$\mathcal{D}_X := \mathbb{R}\mathcal{H}om\big(\mathbb{C}_\Delta, \mathcal{O}_{X\times X}^{t,(0,d_X)}\big),$$

where $\mathcal{O}_{X\times X}^{t,(0,d_X)} := \mathcal{O}_{X\times X}^{t} \otimes^{\mathbb{L}}_{p_2^{-1}\mathcal{O}_X} p_2^{-1}\Omega_X$.

11.5 Microlocalization

We will define microlocalization as the application of a subanalytic sheaf kernel attached to a particular 1-form, as in [KSIW06]. Convolution of subanalytic sheaf kernels is defined in the following way. Consider a diagram

$$\begin{array}{ccccc} & & X\times Y\times Z & & \\ & \swarrow^{q_{12}} & \downarrow{\scriptstyle q_{13}} & \searrow^{q_{23}} & \\ X\times Y & & X\times Z & & Y\times Z \end{array}$$

of manifolds regarded as subanalytic spaces. The composition of a kernel $\mathcal{F} \in D^b(k_{X\times Y})$ and $\mathcal{G} \in D^b(k_{Y\times Z})$ is defined by

$$\mathcal{F} \circ \mathcal{G} := \mathbb{R}q_{13!!}\big(q_{12}^{-1}\mathcal{F} \otimes q_{23}^{-1}\mathcal{G}\big).$$

We denote by $\pi_X : T^*X \to X$ the natural projection. For a closed submanifold Z of X, denote by $T_Z X$ its normal bundle, defined by the exact sequence

$$0 \to TZ \to TX \times_X Z \to T_Z X \to 0$$

and $T_Z^* X$ its conormal bundle, defined by the exact sequence

$$0 \to T_Z^* X \to T^*X \times_X Z \to T^*Z \to 0.$$

In particular, $T_X^* X$ denotes the zero section of T^*X. To a differentiable map $f : X \to Y$, one associates the diagram

$$T^*X \xleftarrow{f_d} T^*Y \times_Y X \xrightarrow{f_\pi} T^*Y.$$

The deformation $\widetilde{X}_Z$ to the normal bundle of Z in X is defined (see [KS90], Chap. IV) as the relative affine scheme over X given by the spectrum

$$\widetilde{X}_Z := \operatorname{Spec}_{\mathcal{O}_X}\Big(\bigoplus_{i\in\mathbb{Z}} z^{-i}\mathcal{I}_Z^i\Big),$$

where $\mathcal{I}_Z$ denotes the ideal of Z in $\mathcal{O}_X$ and its negative powers are defined to be $\mathcal{O}_X$. We will also denote by $\widetilde{X}_Z$ the corresponding manifold. The natural projection $p : \widetilde{X}_Z \to \mathbb{A}^1$ has as fiber the normal bundle $t^{-1}(0) \cong T_Z X$ and $t^{-1}(\mathbb{G}_m) \cong X \times \mathbb{G}_m$, where $\mathbb{G}_m = \mathbb{A}^1 \backslash \{0\}$. One has the following commutative diagram with cartesian squares:

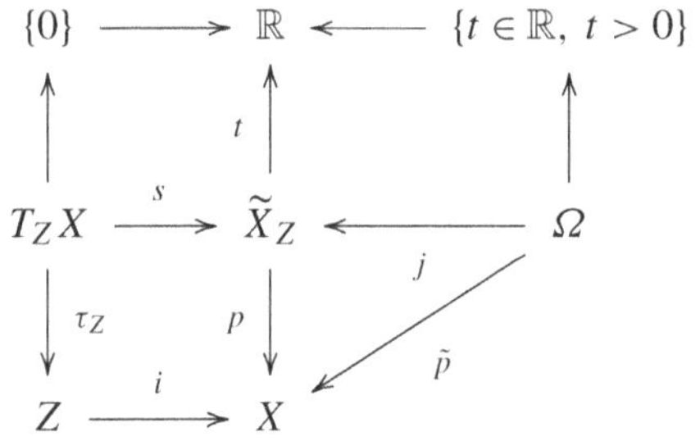

Note that $p : \widetilde{X}_Z \to X$ is not smooth but its relative dualizing complex is $\omega_{\widetilde{X}_Z/X} = k_{\widetilde{X}_Z}[1]$.

Definition 11.5.1 Let S be a subset of X. The *normal cone to S along Z* is the set

$$C_Z(S) := T_M X \cap \overline{\tilde{p}^{-1}(S)}.$$

If $X = T^*Y$ is already the cotangent bundle of a manifold Y, we use the canonical symplectic form on T^*Y to identify T^*X with TX and consider $C_Z(S)$ as a closed subset of $T^*_M X$.

Now let M be a real analytic manifold. The deformation to the normal bundle of the diagonal

$$\Delta_{T^*M} : Z = T^*M \to T^*M \times T^*M = X$$

gives a diagram

$$\begin{array}{ccccccccc}
P & \longrightarrow & TT^*M & \xrightarrow{\sim} & T_{\Delta_{T^*M}}(T^*M \times T^*M) & \xrightarrow{s} & \widetilde{T^*M \times T^*M}_{\Delta_{T^*M}} & \xleftarrow{j} & \Omega \\
 & & & & \downarrow{\scriptstyle \tau_{T^*M}} & & \downarrow{\scriptstyle p} & \swarrow{\scriptstyle \tilde{p}} & \\
 & & & & T^*M & \xrightarrow[\Delta_{T^*M}]{} & T^*M \times T^*M & &
\end{array}$$

where $P \subset TT^*M$ is defined by

$$P = \{(x,\xi; v_x, v_\xi),\ \langle v_x, \xi\rangle = 0\}.$$

Definition 11.5.2 The *microlocalization kernel* $\mathcal{K}_{T^*M} \in D^b(k_{T^*M_{sa} \times T^*M_{sa}})$ is defined by

$$\mathcal{K}_{T^*M} := \mathbb{R}p_{!!}\big(k_{\overline{\Omega}} \otimes \rho_!(K_P)\big) \otimes \rho_!\big(\omega^{\otimes -1}_{\Delta_{T^*M}/T^*M \times T^*M}\big).$$

The *microlocalization functor* is defined by

$$\mu : D^b(k_{M_{sa}}) \to D^b(k_{T^*M_{sa}})$$
$$\mathcal{F} \mapsto \mathcal{K}_{T^*M} \circ \pi_M^{-1}(\mathcal{F}),$$

where the convolution is taken over the triple product $M \times T^*M \times \{.\}$ as a subanalytic space.

The microlocalization functor allows straightforward definitions of many important tools of algebraic analysis. Note that our definition on the subanalytic site differs in general from the usual one (see for example [Pre07], Sect. 5.2).

Definition 11.5.3 For $\mathcal{F}, \mathcal{G} \in D^b(k_{M_{sa}})$, the *microlocal homomorphisms* between $\mathcal{F}$ and $\mathcal{G}$ are defined by

$$\mu hom(\mathcal{F}, \mathcal{G}) := \mathbb{R}\mathcal{H}om\big(\mu(\mathcal{F}), \mu(\mathcal{G})\big).$$

For $\mathcal{F} \in D^b(k_M)$, the *microsupport* of $\mathcal{F}$ is defined by

$$\mathrm{SS}(\mathcal{F}) := \mathrm{supp}\big(\mu(\mathcal{F})\big).$$

If $\mathcal{F}, \mathcal{G} \in D^b(k_M)$, we will also define

$$\mu hom(\mathcal{F}, \mathcal{G}) := \rho^{-1}\mathbb{R}\mathcal{H}om\big(\mu(\rho_*\mathcal{F}), \mu(\rho_*\mathcal{G})\big).$$

By definition, for $\mathcal{F}, \mathcal{G} \in D^b(k_M)$, one has

$$\operatorname{supp}\big(\mu hom(\mathcal{F}, \mathcal{G})\big) \subset \mathrm{SS}(\mathcal{F}) \cap \mathrm{SS}(\mathcal{G}).$$

Example 11.5.1 If $Z \subset M$ is a closed submanifold, one has

$$\mathrm{SS}(\mathbb{C}_Z) = T_Z^*M.$$

We give a simpler characterization of the microsupport, which is its original definition (see [KS90]). Roughly speaking, $\mathcal{F}$ has no cohomology supported by the closed half spaces whose conormals do not belong to its microsupport.

Proposition 11.5.1 *For $\mathcal{F} \in D^b(k_M)$ and $U \subset T^*M$, one has $U \cap \mathrm{SS}(\mathcal{F})$ if and only if for any $x_0 \in M$ and any real C^∞-function $\varphi : X \to \mathbb{R}$ such that $\varphi(x_0) = 0$ and $d\varphi(x_0) \in U$, one has*

$$\big(\mathbb{R}\Gamma_{\varphi \geq 0}(\mathcal{F})\big)_{x_0} = 0.$$

Proof We know from [KS90], Corollary 6.1.3, that, if SS' denotes the above characterization of the microsupport,

$$\mathrm{SS}'(\mathcal{F}) = \operatorname{supp}\big(\mu hom(\mathcal{F}, \mathcal{F})\big).$$

The result then follows from the sequence of equalities

$$\begin{aligned} SS'(\mathcal{F}) &= \operatorname{supp}\big(\mu hom(\mathcal{F}, \mathcal{F})\big) = \operatorname{supp}\big(\mathbb{R}\mathcal{H}om\big(\mu(\mathcal{F}), \mu(\mathcal{F})\big)\big) \\ &= \operatorname{supp}\big(\mu(\mathcal{F})\big) = \mathrm{SS}(\mathcal{F}). \end{aligned}$$

□

The main property of the microsupport is the following.

Theorem 11.5.1 *For $\mathcal{F} \in D^b(k_M)$, the microsupport $\mathrm{SS}(\mathcal{F}) \subset T^*M$ is a closed involutive subset, i.e., contains its own orthogonal for the standard symplectic form.*

Proof See [KS90], Theorem 6.5.4. □

We now define, in analogy with the notion of an infinite order differential operator, presented in Definition 11.4.3, various sheaves of microlocal functions and microlocal operators. The main use of microlocal operators is to define inverses for differential operators in some subspaces of the cotangent bundle.

Definition 11.5.4 Let X be a complex analytic manifold. Let $\dot{T}^*X := T^*X \backslash T^*_X X$ and let $\gamma : \dot{T}^*X \to P_X := \dot{T}^*X/\mathbb{C}^*$ be the canonical projection. The ring of *microlocal operators* is defined as

$$\mathcal{E}_X^{\mathbb{R}} := \mu hom(\mathbb{C}_\Delta, \mathcal{O}_{X\times X}^{(0,d_X)}),$$

where $\mathcal{O}_{X\times X}^{(0,d_X)} := \mathcal{O}_{X\times X} \otimes^{\mathbb{L}}_{p_2^{-1}\mathcal{O}_X} p_2^{-1}\Omega_X$. The ring of *tempered microlocal operators* is defined as

$$\mathcal{E}_X^{\mathbb{R},f} := \mu hom(\mathbb{C}_\Delta, \mathcal{O}_{X\times X}^{t,(0,d_X)}),$$

where $\mathcal{O}_{X\times X}^{t,(0,d_X)} := \mathcal{O}^t_{X\times X} \otimes^{\mathbb{L}}_{p_2^{-1}\mathcal{O}_X} p_2^{-1}\Omega_X$. The ring $\mathcal{E}_X$ of *microdifferential operators* is the corresponding conic sheaf, defined as

$$\mathcal{E}_X := \gamma^{-1}\mathbb{R}\gamma_* \mathcal{E}_X^{\mathbb{R},f}.$$

The ring $\mathcal{E}_X$ is supported on $T^*X \cong T^*_\Delta(X \times X)$, and one has a natural isomorphism

$$\mathrm{sp} : \mathcal{D}_X \xrightarrow{\sim} (\mathcal{E}_X)_{|T^*_\Delta \Delta}$$

where $X \cong T^*_\Delta \Delta \subset T^*_\Delta(X \times X)$ is the zero section.

There is a natural morphism $\pi^{-1}\mathcal{D}_X \to \mathcal{E}_X$. Moreover, the composition of microdifferential operators can be computed locally in an easy way. When X is affine and U is open in T^*X, a microdifferential operator $P \in \mathcal{E}_X(U)$ is described by its "total symbol"

$$\sigma_{tot}(P)(x,\xi) = \sum_{-\infty<j\le m} p_j(x,\xi), \quad m \in \mathbb{Z},\ p_j \in \Gamma(U, \mathcal{O}_{T^*X}(j)),$$

with the condition:

$$\begin{cases} \text{for any compact subset } K \text{ of } U, \text{ there exists a positive constant} \\ C_K \text{ such that } \sup_K |p_j| \le C_K^{-j}(-j)! \text{ for all } j < 0, \end{cases}$$

(where $\mathcal{O}_{T^*X}(j) \subset \mathcal{O}_{T^*X}$ denotes functions on T^*X that are homogeneous of degree j in the fiber variable). The total symbol of the product is given by the Leibniz rule: if Q is another operator, one has

$$\sigma_{tot}(P \circ Q) = \sum_{\alpha\in\mathbb{N}^n} \frac{1}{\alpha!} \partial_\xi^\alpha \sigma_{tot}(P) \partial_x^\alpha \sigma_{tot}(Q).$$

Definition 11.5.5 If $\mathcal{M}$ is a coherent $\mathcal{E}_X$-module, its *characteristic variety* $\mathrm{char}(\mathcal{M})$ is defined to be the support of $\mathcal{M}$.

Proposition 11.5.2 *If $\mathcal{M}$ is a $\mathcal{D}$-module and $\mathcal{M}' = \mathcal{E}_X \otimes_{\pi^{-1}\mathcal{D}_X} \pi^{-1}\mathcal{M}$ is its microlocalization, the characteristic variety of $\mathcal{M}$ equals the characteristic variety of $\mathcal{M}'$, i.e.,*

$$\mathrm{char}(\mathcal{M}) = \mathrm{supp}(\mathcal{M}').$$

Proof See [Kas03], Theorem 7.27. □

Microfunctions are a microlocal version of hyperfunctions. They are naturally equipped with a right $\mathcal{E}_X$-module structure.

Definition 11.5.6 Let M be a real analytic manifold and X be a complexification of M. The sheaf of *microfunctions* on M is defined by

$$\mathcal{C}_M := \mu hom\big(D'_X(\mathbb{C}_M), \mathcal{O}_X\big).$$

The sheaf of *tempered microfunctions* on M is defined by

$$\mathcal{C}^t_M := \mu hom\big(D'_X(\mathbb{C}_M), \mathcal{O}^t_X\big).$$

There is a natural isomorphism

$$\mathrm{sp} : \mathcal{B}_M \xrightarrow{\sim} (\mathcal{C}_M)_{|T^*_M M},$$

called the specialization, where $M \cong T^*_M M \subset T^*_M X$ is the zero section. Since $\mathcal{C}_M$ is a conic sheaf, this also induces a morphism

$$\mathrm{sp} : \mathcal{B}_M(M) \longrightarrow \mathcal{C}_M\big(T^*_M X\big)$$

on global sections.

Definition 11.5.7 Let $u \in \mathcal{B}_M$ be a hyperfunction. The support of $\mathrm{sp}(u)$ in $T^*_M X$ is called the analytic wave front set of u and is denoted by $\mathrm{WF}(u)$.

Proposition 11.5.3 *Let Ω be a subanalytic open subset of X, $\overline{\Omega}$ its closure and suppose that $D'_X(\mathbb{C}_\Omega) \cong \mathbb{C}_{\overline{\Omega}}$ and $M \subset \overline{\Omega}$. Let $\varphi \in \mathcal{H}om_{\mathcal{D}}(\mathcal{M}, \mathcal{O})(\Omega)$ be a solution to a $\mathcal{D}_X$-module $\mathcal{M}$. Then one has*

$$\mathrm{WF}\big(b(\varphi)\big) \subset T^*_M X \cap \mathrm{SS}(\mathbb{C}_\Omega) \cap \mathrm{char}(\mathcal{M}).$$

Proof The morphism $\mathbb{C}_{\overline{\Omega}} \to \mathbb{C}_M$ defines by duality a morphism $D'_X(\mathbb{C}_M) \to D'_X(\mathbb{C}_{\overline{\Omega}}) \cong \mathbb{C}_\Omega$. Applying $\mu hom(-, \mathcal{O}_X)$, one gets that the boundary value morphism

$$b : \mathcal{O}(\Omega) \to \mathcal{C}\big(T^*M\big) \cong \mathcal{B}(M)$$

factors through $\mu hom(\mathbb{C}_\Omega, \mathcal{H}om_{\mathcal{D}_X}(\mathcal{M}, \mathcal{O}_X))$, and the result follows by using the fact that $\mathrm{SS}(\mathcal{H}om_{\mathcal{D}_X}(\mathcal{M}, \mathcal{O}_X)) \subset \mathrm{char}(\mathcal{M})$. □

Proposition 11.5.4 *Let N be a submanifold of M and u be a hyperfunction on M. Suppose that*

$$\mathrm{WF}(u) \cap T_N^* M = \emptyset.$$

Then there is a well defined restriction $u_{|N}$ of u to N. In particular, if $u_1 \in \mathcal{B}(M)$ and $u_2 \in \mathcal{B}(M)$ are two hyperfunctions, the pullback $u_1 \cdot u_2$ of the tensor product $u_1 \otimes u_2 \in \mathcal{B}(M \times M)$ to the diagonal $M \subset M \times M$ is well defined whenever

$$\mathrm{WF}(u) \cap \mathrm{WF}(v)^o = \emptyset,$$

where $\mathrm{WF}(v)^o = \{(x, -\xi) | \, (x, \xi) \in \mathrm{WF}(v)\}$.

Proof See [KKK86], Theorem 3.1.3 and Theorem 3.1.5. □

Microfunctions were designed to analyze the directions of the propagation of singularities of hyperfunctions in the cotangent bundle. We denote by $\dot{\pi} : \dot{T}_M^* X \to M$ the conormal bundle without its zero section. The following proposition shows that microfunctions outside of the zero section describe exactly the singular part of hyperfunctions, i.e., hyperfunctions up to (non-singular) analytic functions.

Proposition 11.5.5 *There is a natural exact sequence*

$$0 \to \mathcal{A}_M \to \mathcal{B}_M \to \dot{\pi}_* \mathcal{C}_M \to 0$$

of sheaves that is also exact at the level of global sections.

Proof See [KS90], Proposition 11.5.2. □

11.6 Some Key Results of Linear Algebraic Analysis

We refer to Schapira's surveys [Sch10] and [Sch12] for refined introductions to the general methods and results of linear algebraic analysis. Let M be a real analytic manifold with complexification X. Recall that if $\mathcal{M}$ is a $\mathcal{D}$-module and $\mathcal{S}$ is another $\mathcal{D}$-module, we denote by

$$\mathbb{R}\mathrm{Sol}_{\mathcal{M}}(\mathcal{S}) := \mathbb{R}\mathcal{H}om_{\mathcal{D}}(\mathcal{M}, \mathcal{S})$$

the space of (derived) solutions of $\mathcal{M}$ with values in $\mathcal{S}$.

Theorem 11.6.1 (Propagation of Singularities) *Let $\mathcal{M}$ be a coherent $\mathcal{E}_X$-module. Then one has an inclusion*

$$\mathrm{SS}\big(\mathbb{R}\mathcal{H}om_{\mathcal{E}_X}(\mathcal{M}, \mathcal{C}_M)\big) \subset C_{T_M^* X}\big(\mathrm{char}(\mathcal{M})\big)$$

of closed subsets of $T^* T^* X$. In particular, one has

$$\mathrm{supp}\big(\mathbb{R}\mathcal{H}om_{\mathcal{E}}(\mathcal{M}, \mathcal{C}_M)\big) \subset \mathrm{char}(\mathcal{M}) \cap T_M^* X.$$

Proof See [KS90], Proposition 11.5.4. □

Definition 11.6.1 A coherent $\mathcal{D}_X$-module is called *elliptic* if

$$\mathrm{char}(\mathcal{M}) \cap T_M^* X \subset T_X^* X.$$

Theorem 11.6.2 (Elliptic Regularity) *Let $\mathcal{M}$ be an elliptic $\mathcal{D}$-module. The natural inclusion of analytic solutions in hyperfunction solutions is an isomorphism:*

$$\mathbb{R}\mathrm{Sol}_{\mathcal{M}}(\mathcal{A}_M) \xrightarrow{\sim} \mathbb{R}\mathrm{Sol}_{\mathcal{M}}(\mathcal{B}_M).$$

Proof This follows from Theorem 11.6.1, by applying the functor $\mathbb{R}\mathcal{H}om_{\mathcal{D}}(\mathcal{M}, -)$ to the exact sequence

$$0 \to \mathcal{A}_M \to \mathcal{B}_M \to \dot{\pi}_* \mathcal{C}_M \to 0$$

of Proposition 11.5.5. □

Definition 11.6.2 Let $f : X \to Y$ be a morphism of complex manifolds. Consider the associated diagram

$$T^* X \xleftarrow{f_d} T^* Y \times_Y X \xrightarrow{f_\pi} T^* Y.$$

Let $\mathcal{N}$ be a $\mathcal{D}_Y$-module. One says that f is *non-characteristic* for $\mathcal{N}$ if

$$f_\pi^{-1} \mathrm{char}(\mathcal{N}) \cap T_X^* Y \subset X \times_Y T_Y^* Y.$$

We now give Kashiwara's generalization of the classical Cauchy-Kowalewskaya theorem, which essentially asserts that, given some nice initial data, one can globally solve analytic partial differential equations.

Theorem 11.6.3 (Cauchy-Kowalewskaya-Kashiwara) *Let $f : X \to Y$ be a morphism of complex manifolds and $\mathcal{N}$ be a $\mathcal{D}_Y$-module. If f is non characteristic for $\mathcal{N}$, the $\mathcal{D}_X$-module $f_{\mathcal{D}}^*(\mathcal{N})$ is coherent and there is a natural isomorphism*

$$f^{-1} \mathbb{R}\mathrm{Sol}_{\mathcal{N}}(\mathcal{O}_Y) \xrightarrow{\sim} \mathbb{R}\mathrm{Sol}_{f_{\mathcal{D}}^* \mathcal{N}}(\mathcal{O}_X).$$

Proof This follows by dévissage from the classical solution of the holomorphic Cauchy problem. We refer to [KS90], Theorem 11.3.5 for a proof. □

A related result gives a relation between the characteristic variety of a $\mathcal{D}$-module and its holomorphic solution space (see [KS90], Theorem 11.3.3).

Theorem 11.6.4 (Kashiwara-Schapira) *Let X be a complex analytic manifold and $\mathcal{M}$ be a coherent $\mathcal{D}$-module. There is a natural bijection*

$$\mathrm{char}(\mathcal{M}) = \mathrm{SS}\big(\mathbb{R}\mathrm{Sol}_{\mathcal{M}}(\mathcal{O}_X)\big).$$

A very important structural result for microlocal analysis is that coherent $\mathcal{E}$-modules have a very simple form at generic points of their characteristic variety. The full statement of this theorem is too long for our presentation and we refer to [KKK86] for more precision. The proof is based on the extension of local canonical (i.e. symplectic) changes of variables in the cotangent bundle to transformations on microdifferential operators, called quantized contact transformations.

We will now state a simplified formulation of the regular Riemann-Hilbert correspondence between derived categories of regular holonomic $\mathcal{D}$-modules and $\mathbb{C}$-constructible sheaves. We start by recollecting some basic facts on holonomic $\mathcal{D}$-modules.

Definition 11.6.3 Let X be a complex analytic manifold. A $\mathcal{D}$-module $\mathcal{M} \in D^b(\mathcal{D}_X)$ is called *holonomic* if $\mathrm{char}(\mathcal{M})$ is a Lagrangian subspace of T^*X. The category of holonomic $\mathcal{D}$-modules is denoted by $D^b_{hol}(\mathcal{D}_X)$.

The category of holonomic $\mathcal{D}$-modules gives the simplest category containing the $\mathcal{D}$-module $\mathcal{O}$ of functions and which is stable under the six Grothendieck operations. It plays an important role in the study of some topological quantum field theories described in Sects. 24.2 and 24.5. We refer to Ayoub [Ayo07] for a general formalization of the four operations, called crossed functors, and to [FHM02] for a presentation of the Grothendieck-Verdier duality formalism of [Har70] in the language of adjunctions between symmetric monoidal categories. We also recommend [KS90] as a general reference on Verdier duality for sheaves.

Theorem 11.6.5 *Let $f : X \to Y$ be a morphism of smooth analytic manifolds. The adjoint pair $(f^*, f_*) := (f^*_{\mathcal{D}}, f^{\mathcal{D}}_*)$ of functors described in Definition 11.2.2, and the duality functor $\mathbb{D}$, can be completed with an adjoint pair $(f_!, f^!)$ of proper supported inverse and direct image, to give the six Grothendieck operations*

$$(f^*, f_*, f^!, f_!, \mathbb{D}, \boxtimes)$$

on the derived category of bounded complexes of holonomic $\mathcal{D}$-modules.

Definition 11.6.4 Let X be a complex manifold and $Z \subset X$ be a closed submanifold. We denote by $\mathcal{O}_X(*Z)$ the subsheaf of the sheaf $\mathcal{K}_X$ of meromorphic functions on X given by sections that have poles on Z. We denote by $\mathcal{D}(*Z) = \mathcal{D}_X(*Z)$ the ring of differential operators on X with coefficients in $\mathcal{O}(*Z) = \mathcal{O}_X(*Z)$. A $\mathcal{D}(*Z)$-module $(\mathcal{E}, \nabla)$ on X that is locally free of finite rank on $\mathcal{O}(*Z)$ is called a meromorphic connection on X with poles on Z.

We first recalling the standard dévissage of holonomic $\mathcal{D}$-modules (see Kashiwara [Kas03]) in meromorphic differential equations.

Proposition 11.6.1 *For X a manifold and $\mathcal{M}$ a holonomic $\mathcal{D}$-module on X, there exists a stratification $Z_0 = \emptyset \subset Z_1 \subset \cdots \subset Z_n = X$ of X by closed analytic subsets such that for all $i = 1, \ldots, n$, the restriction of $\mathcal{M}$ on Z_i is a meromorphic connection on Z_i with poles on Z_{i-1}.*

Definition 11.6.5 Let X be a manifold and Z be a closed submanifold. A meromorphic connection $(\mathcal{E}, \nabla)$ on X with poles on Z is called *regular singular* if the natural restriction morphism

$$\mathbb{R}\mathrm{Sol}_{\mathcal{E}}(\mathcal{O}_X(*Z)) \to \mathbb{R}\mathrm{Sol}_{\mathcal{E}}(\mathcal{O}_{X\setminus Z})$$

is an isomorphism.

Definition 11.6.6 A complex of sheaves $F \in D^b(\mathbb{C}_X)$ is called *constructible* if it is the derived solution sheaf

$$F = \mathbb{R}\mathrm{Sol}_{\mathcal{M}}(\mathcal{O}_X) := \mathbb{R}\mathcal{H}om_{\mathcal{D}}(\mathcal{M}, \mathcal{O}_X)$$

of a holonomic $\mathcal{D}$-module $\mathcal{M}$. The category of constructible sheaves is denoted by $D^b_{\mathbb{C}-c}(\mathbb{C}_X)$. A $\mathcal{D}$-module is called *regular holonomic* if it is of the form

$$\mathcal{M} = \mathbb{R}\mathcal{H}om_{\mathbb{C}_X}(F, \mathcal{O}^t_X)$$

where F is constructible and $\mathcal{O}^t_X$ is the subanalytic sheaf of tempered holomorphic functions (see Definition 11.3.6). The category of regular holonomic $\mathcal{D}$-modules is denoted by $D^b_{reghol}(\mathcal{D}_X)$.

Theorem 11.6.6 (Riemann-Hilbert Correspondence) *The pair of functors*

$$\mathbb{R}\mathcal{H}om_{\mathcal{D}}(-, \mathcal{O}_X) : D^b_{reghol}(\mathcal{D}_X)^{op} \leftrightarrows D^b_{\mathbb{C}-c}(\mathbb{C}_X) : \mathbb{R}\mathcal{H}om_{\mathbb{C}_X}(-, \mathcal{O}^t_X)$$

is an equivalence of categories. Moreover, one has the following concrete characterizations of both categories:

1. *A sheaf F is constructible if and only if it is locally constant of finite rank on an analytic stratification, i.e., there exist a finite sequence $Z_0 = \emptyset \subset Z_1 \subset \cdots \subset Z_n = X$ of closed analytic subsets such that $F_{|Z_{i+1}\setminus Z_i}$ is locally constant.*
2. *A holonomic $\mathcal{D}$-module $\mathcal{M}$ is regular if and only if one of the following equivalent properties is fulfilled:*

   (a) *There exists locally a coherent filtration $F^\bullet\mathcal{M}$ whose characteristic ideal $\mathcal{I}_{F^\bullet\mathcal{M}}$ is radical (see Definition* 11.1.2*), i.e., such that the radical ideal $\mathcal{I}_{\mathcal{M}}$ defining the characteristic variety* char$(\mathcal{M})$ *annihilates* $\mathrm{gr}^F(\mathcal{M})$.

   (b) *There exists a dévissage of $\mathcal{M}$ into meromorphic connections on a stratification $Z_0 = \emptyset \subset Z_1 \subset \cdots \subset Z_n = X$ by closed analytic subsets, such that for $i = 1, \ldots, n$, the restriction of $\mathcal{M}$ to Z_i is a regular singular meromorphic connection with poles on Z_{i-1}.*

Note for the applications to diffeographism groups in renormalization à la Connes-Kreimer of Chap. 19, one needs a more general result, called the irregular Riemann-Hilbert correspondence: a classification of all holonomic $\mathcal{D}$-modules by constructible sheaves with additional combinatorial data. This general result is only known in lower dimensional settings (see [Sab09] for a description of the kind of structures necessary to prove instances of the irregular Riemann-Hilbert correspondence), but the results of Mochizuki [Moc08] and Kedlaya [Ked10] (which give a fine dévissage for general holonomic $\mathcal{D}$-modules) give interesting hints to the definition of a convenient category of irregular (ramified) constructible coefficients that will give an irregular Riemann-Hilbert correspondence (see also [DK12]).

References

[And07] Andronikof, E.: Interview with Mikio Sato. Not. Am. Math. Soc. **54**(2), 243–245 (2007)

[Ayo07] Ayoub, J.: Les Six Opérations de Grothendieck et Le Formalisme des Cycles Évanescents dans Le Monde Motivique. I (2007). Astérisque, 314:x+466 pp. (2008)

[DK12] D'Agnolo, A., Kashiwara, M.: On a reconstruction theorem for holonomic systems. arXiv e-prints (2012). arXiv:1208.6104

[DM99] Deligne, P., Morgan, J.W.: Notes on supersymmetry (following Joseph Bernstein). In: Quantum Fields and Strings: A Course for Mathematicians, Vol. 1, 2 (Princeton, NJ, 1996/1997), pp. 41–97. Am. Math. Soc., Providence (1999)

[FG11] Francis, J., Gaitsgory, D.: Chiral Koszul duality. arXiv e-prints (2011). arXiv:1103.5803

[FHM02] Fausk, H., Hu, P., May, J.P.: Isomorphisms between left and right adjoints. arXiv Mathematics e-prints (2002). arXiv:math/0206079

[Gab96] Gabrielov, A.: Complements of subanalytic sets and existential formulas for analytic functions. Invent. Math. **125**(1), 1–12 (1996). doi:10.1007/s002220050066

[Har70] Hartshorne, R.: Residues and Duality. Springer, Berlin (1970)

[Hör03] Hörmander, L.: The Analysis of Linear Partial Differential Operators. i. Classics in Mathematics, p. 440. Springer, Berlin (2003). Distribution theory and Fourier analysis. Reprint of the second (1990) edition [Springer, Berlin; MR1065993 (91m:35001a)]. ISBN 3-540-00662-1

[Hör07] Hörmander, L.: The Analysis of Linear Partial Differential Operators. III. Classics in Mathematics, p. 525. Springer, Berlin (2007). Pseudo-differential operators. Reprint of the 1994 edition. ISBN 978-3-540-49937-4

[Kas03] Kashiwara, M.: D-Modules and Microlocal Calculus. Translations of Mathematical Monographs, vol. 217, p. 254. Am. Math. Soc., Providence (2003). Translated from the 2000 Japanese original by Mutsumi Saito. Iwanami Series in Modern Mathematics. ISBN 0-8218-2766-9

[Ked10] Kedlaya, K.S.: Good formal structures for flat meromorphic connections, II: excellent schemes. arXiv e-prints (2010). arXiv:1001.0544

[KKK86] Kashiwara, M., Kawai, T., Kimura, T.: Foundations of Algebraic Analysis. Princeton Mathematical Series, vol. 37, p. 255. Princeton University Press, Princeton (1986). Translated from the Japanese by Goro Kato. ISBN 0-691-08413-0

[KS90] Kashiwara, M., Schapira, P.: Sheaves on Manifolds. Grundlehren der Mathematischen Wissenschaften [Fundamental Principles of Mathematical Sciences], vol. 292, p. 512. Springer, Berlin (1990). With a chapter in French by Christian Houzel. ISBN 3-540-51861-4

[KS99] Kashiwara, M., Schapira, P.: Ind-sheaves, distributions, and microlocalization. In: Seminaire: Équations aux Dérivées Partielles, 1998–1999 (1999). Sémin. Équ. Dériv. Partielles, pages Exp. No. XXII, 13. École Polytech., Palaiseau

[KS01] Kashiwara, M., Schapira, P.: Ind-sheaves. Astérisque **271**, 136 (2001)

[KSIW06] Kashiwara, M., Schapira, P., Ivorra, F., Waschkies, I.: Microlocalization of ind-sheaves. In: Studies in Lie Theory. Progr. Math., vol. 243, pp. 171–221. Birkhäuser Boston, Boston (2006). doi:10.1007/0-8176-4478-4_9

[Lan93] Lang, S.: Algebra, 3rd edn., p. 914. Addison-Wesley, New York (1993). ISBN 0-201-55540-9

[Man97] Manin, Y.I.: Gauge Field Theory and Complex Geometry, 2nd edn. Grundlehren der Mathematischen Wissenschaften [Fundamental Principles of Mathematical Sciences], vol. 289, p. 346. Springer, Berlin (1997). Translated from the 1984 Russian original by N. Koblitz and J.R. King, with an appendix by Sergei Merkulov. ISBN 3-540-61378-1

[Moc08] Mochizuki, T.: Good formal structure for meromorphic flat connections on smooth projective surfaces. arXiv e-prints (2008). arXiv:0803.1346

[Pen83] Penkov, I.B.: $\mathcal{D}$-Modules on supermanifolds. Invent. Math. **71**(3), 501–512 (1983). doi:10.1007/BF02095989

[Pre07] Prelli, L.: Microlocalization of subanalytic sheaves. arXiv Mathematics e-prints (2007). arXiv:math/0702459

[Sab09] Sabbah, C.: Introduction to Stokes structures. arXiv e-prints (2009). arXiv:0912.2762

[Sch94a] Schneiders, J.-P.: An introduction to $\mathcal{D}$-modules. Bull. Soc. R. Sci. Liège **63**(3–4), 223–295 (1994). Algebraic Analysis Meeting (Liège, 1993)

[Sch07] Schapira, P.: Mikio Sato, a visionary of mathematics. Not. Am. Math. Soc. **54**(2), 243–245 (2007)

[Sch09a] Schapira, P.: Masaki Kashiwara and algebraic analysis. In: Algebraic Analysis and Around. Adv. Stud. Pure Math., vol. 54, pp. 1–9. Mathematical Society of Japan, Tokyo (2009)

[Sch10] Schapira, P.: Triangulated categories for the analysis. In: Triangulated Categories. London Math. Soc. Lecture Note Ser., vol. 375, pp. 371–388. Cambridge University Press, Cambridge (2010)

[Sch12] Schapira, P.: Three lectures on algebraic microlocal analysis. arXiv e-prints (2012). arXiv:1206.1435

[SKK73] Sato, M., Kawai, T., Kashiwara, M.: Microfunctions and pseudo-differential equations. In: Hyperfunctions and Pseudo-Differential Equations (Proc. Conf., Katata, 1971; Dedicated to the Memory of André Martineau). Lecture Notes in Math., vol. 287, pp. 265–529. Springer, Berlin (1973)

Chapter 12
Algebraic Analysis of Non-linear Partial Differential Equations

This chapter introduces the necessary tools for a differential algebraic coordinate-free study of non-linear partial differential equations in general, and Lagrangian mechanics (see Sect. 8.1 for an introduction compatible with our general approach) in particular.

The algebraic study of non-linear partial differential equations was initiated by E. Cartan. This section expands on the article [Pau11a]. We use the viewpoint of functorial analysis, developed in Chap. 4, meaning that functionals are defined as partially defined functions between spaces in the sense of Chap. 3. We use Beilinson and Drinfeld's functorial approach [BD04] to non-linear algebraic partial differential equations (with non-algebraic coefficients), and we relate this approach to ours by extending it to non-algebraic partial differential equations. We are also inspired by the work of Nestruev's diffiety school, and in particular Vinogradov [Vin01] and Krasilshchik and Verbovetsky [KV98].

12.1 Differential Algebras and Non-linear Partial Differential Equations

In this section, we will systematically use the language of differential calculus in symmetric monoidal categories, and the functor of points approach to spaces of fields, described in Chaps. 3 and 4. We first restrict our presentation to relatively algebraic partial differential equations (with non-algebraic coefficients), but our results also apply to smooth partial differential equations, using the formalism of Sect. 2.5. This extension of our approach to general partial differential equations will be described in Sect. 12.5, but it is pedagogically helpful to first consider the relatively algebraic situation.

We specialize the situation to the symmetric monoidal category

$$\left(\textsc{Mod}(\mathcal{D}_M), \otimes_{\mathcal{O}_M}\right)$$

F. Paugam, *Towards the Mathematics of Quantum Field Theory*,
Ergebnisse der Mathematik und ihrer Grenzgebiete. 3. Folge / A Series of Modern Surveys in Mathematics 59, DOI 10.1007/978-3-319-04564-1_12,

of left $\mathcal{D}_M$-modules on a given (super-)manifold M. Recall that there is an equivalence

$$(\text{Mod}(\mathcal{D}_M), \otimes) \to (\text{Mod}(\mathcal{D}_M^{op}), \otimes^!)$$

given by tensoring with the $(\mathcal{D}, \mathcal{D}^{op})$-modules Ber_M and $\text{Ber}_M^{-1} = \mathcal{H}om_{\mathcal{O}}(\text{Ber}_M, \mathcal{O})$. The unit objects for the two monoidal structures are $\mathcal{O}$ and Ber_M respectively. If $\mathcal{M}$ is a $\mathcal{D}$-module (resp. a $\mathcal{D}^{op}$-module), we let $\mathcal{M}^r := \mathcal{M} \otimes \text{Ber}_M$ (resp. $\mathcal{M}^\ell := \mathcal{M} \otimes \text{Ber}_M^{-1}$) be the corresponding $\mathcal{D}^{op}$-module (resp. $\mathcal{D}$-module).

Recall that if $P \in \mathbb{Z}[X]$ is a polynomial, one can study the solution space

$$\underline{\text{Sol}}_{P=0}(A) = \{x \in A, P(x) = 0\}$$

of P with values in any commutative unital ring. Indeed, in any such ring, one has a sum, a multiplication, a zero and a unit that fulfill the necessary compatibilities to be able to write down the polynomial. One can thus think of the mathematical object given by the category of commutative unital rings as solving the mathematical problem of giving a natural setting for a coordinate-free study of polynomial equations. This solution space is representable, meaning that there is a functorial isomorphism

$$\underline{\text{Sol}}_{P=0}(-) \cong \text{Hom}_{\text{Rings}_{cu}}(\mathbb{Z}[X]/(P), -).$$

This shows that the solution space of an equation essentially determines the equation itself. Note that the polynomial P lives in the free algebra $\mathbb{Z}[X]$ on the given variable that was used to write it.

Suppose now given the bundle $\pi_1 : C = \mathbb{R} \times \mathbb{R} \to \mathbb{R} = M$ of smooth manifolds. We would like to study an algebraic non-linear partial differential equation

$$F(t, \partial_t^i x) = 0$$

that applies to sections $x \in \Gamma(M, C)$, that are functions $x : \mathbb{R} \to \mathbb{R}$. It is given by a polynomial $F(t, x_i) \in \mathbb{R}[t, \{x_i\}_{i \geq 0}]$. The solution space of such an equation can be studied with values in any $\mathcal{O}$-algebra $\mathcal{A}$ equipped with an action of the differentiation ∂_t (which satisfies a Leibniz rule for multiplication), the basic example being given by the algebra $\text{Jet}(\mathcal{O}_C) := \mathbb{R}[t, \{x_i\}_{i \geq 0}]$ above with the action $\partial_t x_i = x_{i+1}$. The solution space of the given partial differential equation is then given by the functor

$$\underline{\text{Sol}}_{\mathcal{D}, F=0}(\mathcal{A}) := \{x \in \mathcal{A}, \ F(t, \partial_t^i x) = 0\}$$

defined on all $\mathcal{O}_C$-algebras equipped with an action of ∂_t. To be more precise, we define the category of $\mathcal{D}$-algebras, that solves the mathematical problem of finding a natural setting for a coordinate-free study of polynomial non-linear partial differential equations with smooth super-function coefficients.

Definition 12.1.1 Let M be a supermanifold. A *$\mathcal{D}_M$-algebra* is an algebra $\mathcal{A}$ in the monoidal category of $\mathcal{D}_M$-modules. More precisely, it is an $\mathcal{O}_M$-algebra equipped with an action

$$\Theta_M \otimes \mathcal{A} \to \mathcal{A}$$

of vector fields on M such that the product in $\mathcal{A}$ satisfies Leibniz's rule

$$\partial(fg) = \partial(f)g + f\partial(g).$$

Recall from Sect. 3.3 that one can extend the jet functor to the category of smooth $\mathcal{D}$-algebras (and even to smooth superalgebras), to extend the forthcoming results to the study of non-polynomial smooth partial differential equations. The forgetful functor

$$\text{Forget} : \text{ALG}_{\mathcal{D}} \to \text{ALG}_{\mathcal{O}}$$

has an adjoint (free $\mathcal{D}$-algebra on a given $\mathcal{O}$-algebra)

$$\text{Jet} : \text{ALG}_{\mathcal{O}} \to \text{ALG}_{\mathcal{D}}$$

called the (infinite) jet functor. It satisfies the universal property that for every $\mathcal{D}$-algebra $\mathcal{B}$, the natural map

$$\text{Hom}_{\text{ALG}_{\mathcal{O}}}(\mathcal{O}_C, \mathcal{B}) \cong \text{Hom}_{\text{ALG}_{\mathcal{D}}}\big(\text{Jet}(\mathcal{O}_C), \mathcal{B}\big)$$

induced by the natural map $\mathcal{O}_C \to \text{Jet}(\mathcal{O}_C)$ is a bijection.

Using the jet functor, one can show that the solution space of the non-linear partial differential equation

$$F\big(t, \partial_t^i x\big) = 0$$

of the above example is representable, meaning that there is a natural isomorphism of functors on $\mathcal{D}$-algebras

$$\underline{\text{Sol}}_{\mathcal{D}, F=0}(-) \cong \text{Hom}_{\text{ALG}_{\mathcal{D}}}\big(\text{Jet}(\mathcal{O}_C)/(F), -\big)$$

where (F) denotes the $\mathcal{D}$-ideal generated by F. This shows that the jet functor plays the role of the polynomial algebra in the differential algebraic setting. If $\pi : C \to M$ is a bundle, we define

$$\text{Jet}(C) := \underline{\text{Spec}}\big(\text{Jet}(\mathcal{O}_C)\big).$$

One can summarize the above discussion in Table 12.1.

Example 12.1.1 If $\pi : C = \mathbb{R}^{n+m} \to \mathbb{R}^n = M$ is a trivial bundle of dimension $m + n$ over M of dimension n, with algebra of coordinates $\mathcal{O}_C := \mathbb{R}[\underline{t}, \underline{x}]$ for $\underline{t} = \{t^i\}_{i=1,\dots,n}$ and $\underline{x} = \{x^j\}_{j=1,\dots,m}$ given in multi-index notation, its jet algebra is

$$\text{Jet}(\mathcal{O}_C) := \mathbb{R}[\underline{t}, \underline{x}_\alpha]$$

Table 12.1 Analogy between polynomial and partial differential equations

Equation	Polynomial	Partial differential
Formula	$P(x)=0$	$F(t,\partial^\alpha x)=0$
Naive variable	$x\in\mathbb{R}$	$x\in\mathrm{Hom}(\mathbb{R},\mathbb{R})$
Algebraic structure	Commutative unitary ring A	$\mathcal{D}_M$-algebra A
Free structure	$P\in\mathbb{R}[x]$	$F\in\mathrm{Jet}(\mathcal{O}_C)$
Solution space	$\{x\in A,\ P(x)=0\}$	$\{x\in A,\ F(t,\partial^\alpha x)=0\}$

where $\alpha\in\mathbb{N}^m$ is a multi-index representing the derivation index. The $\mathcal{D}$-module structure is given by making $\frac{\partial}{\partial t^i}$ act through the total derivative

$$D_i:=\frac{\partial}{\partial t^i}+\sum_{\alpha,k}x^k_{i\alpha}\frac{\partial}{\partial x^k_\alpha}$$

where $i\alpha$ denotes the multi-index α increased by one at the i-th coordinate. For example, if $\pi: C=\mathbb{R}\times\mathbb{R}\to\mathbb{R}=M$, one gets

$$D_1=\frac{\partial}{\partial t}+x_1\frac{\partial}{\partial x}+x_2\frac{\partial}{\partial x_1}+\cdots.$$

Definition 12.1.2 Let $\pi: C\to M$ be a bundle. A *partial differential equation* on the space $\Gamma(M,C)$ of sections of π is given by a quotient $\mathcal{D}_M$-algebra

$$p:\mathrm{Jet}(\mathcal{O}_C)\twoheadrightarrow\mathcal{A}$$

of the jet algebra of the $\mathcal{O}_M$-algebra $\mathcal{O}_C$. Its *local solution space* is the $\mathcal{D}$-space whose points with values in $\mathrm{Jet}(\mathcal{O}_C)$-$\mathcal{D}$-algebras $\mathcal{B}$ are given by

$$\underline{\mathrm{Sol}}_{\mathcal{D},(\mathcal{A},p)}:=\{x\in\mathcal{B}|\ f(x)=0,\ \forall f\in\mathrm{Ker}(p)\}.$$

The *non-local solution space* of the partial differential equation $(\mathcal{A},p)$ is the subspace of $\underline{\Gamma}(M,C)$ given by

$$\underline{\mathrm{Sol}}_{(\mathcal{A},p)}:=\{x\in\underline{\Gamma}(M,C)|\ (j_\infty x)^*L=0\text{ for all }L\in\mathrm{Ker}(p)\}$$

where $(j_\infty x)^*:\mathrm{Jet}(\mathcal{O}_C)\to\mathcal{O}_M$ is (dual to) the Jet of x. Equivalently, $x\in\underline{\mathrm{Sol}}_{(\mathcal{A},p)}$ if and only if there is a natural factorization

$$\begin{array}{ccc}\mathrm{Jet}(\mathcal{O}_C) & \xrightarrow{(j_\infty x)^*} & \mathcal{O}_M\\ & \searrow^{p} & \uparrow\\ & & \mathcal{A}\end{array}$$

of the jet of x through p.

12.2 Local Functionals and Local Differential Forms

The natural functional invariant associated to a given $\mathcal{D}$-algebra $\mathcal{A}$ is given by the de Rham complex

$$\mathrm{DR}(\mathcal{A}) := \left(I_{*,M} \otimes_{\mathcal{O}} \mathcal{D}[n]\right) \overset{\mathbb{L}}{\otimes}_{\mathcal{D}} \mathcal{A}$$

of its underlying $\mathcal{D}$-module with coefficient in the universal complex of integral forms $I_{*,M} \otimes_{\mathcal{O}} \mathcal{D}[n]$, and its cohomology $h^*(\mathrm{DR}(\mathcal{A}))$. We will define

$$h(\mathcal{A}) := h^0\left(\mathrm{DR}(\mathcal{A})\right) = \mathrm{Ber}_M \otimes_{\mathcal{D}} \mathcal{A}$$

where Ber_M denotes the Berezinian object (and not only the complex concentrated in degree 0). If M is a non-super manifold, one gets

$$\mathrm{DR}(\mathcal{A}) = \Omega^n_M \overset{\mathbb{L}}{\otimes}_{\mathcal{D}} \mathcal{A} \quad \text{and} \quad h(\mathcal{A}) = \Omega^n_M \otimes_{\mathcal{D}} \mathcal{A}.$$

The de Rham cohomology is given by the cohomology of the complex

$$\mathrm{DR}(\mathcal{A}) = I_{*,M}[n] \otimes_{\mathcal{O}_M} \mathcal{A},$$

which gives

$$\mathrm{DR}(\mathcal{A}) = \wedge^* \Omega^1_M[n] \otimes_{\mathcal{O}_M} \mathcal{A}$$

in the non-super case.

If $\mathcal{A}$ is a jet algebra for sections of a bundle $\pi : C \to M$ with basis a classical manifold, the de Rham complex identifies with a sub-complex of the usual de Rham complex of $\wedge^* \Omega^1_{\mathcal{A}/\mathbb{R}}$ of $\mathcal{A}$ viewed as an ordinary ring. One can think of classes in $h^*(\mathrm{DR}(\mathcal{A}))$ as defining a special class of (partially defined) functionals on the space $\underline{\Gamma}(M, C)$, by integration along singular homology cycles with compact support.

Recall that a functional $f \in \mathrm{Hom}(\underline{\Gamma}(M, C), \mathbb{A}^1)$ on a space of fields denotes in general, by definition, only a partially defined function (with a well-chosen domain of definition).

Proposition 12.2.1 *Let M be a manifold. Let $\pi : C \to M$ be a bundle and $\mathcal{A}$ be the $\mathcal{D}_M$-algebra* $\mathrm{Jet}(\mathcal{O}_C)$. *There is a natural integration pairing*

$$\begin{aligned} H_{*,c}(M) \times h^{*-n}\left(\mathrm{DR}(\mathcal{A})\right) &\to \mathrm{Hom}\left(\underline{\Gamma}(M, C), \mathbb{A}^1\right) \\ (\Sigma, \omega) &\mapsto \left[x \mapsto \int_\Sigma (j_\infty x)^* \omega\right] \end{aligned}$$

where $j_\infty x : M \to \mathrm{Jet}(C)$ *is the Taylor series of a given section* x. *If* $p : \mathrm{Jet}(\mathcal{O}_C) \to \mathcal{A}$ *is a given partial differential equation (such that* $\mathcal{A}$ *is* $\mathcal{D}$*-smooth) on* $\Gamma(M,C)$ *one also gets an integration pairing*

$$H_{*,c}(M) \times h^{*-n}\big(\mathrm{DR}(\mathcal{A})\big) \to \mathrm{Hom}\big(\underline{\mathrm{Sol}}_{(\mathcal{A},p)}, \mathbb{A}^1\big)$$

$$(\Sigma, \omega) \mapsto S_{\Sigma,\omega} : \left[x(t,u) \mapsto \int_\Sigma (j_\infty x)^* \omega\right].$$

Proof Note that the values of the above pairing are given by partially defined functions, with a domain of definition given by Lebesgue's domination condition to make $t \mapsto \int_\Sigma (j_\infty x_t)^* \omega$ a smooth function of t if x_t is a parametrized trajectory. The only point to check is that the integral is independent of the chosen cohomology class. This follows from the fact that the integral of a total divergence on a closed subspace is zero, by Stokes' formula. □

Definition 12.2.1 A functional $S_{\Sigma,\omega} : \underline{\Gamma}(M,C) \to \mathbb{A}^1$ or $S_{\Sigma,\omega} : \underline{\mathrm{Sol}}_{(\mathcal{A},p)} \to \mathbb{A}^1$ obtained by the above constructed pairing is called a *quasi-local functional.* We denote by

$$\mathcal{O}^{qloc} \subset \mathcal{O} := \underline{\mathrm{Hom}}\big(\underline{\Gamma}(M,C), \mathbb{A}^1\big)$$

the space of quasi-local functionals. A *local functional* is a quasi-local functional that comes from a cohomology class in $h(\mathcal{A}) = h^0(\mathrm{DR}(\mathcal{A}))$.

Note that for $k \le n$, the classes in $h^{*-k}(\mathrm{DR}(\mathcal{A}))$ are usually called (higher) conservation laws for the partial differential equation $p : \mathrm{Jet}(\mathcal{O}_C) \to \mathcal{A}$.

If $\mathcal{A}$ is a $\mathcal{D}$-algebra, i.e., an algebra in $(\mathrm{MOD}(\mathcal{D}), \otimes_\mathcal{O})$, one defines, as usual, an $\mathcal{A}$-module of differential forms as an $\mathcal{A}$-module $\Omega^1_\mathcal{A}$ in the monoidal category $(\mathrm{MOD}(\mathcal{D}), \otimes)$, equipped with a ($\mathcal{D}$-linear) derivation $d : \mathcal{A} \to \Omega^1_\mathcal{A}$ such that for every $\mathcal{A}$-module $\mathcal{M}$ in $(\mathrm{MOD}(\mathcal{D}), \otimes)$, the natural map

$$\mathrm{Hom}_{\mathrm{MOD}(\mathcal{A})}\big(\Omega^1_\mathcal{A}, \mathcal{M}\big) \to \mathrm{Der}_{\mathrm{MOD}(\mathcal{A})}(\mathcal{A}, \mathcal{M})$$

given by $f \mapsto f \circ d$ is a bijection.

Note that the natural $\mathcal{O}$-linear map

$$\Omega^1_{\mathcal{A}/\mathcal{O}} \to \Omega^1_\mathcal{A}$$

is an isomorphism of $\mathcal{O}$-modules. The $\mathcal{D}$-module structure on $\Omega^1_\mathcal{A}$ can be regarded as an Ehresman connection, i.e., a section of the natural projection

$$\Omega^1_{\mathcal{A}/\mathbb{R}} \to \Omega^1_{\mathcal{A}/M}.$$

Example 12.2.1 In the case of the jet space algebra $\mathcal{A} = \mathrm{Jet}(\mathcal{O}_C)$ for $C = \mathbb{R}^{n+m} \to \mathbb{R}^n = M$, a basis of $\Omega^1_{\mathcal{A}/M}$ compatible with this section is given by the Cartan forms

$$\theta^i_\alpha = dx^i_\alpha - \sum_{j=1}^{n} x^i_{j\alpha} dt^j.$$

The de Rham differential $d : \mathcal{A} \to \Omega^1_{\mathcal{A}}$ in the $\mathcal{D}$-algebra setting and its de Rham cohomology (often denoted by d^V in the literature), can then be computed by expressing the usual de Rham differential $d : \mathcal{A} \to \Omega^1_{\mathcal{A}/M}$ in the basis of Cartan forms.

As explained in Sect. 11.1, the right notion of finiteness and duality in the monoidal category of $\mathcal{D}$-modules is not the $\mathcal{O}$-finite presentation and duality but the $\mathcal{D}$-finite presentation and duality. This extends to the category of $\mathcal{A}$-modules in $(\mathrm{Mod}(\mathcal{D}), \otimes)$. The following notion of smoothness differs from the usual one over a ring, because, in order to ensure good duality properties, we impose the $\mathcal{A}[\mathcal{D}]$-finite presentation, where $\mathcal{A}[\mathcal{D}] := \mathcal{A} \otimes_{\mathcal{O}} \mathcal{D}$. This is however equivalent to the usual categorical finite presentation (in terms of commutation to directed colimits).

Definition 12.2.2 The $\mathcal{D}$-algebra $\mathcal{A}$ is called *$\mathcal{D}$-smooth* if $\Omega^1_{\mathcal{A}}$ is a projective $\mathcal{A}$-module of finite $\mathcal{A}[\mathcal{D}]$-presentation in the category of $\mathcal{D}$-modules, and $\mathcal{A}$ is a (geometrically) finitely presented $\mathcal{D}$-algebra, meaning that there exists a finitely presented $\mathcal{O}$-module $\mathcal{M}$, a $\mathcal{D}$-ideal $\mathcal{I} \subset \mathrm{Jet}(\mathrm{Sym}_{\mathcal{O}}(\mathcal{M}))$, and a surjective morphism

$$\mathrm{Jet}\big(\mathrm{Sym}_{\mathcal{O}}(\mathcal{M})\big)/\mathcal{I} \twoheadrightarrow \mathcal{A}.$$

Proposition 12.2.2 *If $\mathcal{A} = \mathrm{Jet}(\mathcal{O}_C)$ for $\pi : C \to M$ a smooth map of varieties, then $\mathcal{A}$ is $\mathcal{D}$-smooth and the $\mathcal{A}[\mathcal{D}]$-module $\Omega^1_{\mathcal{A}}$ is isomorphic to*

$$\Omega^1_{\mathcal{A}} \cong \Omega^1_{C/M} \otimes_{\mathcal{O}_C} \mathcal{A}[\mathcal{D}].$$

In particular, if $\pi : C = \mathbb{R} \times M \to M$ is the trivial bundle with fiber coordinate u, one gets the free $\mathcal{A}[\mathcal{D}]$-module of rank one

$$\Omega^1_{\mathcal{A}} \cong \mathcal{A}[\mathcal{D}]^{(\{du\})}$$

generated by the form du.

Remark 12.2.1 One may identify local functionals on $\underline{\Gamma}(M, C)$ with cohomology classes in $h(\mathcal{A})$, where $\mathcal{A} := \mathrm{Jet}(\mathcal{O}_C)$ is the jet algebra, and

$$h(\mathcal{A}) = h^0\big(\mathrm{DR}(\mathcal{A})\big) := h^0\big(\mathrm{Ber}_M \overset{\mathbb{L}}{\otimes}_{\mathcal{D}} \mathcal{A}\big).$$

Up to now, we have been using this degree convention, because we were working mostly with left $\mathcal{D}$-modules, and $\mathbb{R}$-valued action functionals. We may also see this from the right $\mathcal{D}$-module viewpoint as

$$h(\mathcal{A}) = h^n(\mathcal{A}^r \otimes_{\mathcal{D}} \mathcal{O}),$$

where $\mathcal{A}^r := \mathrm{Ber}_M \otimes_{\mathcal{O}} \mathcal{A}$ is the right $\mathcal{D}$-algebra associated to $\mathcal{A}$. From this point of view, it is desirable to think of volume forms in Ber_M as living in cohomological degree n, where n is the dimension of the supermanifold in play, so that $\mathcal{A}^r$, and its cohomology, is in degree n. It is possible to think of a local action functional $S \in h(\mathcal{A})$ as a space morphism

$$\underline{\Gamma}(M, C) \to \underline{\Gamma}(M, \mathrm{Ber}_M)$$

whose value can be integrated on compact subspaces of M, to get a real number. If not integrated, a density must be thought as living in cohomological degree n. This is what will explain the degree shifts that one finds in computations on the Ran space.

12.3 $\mathcal{D}$-Modules on the Ran Space and Operations

We refer to Beilinson-Drinfeld's book [BD04] and to Francis-Gaitsgory's article [FG11] for a systematic study of $\mathcal{D}$-modules on the Ran space. We will give here a short presentation of local operations (also called $*$-operations by Beilinson and Drinfeld), and chiral operations, since these are at the basis of the deformation quantization approach to Euclidean quantum field theory. Local operations are very important because they allow the algebraic manipulation of covariant objects such as local vector fields or local Poisson brackets. The main idea of Beilinson and Drinfeld's approach to geometry of $\mathcal{D}$-spaces is that functions and differential forms usually multiply by using ordinary tensor product of $\mathcal{D}$-modules, but that local vector fields and local differential operators multiply by using local operations. This gives the complete toolbox to do differential geometry on $\mathcal{D}$-spaces in a way that is very similar to ordinary differential geometry.

One defines the ∞-category (we may also use dg-categories here) of $\mathcal{D}$-modules on the Ran space as the ∞-limit (or the homotopy limit of dg-categories)

$$\mathcal{D}(\mathrm{Ran}(M)) := \lim \mathcal{D}(M^I)$$

of the categories $\mathcal{D}(M^I)$ of $\mathcal{D}$-modules on finite powers of M, indexed by the category of finite sets I with surjective maps between them. We refer to Sect. 12.5 for a refined geometric explanation of this definition.

By definition, a $\mathcal{D}$-module on the Ran space is a collection of $\mathcal{D}$-modules $\mathcal{M}^I$ on the finite powers M^I with homotopy equivalences

$$\Delta(\pi)(\mathcal{M}^I) \cong \mathcal{M}^J$$

for every surjection $\pi : I \twoheadrightarrow J$, with $\Delta(\pi) : M^J \to M^I$ the associated diagonal map.

The canonical projection functor

$$\Delta(\pi)^! : \mathcal{D}(\mathrm{Ran}(M)) \to \mathcal{D}(M^I)$$

has a left adjoint

$$\Delta(\pi)_* : \mathcal{D}(M^I) \to \mathcal{D}(\mathrm{Ran}(M))$$

that may be described explicitly by the following construction (see [Roz10] for a more precise description): Let $\mathcal{M}$ be a $\mathcal{D}$-module on M^I and denote by $\{\mathcal{M}^I\} := \Delta_*\mathcal{M}$ the corresponding $\mathcal{D}$-module on the Ran space. Let $Z_{I,J} \subset M^{I \cup J}$ be the (singular) subvariety given by the union of all diagonals corresponding to diagrams

$$I \twoheadrightarrow K \twoheadleftarrow J$$

with K an arbitrary finite set. Then we have (inverse and direct images are defined through the de Rham space construction for $Z_{I,J}$)

$$\mathcal{M}^J \cong (p_J)_*(p_I)^!\mathcal{M}.$$

In the case where $\mathcal{M}$ lives on M, there is a unique diagram

$$I = \{1\} \twoheadrightarrow K = \{1\} \twoheadleftarrow J : \pi$$

that corresponds to the natural projection $J \to \{1\}$, $Z_{I,J} \subset M \times M^J$ is the main diagonal and

$$\mathcal{M}^J \cong \Delta(\pi)_*\mathcal{M}$$

is the $\mathcal{D}$-module supported on the main diagonal with value $\mathcal{M}$.

There are two natural monoidal structures on $\mathcal{D}(\mathrm{Ran}(M))$. The local tensor product is given by

$$\mathcal{M} \otimes^* \mathcal{N} := \mathrm{union}_*(\mathcal{M} \boxtimes \mathcal{N}),$$

where

$$\mathrm{union} : \mathrm{Ran}(M) \times \mathrm{Ran}(M) \to \mathrm{Ran}(M)$$

is given by union of finite sets. The chiral tensor product is given by

$$\mathcal{M} \otimes^{ch} \mathcal{N} := \mathrm{union}_* \circ j_* \circ j^*(\mathcal{M} \boxtimes \mathcal{N}),$$

where j is the open embedding of the locus

$$\big(\mathrm{Ran}(M) \times \mathrm{Ran}(M)\big)_{disj} \subset \mathrm{Ran}(M) \times \mathrm{Ran}(M)$$

corresponding to pairs of finite subsets of M that are disjoint.

We now describe the relation between the two descriptions of local operations given in the literature (see [BD04] and [FG11]).

Proposition 12.3.1 *Let $\{\mathcal{M}_1, \ldots, \mathcal{M}_n, \mathcal{N}\}$ be $\mathcal{D}$-modules on M, viewed as $\mathcal{D}$-modules on the Ran space supported on the main diagonal. There is a natural isomorphism*

$$\mathbb{R}\mathcal{H}om_{\mathcal{D}(\mathrm{Ran}(M))}(\mathcal{M}_1 \otimes^* \cdots \otimes^* \mathcal{M}_n, \mathcal{N}) \cong \mathbb{R}\mathcal{H}om_{\mathcal{D}_{M^n}}(\mathcal{M}_1 \boxtimes \cdots \boxtimes \mathcal{M}_n, \Delta_* \mathcal{N}).$$

Another way to explain the interest of $*$-operations is that they induce ordinary operations in de Rham cohomology. Since de Rham cohomology is the main tool of local functional calculus (it gives an algebraic presentation of local functions, differential forms and vector fields on the space $\underline{\Gamma}(M, C)$ of trajectories of a given field theory), we will make systematic use of these operations. We refer to Beilinson-Drinfeld [BD04] and [FG11] for a proof of the following result, which roughly says that the de Rham functor is a monoidal functor. This means, in simpler terms, that local, i.e., $*$-operations can be used to define the usual operations on quasi-local functionals.

Proposition 12.3.2 *The de Rham functor is an ∞-monoidal functor*

$$\mathrm{DR} : (\mathcal{D}(\mathrm{Ran}(M)), \otimes^*) \to (\mathrm{SH}(\mathrm{Ran}(M)), \otimes^!).$$

In particular, the central de Rham cohomology functor $h : \mathrm{MOD}(\mathcal{D}_M^{op}) \to \mathrm{MOD}(\mathbb{R}_M)$ given by $h(\mathcal{M}) := h^0(\mathrm{DR}^r(\mathcal{M})) := \mathcal{M} \otimes_{\mathcal{D}} \mathcal{O}$ induces a natural map

$$h : \mathrm{Hom}_{\mathcal{D}(\mathrm{Ran}(M))}(\otimes^*_{i \in I} \mathcal{L}_i, \mathcal{M}) \to \mathrm{Hom}(\otimes_i h(\mathcal{L}_i), h(\mathcal{M}))$$

from $*$-operations to multilinear operations.

12.4 Local Vector Fields and Their Local Bracket

We now explain, on the concrete example of local functional calculus, the usefulness of the general theory of local operations when doing differential calculus on $\mathcal{D}$-spaces. The following construction is based on the notion of a local dual for a module $\mathcal{M}$ over a $\mathcal{D}$-algebra, which is defined as the dual for the monoidal structure $\otimes^*$, i.e., by the isomorphism of functors

$$\mathrm{Hom}_{\mathcal{A}}(-, (\mathcal{M}^\circ)^\ell) \cong \mathrm{Hom}_{\mathcal{D}(\mathrm{Ran}(M))}(- \otimes^* \mathcal{M}, \mathcal{A}).$$

Concretely, if $\mathcal{M}$ is a projective $\mathcal{A}[\mathcal{D}]$-module of finite rank, one has

$$\mathcal{M}^\circ \cong \mathcal{H}om_{\mathcal{A}[\mathcal{D}]}(\mathcal{M}, \mathcal{A}[\mathcal{D}]).$$

We now define the notion of local vector fields.

Definition 12.4.1 Let $\mathcal{A}$ be a smooth $\mathcal{D}$-algebra. The $\mathcal{A}^r[\mathcal{D}^{op}]$-module of *local vector fields* is defined by

$$\Theta_{\mathcal{A}} = (\Omega^1_{\mathcal{A}})^{\circ} := \mathcal{H}om_{\mathcal{A}[\mathcal{D}]}(\Omega^1_{\mathcal{A}}, \mathcal{A}[\mathcal{D}]),$$

where $\mathcal{A}^r[\mathcal{D}^{op}] := \mathcal{A}^r \otimes_{\mathrm{Ber}_M} \mathcal{D}^{op}$ acts on the right via the isomorphism

$$\mathcal{A}^r[\mathcal{D}^{op}] \cong (\mathcal{A} \otimes_{\mathcal{O}} \mathrm{Ber}_M) \otimes_{\mathrm{Ber}_M} \mathcal{D}^{op} \cong \mathcal{A}[\mathcal{D}^{op}].$$

Note now that in ordinary differential geometry, one way to define vector fields on a manifold M is to take the $\mathcal{O}_M$-dual

$$\Theta_M := \mathcal{H}om_{\mathcal{O}_M}(\Omega^1_M, \mathcal{O}_M)$$

of the module of differential forms. The Lie bracket

$$[\cdot,\cdot] : \Theta_M \otimes \Theta_M \to \Theta_M$$

of two vector fields X and Y can then be defined from the universal derivation $d : \mathcal{O}_M \to \Omega^1_M$ as the only vector field $[X, Y]$ on M such that for every function $f \in \mathcal{O}_M$, one has the equality of derivations

$$[X, Y] \cdot f = X \cdot i_Y(df) - Y \cdot i_X(df).$$

In the case of a $\mathcal{D}$-algebra $\mathcal{A}$, this construction does not work directly because the duality used to define local vector fields is not the $\mathcal{A}$-linear duality (because it doesn't have good finiteness properties) but the $\mathcal{A}[\mathcal{D}]$-linear duality. This explains why the Lie bracket of local vector fields and their action on $\mathcal{A}$ are new kinds of operations of the form

$$[\cdot,\cdot] : \Theta_{\mathcal{A}} \boxtimes \Theta_{\mathcal{A}} \to \Delta_* \Theta_{\mathcal{A}}$$

and

$$L : \Theta_{\mathcal{A}} \boxtimes \mathcal{A} \to \Delta_* \mathcal{A}$$

where $\Delta : M \to M \times M$ is the diagonal map and the box product is defined by

$$\mathcal{M} \boxtimes \mathcal{N} := p_1^* \mathcal{M} \otimes p_2^* \mathcal{N}$$

where $p_1, p_2 : M \times M \to M$ are the two projections. One way to understand these construction is by looking at the natural injection

$$\Theta_{\mathcal{A}} \hookrightarrow \mathcal{H}om_{\mathcal{D}}(\mathcal{A}, \mathcal{A}[\mathcal{D}])$$

given by sending $X : \Omega^1_{\mathcal{A}} \to \mathcal{A}[\mathcal{D}]$ to $X \circ d : \mathcal{A} \to \mathcal{A}[\mathcal{D}]$. The theory of $\mathcal{D}$-modules tells us that the datum of this map is equivalent to the datum of a $\mathcal{D}^{op}_{M \times M}$-linear map

$$L : \Theta_{\mathcal{A}} \boxtimes \mathcal{A}^r \to \Delta_* \mathcal{A}^r.$$

Similarly, the above formula

$$[X,Y]\cdot f = X\cdot i_Y(df) - Y\cdot i_X(df)$$

of ordinary differential geometry makes sense in local computations only if we regard $\Theta_{\mathcal{A}}$ as contained in $\mathcal{H}om_{\mathcal{D}}(\mathcal{A},\mathcal{A}[\mathcal{D}])$ (and not in $\mathcal{H}om_{\mathcal{D}}(\mathcal{A},\mathcal{A})$, as is usually done), so that we must think of the bracket as a morphism of sheaves

$$\Theta_{\mathcal{A}} \to \mathcal{H}om_{\mathcal{D}^{op}}\big(\Theta_{\mathcal{A}}, \Theta_{\mathcal{A}}\otimes\mathcal{D}^{op}\big)^r.$$

This is better formalized by a morphism of $\mathcal{D}^{op}_{M\times M}$-modules

$$[\cdot,\cdot] : \Theta_{\mathcal{A}}\boxtimes\Theta_{\mathcal{A}} \to \Delta_*\Theta_{\mathcal{A}}$$

as above. Another way to understand these local operations is to make an analogy with multilinear operations on $\mathcal{O}_M$-modules. Indeed, if $\mathcal{F}$, $\mathcal{G}$ and $\mathcal{H}$ are three quasi-coherent $\mathcal{O}_M$-modules, one has a natural adjunction isomorphism

$$\mathcal{H}om_{\mathcal{O}_M}(\mathcal{F}\otimes\mathcal{G},\mathcal{H}) \cong \mathcal{H}om_{\mathcal{O}_M}\big(\Delta^*(\mathcal{F}\boxtimes\mathcal{G}),\mathcal{H}\big) \cong \mathcal{H}om_{\mathcal{O}_{M\times M}}(\mathcal{F}\boxtimes\mathcal{G},\Delta_*\mathcal{H}),$$

and local operations are given by a $\mathcal{D}$-linear version of the right part of the above equality. It is better to work with this expression because of the finiteness properties of the $\mathcal{D}$-modules in play.

12.5 Non-algebraic Partial Differential Equations

We now use the general approach to differential calculus, given in Sect. 2.5, to describe a formulation of the jet space formalism that also applies to nonlinear partial differential equations on general types of varieties (e.g., smooth, analytic, super or graded manifolds). For example, this formalism is adapted to the study of a partial differential equation of the form

$$F\big(x,\phi(x),\partial_\alpha\phi(x)\big) = 0$$

for $F \in \mathcal{C}^\infty(\mathrm{Jet}(C))$ a smooth function on the jet space of a bundle $C \to M$ over a smooth manifold M. The formalism of $\mathcal{D}$-algebras described before is only adapted to the study of differential equations of this form, where F is a function that is polynomial in the higher jet coordinates, but the generalization is not that hard from our viewpoint, so we shall present it shortly.

Recall from Sect. 2.5 that, if M is a space (meaning a sheaf on a differentially convenient site, see Definition 3.1.3), one may define its de Rham space $\pi : M \to M_{DR}$ as the quotient space

$$M_{DR} := \pi_0\big(\Pi_\infty^{inf}(M)\big)$$

of M by the (free) action of the ∞-groupoid $\Pi_\infty^{inf}(M)$ of its infinitesimal paths.

The datum of an action of $\Pi_\infty^{inf}(M)$ on a space $Z \to M$ is the same as the datum of a space $Z_{DR} \to M_{DR}$ such that

$$Z \cong \pi^* Z_{DR} := Z_{DR} \times_{M_{DR}} M \to M,$$

which is the same as a flat Grothendieck connection on Z relative to the morphism $Z \to M$.

In particular, a $\mathcal{D}$-module on a space M may be defined as a module over M_{DR}, which is the same as a space thickening

$$M_{DR} \to Z$$

that is equipped with an abelian cogroup object structure.

There is an adjunction

$$\pi^* : \text{SPACES}/M_{DR} \leftrightarrows \text{SPACES}/M : \pi_* =: \text{Jet}_{DR}$$

between spaces over M_{DR} and spaces over M, and we define

$$\text{Jet} := \pi^* \circ \text{Jet}_{DR} : \text{SPACES}/M \to \text{SPACES}/M.$$

By definition, the jet space $\text{Jet}(C) \to M$ of a space $C \to M$ is equipped with a flat connection, given by the corresponding morphism $\text{Jet}_{DR}(C) \to M_{DR}$.

There is a natural inclusion

$$\underline{\Gamma}\big(M_{DR}, \text{Jet}_{DR}(C)\big) \hookrightarrow \underline{\Gamma}\big(M, \text{Jet}(C)\big)$$

that identifies horizontal sections of the jet bundle, and an isomorphism

$$j_\infty : \underline{\Gamma}(M, C) \xrightarrow{\sim} \underline{\Gamma}\big(M_{DR}, \text{Jet}(C)_{DR}\big)$$

that sends a section to its infinite jet, which is horizontal for the canonical connection on $\text{Jet}(C)$.

These considerations also work when C is a derived stack, such as BG for G a Lie group.

Definition 12.5.1 The Ran space is the space

$$\text{Ran}(M) := \operatorname*{colim}_{I \twoheadrightarrow J} M^I$$

given by the colimit along diagonals $\Delta(\pi) : M^J \hookrightarrow M^I$ associated to surjections $\pi : I \twoheadrightarrow J$ of finite sets.

One may define a $\mathcal{D}$-module on the Ran space as an object $\mathcal{M}$ in the tangent category

$$\text{MOD}(\mathcal{D}_{\text{Ran}(M)}) := T\big(\text{Ran}(M)_{DR}\big)$$

to the de Rham space of the Ran space. Concretely, it is given by a family of $\mathcal{D}$-modules on powers M^I of M that are strictly compatible with pullbacks along the various diagonal embeddings. One similarly defines a $\mathcal{D}$-space over the Ran space as a morphism

$$Z \to \mathrm{Ran}(M)_{DR},$$

which is identified with a family of $\mathcal{D}$-spaces $Z^I \to M^I_{DR}$, compatible with pullbacks along diagonals.

Note that the quantization problem for higher dimensional gauge theories (to be described in Chap. 24), does not work well in this non-homotopical setting, essentially because quantizations are not strict geometric objects, but homotopical in nature (i.e., strict quantizations do not exist in higher dimensions in general, as one can see in the example of quantum groups, described in Chap. 6; this may be intuitively explained by a Heckman-Hilton type argument that says that two commuting monoid structures on a set are equal and commutative). Note also that the Batalin-Vilkovisky formalism for homotopical Poisson reduction of gauge theories (see Chap. 13) naturally gives a derived Poisson $\mathcal{D}$-space as output, so that it is very natural, also from the physicist's viewpoint, to quantize in the derived setting.

These remarks imply that it is profitable to view the Ran space as an ∞-colimit

$$\mathbb{R}\mathrm{Ran}(M) := \operatorname*{colim}_{I \twoheadrightarrow J} M^I$$

of derived stacks, and to work with the stable (tangent) ∞-category

$$\mathcal{D}\big(\mathrm{Ran}(M)\big) := \mathrm{M{\scriptsize OD}}\big(\mathbb{R}\mathrm{Ran}(M)_{DR}\big)$$

of differential modules over it.

This category identifies canonically with the ∞-limit

$$\mathcal{D}\big(\mathrm{Ran}(M)\big) \cong \lim_{I \twoheadrightarrow J} \mathcal{D}\big(M^I\big).$$

We will simply denote by $\mathrm{Ran}(M)$ the derived Ran space $\mathbb{R}\mathrm{Ran}(M)$, since both spaces are isomorphic as derived spaces.

In this setting, one also defines the $\otimes^*$ and $\otimes^{ch}$ monoidal structures on $\mathcal{D}(\mathrm{Ran}(M))$ as before. To work with general partial differential equations and the corresponding $\mathcal{D}$-spaces, it is sometimes convenient to avoid working with algebras, and to use directly spaces over $\mathrm{Ran}(M)_{DR}$. This geometric approach will be used in Chap. 24 to describe the general quantization problem for $\mathcal{D}$-spaces.

References

[BD04] Beilinson, A., Drinfeld, V.: Chiral Algebras. American Mathematical Society Colloquium Publications, vol. 51, p. 375. Am. Math. Soc., Providence (2004). ISBN 0-8218-3528-9

[FG11] Francis, J., Gaitsgory, D.: Chiral Koszul duality. arXiv e-prints (2011). arXiv:1103.5803

[KV98] Krasil'shchik, J., Verbovetsky, A.: Homological methods in equations of mathematical physics. arXiv (1998)

[Pau11a] Paugam, F.: Histories and observables in covariant field theory. J. Geom. Phys. **61**(9), 1675–1702 (2011). doi:10.1016/j.geomphys.2010.11.002

[Roz10] Rozenblyum, N.: Modules over a chiral algebra. arXiv e-prints (2010). arXiv:1010.1998

[Vin01] Vinogradov, A.M.: Cohomological Analysis of Partial Differential Equations and Secondary Calculus. Translations of Mathematical Monographs, vol. 204, p. 247. Am. Math. Soc., Providence (2001). Translated from the Russian manuscript by Joseph Krasil'shchik. ISBN 0-8218-2922-X

Chapter 13
Gauge Theories and Their Homotopical Poisson Reduction

In this chapter, we define general gauge theories and study their classical aspects. These may also be called *local variational problems*, because they fit with the general Lagrangian formalism of Sect. 8.1, but have the additional property of having as action functional

$$S : H \to R$$

a local functional (in the sense of Definition 12.2.1). The corresponding equations of motion are given by an Euler-Lagrange partial differential equation, which we shall study in the setting of non-linear algebraic analysis, presented in Chap. 12.

This chapter is based and expands (by adding the gauge fixing procedure) on the author's precise and coordinate-free mathematical formalization in [Pau11b] of the theory of homotopical Poisson reduction of gauge theories. The main interest of our approach over the previous ones is that it is fully expressed in terms of functors of points. This allows us to always clearly state the nature of the various spaces and functions in play in the gauge fixing (and later in the functional integral) procedure. We also always give the most general finiteness hypothesis in which the formalism can be applied, which differs from the usual approach given in the physical literature, based on explicit examples, but also very often, on finite dimensional toy models.

We are inspired, when discussing the Batalin-Vilkovisky formalism, by a huge physical literature, starting with Peierls [Pei52] and DeWitt [DeW03] for the covariant approach to quantum field theory, and with [HT92] and [FH90] as general references for the BV formalism. More specifically, we also use Stasheff's work [Sta97] and [Sta98] as homotopical inspiration, and [FLS02, Bar10] and [CF01] for explicit computations. We refer to the author's work loc. cit. for a more complete list of references on this subject.

13.1 A Finite Dimensional Toy Model

In this section, we will describe a new kind of differential geometry on spaces of the form $X = \underline{\mathrm{Spec}}_{\mathcal{D}}(\mathcal{A})$ given by spectra of $\mathcal{D}$-algebras, that encode solution spaces of

F. Paugam, *Towards the Mathematics of Quantum Field Theory*,
Ergebnisse der Mathematik und ihrer Grenzgebiete. 3. Folge / A Series of Modern Surveys in Mathematics 59, DOI 10.1007/978-3-319-04564-1_13,

non-linear partial differential equations in a coordinate-free fashion (to be explained in the next section).

Before starting this general description, that is entailed of technicalities, we present a finite dimensional analog that can be used as a reference to better understand the constructions appearing in the setting of $\mathcal{D}$-spaces.

Let H be a finite dimensional smooth manifold (analogous to the space of histories $H \subset \underline{\Gamma}(M, C)$ of a given Lagrangian variational problem) and $S : H \to \mathbb{R}$ be a smooth function (analogous to an action functional on the space of histories). Let $d : \mathcal{O}_H \to \Omega_H^1$ be the de Rham differential and

$$\Theta_H := \mathcal{H}om_{\mathcal{O}_H}(\Omega_H^1, \mathcal{O}_H).$$

There is a natural biduality isomorphism

$$\Omega_H^1 \cong \mathcal{H}om_{\mathcal{O}_H}(\Omega_H^1, \mathcal{O}_H).$$

Let $i_{dS} : \Theta_H \to \mathcal{O}_H$ be given by the insertion of vector fields in the differential $dS \in \Omega_H^1$ of the given function $S : H \to \mathbb{R}$.

The claim is that there is a natural homotopical Poisson structure on the space T of critical points of $S : H \to \mathbb{R}$, defined by

$$T = \{x \in H,\ d_x S = 0\}.$$

Define the algebra of functions $\mathcal{O}_T$ as the quotient of $\mathcal{O}_H$ by the ideal $\mathcal{I}_S$ generated by the equations $i_{dS}(\mathbf{v}) = 0$ for all $\mathbf{v} \in \Theta_H$. Note that $\mathcal{I}_S$ is the image of the insertion map $i_{dS} : \Theta_H \to \mathcal{O}_H$ and is thus locally finitely generated by the image of the basis vector fields $\mathbf{x}_i := \frac{\partial}{\partial x_i}$ that correspond to the local coordinates x_i on H. Now let $\mathcal{N}_S$ be the kernel of i_{dS}. It describes the relations between the generating equations $i_{dS}(\mathbf{x}_i) = \frac{\partial S}{\partial x_i}$ of $\mathcal{I}_S$.

The differential graded $\mathcal{O}_H$-algebra

$$\mathcal{O}_P := \mathrm{Sym}_{dg}\left(\left[\underset{-1}{\Theta_H} \xrightarrow{i_{dS}} \underset{0}{\mathcal{O}_H}\right]\right)$$

is isomorphic, as a graded algebra, to the algebra of multi-vectors

$$\wedge^*_{\mathcal{O}_H} \Theta_H.$$

This graded algebra is equipped with an odd, so-called Schouten bracket, given by extending the Lie derivative

$$L : \Theta_H \otimes \mathcal{O}_H \to \mathcal{O}_H$$

and Lie bracket

$$[\cdot, \cdot] : \Theta_H \otimes \Theta_H \to \Theta_H$$

by Leibniz's rule.

Proposition 13.1.1 *The Schouten bracket is compatible with the insertion map i_{dS} and makes $\mathcal{O}_P$ a differential graded odd Poisson algebra. The Lie bracket on Θ_H induces a Lie bracket on $\mathcal{N}_S$.*

Now, we will compute, following Tate [Tat57], a cofibrant resolution

$$\mathcal{B} \overset{\sim}{\to} \mathcal{O}_H/\mathcal{I}_S =: \mathcal{O}_T$$

of the algebra of functions on the critical space, as an $\mathcal{O}_H$-algebra. This proceeds by adding inductively to the algebra $\mathcal{O}_P$ higher degree generators to annihilate its cohomology. More precisely, if $H^1(\mathcal{O}_P, i_{dS}) \neq 0$, we choose a submodule

$$\mathfrak{g}_S^0 \subset \mathrm{Ker}(i_{dS}) \subset \Theta_H$$

such that

$$H^1\left(\mathrm{Sym}_{dg-\mathcal{O}_H}\left(\left[\underset{-2}{\mathfrak{g}_S^0} \to \underset{-1}{\Theta_H} \to \underset{0}{\mathcal{O}_H}\right]\right)\right) = 0.$$

We apply the same procedure by adding a chosen $\mathfrak{g}_S^1$ to kill the H^2 of the algebra

$$\mathrm{Sym}_{dg-\mathcal{O}_H}\left(\left[\underset{-2}{\mathfrak{g}_S^0} \to \underset{-1}{\Theta_H} \to \underset{0}{\mathcal{O}_H}\right]\right),$$

and so on... The graded module $\mathfrak{g}_S$ is the finite dimensional analog of the space of gauge (and higher gauge) symmetries. Suppose that $\mathfrak{g}_S$ is bounded with projective components of finite rank.

Definition 13.1.1 The *finite dimensional BV algebra* associated to $S : H \to \mathbb{R}$ is the bigraded $\mathcal{O}_H$-algebra

$$\mathcal{O}_{BV} := \mathrm{Sym}_{bigrad}\left(\begin{bmatrix} \mathfrak{g}_S[2] \oplus \Theta_H[1] \oplus & \mathcal{O}_H \\ & \oplus \\ & {}^t\mathfrak{g}_S^*[-1] \end{bmatrix}\right),$$

where ${}^t\mathfrak{g}_S^*$ is the $\mathcal{O}_H$-dual of the graded module $\mathfrak{g}_S$ transposed to become a vertical ascending graded module.

The main theorem of the Batalin-Vilkovisky formalism, which is the aim of this section, is the following:

Theorem 13.1.1 *There exists a non-trivial extension*

$$S_{cm} = S_0 + \sum_{i \geq 1} S_i \in \mathcal{O}_{BV}$$

of the classical function S that satisfies the classical master equation

$$\{S_{cm}, S_{cm}\} = 0.$$

The differential $D = \{S_{cm}, \cdot\}$ *gives* $\mathcal{O}_{BV}$ *the structure of a differential graded odd Poisson algebra.*

The corresponding derived space $\underline{\mathbb{R}\mathrm{Spec}}(\mathcal{O}_{BV}, D)$ can be regarded as a kind of derived Poisson reduction

$$\underline{\mathbb{R}\mathrm{Spec}}(\mathcal{O}_{BV}, D) \cong \underline{\mathbb{R}\mathrm{Spec}}(\mathcal{O}_H/\mathcal{I}_H)/_{\mathbb{R}}\mathcal{N}_S$$

that corresponds to taking the quotient of the homotopical critical space of S (cofibrant replacement of $\mathcal{O}_T = \mathcal{O}_H/\mathcal{I}_H$) by the foliation induced by the Noether relations $\mathcal{N}_S$.

Example 13.1.1 To be even more explicit, let us treat a simple example of the above construction. Let $H = \mathbb{R}^2$ be equipped with the polynomial function algebra $\mathcal{O}_H = \mathbb{R}[x, y]$. The differential one forms on H are given by the free $\mathcal{O}_H$-modules

$$\Omega_H^1 = \mathbb{R}[x, y]^{(dx,dy)}.$$

Let $S \in \mathcal{O}_H$ be the function $F(x, y) = \frac{x^2}{2}$. One then has $dS = xdx$. The module Θ_H of vector fields is the free module

$$\Theta_H = \mathbb{R}[x, y]^{(\frac{\partial}{\partial x}, \frac{\partial}{\partial y})}$$

and the insertion map is given by the $\mathbb{R}[x, y]$-module morphism

$$\begin{aligned} i_{dS} : \Theta_H &\to \mathcal{O}_H \\ \mathbf{v} &\mapsto \langle dS, \mathbf{v}\rangle, \end{aligned}$$

and, in particular, $i_{dS}(\frac{\partial}{\partial x}) = x$, and $i_{dS}(\frac{\partial}{\partial y}) = 0$. The image of the insertion map is given by the ideal $\mathcal{I}_S = (x)$ in $\mathcal{O}_H = \mathbb{R}[x, y]$. The kernel $\mathcal{N}_S$ of the insertion map is the free submodule

$$\mathcal{N}_S = \mathfrak{g}_S = \mathbb{R}[x, y]^{(\frac{\partial}{\partial y})}$$

of Θ_H. We apply the above inductive construction of a Koszul-Tate resolution to get a graded module $\mathfrak{g}_S$. The corresponding cofibrant resolution is given by a quasi-isomorphism

$$\mathcal{B}_S \xrightarrow{\sim} \mathcal{O}_H/\mathcal{I}_S.$$

The obtained algebra $\mathcal{B}$ is called the Koszul-Tate resolution of $\mathcal{O}_H/\mathcal{I}_S$. Now the bigraded BV algebra $\mathcal{O}_{BV}$ is given by

$$\mathcal{O}_{BV} := \mathrm{Sym}_{bigrad}\left(\begin{bmatrix} \mathfrak{g}_S[2] \oplus \Theta_H[1] \oplus & \mathcal{O}_H \\ & \oplus \\ & {}^t\mathfrak{g}_S^*[-1] \end{bmatrix}\right).$$

The graded version is the $\mathcal{O}_H$-algebra

$$\mathcal{O}_{BV} = \mathrm{Sym}^*(\mathfrak{g}_S) \otimes \wedge^* \Theta_H \otimes \wedge^* \mathfrak{g}_S^*$$

on $4 = 1 + 2 + 1$ variables. The BV formalism gives a way to combine the Koszul-Tate differential with the Chevalley-Eilenberg differential by constructing an $S_{cm} \in \mathcal{O}_{BV}$ such that some components of the bracket $\{S_{cm}, \cdot\}$ induce both differentials on the corresponding generators of the BV algebra.

The aim of this section is to generalize the above construction to local variational problems, where $H \subset \underline{\Gamma}(M, C)$ is a space of histories (subspace of the space of sections of a bundle $\pi : C \to M$) and $S : H \to \mathbb{R}$ is given by the integration of a Lagrangian density. The main difficulties that we will encounter and overcome trickily in this generalization are that:

1. The $\mathcal{D}$-module $\mathcal{D} \otimes_{\mathcal{O}} \mathcal{D}$ is not $\mathcal{D}$-coherent, so that a projective resolution of $\mathcal{N}_S$ will not be dualizable in practical cases. This will force us to use finer resolutions of the algebra $\mathcal{O}_T$ of functions on the critical space.
2. Taking the bracket between two *densities* of vector fields on the space H of histories is not an $\mathcal{O}_M$-bilinear operation but a new kind of operation, called a locally bilinear operation and described in Sect. 12.4.
3. The ring $\mathcal{D}$ is not commutative. This will be overcome by using the equivalence between $\mathcal{D}$-modules and $\mathcal{D}^{op}$-modules given by tensoring by Ber_M.

Besides the above technical points, the rest of the constructions of this section are completely parallel to those on the finite dimensional toy model.

13.2 General Gauge Theories

Proposition 13.2.1 *Let M be a supermanifold and $\mathcal{A}$ be a smooth $\mathcal{D}$-algebra. Then there is a natural isomorphism, called the local interior product*

$$\begin{aligned} i : h(\Omega^1_{\mathcal{A}}) := \mathrm{Ber}_M \otimes_{\mathcal{D}} \Omega^1_{\mathcal{A}} &\longrightarrow \mathcal{H}om_{\mathcal{A}[\mathcal{D}]}(\Theta^{\ell}_{\mathcal{A}}, \mathcal{A}) \\ \omega &\longmapsto [X \mapsto i_X \omega]. \end{aligned}$$

Proof By definition, one has

$$\Theta_{\mathcal{A}} := \mathcal{H}om_{\mathcal{A}[\mathcal{D}]}(\Omega^1_{\mathcal{A}}, \mathcal{A}[\mathcal{D}])$$

and since $\mathcal{A}$ is $\mathcal{D}$-smooth, the biduality map

$$\Omega^1_{\mathcal{A}} \to \mathcal{H}om_{\mathcal{A}^r[\mathcal{D}^{op}]}(\Theta_{\mathcal{A}}, \mathcal{A}^r[\mathcal{D}^{op}])$$

is an isomorphism. Tensoring this map with Ber_M over $\mathcal{D}$ gives the desired result. □

Definition 13.2.1 If $\mathcal{A}$ is a smooth $\mathcal{D}$-algebra and $\omega \in h(\Omega^1_{\mathcal{A}})$, the $\mathcal{A}[\mathcal{D}]$-linear map

$$i_\omega : \Theta^\ell_{\mathcal{A}} \to \mathcal{A}$$

is called the *insertion map*. Its kernel $\mathcal{N}_\omega$ is called the $\mathcal{A}[\mathcal{D}]$-module of *Noether identities* and its image $\mathcal{I}_\omega$ is called the *Euler-Lagrange ideal.*

If $\mathcal{A} = \mathrm{Jet}(\mathcal{O}_C)$ for $\pi : C \to M$ a bundle and $\omega = dS$, the Euler-Lagrange ideal $\mathcal{I}_{dS}$ is locally generated as an $\mathcal{A}[\mathcal{D}]$-module by the image of the local basis of vector fields in

$$\Theta^\ell_{\mathcal{A}} \cong \mathcal{A}[\mathcal{D}] \otimes_{\mathcal{O}_C} \Theta_{C/M}.$$

If M is of dimension n and the relative dimension of C over M is m, this gives n equations (indexed by $i = 1, \ldots, n$, one for each generator of $\Theta_{C/M}$) given in local coordinates by

$$\sum_\alpha (-1)^{|\alpha|} D_\alpha \left(\frac{\partial L}{\partial x_{i,\alpha}} \right) \circ (j_\infty x)(t) = 0,$$

where $S = [Ld^n t] \in h(\mathcal{A})$ is the local description of the Lagrangian density.

We now define the notion of a local variational problem with nice histories. This type of variational problem can be studied completely by using only geometry of $\mathcal{D}$-spaces. This gives powerful finiteness and biduality results that are necessary in the study of general gauge theories.

Definition 13.2.2 Let $\pi : C \to M$, $H \subset \underline{\Gamma}(M, C)$ and $S : H \to \mathbb{R}$ be a Lagrangian variational problem (in the sense of Definition 8.1.1), and suppose that S is a local functional, i.e., if $\mathcal{A} = \mathrm{Jet}(\mathcal{O}_C)$, there exists $[L\omega] \in h(\mathcal{A}) := \mathrm{Ber}_M \otimes_{\mathcal{D}} \mathcal{A}$ and $\Sigma \in H_{c,n}(M)$ such that $S = S_{\Sigma, L\omega}$. The variational problem is called a *local variational problem with nice histories* if the space of critical points $T = \{x \in H,\ d_x S = 0\}$ identifies with the space $\underline{\mathrm{Sol}}(\mathcal{A}/\mathcal{I}_{dS})$ of solutions to the Euler-Lagrange equation.

The notion of a variational problem with nice histories can be explained in simple terms by looking at the following example. The point is to define H by adding boundary conditions to elements in $\underline{\Gamma}(M, C)$, so that the boundary terms of the integration by parts, which we do to compute the variation $d_x S$ of the action, vanish.

Note that one may refine the above notions, improving on the use of histories to kill boundary terms, by the use of manifolds with corners, based on the formalism described in Example 3.2.4 and the general differential calculus formalism of Sect. 2.5.

Example 13.2.1 Let $\pi : C = \mathbb{R}^3 \times [0,1] \to [0,1] = M$, $\mathcal{A} = \mathbb{R}[t, x_0, x_1, \ldots]$ be the corresponding $\mathcal{D}_M$-algebra with action of ∂_t given by $\partial_t x_i = x_{i+1}$, and $S = \frac{1}{2} m (x_1)^2 dt \in h(\mathcal{A})$ be the local action functional for the variational problem of

Newtonian mechanics for a free particle in $\mathbb{R}^3$. The differential of $S : \underline{\Gamma}(M,C) \to \underline{\mathbb{R}}$ at $u : U \to \underline{\Gamma}(M,C)$ along the vector field $\mathbf{u} \in \Theta_U$ is given by integrating by parts

$$\langle d_x S, \mathbf{u}\rangle = \int_M \left\langle -m\partial_t^2 x, \frac{\partial x}{\partial \mathbf{u}}\right\rangle dt + \left[\left\langle \partial_t x, \frac{\partial x}{\partial \mathbf{u}}\right\rangle\right]_0^1.$$

The last term of this expression is called the boundary term and we define nice histories for this variational problem by fixing the starting and ending point of trajectories to annihilate this boundary term:

$$H = \left\{x \in \underline{\Gamma}(M,C),\ x(0) = x_0,\ x(1) = x_1\right\}$$

for x_0 and x_1 some given points in $\mathbb{R}^3$. In this case, one has

$$T = \{x \in H,\ d_x S = 0\} \cong \underline{\mathrm{Sol}}(\mathcal{A}/\mathcal{I}_{dS})$$

where $\mathcal{I}_{dS}$ is the $\mathcal{D}$-ideal in $\mathcal{A}$ generated by $-mx_2$, i.e., by Newton's differential equation for the motion of a free particle in $\mathbb{R}^3$. The critical space is thus given by

$$T = \left\{x \in H,\ \partial_t x \text{ is constant on } [0,1]\right\},$$

i.e., the free particle is moving on the line from x_0 to x_1 with constant speed.

Definition 13.2.3 A *general gauge theory* is a local variational problem with nice histories.

13.3 Regularity Conditions and Higher Noether Identities

We now describe regularity properties of gauge theories, basing our exposition on the article [Paul1a]. We will moreover use the language of homotopical and derived geometry in the sense of Toen-Vezzosi [TV08] to get geometric insights on the spaces in play in this section (see Chap. 10 for an overview). We denote by $\mathcal{A} \mapsto Q\mathcal{A}$ a cofibrant replacement functor in a given model category. Recall that all differential graded algebras are fibrant with respect to their standard model structure.

Throughout this section, we set $\pi : C \to M$, $H \subset \underline{\Gamma}(M,C)$, $\mathcal{A} = \mathrm{Jet}(\mathcal{O}_C)$ and $S \in h(\mathcal{A})$ a gauge theory. The kernel of its insertion map

$$i_{dS} : \Theta_{\mathcal{A}}^{\ell} \to \mathcal{A}$$

is called the space $\mathcal{N}_S$ of Noether identities. Its right version

$$\mathcal{N}_S^r = \mathrm{Ber}_M \otimes \mathcal{N}_S \subset \Theta_{\mathcal{A}}$$

is called the space of Noether gauge symmetries.

Definition 13.3.1 The *derived critical space* of a gauge theory is the differential graded $\mathcal{A}$-space

$$P := \mathrm{Spec}(\mathcal{A}_P) : dg - \mathcal{A} - \mathrm{ALG} \to \mathrm{SSETS}$$
$$\mathcal{R} \mapsto s\mathrm{Hom}_{dg-Alg_{\mathcal{D}}}(\mathcal{A}_P, \mathcal{R})$$

whose coordinate differential graded algebra is

$$\mathcal{A}_P := \mathrm{Sym}_{dg}\left([\Theta^\ell_{\mathcal{A}}[1] \xrightarrow{i_{dS}} \mathcal{A}]\right).$$

A *non-trivial Noether identity* is a class in $H^1(\mathcal{A}_P, i_{dS})$.

We refer to Beilinson and Drinfeld's book [BD04] for the following proposition.

Proposition 13.3.1 *The local Lie bracket of vector fields extends naturally to an odd local (so-called Schouten) Poisson bracket on the dg-$\mathcal{A}$-algebra $\mathcal{A}_P$ of coordinates on the derived critical space.*

The following corollary explains why we called $\mathcal{N}^r_S$ the space of Noether gauge symmetries.

Corollary 13.3.1 *The natural map*

$$\mathcal{N}^r_S \boxtimes \mathcal{N}^r_S \to \Delta_* \Theta_{\mathcal{A}}$$

induced by the local bracket on local vector fields always factors through $\Delta_* \mathcal{N}^r_S$ and the natural map

$$\mathcal{N}^r_S \boxtimes \mathcal{A}^r/\mathcal{I}^r_S \to \Delta_* \mathcal{A}^r/\mathcal{I}^r_S$$

is a local Lie $\mathcal{A}$-algebroid action.

The *trivial Noether identities* are those in the image of the natural map

$$\wedge^2 \Theta^\ell_{\mathcal{A}} \to \Theta_{\mathcal{A}},$$

and these usually don't give a finitely generated $\mathcal{A}[\mathcal{D}]$-module because of the simple fact that $\mathcal{D} \otimes_{\mathcal{O}} \mathcal{D}$ is not $\mathcal{D}$-coherent. This is a very good reason to consider only non-trivial Noether identities, because these can usually (i.e., in all the applications we have in mind) be given by a finitely generated $\mathcal{A}[\mathcal{D}]$-module.

Definition 13.3.2 The *proper derived critical space* of a gauge theory is the differential graded space

$$\mathbb{R}\mathrm{Spec}(\mathcal{A}/\mathcal{I}_S) : dg - \mathcal{A} - \mathrm{ALG} \to \mathrm{SSETS}$$
$$\mathcal{R} \mapsto s\mathrm{Hom}_{dg-Alg_{\mathcal{D}}}(\mathcal{B}, \mathcal{R})$$

where $\mathcal{B} \xrightarrow{\sim} \mathcal{A}/\mathcal{I}_{dS}$ is a cofibrant resolution of $\mathcal{A}/\mathcal{I}_S$ as a dg-$\mathcal{A}$-algebra in degree 0.

From the point of view of derived geometry, differential forms on the cofibrant resolution $\mathcal{B}$ give a definition of the cotangent complex of the $\mathcal{D}$-space morphism

$$i : \underline{\mathrm{Spec}_{\mathcal{D}}}(\mathcal{A}/\mathcal{I}_S) \to \underline{\mathrm{Spec}_{\mathcal{D}}}(\mathcal{A})$$

of inclusion of critical points of the action functional in the $\mathcal{D}$-space of general trajectories. This notion of cotangent complex gives a well behaved way to study infinitesimal deformations of the above inclusion map i (see Illusie [Ill71]), even if it is not a smooth morphism (i.e., even if the critical space is singular).

We will see how to define a finer, so-called Koszul-Tate resolution, that will have better finiteness properties, by using generating spaces of Noether identities. These can be defined by adapting Tate's construction [Tat57] to the local context. We are inspired here by Stasheff's paper [Sta97].

Definition 13.3.3 A *generating space of Noether identities* is a tuple $(\mathfrak{g}_S, \mathcal{A}_n, i_n)$ composed of:

1. a negatively graded projective $\mathcal{A}[\mathcal{D}]$-module $\mathfrak{g}_S$;
2. a negatively indexed family $\mathcal{A}_n$ of dg-$\mathcal{A}$-algebras with $\mathcal{A}_0 = \mathcal{A}$; and
3. for each $n \leq -1$, an $\mathcal{A}[\mathcal{D}]$-linear morphism $i_n : \mathfrak{g}_S^{n+1} \to Z^n\mathcal{A}_n$ to the n-cycles of $\mathcal{A}_n$,

such that if one extends $\mathfrak{g}_S$ by setting $\mathfrak{g}_S^1 = \Theta_{\mathcal{A}}^{\ell}$ and if one sets

$$i_0 = i_{dS} : \Theta_{\mathcal{A}}^{\ell} \to \mathcal{A},$$

1. one has for all $n \leq 0$ an equality

$$\mathcal{A}_{n-1} = \mathrm{Sym}_{\mathcal{A}_n}\left(\left[\mathfrak{g}_S^{n+1}[-n+1] \otimes_{\mathcal{A}} \mathcal{A}_n \xrightarrow{i_n} \underset{0}{\mathcal{A}_n}\right]\right),$$

2. the natural projection map

$$\mathcal{A}_{KT} := \varinjlim \mathcal{A}_n \to \mathcal{A}/\mathcal{I}_S$$

is a cofibrant resolution, called the *Koszul-Tate algebra*, whose differential is denoted by d_{KT}.

We are now able to define the right regularity properties for a given gauge theory. These finiteness properties are imposed to make the generating space of Noether identities dualizable as an $\mathcal{A}[\mathcal{D}]$-module (resp. as a graded $\mathcal{A}[\mathcal{D}]$-module). Without any regularity hypothesis, the constructions given by homotopical Poisson reduction of gauge theories, the so-called derived covariant phase space, don't give $\mathcal{A}$-algebras, but only $\mathbb{R}$-algebras, that are too poorly behaved and infinite dimensional to be of any (even theoretical) use. We thus don't go through the process of their definition, which is left to the interested reader.

We now recall the language used by physicists (see for example [HT92]) to describe the situation. This can be useful to relate our constructions to those described in physics books.

Definition 13.3.4 A gauge theory is called *regular* if there exists a generating space of Noether identities $\mathfrak{g}_S$ whose components are finitely generated and projective. It is called *strongly regular* if this regular generating space is a bounded graded module. Suppose given a regular gauge theory. Consider the inner dual graded space (well-defined because of the regularity hypothesis)

$$\mathfrak{g}_S^\circ := \mathcal{H}om_{\mathcal{A}[\mathcal{D}]}(\mathfrak{g}_S, \mathcal{A}[\mathcal{D}])^\ell.$$

1. The generators of $\Theta_{\mathcal{A}}^\ell$ are called *antifields* of the theory.
2. The generators of $\mathfrak{g}_S$ of higher degree are called *antifields of ghosts*, or (non-trivial) *higher Noether identities* of the theory.
3. The generators of the graded $\mathcal{A}^r[\mathcal{D}^{op}]$-module $\mathfrak{g}_S^r$ are called (non-trivial) *higher gauge symmetries* of the theory.
4. The generators of the graded $\mathcal{A}[\mathcal{D}]$-module $\mathfrak{g}_S^\circ$ are called *ghosts* of the theory.

Note that the natural map $\mathfrak{g}_S^{0,r} \to \mathcal{N}_S^r \subset \Theta_{\mathcal{A}}$ identifies order zero gauge symmetries with (densities of) local vector fields that induce tangent vectors to the $\mathcal{D}$-space $\underline{\mathrm{Spec}}_{\mathcal{D}}(\mathcal{A}/\mathcal{I}_S)$ of solutions to the Euler-Lagrange equation. This explains the denomination of higher gauge symmetries for $\mathfrak{g}_S^r$.

We now define an important invariant of gauge theories, called the Batalin-Vilkovisky bigraded algebra. This will be used in next section on the derived covariant phase space.

Definition 13.3.5 Let $\mathfrak{g}_S$ be a regular generating space of the Noether gauge symmetries. The bigraded $\mathcal{A}[\mathcal{D}]$-module

$$\mathcal{V}_{BV} := \begin{bmatrix} \mathfrak{g}_S[2] \oplus \Theta_{\mathcal{A}}^\ell[1] \oplus & 0 \\ & \oplus \\ & {}^t\mathfrak{g}_S^\circ[-1] \end{bmatrix},$$

where ${}^t\mathfrak{g}_S^\circ$ is the vertical chain graded space associated to $\mathfrak{g}_S^\circ$, is called the module of *auxiliary fields*. The completed bigraded symmetric algebra

$$\hat{\mathcal{A}}_{BV} := \widehat{\mathrm{Sym}}_{\mathcal{A}\text{-bigraded}}(\mathcal{V}_{BV})$$

is called the *completed Batalin-Vilkovisky algebra* of the given gauge theory. The corresponding symmetric algebra

$$\mathcal{A}_{BV} := \mathrm{Sym}_{\mathcal{A}\text{-bigraded}}(\mathcal{V}_{BV})$$

is called the *Batalin-Vilkovisky algebra*.

In practical situations, physicists usually think of ghosts and antifields as sections of an ordinary graded bundle on spacetime itself (and not only on jet space). This idea can be formalized by the following.

Definition 13.3.6 Let $\mathfrak{g}$ be a regular generating space of Noether symmetries for $S \in h(\mathcal{A})$. Suppose that all the $\mathcal{A}[\mathcal{D}]$-modules $\mathfrak{g}^i$ and $\Theta^\ell_\mathcal{A}$ are locally free on M. A *Batalin-Vilkovisky bundle* is a bigraded vector bundle

$$E_{BV} \to C$$

with an isomorphism of $\mathcal{A}[\mathcal{D}]$-modules

$$\mathcal{A}[\mathcal{D}] \otimes_{\mathcal{O}_C} \mathcal{E}^*_{BV} \to \mathcal{V}_{BV},$$

where $\mathcal{E}_{BV}$ are the sections of $E_{BV} \to C$. The sections of the graded bundle $E_{BV} \to M$ are called the *fields-antifields variables* of the theory.

Recall that neither $C \to M$ nor $E_{BV} \to M$ are vector bundles in general. To illustrate the above general construction by a simple example, suppose that the action $S \in h(\mathcal{A})$ has no non-trivial Noether identities, meaning that for all $k \geq 1$, one has $H^k(\mathcal{A}_P) = 0$. In this case, one gets

$$\mathcal{V}_{BV} = \Theta^\ell_\mathcal{A}[1]$$

and the relative cotangent bundle $E_{BV} := T^*_{C/M} \to C$ gives a BV bundle because

$$\Theta^\ell_\mathcal{A} \cong \mathcal{A}[\mathcal{D}] \otimes_{\mathcal{O}_C} \Theta_{C/M}.$$

The situation simplifies further if $C \to M$ is a vector bundle because then, the vertical bundle $VC \subset TC \to M$, given by the kernel of $TC \to \pi^*TM$, is isomorphic to $C \to M$. Since one has $T^*_{C/M} \cong (VC)^*$, one gets a natural isomorphism

$$E_{BV} \cong C \oplus C^*$$

of bundles over M. This linear situation is usually used as a starting point for the definition of a BV theory (see for example Costello's book [Cos11]). Starting from a non-linear bundle $C \to M$, one can linearize the situation by working with the bundle

$$C^{linear}_{x_0} := x_0^* T_{C/M} \to M$$

with $x_0 : M \to C$ a given solution of the equations of motion (sometimes called the vacuum).

Proposition 13.3.2 *Let $E_{BV} \to C$ be a BV bundle. There is a natural isomorphism of bigraded algebras*

$$\mathrm{Jet}(\mathcal{O}_{E_{BV}}) \xrightarrow{\sim} \mathcal{A}_{BV} = \mathrm{Sym}_{bigrad}(\mathcal{V}_{BV}).$$

Proof Since $E_{BV} \to C$ is a graded vector bundle concentrated in non-zero degrees, one has

$$\mathcal{O}_{E_{BV}} = \mathrm{Sym}_{\mathcal{O}_C}(\mathcal{E}_{BV}^*).$$

The natural map

$$\mathcal{E}_{BV}^* \to \mathcal{V}_{BV}$$

induces a morphism

$$\mathcal{O}_{E_{BV}} = \mathrm{Sym}_{\mathcal{O}_C}(\mathcal{E}_{BV}^*) \to \mathcal{A}_{BV}.$$

Since $\mathcal{A}_{BV}$ is a $\mathcal{D}$-algebra, one gets a natural morphism

$$\mathrm{Jet}(\mathcal{O}_{E_{BV}}) \to \mathcal{A}_{BV} = \mathrm{Sym}_{bigrad}(\mathcal{V}_{BV}).$$

Conversely, the natural map $\mathcal{E}_{BV}^* \to \mathcal{O}_{E_{BV}}$ extends to an $\mathcal{A}[\mathcal{D}]$-linear map

$$\mathcal{V}_{BV} \to \mathrm{Jet}(\mathcal{O}_{E_{BV}}),$$

that gives a morphism

$$\mathcal{A}_{BV} \to \mathrm{Jet}(\mathcal{O}_{E_{BV}}).$$

The two constructed maps are the inverse of each other. □

The main interest of the datum of a BV bundle is that it allows us to work with non-local functionals of the fields and antifields variables. This is important for the effective renormalization of gauge theories, which involves non-local functionals.

Definition 13.3.7 Let $E_{BV} \to C$ be a BV bundle. Denote by $\mathbb{A}^1(A) := A$ the graded affine space. The space of *non-local functionals of the fields-antifields* is defined by

$$\mathcal{O}_{BV} := \underline{\mathrm{Hom}}(\underline{\Gamma}(M, E_{BV}), \mathbb{A}^1)$$

of (non-local) functionals on the space of sections of E_{BV}. The image of the natural map

$$h(\mathcal{A}_{BV}) \cong h(\mathrm{Jet}(\mathcal{O}_{E_{BV}})) \longrightarrow \mathcal{O}_{BV}$$

is called the space of *local functionals of the fields-antifields* and is denoted by $\mathcal{O}_{BV}^{loc} \subset \mathcal{O}_{BV}$. The image of the period map

$$h^*(\mathrm{DR}(\mathcal{A}_{BV})) \times H_{*,c}(M) \to \mathcal{O}_{BV}$$

is called the space of *quasi-local functionals* and is denoted by $\mathcal{O}_{BV}^{qloc} \subset \mathcal{O}_{BV}$.

13.4 The Derived Covariant Phase Space

Throughout this section, we set $\pi : C \to M$, $H \subset \underline{\Gamma}(M, C)$, $\mathcal{A} = \mathrm{Jet}(\mathcal{O}_C)$ and $S \in h(\mathcal{A})$ a gauge theory. Suppose given a strongly regular generating space of Noether symmetries $\mathfrak{g}_S$ for S, in the sense of Definitions 13.3.3 and 13.3.4.

The idea of the BV formalism is to define a (local and odd) Poisson dg-$\mathcal{A}$-algebra $(\mathcal{A}_{BV}, D, \{\cdot,\cdot\})$ whose spectrum $\mathbb{R}\underline{\mathrm{Spec}}_{\mathcal{D}}(\mathcal{A}_{BV}, D)$ can be though as a kind of homotopical space of leaves

$$\mathbb{R}\mathrm{Spec}(\mathcal{A}/\mathcal{I}_S)\underset{\mathbb{R}}{/}\mathcal{N}_S^r$$

of the foliation induced by the action (described in Corollary 13.3.1) of Noether gauge symmetries $\mathcal{N}_S^r$ on the derived critical space $\mathbb{R}\underline{\mathrm{Spec}}_{\mathcal{D}}(\mathcal{A}/\mathcal{I}_S)$. It is naturally equipped with a homotopical Poisson structure, which gives a nice starting point for quantization. From this point of view, the above space is a wide generalization of the notion extensively used by DeWitt in his covariant approach to quantum field theory [DeW03] called the covariant phase space. This explains the title of this section.

We will first define the BV Poisson dg-algebra by using only a generating space for Noether identities, and explain in more detail in the next section how this relates to the above intuitive statement.

Proposition 13.4.1 *The local Lie bracket and local duality pairings*

$$[\cdot,\cdot] : \Theta_{\mathcal{A}} \boxtimes \Theta_{\mathcal{A}} \to \Delta_* \Theta_{\mathcal{A}} \quad \text{and} \quad \langle\cdot,\cdot\rangle : \left(\mathfrak{g}_S^n\right)^r \boxtimes \left(\mathfrak{g}_S^{n\circ}\right)^r \to \Delta_* \mathcal{A}^r, \qquad n \geq 0,$$

induce an odd local Poisson bracket

$$\{\cdot,\cdot\} : \hat{\mathcal{A}}_{BV}^r \boxtimes \hat{\mathcal{A}}_{BV}^r \to \Delta_* \hat{\mathcal{A}}_{BV}^r$$

called the BV-antibracket on the completed BV algebra

$$\hat{\mathcal{A}}_{BV} = \widehat{\mathrm{Sym}}_{bigrad}\left(\begin{bmatrix} \mathfrak{g}_S[2] \oplus \Theta_{\mathcal{A}}^{\ell}[1] \oplus & \mathcal{A} \\ & \oplus \\ & {}^t\mathfrak{g}_S^{\circ}[-1] \end{bmatrix}\right)$$

and on the BV algebra $\mathcal{A}_{BV}$.

Definition 13.4.1 Let $\mathfrak{g}_S$ be a regular generating space of Noether identities. A *formal solution to the classical master equation* is an $S_{cm} \in h(\hat{\mathcal{A}}_{BV})$ such that:

1. the degree $(0, 0)$ component of S_{cm} is S;
2. the component of S_{cm} that is linear in the ghost variables, denoted by S_{KT}, induces the Koszul-Tate differential $d_{KT} = \{S_{KT}, \cdot\}$ on antifields of degrees $(k, 0)$; and
3. the *classical master equation*

$$\{S_{cm}, S_{cm}\} = 0,$$

(meaning $D^2 = 0$ for $D = \{S_{cm}, \cdot\}$) is satisfied in $h(\hat{\mathcal{A}}_{BV})$.

A *solution to the classical master equation* is a formal solution that comes from an element in $h(\mathcal{A}_{BV})$.

Definition 13.4.2 Let $S \in h(\mathcal{A})$ be a gauge theory, and suppose given a solution $S_{cm} \in h(\widehat{\mathcal{A}}_{BV})$ of the classical master equation. The *anomalies of the gauge theory* are defined as the degree 1 cohomology classes of the complex

$$\big(h(\widehat{\mathcal{A}}_{BV}), \{S_{CM}, -\}\big).$$

The main theorem of homological perturbation theory, given in a physical language in Henneaux-Teitelboim [HT92], Chap. 17 (DeWitt indices), can be formulated in our language as follows.

Theorem 13.4.1 *Let $\mathfrak{g}_S$ be a regular generating space of Noether symmetries. There exists a formal solution to the corresponding classical master equation, constructed via an inductive method. If $\mathfrak{g}_S$ is further strongly regular and the inductive method ends after finitely many steps, then there exists a solution to the classical master equation.*

Proof One can attack this theorem conceptually using the general setting of homotopy transfer for curved local L_∞-algebroids (see Schaetz's paper [Sch09b] for a finite dimensional analog). We only need to prove the theorem when $\mathfrak{g}$ has all $\mathfrak{g}^i$ given by free $\mathcal{A}[\mathcal{D}]$-modules of finite rank since this is true locally on M. We start by extending S to a generator of the Koszul-Tate differential $d_{KT} : \mathcal{A}_{KT} \to \mathcal{A}_{KT}$. Note that the BV bracket with S on $\mathcal{A}_{BV}$ already identifies with the insertion map

$$\{S, \cdot\} = i_{dS} : \Theta^{\ell}_{\mathcal{A}} \to \mathcal{A}.$$

We want to define $S_{KT} := \sum_{k\geq 0} S_k$ with $S_0 = S$ such that

$$\{S_{KT}, \cdot\} = d_{KT} : \mathcal{A}_{KT} \to \mathcal{A}_{KT}.$$

Let $C^*_{\alpha_i}$ be generators of the free $\mathcal{A}[\mathcal{D}]$-modules $\mathfrak{g}^i$ and C^{α_i} be the dual generators of the free $\mathcal{A}[\mathcal{D}]$-modules $(\mathfrak{g}^i)^\circ$. We suppose further that all these generators correspond to closed elements for the de Rham differential. Let $n_{\alpha_i} := d_{KT}(C^*_{\alpha_i})$ in $\mathcal{A}_{KT}$. Then setting $S_k = \sum_{\alpha_k} n_{\alpha_k} C^{\alpha_k}$, one gets

$$\begin{aligned}\{S_i, C^*_{\alpha_i}\} &= \{n_{\alpha_i} C^{\alpha_i}, C^*_{\alpha_i}\}\\ &= n_{\alpha_i}\\ &= d_{KT}(C^*_{\alpha_i})\end{aligned}$$

so that $\{S_{KT}, \cdot\}$ identifies with d_{KT} on $\mathcal{A}_{KT}$. Now let m_{α_j} denote the coordinates of n_{α_i} with respect to the basis $C^*_{\alpha_i}$, so that

$$n_{\alpha_i} = \sum_j m_{\alpha_j} C^*_{\alpha_j}.$$

One gets in these coordinates

$$S_i = \sum_{\alpha_i,\alpha_j} C^*_{\alpha_j} m_{\alpha_j} C^{\alpha_i}.$$

The next terms in $S = \sum_{k\geq 0} S_k$ are determined by the recursive equation

$$2d_{KT}(S_k) + D_{k-1} = 0$$

where D_{k-1} is the component of Koszul-Tate degree (i.e., degree in the variables $C^*_{\alpha_i}$) $k-1$ in $\{R_{k-1}, R_{k-1}\}$, with

$$R_{k-1} = \sum_{j\leq k-1} S_j.$$

These equations have a solution because D_{k-1} is d_{KT}-closed, because of Jacobi's identity for the odd bracket $\{\cdot,\cdot\}$ and since d_{KT} is exact on the Koszul-Tate components (because it gives, by definition, a resolution of the critical ideal), these are also exact. If we suppose that the generating space $\mathfrak{g}_S$ is strongly regular (i.e., bounded) and the inductive process ends after finitely many steps, one can choose the solution S in $h(\mathcal{A}_{BV})$. □

13.5 General Gauge Fixing Procedure

We refer to Sect. 8.4, and particularly Example 8.4.2 for some background on canonical transformations in classical Hamiltonian mechanics. We are giving here a generalization of this notion to the $\mathcal{D}$-geometrical setting. Note that in some particular cases, it may also be interesting to use derived Lagrangian subspaces (see Definition 10.7.7) in derived critical spaces of action functionals on spaces of stacky fields, which have a more intrinsic flavour. We however want to present the physicists' approach, which is also appealing because it is entirely based on the detailed study of the action functional's properties, and because it is at the heart of the functional integral quantization method.

Let $S \in h(\mathcal{A})$ be a strongly regular gauge theory and $S_{cm} \in h(\mathcal{A}_{BV})$ be a classical master action. Recall that the algebra $\mathcal{A}_{BV}$ may be defined as the bigraded symmetric algebra

$$\mathcal{A}_{BV} := \mathrm{Sym}_{\mathcal{A}\text{-}bigraded}(\mathcal{V}_{BV})$$

where

$$\mathcal{V}_{BV} := \begin{bmatrix} \mathfrak{g}_S[2] \oplus \Theta^{\ell}_{\mathcal{A}}[1] \oplus & 0 \\ & \oplus \\ & {}^t\mathfrak{g}_S^{\circ}[-1] \end{bmatrix}$$

and $\mathfrak{g}_S$ is the graded $\mathcal{A}[\mathcal{D}]$-module of Noether identities. We denote by

$$X_{BV} := \underline{\mathrm{Spec}}_{\mathcal{D}}(\mathcal{A}_{BV})$$

the associated $\mathcal{D}$-space.

Definition 13.5.1 The *minimal space of extended fields* is the $\mathcal{D}$-space

$$X^{min}_{fields} := \underline{\mathrm{Spec}}_{\mathcal{D}}\left(\mathcal{A}^{min}_{fields}\right)$$

where

$$\mathcal{A}^{min}_{fields} := \mathrm{Sym}_{\mathcal{A}}\left(\mathfrak{g}^{\circ}[-1]\right)$$

is the *algebra of minimal fields.*

Proposition 13.5.1 *There is a graded symplectic isomorphism*

$$X_{BV} \cong T^*[-1]X^{min}_{fields}.$$

Proof This follows from the graded biduality isomorphism

$$\left(\mathfrak{g}_S^{\circ}[-1]\right)^{\circ}[1] \cong \mathfrak{g}_S[2],$$

due to the finiteness hypothesis imposed on $\mathfrak{g}_S$ and the symplectic isomorphism

$$T^*[-1]\underline{\mathrm{Spec}}_{\mathcal{D}}(\mathcal{A}) \cong \underline{\mathrm{Spec}}_{\mathcal{D}}\left(\mathrm{Sym}_g\left(\Theta^{\ell}_{\mathcal{A}}[1] \oplus \mathcal{A}\right)\right). \qquad \square$$

Definition 13.5.2 A *graded gauge fixing* is the datum of a triple $(\mathcal{A}^{nm}_{BV}, i, L)$ composed of:

1. a dg-$\mathcal{A}$-Poisson algebra $\mathcal{A}^{nm}_{BV}$ called the *non-minimal Batalin-Vilkovisky algebra*;
2. a quasi-isomorphism

$$i : \mathcal{A}^{nm}_{BV} \xrightarrow{\sim} \mathcal{A}_{BV}$$

   of Poisson dg-$\mathcal{D}$-algebras; and
3. a graded Lagrangian $\mathcal{D}$-subspace

$$\psi : L \hookrightarrow \underline{\mathrm{Spec}}_{\mathcal{D}}\left(\mathcal{A}^{nm}_{BV}\right)$$

   called the *gauge fixing condition.*

An *admissible gauge fixing* is a gauge fixing such that the associated *gauge fixed action* $S_\psi \in h(\mathcal{O}_L)$ defined by

$$S_\psi := S_{cm} \circ i^{-1} \circ \psi : L \to \underline{\mathbb{R}},$$

as a degree zero graded map, has an invertible second order derivative.

The original (so-called minimal) algebra $\mathcal{A}_{BV}$ is usually not big enough to have a gauge fixing, and it is necessary to extend the original bundle $\mathcal{V}_{BV}$ by additional non-minimal generators, which form cohomologically trivial pairs, to get an admissible gauge fixing condition. We will describe here the original recipe of Batalin and Vilkovisky [BV81] to get a gauge fixing Lagrangian, which will be given by a Hamiltonian perturbation of the zero section of an odd cotangent bundle.

Definition 13.5.3 The *(non-minimal) bundle of extended fields* is the bundle

$$\mathcal{V}_{fields} := \mathfrak{g}_S^\circ[-1] \oplus \mathfrak{a}_S[1] \oplus \mathfrak{l}_S,$$

where $\mathfrak{g}_S$ is called the space of *ghosts*, $\mathfrak{a}_S$ is the space of *antighosts*, $\mathfrak{l}_S$ the space of *Lagrange multipliers*, and one has

$$\mathfrak{a}_S \cong \mathfrak{l}_S \cong \mathfrak{g}_S.$$

The *standard non-minimal Batalin-Vilkovisky space* is the space

$$X_{BV}^{nm} := T^*[-1]X_{fields},$$

with algebra of functions $\mathcal{A}_{BV}^{nm}$, where

$$X_{fields} := \underline{\mathrm{Spec}}_{\mathcal{D}}(\mathrm{Sym}_{dg}(\mathcal{V}_{fields}))$$

is the *space of non-minimal fields* with algebra of functions $\mathcal{A}_{fields}$.

Proposition 13.5.2 *There is a natural structure of a dg-$\mathcal{D}$-symplectic space on X_{BV}^{nm} such that the natural injection*

$$i : X_{BV} \to X_{BV}^{nm}$$

is an equivalence.

Proof The graded symplectic structure is given by the fact that X_{BV}^{nm} is a shifted cotangent space. One extends the classical master function $S_{cm} \in h(\mathcal{A}_{BV})$ by adding a trivial pairing

$$\sum_i l_i \cdot a_i^*$$

where l_i are generators of the space $\mathfrak{l}_S$ of Lagrange multipliers and a_i^* are duals to the generators a_i of the space $\mathfrak{a}_S$ of antighosts. This addition does not change the cohomology because it makes the pair (a_i, l_i) cohomologically trivial. This means that the projection graded morphism

$$i : \mathcal{A}_{BV}^{nm} \to \mathcal{A}_{BV}$$

is then a dg-Poisson quasi-isomorphism. □

Note that the solution spaces of the $\mathcal{D}$-spaces X^{nm}_{BV} and X_{fields} may be regarded as the smooth spaces $\underline{\Gamma}(M, E_{BV})$ and $\underline{\Gamma}(M, E_{fields})$ of bundles $E^{nm}_{BV} \to C \to M$ and $E_{fields} \to C \to M$. We use the same notation for these underlying smooth spaces.

By definition, the classical master action $S_{cm} \in h(\mathcal{A}_{BV})$, being of total degree zero, induces a (partially defined) real valued functional

$$S_{cm} : X^{nm}_{BV} \to \underline{\mathbb{R}}.$$

If $\mathbb{A}$ is the graded smooth affine space and $\widehat{\mathbb{A}^0}$ is the subspace with zero real coordinate, the value of S_{cm} on $\widehat{\mathbb{A}^0}$ is a function

$$S_{cm}(\widehat{\mathbb{A}^0}) : X^{nm}_{BV}(\widehat{\mathbb{A}^0}) \to \underline{\mathbb{R}}(\widehat{\mathbb{A}^0}) = \mathbb{R},$$

which gives the value of the above functional on a non-trivial family of fields

$$(\varphi, \varphi^\dagger) : M \times \widehat{\mathbb{A}^0} \to E^{nm}_{BV},$$

that is concretely given by a true family of sections of the underlying bundle E^{nm}_{BV}. This gives a clear relation between our formalism and DeWitt's global approach [DeW03].

There is a natural projection

$$\pi : X^{nm}_{BV} := T^*[-1]X_{fields} \to X_{fields}$$

and a zero section 0 for π. The functional

$$S_{cm}(-, 0) : X_{fields} \to \mathbb{R}$$

has a highly degenerate derivative, and the aim of the gauge fixing procedure is to move the zero section $0 : X_{fields} \to X_{BV}$ along a Hamiltonian vector field in the antifield direction to get a better behaved function, without changing the differential graded algebra of extended fields, i.e., without changing classical local observables.

Definition 13.5.4 A *gauge fixing fermion* is a function $\psi \in \mathcal{A}_{fields}$ whose differential induces a *vertical canonical transformation*

$$f_\psi : T^*[-1]X_{fields} \to T^*[-1]X_{fields}$$

given by the formula

$$f_\psi = \mathrm{id} + d\psi \circ \pi$$

where the addition is given relative to the projection π (i.e., in the odd cotangent fiber). This transformation is local, i.e., given by a local Poisson $\mathcal{D}$-algebra morphism

$$f_\psi^* : (\mathcal{A}^{nm}_{BV}, \{-,-\}) \to (\mathcal{A}^{nm}_{BV}, \{-,-\}).$$

The *gauge fixed action* associated to a gauge fixing ψ is the function

$$S_\psi := (S_{cm}) \circ h(f_\psi) \in h(\mathcal{A}_{BV}^{nm}) : X_{BV}^{nm} \to \mathbb{R}.$$

The fact that the gauge fixing can be made an $\mathbb{R}$-valued transformation is due to the presence of fields of negative degrees, called the antighosts. Without them, the composition of S_ψ with the zero section

$$0 : X_{fields} \to T^*[-1]X_{fields}$$

would give a function $S_\psi(-,0) : X_{fields} \to \mathbb{R}$ with only non-trivial component the degree zero component, defined on the space $\Gamma(M,C)$ of fields we started from, and the ghost components would be zero.

We now explain that fixing the gauge does not change the classical local observables.

Theorem 13.5.1 *The vertical canonical transformation f_ψ induces an isomorphism of differential graded $\mathcal{D}$-algebras*

$$f_\psi : (\mathcal{A}_{BV}^{nm}, \{S_{cm}, -\}, \{-,-\}) \xrightarrow{\sim} (\mathcal{A}_{BV}^{nm}, \{S_\psi, -\}, \{-,-\}).$$

Proof This follows from the fact that f_ψ is a canonical transformation and from the definition of S_ψ. □

We may informally define an admissible gauge fixing fermion as a gauge fixing fermion such that, when the antifield variable $\varphi^\dagger$ is fixed to zero, the second order derivative of the function

$$S_\psi(-,0) : X_{fields} \to \mathbb{R}$$

is non-zero at its critical points (and thus allows us to initiate a perturbative construction of the functional integral, via a renormalization procedure). We will give a more abstract definition, which takes into account the higher gauge symmetries, and thus ensures that we shall avoid any possible pathology.

Note that one may also view $S_\psi(-,0)$ as a functional on X_{BV} by composing it with π. We will use the same notation for this composition, as an element of $h(\mathcal{A}_{BV})$. Observe now that there is a natural isomorphism

$$(\mathcal{A}_{BV}, \{S_\psi(-,0), -\}) \xrightarrow{\sim} \mathrm{Sym}_{dg-\mathcal{A}_{fields}}\left(\Theta^\ell_{\mathcal{A}_{fields}}[1] \xrightarrow{i_{dS_\psi(-,0)}} \mathcal{A}_{fields}\right)$$

between the Batalin-Vilkovisky algebra with its differential given by bracketing with $S_\psi(-,0) \in h(\mathcal{A}_{BV})$, and the derived critical space of the functional $S_\psi(-,0) \in h(\mathcal{A}_{fields})$.

We will use the above isomorphism to give a natural admissibility condition on the gauge fixing.

Definition 13.5.5 An *admissible gauge fixing fermion* is a gauge fixing fermion $\psi \in \mathcal{A}_{fields}$ such that, if 0 denotes the zero section of the natural projection

$$\pi : T^*[-1]X_{fields} \to X_{fields},$$

the associated gauge fixed differential graded algebra

$$\left(\mathcal{A}_{BV}, \{S_\psi(-,0), -\}\right)$$

has no non-zero cohomology.

Note that the above definition is compatible with the fact that $(\mathcal{A}_{BV}, \{S_\psi, -\})$ is quasi-isomorphic to the usual Batalin-Vilkovisky algebra: the differential is not the same.

Recall that the $\mathcal{D}$-space X_{fields} may be viewed as corresponding to the smooth space (which we will denote by the same name)

$$X_{fields} := \underline{\Gamma}(M, E_{fields})$$

of sections of the bundle $E_{fields} \to C \to M$ of field variables. The function

$$S_\psi(-,0) : X_{fields} \to \mathbb{R}$$

may and will then be used to define a perturbative functional integral because it does not have any gauge symmetries left.

References

[Bar10] Barnich, G.: A note on gauge systems from the point of view of Lie algebroids. arXiv e-prints (2010). arXiv:1010.0899

[BD04] Beilinson, A., Drinfeld, V.: Chiral Algebras. American Mathematical Society Colloquium Publications, vol. 51, p. 375. Am. Math. Soc., Providence (2004). ISBN 0-8218-3528-9

[BV81] Batalin, I.A., Vilkovisky, G.A.: Gauge algebra and quantization. Phys. Lett. B **102**(1), 27–31 (1981). doi:10.1016/0370-2693(81)90205-7

[CF01] Cattaneo, A.S., Felder, G.: Poisson sigma models and deformation quantization. Mod. Phys. Lett. A **16**(4–6), 179–189 (2001). doi:10.1142/S0217732301003255. Euroconference on brane new world and noncommutative geometry (Torino, 2000)

[Cos11] Costello, K.: Renormalization and Effective Field Theory. Mathematical Surveys and Monographs, vol. 170, p. 251. Am. Math. Soc., Providence (2011). ISBN 978-0-8218-5288-0

[DeW03] DeWitt, B.: The Global Approach to Quantum Field Theory. Vol. 1, 2. International Series of Monographs on Physics, vol. 114. Clarendon/Oxford University Press, Oxford/New York (2003). ISBN 0-19-851093-4

[FH90] Fisch, J.M.L., Henneaux, M.: Homological perturbation theory and the algebraic structure of the antifield-antibracket formalism for gauge theories. Commun. Math. Phys. **128**(3), 627–640 (1990)

[FLS02] Fulp, R., Lada, T., Stasheff, J.: Noether's variational theorem II and the BV formalism. arXiv (2002)

[HT92] Henneaux, M., Teitelboim, C.: Quantization of Gauge Systems, p. 520. Princeton University Press, Princeton (1992). ISBN 0-691-08775-X; 0-691-03769-8

[Ill71] Illusie, L.: Complexe Cotangent et Déformations. I. Lecture Notes in Mathematics, vol. 239, p. 355. Springer, Berlin (1971)

[Pau11a] Paugam, F.: Histories and observables in covariant field theory. J. Geom. Phys. **61**(9), 1675–1702 (2011). doi:10.1016/j.geomphys.2010.11.002

[Pau11b] Paugam, F.: Homotopical Poisson reduction of gauge theories. In: Mathematical Foundations of Quantum Field Theory and Perturbative String Theory. Proc. Sympos. Pure Math., vol. 83, pp. 131–158. Am. Math. Soc., Providence (2011)

[Pei52] Peierls, R.E.: The commutation laws of relativistic field theory. Proc. R. Soc. Lond. Ser. A **214**, 143–157 (1952)

[Sch09b] Schätz, F.: BFV-complex and higher homotopy structures. Commun. Math. Phys. **286**(2), 399–443 (2009). doi:10.1007/s00220-008-0705-0

[Sta97] Stasheff, J.: Deformation theory and the Batalin-Vilkovisky master equation. In: Deformation Theory and Symplectic Geometry (Ascona, 1996). Math. Phys. Stud., vol. 20, pp. 271–284. Kluwer Academic, Dordrecht (1997)

[Sta98] Stasheff, J.: The (secret?) homological algebra of the Batalin-Vilkovisky approach. In: Secondary Calculus and Cohomological Physics (Moscow, 1997). Contemp. Math., vol. 219, pp. 195–210. Am. Math. Soc., Providence (1998)

[Tat57] Tate, J.: Homology of Noetherian rings and local rings. Ill. J. Math. **1**, 14–27 (1957)

[TV08] Toën, B., Vezzosi, G.: Homotopical Algebraic Geometry. II. Geometric Stacks and Applications. Mem. Am. Math. Soc. 193(902):x+224 (2008)

Part II
Classical Trajectories and Fields

Chapter 14
Variational Problems of Experimental Classical Physics

We now define the basic Lagrangian variational problems of experimental classical physics and provide a brief description of their physical interpretation.

14.1 Newtonian Mechanics

In Newtonian mechanics, trajectories are given by smooth maps

$$x : [0, 1] \to \mathbb{R}^3,$$

which are the same as sections of the fiber bundle

$$\boxed{\pi : C = \mathbb{R}^3 \times [0, 1] \to [0, 1] = M.}$$

The trajectories represent one material point moving in $\mathbb{R}^3$. The space of histories is given by fixing a pair of starting and ending points for trajectories $\{x_0, x_1\}$, i.e.,

$$H = \big\{x \in \Gamma(M, C),\ x(0) = x_0, x(1) = x_1\big\}.$$

If $\langle\cdot,\cdot\rangle$ is the standard metric on $\mathbb{R}^3$ and $V : \mathbb{R}^3 \to \mathbb{R}$ is a given "potential" function, one defines the action functional by

$$S(x) = \int_M \frac{1}{2} m \|\partial_t x\|^2 - V\big(x(t)\big)dt.$$

The tangent space to H is given by smooth functions $\mathbf{h} : \mathbb{R} \to \mathbb{R}^3$ that satisfy $\mathbf{h}(0) = \mathbf{h}(1) = 0$. Defining

$$d_x S(\mathbf{h}) := \lim_{\epsilon \to 0} \frac{S(x + \epsilon h) - S(x)}{\epsilon},$$

one gets

$$d_x S(\mathbf{h}) = \int_M \langle m\partial_t x, \partial_t \mathbf{h}\rangle - \big\langle d_x V(x), \mathbf{h}\big\rangle dt$$

F. Paugam, *Towards the Mathematics of Quantum Field Theory*,
Ergebnisse der Mathematik und ihrer Grenzgebiete. 3. Folge / A Series of Modern Surveys in Mathematics 59, DOI 10.1007/978-3-319-04564-1_14,

and by integrating by parts using that $h(0) = h(1) = 0$, finally,

$$d_x S(\mathbf{h}) = \int_M \langle m\partial_t^2 x - d_x V(x), \mathbf{h} \rangle dt.$$

The space of physical trajectories is thus the space of maps $x : [0, 1] \to \mathbb{R}^3$ such that

$$m\partial_t^2 x = -V'(x).$$

This is the standard law of Newtonian mechanics. For example, if $V = 0$, the physical trajectories are those with constant speed, which corresponds to free Galilean inertial bodies. Their trajectories are also the geodesics in $\mathbb{R}^3$, i.e., lines.

Note that in this simple case, there are no gauge symmetries. Here we work in the algebraic setting. If

$$\mathcal{A} = \mathrm{Jet}(\mathcal{O}_C) = \mathbb{R}[t, x_0, x_1, \dots]$$

is the jet $\mathcal{D}$-algebra of the given bundle, the action functional is given by the class

$$S = \frac{1}{2} m \|x_1\|^2 - V\big(x(t)\big)dt \in h(\mathcal{A})$$

of the usual Lagrangian density modulo the subspace $\{D_t F dt\}_{F \in \mathcal{A}}$ of total derivatives. The critical $\mathcal{D}$-algebra is given by $\mathcal{A}/(mx_2 + V'(x_0))$. The tangent algebroid to the jet space is the rank one module

$$\Theta_{\mathcal{A}} \cong \mathcal{A}^r\big[\mathcal{D}^{op}\big]\langle \partial_x \rangle.$$

The insertion map

$$i_{dS} : \Theta_{\mathcal{A}}^{\ell} \cong \mathcal{A}[\mathcal{D}]\langle \partial_x \rangle \to \mathcal{A}$$

sends the generator $\langle \partial_x \rangle$ to the Euler-Lagrange equation $mx_2 + V'(x_0)$, which is the generator of the Euler-Lagrange $\mathcal{D}$-ideal $\mathcal{I}_S \subset \mathcal{A}$. The derived critical algebra

$$\mathrm{Sym}_{dg}\big(\big[\Theta_{\mathcal{A}}^{\ell}[1] \xrightarrow{i_{dS}} \mathcal{A}\big]\big)$$

has no cohomology at all, and is thus quasi-isomorphic to the usual critical algebra $\mathcal{A}/\mathcal{I}_S$. So in this case, there are no gauge symmetries and the BV action functional S_{CM} is simply given by $S_{CM} = S$. The (Schouten-Nijenhuis) bracket with S indeed induces the Koszul-Tate differential S_{KT} on $\mathcal{A}_P$ because

$$i_{dS}(\mathbf{v}) = \{S, \mathbf{v}\}$$

and the classical master equation $\{S, S\}$ is satisfied by definition.

This setting can easily be generalized to a configuration space $C = [0, 1] \times X$ where X is a manifold equipped with a Euclidean metric

$$g : TX \times_X TX \to \mathbb{R}_X.$$

If $V : X \to \mathbb{R}$ is a smooth "potential" function, the action functional on $x : M \to C$ is given by

$$S(x) = \int_M \frac{1}{2} m.x^* g(Dx, Dx) - x^* V dt$$

where $Dx : TM \to x^*TX$ is the differential and $x^*g : x^*TX \times x^*TX \to \mathbb{R}_M$ is the induced metric. This generalization has to be done if one is working with constrained trajectories, for example with a pendulum.

Note that the free action functional above may be formally replaced by

$$S(x) = \int_M \frac{1}{2} \sqrt{m} \|\partial_t x\| dt,$$

and this gives the same drawings in C but there is then a natural (gauge) symmetry of S by reparametrizations, i.e., diffeomorphisms of M. This means that the trajectories are not necessarily of constant speed. However, they correspond to the smallest path in C between two given points for the given metric on C, also called the geodesics.

14.2 Relativistic Mechanics

Special relativity is based on the following simple Lagrangian variational problem, which one can take as axiom of this theory. The space of configurations for trajectories is not ordinary space $\mathbb{R}^3$, but spacetime $\mathbb{R}^{3,1} := \mathbb{R}^4$, equipped with the Minkowski quadratic form

$$g(t, x) = -c^2 t^2 + \|x\|^2.$$

The space of parameters for trajectories is $M = [0, 1]$, and its variable is called the proper time of the given particle and is denoted by τ to distinguish it from the usual time of Newtonian mechanics t, which is one of the coordinates in spacetime. The fiber bundle in play is thus given by

$$\boxed{\pi : C = \mathbb{R}^{3,1} \times [0, 1] \to [0, 1] = M.}$$

The action functional is exactly the same as the action of Newtonian mechanics, except that the quadratic form used to write it is Minkowski's metric g. It is given on a section $x : M \to C$ by

$$S(x) = \int_M \frac{1}{2} mg(\partial_\tau x, \partial_\tau x) - V\big(x(\tau)\big)$$

for $V : \mathbb{R}^{3,1} \to \mathbb{R}$ a given potential function.

The equations of motion are given as before by

$$m\partial_\tau^2 x = -V'(x),$$

but they are now valid only if x is in the light cone (the domain where the Minkowski quadratic form is positive)

$$g(t,x) \geq 0.$$

One can think of the free action functional

$$S(x) = \int_M \frac{1}{2} m g(\partial_t x, \partial_t x)$$

as measuring the (square of) Minkowski length of the given trajectory, which is called proper time by physicists. Minimizing this action functional then means making your proper time the smallest possible to go from one point to another in space-time.

In physics, one usually replaces the above (free) action functional by

$$S(x) = \int_M \frac{1}{2} m \sqrt{g(\partial_\tau x, \partial_\tau x)}.$$

It has a (gauge) symmetry by reparametrization of $[0,1]$, i.e., by diffeomorphisms of $[0,1]$ that fix 0 and 1. This means that the free trajectories are following geodesics (smallest proper time, i.e., length of the path in Minkowski's metric), but they can have varying speed. It is in some sense problematic to allow varying speed because the true theory of accelerated particles is not special relativity but general relativity: special relativity only treats inertial frames (with no acceleration). However, for a general relativistic particle, which can be accelerated, this is the correct version of the action functional.

14.3 Electromagnetism

A simple physical example of a Lagrangian variational problem whose parameter space for trajectories is, unlike Newtonian and special relativistic mechanics, not a segment, is given by Maxwell's theory of electromagnetism. We refer to Landau and Lifchitz's book [LL66] for more details and to Derdzinski's book [Der92] for the covariant formulation.

14.3.1 Flat Space Formulation

Let M be an oriented 4-dimensional manifold. A Minkowski metric on M is a symmetric non-degenerate bilinear form

$$g : TM \times_M TM \to \mathbb{R}_M$$

that is locally of signature (3, 1), i.e., isomorphic to the Minkowski metric in the fiber at every point $x \in M$. We can view g as a bundle isomorphism $g : TM \overset{\sim}{\to} T^*M$, which can be extended to a bundle isomorphism

$$g : \wedge^* TM \to \wedge^* T^* M.$$

There is a natural operation of contraction of multi-vectors by the volume form

$$i_\bullet \omega_{\text{vol}} : \wedge^* TM \overset{\sim}{\longrightarrow} \wedge^* T^* M$$

and the Hodge $*$-operator is defined by

$$* := i_\bullet \text{vol} \circ g^{-1} : \wedge^* T^* M \to \wedge^* T^* M.$$

If $d : \Omega^i \to \Omega^{i+1}$ is the de Rham differential, we denote by $d^* : \Omega^3 \to \Omega^2$ the adjoint of d for the given metric g given by $g(d^*\omega, \nu) = g(\omega, d\nu)$. The bundle $\pi : C \to M$ underlying classical electromagnetism in a flat space is the bundle

$$\boxed{\pi : C = T^* M \to M}$$

of differential one forms on M. Let $J \in \Omega^3_M$ be a fixed 3-form (called the charge-current density or the source) that satisfies the compatibility condition

$$dJ = 0.$$

The Lagrangian density is the section L_J of

$$\text{Jet}^1 T^* M \otimes \wedge^4 T^* M$$

given in jet coordinates (A, A_1) by

$$L_J(A, A_1) = \frac{1}{2} A_1 \wedge *A_1 + A \wedge J.$$

It gives the action functional $S : \Gamma(M, C) \to \mathbb{R}$ defined by

$$S_J(A) = \int_M L_J(A, dA).$$

Its variable A is called the electromagnetic potential and the derivative $F = dA$ is called the electromagnetic field. If one decomposes the spacetime manifold M as a product $T \times N$ of time and space, one can define the electric field $E_A \in \Omega^1(N)$ and magnetic field $B_A \in \Omega^2(N)$ by

$$F = B_A + dt \wedge E_A.$$

One then gets the Lagrangian density

$$L(A, dA) = \frac{1}{2}\left(\frac{|E_A|^2}{c^2} - |B_A|^2\right)$$

if the source J is zero. Minimizing the corresponding action functional means that variations of the electric field induce variation of the magnetic field and vice versa, which is the induction law of electromagnetism.

Proposition 14.3.1 *The equations of motion for the electromagnetic field Lagrangian $S_J(A)$ are given by*

$$d * dA = J.$$

Proof As explained in Chap. 13, one can compute the equations of motion from the Lagrangian density in terms of differential forms by

$$\frac{\partial L}{\partial A} - d\left(\frac{\partial L}{\partial A_1}\right) = 0.$$

Recall that the Lagrangian density is

$$L_J(A, A_1) = \frac{1}{2} A_1 \wedge *A_1 + A \wedge J.$$

On has $\frac{\partial L}{\partial A} = J$ and $\frac{\partial L}{\partial A_1} = *A_1$ because of the following equalities, which are true for $\epsilon^2 = 0$:

$$\begin{aligned}\left(A_1 + \epsilon A_1'\right) \wedge *\left(A_1 + \epsilon A_1'\right) - A_1 \wedge *A_1 &= \epsilon\left(A_1' \wedge *A_1 + A_1 \wedge *A_1'\right) \\ &= \epsilon\left(A_1' \wedge *A_1 - *A_1' \wedge A_1\right) \\ &= 2\epsilon\left(A_1' \wedge *A_1\right).\end{aligned}$$

This gives

$$d * dA = J.$$ □

The equation of motion is equivalent to

$$\Box A = J$$

where $\Box := d * d$ is the d'Alembertian (or wave) operator. On Minkowski's space, this coincides with $d^* d + dd^*$, which gives in local coordinates

$$\Box = -\frac{1}{c^2}\partial_t^2 + \partial_x^2.$$

In classical physics, $F = dA$ is the object of interest and the solution to $\Box A = J$ is not unique because for any function α, one also has $\Box(A + d\alpha) = 0$ since $d^2 = 0$. One can however fix a section of the projection $A \mapsto F := dA$ (called a Gauge fixing) by setting for example $\sum_\mu A_\mu = 0$. This implies the unicity of the solution A to the equation of motion $\Box A = J$.

In quantum physics, the electromagnetic potential itself can be thought of as the wave function of the photon light particle.

If one wants to study the motion of a charged particle in a given fixed electromagnetic potential $A \in \Omega^1_X$ on a given Lorentzian manifold (X, g), one uses the action functional on relativistic trajectories $x : [0, 1] \to X$ defined by

$$S(x) = \int_M \frac{1}{2} mc \sqrt{g(\partial_\tau x, \partial_\tau x)}$$

and adds to it a so-called Coulomb potential term of the form

$$\int_M \frac{e}{c} x^* A$$

where the constant e, called the charge of the particle, can be of positive or negative sign.

14.3.2 Generally Covariant Formulation

Let M be an oriented manifold of dimension n equipped with a non-degenerate metric $g : TM \times_M TM \to \mathbb{R}_M$ (the corresponding volume form is denoted by ω_g), and a connection ∇ on the tangent bundle given on sections by a map

$$\nabla : \mathcal{T}_M \to \mathcal{T}_M \otimes \Omega^1_M$$

that satisfies Leibniz's rule. The bundle $\pi : C \to M$ underlying generalized electromagnetism is one of the following linear bundles:

$$\mathbb{R}_M, \qquad \wedge^n T^* M, \qquad T^* M,$$

(the case of interest for electromagnetism is $C = T^* M$ but other cases can be useful in particle physics). We will now write down the Lagrangian density for a generalized electromagnetism action functional

$$S : \Gamma(M, C) \to \mathbb{R}.$$

Since the bundle C is linear, we will denote it by F. The metric and connection on TM induces a metric $g : F \times_M F \to \mathbb{R}_M$ and a connection

$$\nabla : \mathcal{F} \to \mathcal{F} \otimes \Omega^1_M.$$

Definition 14.3.1 Fix an element $J \in \mathcal{F}$ called the *source*. The generalized *electromagnetism Lagrangian* (F, g, ∇) with source J sends a section $A \in \mathcal{F}$ to the differential form of maximal degree

$$L_J(A, \nabla A) := g(\nabla A, \nabla A)\omega_g + g(J, A)\omega_g$$

where g is here extended naturally to $\mathcal{F} \otimes \Omega^1_M$.

The usual electromagnetic Lagrangian (in not necessarily flat space) is obtained by using for F the cotangent bundle T^*M.

Definition 14.3.2 The *d'Alembertian operator* of the pair (g, ∇) is the operator $\square : \mathcal{F} \to \mathcal{F}$ defined by contracting the second order derivative with the metric

$$\square := \mathrm{Tr}\big((\mathrm{id}_{\mathcal{F}\otimes\Omega^1} \otimes g^{-1}) \circ \nabla_1 \circ \nabla\big).$$

Let us explain this in detail. The inverse of the metric gives an isomorphism $g^{-1} : \Omega^1_M \xrightarrow{\sim} \mathcal{T}_M$ and thus an isomorphism

$$\mathrm{id} \otimes g^{-1} : \Omega^1 \otimes \Omega^1_M \to \Omega^1 \otimes \mathcal{T}_M \cong \mathrm{End}(\mathcal{T}_M).$$

If we denote by

$$\nabla_1 : \mathcal{F} \otimes \Omega^1 \to Fc \otimes \Omega^1 \otimes \Omega^1$$

the extension $\nabla_1 := \nabla \otimes \mathrm{id} + \mathrm{id} \otimes \nabla$ of ∇ to differential 2-forms, one gets by composition

$$\mathcal{F} \xrightarrow{\nabla} \mathcal{F} \otimes \Omega^1 \xrightarrow{\nabla_1} \mathcal{F} \otimes \Omega^1 \otimes \Omega^1 \xrightarrow{\mathrm{id}\otimes\mathrm{id}\otimes g^{-1}} \mathcal{F} \otimes \mathrm{End}(\mathcal{T})$$

a map $\mathcal{F} \to \mathcal{F} \otimes \mathrm{End}(\mathcal{T})$ of which we can compute the trace to get the d'Alembertian.

If we suppose that $F = T^*M$, $\nabla g = 0$ and ∇ is torsion free, meaning that

$$\nabla_x y - \nabla_y x = [x, y],$$

we get as equations of motion for the above Lagrangian the equation

$$\square A = J$$

which is the wave equation on the given metric manifold with connection.

14.3.3 Gauge Theoretic Formulation

Let M be an oriented manifold of dimension n equipped with a non-degenerate metric $g : TM \times_M TM \to \mathbb{R}_M$ (the corresponding volume form is denoted by ω_g) and a connection $\nabla : \mathcal{T}_M \to \mathcal{T}_M \otimes \Omega^1_M$ (the Levi-Civita connection for example). Let $G := \mathrm{SU}(1) \subset \mathbb{C}^*$ be the group of complex numbers of norm 1 and consider a principal G-bundle P. Recall that the Lie algebra $\mathfrak{g}$ of G is identified with $2i\pi\mathbb{R}$ by the exponential map. It is equipped with a natural S^1-equivariant inner product $\langle \cdot, \cdot \rangle$. The variable of the gauge theoretic formulation of electromagnetism is a principal G-connection A on P, i.e., an equivariant differential form

$$A \in \Omega^1(P, \mathfrak{g})^G$$

such that the following condition is satisfied:

- (non-degeneracy) the natural map $\mathfrak{g} \to TP \to \mathfrak{g}$ induced by the action of G on P and ω is an isomorphism.

The curvature of A is defined as the $\mathfrak{g}$-valued 2 form

$$F := dA + [A \wedge A] \in \Omega^2(P, \mathfrak{g})^G.$$

Note that the choice of a section $s : M \to P$ gives an identification

$$s^* : \Omega^1(P, \mathfrak{g})^G \to \Omega^1(M, \mathfrak{g}) = \Omega^1(M)$$

of the space of G-equivariant principal connections on P with differential forms on M by pull-back. This relates this gauge theoretic formulation to the covariant formulation of Sect. 14.3.2.

The Yang-Mills action for electromagnetism is given by

$$S(A) := \int_M \langle F \wedge *F \rangle + \langle A \wedge *J \rangle,$$

where $* : \Omega^*(M, \mathfrak{g}) \to \Omega^*(M, \mathfrak{g})$ is the operator obtained by combining the usual star operator on M with the star operator on $\mathfrak{g}$ induced by the given metric.

The corresponding equation is

$$d_A * F = *J,$$

where d_A is the combination of the gauge derivative and the covariant derivative.

14.4 General Relativity

This section is mainly inspired by the excellent old book of Landau and Lifchitz [LL66].

14.4.1 The Einstein-Hilbert-Palatini Action

Here we give a version of general relativity that is due (in another language) to Palatini. It is the one used by people studying quantization of gravity, and it is better adapted to the introduction of fermionic ("matter") fields in general relativity. It is called the first order formulation of relativity because the equations of motion are of first order, in contrast to the classical Einstein-Hilbert approach that gives an equation of order two.

Let M be an oriented and compact 4-dimensional manifold. A Minkowski metric on M is a symmetric non-degenerate bilinear form

$$g : TM \times_M TM \to \mathbb{R}_M$$

that is locally of signature (3, 1), i.e., isomorphic to the Minkowski metric. We will denote by

$$\mathrm{Sym}^2_{mink}(T^*M) \to M$$

the bundle whose sections are metrics of this kind. It is a sub-bundle of the linear bundle $\mathrm{Sym}^2(T^*M) \to M$ of all bilinear forms. Recall that a connection on a linear bundle $F \to M$ is given by a covariant derivation

$$\nabla : \mathcal{F} \to \Omega^1_M \otimes \mathcal{F}$$

on the space $\mathcal{F} = \Gamma(M, F)$ of section of F, i.e., an $\mathbb{R}$-linear map that satisfies

$$\nabla(gf) = dg \otimes f + g\nabla(f).$$

We will consider the space $\mathrm{Conn}(F)$ of (parametrized families of) connections on M whose points with values in a smooth algebra B are

$$\mathrm{Conn}(F)(B) := \left\{ (x, \nabla) \middle| \begin{array}{l} x : \mathrm{Spec}(B) \to M \\ \nabla : \mathcal{F} \otimes_{C^\infty(M)} B \to \Omega^1_M \otimes (\mathcal{F} \otimes_{C^\infty(M)} B) \end{array} \right\}.$$

There is a natural projection

$$\pi : \mathrm{Conn}(F) \to M$$

that sends the pair (x, ∇) to x. There is also a natural action of sections A of $\Omega^1_M \otimes_{\mathcal{O}} \mathrm{End}(\mathcal{F})$ on the space $\mathrm{Conn}(F)$ of connections on F given by

$$\nabla \mapsto \nabla + A$$

that makes it a principal homogeneous space. In the case $M = \mathbb{R}^4$ and $F = TM$, this means that every connection can be written as

$$\nabla = d + A$$

where d is the de Rham differential extended to the trivial bundle $TM = \mathbb{R}^4 \times \mathbb{R}^4$ and A is a 1-form on M with values in $\mathrm{End}(\mathbb{R}^4)$.

The bundle $\pi : C \to M$ that is at the heart of the Lagrangian formulation of general relativity is the bundle

$$\boxed{\pi : C = \mathrm{Conn}_{tf}(TM) \times_M \mathrm{Sym}^2_{mink}(T^*M) \to M}$$

whose sections are pairs (∇, g) with

- $\nabla : \mathcal{T}_M \to \Omega^1_M \otimes \mathcal{T}_M$ a connection on the tangent bundle (often called an affine connection) that is supposed to be torsion free, i.e., such that

$$\nabla_x y - \nabla_y x = [x, y],$$

and
- $g : TM \times_M TM \to \mathbb{R}_M$ a metric of Minkowski's type.

The Riemann curvature tensor of the connection is given by the composition

$$R_\nabla := \nabla_1 \circ \nabla : \mathcal{T} \to \mathcal{T} \otimes \Omega^2.$$

It measures the local difference between ∇ and the de Rham trivial connection on a trivial bundle. Unlike the original connection, this is a $\mathcal{C}^\infty(M)$-linear operator that can be viewed as a tensor

$$R_\nabla \in \Omega^1 \otimes \mathcal{T} \otimes \Omega^2 \subset \mathrm{End}(\mathcal{T}) \otimes \mathrm{Sym}^2 \Omega^1_M,$$

where we pair the $\mathcal{T}$ component with one of the Ω^1 components of Ω^2 to find $\mathrm{End}(\mathcal{T})$ and the two remaining Ω^1 are paired in the symmetric power. The Ricci curvature is simply the trace

$$\mathrm{Ric}(\nabla) := -\mathrm{Tr}(R_\nabla) \in \mathrm{Sym}^2 \Omega^1_M.$$

One can also interpret it as a morphism

$$\mathrm{Ric} : \mathcal{T} \to \Omega^1_M.$$

The metric g, being non-degenerate, is an isomorphism $g : \mathcal{T} \to \Omega^1$ and its inverse g^{-1} is a morphism $g^{-1} : \Omega^1 \to \mathcal{T}$. One can then define the scalar curvature of the pair (∇, g) as the function on M defined by

$$R(\nabla, g) := \mathrm{Tr}\big(g^{-1} \circ \mathrm{Ric}\big).$$

The given metric on M gives a fundamental class $d\mu_g$ and one defines the Einstein-Hilbert action functional $S : \Gamma(M, C) \to \mathbb{R}$ by

$$S(\nabla, g) := \int_M R(\nabla, g) d\mu_g.$$

One sometimes adds a constant Λ to get an action

$$S(\nabla, g) := \int_M \big[R(\nabla, g) - \Lambda\big] d\mu_g$$

that takes into account the "expansion of universe" experiment (red shift in old/far-away starlight).

The space of histories H is usually fixed to be a subspace of $\Gamma(M, C)$ with fixed value and normal derivative for the metric and ∇ along a starting and ending space-like hyper-surface. The Cauchy problem is well posed in small time once that starting datum is fixed, by the work of Choquet-Bruhat [CB72]. The problem of finding solutions in long time with special asymptotic properties is presently an active domain of research in mathematics.

Proposition 14.4.1 *The space $T \subset H$ of physical trajectories for the Einstein-Hilbert action functional, defined by*

$$T := \{(\nabla, g) \in H,\ d_{(\nabla,g)}S = 0\}$$

is given by the space of solutions of the following partial differential equations

$$\nabla \text{ is the Levi-Civita connection for } g\text{:} \quad \nabla g = 0,$$

and

$$\text{Einstein's equations:} \quad \mathrm{Ric}(\nabla) - \frac{1}{2}R(\nabla, g)g = 0.$$

Proof One may adapt the proof given in Landau and Lifchitz [LL66]. □

14.4.2 Moving Frames and the Cartan Formalism

The main advantage of the Cartan approach is that it allows us to easily combine general relativity with fermionic matter particles, and that it is also the base of the present work on quantum gravity and on the covariant formulation of supersymmetric gravity theory. We use Sect. 7 for background material on connections and, in particular, Sect. 7.4 for background on Cartan geometry.

We just recall the definition of a Cartan connection for the reader's convenience.

Definition 14.4.1 Let M be a manifold and $H \subset G$ be two groups. A *Cartan connection* on M is the data comprising:

1. a principal G-bundle Q on M;
2. a principal G-connection A on Q;
3. a section $s : M \to E$ of the associated bundle $E = Q \times_G G/H$ with fibers G/H,

such that the pullback

$$e = s^* A \circ ds : TM \to VE,$$

called the moving frame (*vielbein*), for $A : TE \to VE$ the associated connection, is a linear isomorphism of bundles.

The role of the section s here is to "break the G/H symmetry". It is equivalent to the choice of a principal H-sub-bundle $P \subset Q$. This section s may also be encoded in a stack morphism

$$s : M \to [G\backslash(G/H)]$$

where (G/H) is the quotient manifold and the left quotient by G is the stacky quotient. The associated principal G-bundle P is given by the composition

$$P : M \xrightarrow{s} \left[G\backslash(G/H)\right] \to \left[G\backslash\{*\}\right] =: BG.$$

In general relativity, we are mostly interested in the case of a four dimensional spacetime M, with the group H being the Lorentz group $O(3,1)$, G being the Poincaré group $\mathbb{R}^4 \rtimes O(3,1)$. There is a canonical GL_4 principal bundle on the manifold M given by the bundle

$$P_{TM} := \mathrm{Isom}\left(TM, \mathbb{R}^4_M\right)$$

of frames in the tangent bundle TM. From a given metric g on TM, one can define a principal $O(3,1)$ bundle by considering

$$P := PO_{(TM,g)} := \mathrm{Isom}_{isom}\left((TM,g), \left(\mathbb{R}^4_M, g_{lorentz}\right)\right)$$

of isometries between (TM,g) and the trivial bundle with its standard Lorentzian metric

$$g(t,x) = -c^2t^2 + \|x\|^2.$$

One then defines the underlying principal G-bundle of the Cartan connection by

$$Q = P \times_H G := PO_{(TM,g)} \times_{O(3,1)} \left(\mathbb{R}^4 \rtimes O(3,1)\right),$$

and the G/H bundle is given by

$$E = Q \times_G (G/H) = P \times_H (G/H) := PO_{(TM,g)} \underset{O(3,1)}{\times} \mathbb{R}^4.$$

The bundle E is a vector bundle equipped with a natural metric $g : E \times_M E \to E$ and the induced connection $A : TE \to VE$ being G-equivariant is also linear, so that it corresponds to an ordinary Koszul connection on E.

In fact, since E is linear, the natural projection $VE \to E$ is an isomorphism. This gives an isomorphism of TM with E, which is actually the identity in this particular case because E was constructed from (TM,g).

We thus retrieve from the Cartan connection $(Q, A, s : M \to E)$ on E the data of a Koszul connection ∇ and a metric on TM, which are the fields of the Einstein-Hilbert-Palatini formulation of gravity. The curvature of the G-equivariant Ehresmann connection A is directly related to the Riemann curvature tensor via this relation.

The bundle underlying the Cartan-Palatini action is the categorical bundle

$$\boxed{\pi : C = \mathrm{Conn}_{\mathrm{Cartan}}\left(M, O(3,1), \mathbb{R}^4 \rtimes O(3,1)\right) \to M}$$

whose sections are Cartan connections $(Q, A, s : M \to E)$ composed of a principal G-bundle, a principal G-connection $A \in \Omega^1(Q, TQ)^G$ on Q and a section $s : M \to$

$E = Q \times_G G/H$ satisfying Cartan's condition: $e = s^* A \circ ds : TM \to VE$ is an isomorphism. The morphisms between these objects are given by isomorphisms of Cartan connections.

Note that the G-connection A on the principal G-bundle Q is equivalent to an equivariant $\mathfrak{g}$-valued differential form

$$A : TQ \to \mathfrak{g},$$

and its restriction to $P \subset Q$ gives an H-equivariant differential form

$$A : TP \to \mathfrak{g}.$$

This is the original notion of a Cartan connection form. One can decompose $\mathfrak{g} = \mathfrak{h} \oplus \mathfrak{g}/\mathfrak{h}$ in an H-equivariant way (the Cartan geometry is called reductive). In fact, $\mathfrak{h}$ is the special orthogonal Lie algebra and $\mathfrak{g}/\mathfrak{h} = \mathbb{R}^4$. The Cartan connection form can thus be decomposed as

$$A = \omega + e$$

for $\omega \in \Omega^1(P, \mathfrak{h})$ and $e \in \Omega^1(P, \mathfrak{g}/\mathfrak{h})$. In this particular case, e can also be seen as

$$e \in \Omega^1(M, \underline{\mathfrak{g}/\mathfrak{h}})$$

for $\underline{\mathfrak{g}/\mathfrak{h}}$ the H-bundle associated to $\mathfrak{g}/\mathfrak{h}$. This form is called the *vielbein* by Cartan. It gives an isomorphism

$$e : TM \to \underline{\mathfrak{g}/\mathfrak{h}}.$$

In terms of these coordinates, the Palatini action

$$S_{pal} : \Gamma(M, C) \to \mathbb{R}$$

is given by

$$S_{pal}(\omega, e) = \int_M *(e \wedge e \wedge R)$$

where:

- $R \in \Omega^2(P, \mathfrak{h})$ is the curvature of $\omega : TP \to \mathfrak{h}$;
- we use the isomorphism $\mathfrak{h} = \mathfrak{so}(3, 1) \cong \wedge^2 \mathbb{R}^{3,1} = \wedge^2 \mathfrak{g}/\mathfrak{h}$ to identify R with an element of $\Omega^2(P, \wedge^2 \underline{\mathfrak{g}/\mathfrak{h}})$;
- the wedge product acts on both space-time indices and internal Lorentz indices; and
- $* : \Omega^4_M(\wedge^4 \underline{\mathfrak{g}/\mathfrak{h}}) \to \Omega^4_M$ is the Hodge $*$-operator.

The equations of motion for this action (obtained by deriving with respect to ω and e) are

$$\begin{cases} e \wedge R = 0 \\ d_\omega e = 0. \end{cases}$$

The first equation is Einstein's equation.

Note that, up to a simple change of variable, this Cartan action for M of dimension 3 and a Riemannian metric is equivalent to the Chern-Simons action functional to be studied in Sect. 16.4. Indeed, these two theories have the same equations of motion. The Chern-Simons theory is a topological field theory (the equations do not depend on the metric and are diffeomorphism invariants). Its quantum partition function gives important topological invariants such as, for example, the Jones polynomial for varieties of dimension 3.

As explained before, one of the main motivations for the Cartan formalism is to treat the case of fermionic variables in general relativity. So let $PSpin$ be a principal $\mathrm{Spin}(3,1)$-bundle on M, $H = \mathrm{Spin}(3,1)$, $V \cong \mathbb{R}^4$ be its orthogonal representation and $G = V \rtimes \mathrm{Spin}(3,1)$. Note that the representation of $\mathrm{Spin}(3,1)$ on V is not faithful. We consider the fiber bundle

$$Q = PSpin \underset{\mathrm{Spin}(3,1)}{\times} G$$

over M and a principal G-connection A on Q. The zero section s of the vector bundle

$$\underline{V} = Q \times_G (G/H) = PSpin \times_{\mathrm{Spin}(3,1)} V$$

gives the principal H-sub-bundle $PSpin \subset Q$. The pull-back of the induced connection gives an isomorphism

$$TM \to \underline{V}$$

of the tangent bundle with $\underline{V}$ which is naturally equipped with a metric (induced by the Minkowski metric on $V \cong \mathbb{R}^4$) and a connection (induced by the given connection A). We thus get a metric g and a connection ∇ on TM. Moreover, if S is the standard spin representation of $\mathrm{Spin}(3,1)$, the G-connection form A induces a connection on the linear spin bundle

$$\underline{S} := PSpin \underset{\mathrm{Spin}(3,1)}{\times} S.$$

Note that, in the construction above, one could use any principal G-bundle Q. This means that one could replace the bundle

$$\pi : C = \mathrm{Conn}_{\mathrm{Cartan}}(Q, P, s) \to M$$

by the categorical bundle

$$\boxed{\pi : C = \mathrm{B{\small UN}Conn}_{G,\mathrm{Cartan}}(M) \to M}$$

whose sections are triples (Q, A, s) of a principal G-bundle, Q, a principal G-connection A on Q and a section $s : M \to E = Q \times_G G/H$ of the associated bundle satisfying Cartan's condition. One may interpret the datum (s, Q, A) of a

Cartan connection as a point of the stack (see Sect. 10.3 for a precise definition of this object)

$$(s, Q, A) \in \underline{\mathrm{Hom}}\big(M, \big[G\backslash(G/H)\big]_{conn}\big)$$

satisfying the additional non-triviality condition.

This shows that even in classical general relativity, the tools of homotopical geometry are mandatory to properly understand the differential geometric properties of the space of gravitational fields.

14.4.3 A Black Hole Solution with Simple Applications

The most important solution of Einstein's equations from the experimental point of view was found by Schwarzschild in 1916. With some symmetry condition on the background manifold M of general relativity, one can explicitly compute particular solutions of the general relativity equations. We refer to Landau-Lifchitz [LL66], 12.100, Besse [Bes08], Chap. III Sect. F and Sachs-Wu [SW77], Example 1.4.2 for a detailed study of the Schwarzschild solution, which we call a black hole solution because it is used to formalize black holes. This solution is also the one used to explain the bending of light by a massive star: if you wait for the moon to make a solar eclipse, and look at a star that is far away behind the sun, the light of the star will be bended by the sun's gravitational field. The bending angle can be computed using Schwarzschild's solution. One can also use this model to explain Mercury's perihelion precession.

Here we give a description of the Schwarzschild solution first by describing the background manifold, and then giving the metric. This is not very satisfactory from the physicist's viewpoint because the geometry of spacetime should be dynamically obtained from Einstein's equations and some additional physical hypothesis. The physical hypothesis here is that:

1. spacetime is static, i.e., there is a non-zero time-like Killing vector field X on M (i.e., a vector field such that $L_X g = 0$);
2. spacetime is spherically symmetric, i.e., there is an action of SO(3) on M whose orbits are either points or space-like hyper-surfaces.

We now construct the most simple static and spherically symmetric spacetime. Let S^2 be the unit two sphere, equipped with its metric h induced by the embedding $S^2 \subset \mathbb{R}^3$ and its fundamental form ω. Let $\mu \in]0, +\infty[$ be given. Consider the space

$$M = \big(]0, \mu[\cup]\mu, +\infty[\big) \times S^2 \times \mathbb{R}$$

with coordinates (r, θ, t) (θ being the vector of $\mathbb{R}^3$ representing a point in S^2). One must *not* think of t as being a time coordinate in all of M. Note that $(1 - \frac{2\mu}{r})$ is a smooth function from M to $]-\infty, 0[\cup]0, 1[$. Denote by $p : M \to S^2$ the natural projection. The Schwarzschild metric on M is defined from the metric h on S^2 by

$$g := \left(1 - \frac{2\mu}{r}\right)^{-1} d^2r + r^2 p^* h - \left(1 - \frac{2\mu}{r}\right) d^2 t.$$

The Schwarzschild spacetime is the subspace N of M defined by

$$N =]\mu, +\infty[\times S^2 \times \mathbb{R},$$

with coordinates (r, θ, t). The complement of N in M is identified with

$$B =]0, \mu[\times S^2 \times \mathbb{R}.$$

If $r_0 \gg 8\pi\mu$, one can interpret the open sub-manifold of N defined by $r > r_0$ as an excellent history of the exterior of a star of radius r_0 and active mass $8\pi\mu$. The interior of the star is not modeled by any sub-manifold of (M, g).

This model is used to explain Mercury's perihelion precession. The space around the sun is modeled by a Schwarzschild spacetime (N, g). Consider the earth turning around the sun, given by a geodesic

$$x : \mathbb{R} \to N$$

lying in a totally geodesic sub-manifold $]\mu, +\infty[\times S^1 \times \mathbb{R}$ where S^1 is the S^2 component of $\mathbb{R}$. If one supposes that the radius r is constant, meaning that the orbit is circular, we get (see [Bes08], III.G) Kepler's law: the orbit of the planet around a star of mass M takes proper time

$$T = \frac{4\pi^2}{M} R$$

where R is the distance between the planet and the center of the star. Using almost circular geodesics $x : \mathbb{R} \to N$, one gets (see [Bes08], III.H) that the period of the Mercury planet (of mass m) turning around the sun (of mass M) differs from the Newtonian period by a factor

$$\left(1 - \frac{6m}{M}\right)^{-1/2}.$$

This is the most famous experimental prediction of general relativity theory.

The study of null geodesics (light rays) in N is used to explain the bending of light coming from a far away star behind the sun to the Earth (see [Bes08], III.J).

The subspace B of M is called the Schwarzschild black hole of active mass $8\pi\mu$. Note that the vector field $(\partial/\partial r)_{|B}$ is time-like, so that an observer close to B in N will tend to B, and he will experience an infinite curvature of spacetime when arriving at the boundary of B. We advise the reader not to try this.

14.4.4 The Big Bang Solution

We refer to Landau-Lifchitz [LL66], Chap. 14, for a very nice account of this theory. This solution, called the Friedmann-Lemaitre-Robertson-Walker solution, or the standard model of cosmology, is a solution that can be obtained by assuming that spacetime M is:

1. homogeneous, i.e., of the form $M = N \times I$ with I an interval in $\mathbb{R}$ and (N, h) a simply connected Riemannian space; and
2. isotropic, meaning that (N, h) is of constant curvature.

We refer to Wolf's book [Wol84] for a complete study of spaces of constant scalar curvature. If $p : M \to N$ is the projection, the Lorentzian metric on M is then supposed to be of the form

$$g = -c^2 dt^2 + a(t)^2 p^* h.$$

The main examples of Riemannian spaces (N, h) like the one above are S^3 (positive curvature, defined by the equation $\|x\|^2 = a^2$ in Euclidean $\mathbb{R}^4$), $\mathbb{R}^3$ (zero curvature) and $\mathbb{H}^3$ (negative curvature, defined as the positive cone in $\mathbb{R}^{3,1}$). The above metric gives a theoretical explanation of the expansion of the universe, which is experimentally explained by observing that the spectrum of light of far away stars is shifted to the red.

One can also give examples that have a Klein geometric description as homogeneous spaces under a group. This allows us to add a geometric cosmological constant in the Lagrangian, by using the Cartan formalism (see [Wis10]).

- The de Sitter space, the universal covering of $SO(4, 1)/SO(3, 1)$, has a positive cosmological constant.
- The Minkowski space $\mathbb{R}^{3,1} \rtimes SO(3, 1)/SO(3, 1)$ has cosmological constant zero.
- The anti de Sitter space, the universal covering of $SO(3, 2)/SO(3, 1)$, has a negative cosmological constant.

Recent fine measurements of the red shift show that the expansion of universe is accelerating. The standard model of cosmology with a cosmological constant is able to explain these two experimental facts by using an FLRW spacetime with positive cosmological constant.

References

[Bes08] Besse, A.L.: Einstein Manifolds. Classics in Mathematics, p. 516. Springer, Berlin (2008). Reprint of the 1987 edition. ISBN 978-3-540-74120-6

[CB72] Choquet-Bruhat, Y.: Stabilité de solutions d'équations hyperboliques non linéaires. Application à l'espace-temps de Minkovski en relativité générale. C. R. Acad. Sci. Paris, Sér. A-B **274**, 843–845 (1972)

[Der92] Derdziński, A.: Geometry of the Standard Model of Elementary Particles. Texts and Monographs in Physics, p. 199. Springer, Berlin (1992). ISBN 3-540-54356-2

[LL66] Landau, L., Lifchitz, E.: Physique Théorique. Tome II: Théorie du Champ. Deuxième Édition Revue. Traduit du Russe Par Edouard Gloukhian, p. 451. Éditions Mir, Moscow (1966)

[SW77] Sachs, R.K., Wu, H.H.: General Relativity for Mathematicians. Graduate Texts in Mathematics, vol. 48, p. 291. Springer, New York (1977). ISBN 0-387-90218-X

[Wis10] Wise, D.K.: MacDowell-Mansouri gravity and Cartan geometry. Class. Quantum Gravity **27**(15), 155010 (2010). 26. doi:10.1088/0264-9381/27/15/155010

[Wol84] Wolf, J.A.: Spaces of Constant Curvature, 5th edn., p. 412. Publish or Perish Inc., Houston (1984). ISBN 0-914098-07-1

Chapter 15
Variational Problems of Experimental Quantum Physics

The following variational problems are of interest only if they are used in the construction of a quantum theory of the corresponding fields. They can't be used in the setting of classical physics. However, their extremal trajectories are often called classical fields. This is because in the very special case of electromagnetism, the classical trajectories correspond to the motion of an electromagnetic wave in space-time.

15.1 The Klein-Gordon Lagrangian

15.1.1 The Classical Particle and the Klein-Gordon Operator

One can view the Klein-Gordon operator as the canonical quantization of the homological symplectic reduction of the classical particle phase space. This relation is explained in Polchinski's book [Pol05], p. 129.

Let X be an ordinary manifold equipped with a metric g and $M = [0, 1]$. The bundle underlying the classical particle is the bundle

$$\boxed{\pi : C = X \times M \to M}$$

of spaces. Its sections correspond to maps $x : [0, 1] \to X$. The free particle action functional is given by

$$S(x) = \int_{[0,1]} \frac{1}{2}\sqrt{x^* g(Dx, Dx)}.$$

The space of initial conditions (the so-called phase space of the corresponding Hamiltonian system) for the corresponding Euler-Lagrange equation is the cotangent bundle T^*X, with algebra of polynomial functions

$$\mathcal{O}_{T^*X} = \mathrm{Sym}^*_{\mathcal{O}_X}(\Theta_X)$$

F. Paugam, *Towards the Mathematics of Quantum Field Theory*,
Ergebnisse der Mathematik und ihrer Grenzgebiete. 3. Folge / A Series of Modern
Surveys in Mathematics 59, DOI 10.1007/978-3-319-04564-1_15,

given by the symmetric algebra on the $\mathcal{O}_X$-module $\Theta_X = \Gamma(X, TX)$ of vector fields on X.

There is a canonical two form $\omega \in \Omega^2_{T^*X}$ that gives a symplectic structure

$$\omega : \Theta_{T^*X} \otimes \Theta_{T^*X} \to \mathcal{O}_{T^*X}.$$

The Weyl algebra $\mathcal{W}_{T^*X}$ is given by the universal property

$$\mathrm{Hom}_{\mathrm{ALG}(\mathcal{O}_{T^*X})}(\mathcal{W}_{T^*X}, B) \cong \big\{ j \in \mathrm{Hom}_{\mathrm{MOD}(\mathcal{O}_{T^*X})}(\Theta_{T^*X}, B) \text{ such that } \\ j(v) \cdot j(w) - j(w) \cdot j(v) = \omega(v, w) \cdot 1_B \big\}.$$

It is defined as a quotient of the tensor algebra $T_{\mathcal{O}_{T^*X}}(\Theta_{T^*X})$. This gives a natural filtration and the corresponding graded algebra is

$$\mathrm{gr}^F \mathcal{W}_{T^*X} \cong \mathrm{Sym}^*_{\mathcal{O}_{T^*X}}(\Theta_{T^*X}) = \mathcal{O}_{T^*(T^*X)}.$$

Note that the algebra $\mathcal{D}_{T^*X}$ of differential operators on T^*X also has a filtration whose graded algebra is

$$\mathrm{gr}^F \mathcal{D}_{T^*X} = \mathcal{O}_{T^*(T^*X)}.$$

The algebra of differential operators $\mathcal{D}_X$ is in some sense a generalization of the Weyl algebra to non-symplectic manifolds, since one always has that its graded algebra

$$\mathrm{gr}^F \mathcal{D}_X \cong \mathrm{Sym}^*_{\mathcal{O}_X}(\Theta_X) = \mathcal{O}_{T^*X}$$

is isomorphic to the algebra of functions on T^*X. The action of $\mathcal{D}_X$ on $L^2(X)$ is regarded by physicists as the canonical quantization of the algebra $\mathcal{O}_{T^*X}$ of functions on phase space.

Because of the symmetries of the Lagrangian of the free particle by reparametrization, the complete explanation of the relation between the classical particle and the Klein-Gordon operator can only be done in the setting of the BRST formalism for fixing the gauge symmetry. This is explained in Polchinski's book [Pol05], p. 129.

15.1.2 The Klein-Gordon Lagrangian

The Klein-Gordon Lagrangian is a slight generalization of the Lagrangian of electromagnetism, which includes a mass for the field in play. Here we use the book of Derdzinski [Der92]. Let (M, g) be a metric manifold, equipped with a connection $\nabla : \Omega^1 \to \Omega^1 \otimes \Omega^1$. The bundle $\pi : C \to M$ underlying the Klein-Gordon Lagrangian is one of the following linear bundles:

1. $\mathbb{R}_M$: the corresponding particle is said to be of spin 0 (because the action of the orthogonal group $O(g)$ on it is trivial) and of parity 1. Its classical state is thus represented simply by a real valued function on M.

2. $\wedge^n T^*M$: the corresponding particle is also of spin 0 but of parity -1.
3. T^*M: the corresponding particle is of spin 1 (the standard representation of $O(g)$). The electromagnetic particles, called Photons, are of this kind.

The metric and connection on TM can be extended to C and to $C \otimes T^*M$. We will now define the Klein-Gordon action functional

$$S : \Gamma(M, C) \to \mathbb{R}$$

by its Lagrangian density.

Definition 15.1.1 The *Klein-Gordon Lagrangian* of mass m is given on sections $\varphi \in \Gamma(M, C)$ by

$$L(\varphi) = -\frac{1}{2}\left[g(\nabla\varphi, \nabla\varphi) + \frac{m^2c^2}{\hbar^2} g(\varphi, \varphi) \right] \omega_g.$$

The equation of motion for this Lagrangian is given by

$$\left(\Box - \frac{m^2c^2}{\hbar^2}\right)\varphi = 0,$$

where the d'Alembertian $\Box$ was introduced in Sect. 14.3.2 on the covariant formulation of electromagnetism.

15.2 The Dirac Lagrangian

We refer to Sect. 5.4 for details on spinors and the Clifford algebra.

15.2.1 Fermionic Particles and the Dirac Operator

The Dirac operator can be viewed as the canonical quantization of the fermionic particle. This relation is explained in Polchinski's book [Pol98] on superstring theory.

Let X be an ordinary manifold equipped with a metric g and $M = \mathbb{R}^{0|1}$. The bundle underlying the fermionic particle is the bundle

$$\boxed{\pi : C = X \times M \to M}$$

of superspaces. Its sections correspond to maps $x : \mathbb{R}^{0|1} \to X$.

The fermionic particle action functional is given by

$$S(x) = \int_{\mathbb{R}^{0|1}} \frac{1}{2} x^* g(Dx, Dx).$$

Note that the superspace $\underline{\Gamma}(M,C)$ of such maps is identified with the odd tangent bundle $T[1]X$ of X, whose super-functions are given by the superalgebra

$$\mathcal{O}_{T[1]X} = \wedge^* \Omega_X^1$$

of differential forms. This space $T[1]X$ plays the role of a phase (i.e. initial condition) space for the fermionic particle and superfunctions on it are classical observables for this system. If one fixes a metric g on X (for example if X is ordinary spacetime), as explained in Sect. 5.4 on Clifford algebras, one can think of $\Gamma(X, \mathrm{Cliff}(TX, g))$ as a canonical quantization of the algebra $\mathcal{O}_{T[1]X}$, and the spinor bundle S on which it acts as the state space for the canonical quantization of the fermionic particle. This means that fermions in the setting of field theory are obtained by canonically quantizing the fermionic particle.

15.2.2 Generally Covariant Formulation

Let (M, g, ∇) be an oriented pseudo-metric manifold with a connection and let $S \subset \mathrm{Cliff}(TM_{\mathbb{C}}, g)$ be an irreducible sub-representation of the action of the complexified Clifford algebra bundle on itself over (M, g). We assume that S is of minimal dimension (equal to the dimension of a column of a fiber). It is then called a Dirac spinor bundle. The connection ∇ is assumed to extend to S (this is the case for the Leci-Civita connection). Denote by $\mathcal{S}$ the space of sections of S. The Clifford multiplication induces a natural map

$$c : \Omega^1_{M,\mathbb{C}} \otimes \mathcal{S} \to \mathcal{S}.$$

The Dirac operator is defined by

$$\not{D} = c \circ \nabla : \mathcal{S} \to \mathcal{S}.$$

Suppose that there exists a real spinor bundle S whose extension of scalars to $\mathbb{C}$ gives the bundle used above. The Clifford multiplication gives a real Dirac operator

$$\not{D} : \mathcal{S} \to \mathcal{S}.$$

There are natural pairings

$$\tilde{\Gamma} : S \times_M S \to TM$$

or equivalently

$$\tilde{\Gamma} : S \times_M \left(S \times_M T^*M\right) \to \mathbb{R}_M,$$

and

$$\epsilon : S \times_M S \to \mathbb{R}_M.$$

If $m \in \mathbb{R}$ is a fixed number, we define the Dirac Lagrangian acting on sections of the super-bundle

$$\boxed{\pi : C = \Pi S \to M}$$

by

$$L(\psi) = \left(\frac{1}{2}\psi \not{D}\psi - \frac{1}{2}\psi m \psi\right) d^n x,$$

where the kinetic term is given by

$$\psi \not{D}\psi := \tilde{\Gamma}(\psi, \nabla\psi)$$

and the mass term is given by

$$\psi m \psi := m\epsilon(\psi, \psi).$$

Let us describe the above construction in the case where $M = \mathbb{R}^{3,1}$ and g is Minkowski's metric, following Derdzinski's book [Der92], Sect. 1.3. Let (V, q) be the corresponding quadratic real vector space. In this case, we have

$$\mathrm{Cliff}(V, q) = \mathrm{Res}_{\mathbb{C}/\mathbb{R}} \mathrm{M}_{4,\mathbb{C}} \quad \text{and} \quad \mathrm{Cliff}^0(V, q) = \mathrm{Res}_{\mathbb{C}/\mathbb{R}} \mathrm{M}_{2,\mathbb{C}},$$

the spinor group is $\mathrm{Spin}(3, 1) = \mathrm{Res}_{\mathbb{C}/\mathbb{R}} \mathrm{SL}_{2,\mathbb{C}}$ and its real spinor representation is $S = \mathrm{Res}_{\mathbb{C}/\mathbb{R}} \mathbb{C}^2$. An element of S is called a Weyl spinor and an element of $S_{\mathbb{C}}$ is called a Dirac spinor. One has a canonical symplectic form ω on S given by

$$\omega(x, y) = x_1 y_2 - x_2 y_1.$$

Now consider the vector space

$$V' := \left\{ f \in \mathrm{End}_{\mathbb{R}}(S),\ f(\lambda x) = \bar{\lambda} f(x), \omega(-, f-) \text{ is hermitian} \right\}.$$

Note that V' is a real vector space of dimension 4 with a Lorentz form q', such that

$$fg + gf = q'(f, g) \cdot \mathrm{id}_S.$$

There is an identification between (V, q) and (V', q') and an identification

$$S \otimes \bar{S} \cong \mathrm{Res}_{\mathbb{C}/\mathbb{R}} V_{\mathbb{C}}.$$

One defines $\epsilon : S \otimes S \to \mathbb{R}$ by

$$\epsilon = \mathrm{Im}(\omega).$$

The Dirac Lagrangian can then be written as

$$L(\psi) = \frac{1}{2} \mathrm{Im}\, \omega\big(\psi, (\not{D} + m)\psi\big).$$

15.2.3 The Dirac Lagrangian in Cartan's Formalism

One can also formulate this from a more gauge theoretical viewpoint by using Cartan's formalism. Let (V, q) be a quadratic space, $H = \mathrm{Spin}(V, q)$ and $G = V \rtimes \mathrm{Spin}(V, q)$. We choose a section of the exact sequence

$$0 \to V \to G \to H \to 1,$$

i.e., an embedding $H \subset G$. Let Q be a principal G-bundle equipped with a principal G-connection and $s : M \to P \times_G G/H$ be a reduction of the structure group of P to H, i.e., a sub-principal H-bundle P of Q. The motivation for doing Cartan geometry is that the pull-back of the connection A along $P \subset Q$ gives an isomorphism

$$e : TM \to \underline{V}$$

between the tangent bundle and the fake tangent bundle

$$\underline{V} := P \times_{\mathrm{SO}(V,q)} V.$$

If S is a spinorial representation of $\mathrm{Spin}(V, q)$, the pairings

$$\tilde{\Gamma} : S \otimes S \to V$$

and

$$\epsilon : S \otimes S \to \mathbb{R},$$

being $\mathrm{SO}(V, q)$-equivariant, induce natural pairings on the associated bundles $\underline{S}$ and $\underline{V}$. The connection A induces a connection on both of them.

In this setting, the underlying bundle of the Dirac Lagrangian is the odd superbundle

$$\boxed{\pi : C := \Pi \underline{S} \to M}$$

and the Dirac Lagrangian is defined as before by

$$L(\psi) = \left(\frac{1}{2}\psi \not{D}_A \psi - \frac{1}{2}\psi m \psi\right) d^n x.$$

We may also allow the gravitational field (P, Q, A) to vary. This corresponds to adding the space of fields

$$\mathrm{B{\small UN}Conn}_{G,\mathrm{Cartan}}(M) \rightarrowtail \underline{\mathrm{Hom}}\left(M, \left[G\backslash(G/H)\right]_{conn}\right)$$

for the Cartan formalism for general relativity.

15.3 Classical Yang-Mills Theory

We already presented the gauge theoretical version of electromagnetism in Sect. 14.3.3.

Let M be a manifold and G be a connected Lie group. Suppose given a principal G-bundle P on M. The bundle underlying Yang-Mills theory is the bundle

$$\boxed{\pi : C = \mathrm{Conn}_G(P) \to M}$$

of principal G-connections on P. Recall that such a connection is given by a (non-degenerate) G-equivariant $\mathfrak{g}$-valued 1-form

$$A \in \Omega^1(P, \mathfrak{g})^G$$

and that its curvature is the G-equivariant $\mathfrak{g}$-valued 2-form

$$F_A := dA + [A \wedge A] \in \Omega^2(P, \mathfrak{g})^G.$$

We assume given a bi-invariant inner product $\langle \cdot, \cdot \rangle$ on $\mathfrak{g}$. The pure Yang-Mills action is given by the formula

$$S(A) := -\int_M \frac{1}{4} \langle F_A \wedge *F_A \rangle.$$

The principal G-connection A induces a covariant derivation

$$d_A : \Omega^*(P, \mathfrak{g}) \to \Omega^*(P, \mathfrak{g}).$$

The equations of motion are given by

$$d_A * F_A = 0.$$

The choice of a section $s : M \to P$, called a gauge fixing, gives an isomorphism

$$s^* : \Omega^1(P, \mathfrak{g})^G \supset \mathrm{Conn}_G(P) \to \Omega^1(M, \mathfrak{g})$$

between G-principal connections on P and Lie algebra valued differential forms on M. However, such a choice is not always possible and we will see later that the obstruction to this choice plays an important role in the quantization of Yang-Mills gauge theories. If we assume given a gauge fixing, the bundle associated to Yang-Mills theory is the bundle

$$\boxed{\pi : C = T^*M \otimes \mathfrak{g} \to M}$$

whose space of sections is $\Omega^1(M, \mathfrak{g})$ and the action functional is

$$S(A) = \int_M \langle dA \wedge *dA \rangle.$$

One can show (by extending [Cos11], Chap. 5, 7.2 to the case of a base manifold M of dimension n) that the corresponding BV bundle is given by the $\mathcal{O}_M$-module map

$$\begin{array}{lcl} \mathcal{W}_{BV} := \Omega^0(M)\otimes\mathfrak{g}[1]\oplus\Omega^1(M)\otimes\mathfrak{g}\oplus\Omega^{n-1}(M)\otimes\mathfrak{g}[-1]\oplus\Omega^n(M)\otimes\mathfrak{g}[-2] \\ \uparrow \qquad\qquad\qquad\qquad\qquad\qquad \uparrow \\ \mathcal{C} \ := \qquad\qquad\qquad\qquad\qquad \Omega^1(M)\otimes\mathfrak{g} \end{array}$$

Note that the BV bundle is defined as the $\mathcal{O}_C$-module

$$\mathcal{V}_{BV} := \left[\Omega^0(M)\otimes\mathfrak{g}[1]\oplus\Omega^{n-1}(M)\otimes\mathfrak{g}[-1]\oplus\Omega^n(M)\otimes\mathfrak{g}[-2]\right]\otimes_{\mathcal{O}_M}\mathcal{O}_C.$$

The BV algebra is the algebra on $\mathcal{A}=\mathrm{Jet}(\mathcal{O}_C)$ given by

$$\mathcal{A}_{BV}=\mathrm{Jet}\big(\mathrm{Sym}_{\mathcal{O}_C}(\mathcal{V}_{BV})\big).$$

One then has identifications of the various terms of the BV algebra with the corresponding local constructions

$$\begin{aligned} \Theta^\ell_{\mathcal{A}} &\cong \Omega^{n-1}(M,\mathfrak{g})\otimes_{\mathcal{O}_M}\mathcal{A}[\mathcal{D}], \\ \mathfrak{g}^0 &\cong \Omega^0(M,\mathfrak{g})\otimes_{\mathcal{O}_M}\mathcal{A}[\mathcal{D}]\cong\mathfrak{g}\otimes\mathcal{A}[\mathcal{D}], \\ \big(\mathfrak{g}^0\big)^\circ &\cong \Omega^n(M,\mathfrak{g})\otimes_{\mathcal{O}_M}\mathcal{A}[\mathcal{D}]. \end{aligned}$$

The BV Lagrangian is then given (see [HT92]) by

$$L\big(A,A^*,C_*,C^*\big)=-\frac{1}{4}\langle F_A\wedge *F_A\rangle+\big\langle A^*\wedge dC_*\big\rangle.$$

Note that in this case, the BV formalism can be formulated purely in terms of the graded linear bundle $\mathcal{W}_{BV}$ on M. One usually does this in perturbative quantum field theory (see for example Costello's book [Cos11]).

There is no reason to choose a particular principal bundle so that the bundle underlying Yang-Mills theory could be the moduli space

$$\pi: C=\mathrm{B\scriptsize{UN}}\mathrm{Conn}_G(M)\to M$$

whose sections are pairs (P,A) composed of a principal bundle P on M and a principal connection. The space of fields is simply the stack of maps (see Sects. 10.2 and 10.4 for a precise definition of this object)

$$\mathrm{B\scriptsize{UN}}\mathrm{Conn}_G(M):=\underline{\mathrm{Hom}}(M,BG_{conn}).$$

This shows that the natural setting of Yang-Mills gauge theory is homotopical geometry.

15.4 Classical Matter and Interaction Particles

One can easily combine the Klein-Gordon, Dirac and Yang-Mills action functionals. We will do this in the Cartan formalism that is easier to use in practice with fermions. The general relativistic version can be found by assuming that the Cartan connection is the Levi-Civita connection. We refer to Derdzinski's book [Der92] for another presentation that avoids the use of Cartan connections.

Let (V, q) be an even dimensional quadratic space and M be a manifold of the same dimension.

Suppose given a Lie group K whose Lie algebra $\mathfrak{k}$ is equipped with an invariant scalar product $\langle\cdot,\cdot\rangle$. Let R be a principal K-bundle on M. We continue to use the notations of Sect. 15.2.3 for the Dirac Lagrangian in Cartan formalism, so that Q is a principal $G = V \rtimes \mathrm{Spin}(V, q)$-bundle and P is a principal $H = \mathrm{Spin}(V, q)$ bundle.

Suppose given a representation S of $K \times \mathrm{Spin}(V, q)$ such that the two components commute. We associate to S the bundle

$$\underline{S} := (R \times P) \times_{K \times \mathrm{Spin}(V,q)} S.$$

If B is a Cartan connection on (P, Q) (which is assumed to be fixed here) and A is a principal K-connection on R, one gets a covariant derivation

$$d_{(A,B)} : \mathcal{S} \to \mathcal{S} \otimes \Omega^1_M$$

on the space $\mathcal{S}$ of sections of $\underline{S}$, which can be combined with the Clifford multiplication map

$$\mathcal{S} \otimes \Omega^1_M \to \mathcal{S}$$

to yield the covariant Dirac operator

$$\not{D}_{(A,B)} : \mathcal{S} \to \mathcal{S}.$$

In the physicist's language, the Cartan connection (which is in practice the combination of the metric and the Levi-Civita connection on spacetime) is called the fixed background, the principal K-connection A on R is called the gauge field and a section ψ of $\Pi\underline{S} \to M$ is called an interacting spinor field.

The fiber bundle of Yang-Mills theory in a fixed gravitational background and with interaction is the bundle

$$\boxed{\pi : C = \mathrm{Conn}_K(R) \times_M \Pi\underline{S} \to M}$$

whose sections are pairs (A, ψ) of a principal K-connection on the gauge bundle R and of a section ψ of the super-bundle $\Pi\underline{S}$ of fermionic interacting particles.

The Yang-Mills theory with fermionic matter Lagrangian is then given by simply adding the pure Yang-Mills and the Dirac Lagrangian in the above setting

$$L(A, \psi) := -\frac{1}{2}\langle F_A \wedge *F_A\rangle + \psi \not{D}_A \psi + \psi m \psi.$$

Note that the Dirac operator now depends on the gauge field A.

For example, in the case of electromagnetism, we can work with the representation given by complex spinors $S_{\mathbb{C}}$ equipped with the action of SU_1 by multiplication and the standard action of $\mathrm{Spin}(V, q)$. This gives the Quantum Electrodynamics Lagrangian.

Note that it is not really natural to fix once and for all the principal bundles used to formalize the Lagrangian. A Cartan connection may be described as a stack morphism (see Sect. 10.3 for more details)

$$(s, Q, A) : M \to \left[G\backslash(G/H)\right]_{conn}.$$

The Yang-Mills field is given by a morphism

$$(R, A) : M \to BK_{conn}.$$

If we give a representation S of $G \times K$, we get a bundle

$$\left[S/(G \times K)\right] \to B(G \times K)$$

such that a morphism

$$M \to \left[S/(G \times K)\right]_{conn}$$

is the same as a principal $(G \times K)$ bundle $Q \times R$ with connection, equipped with a section ψ of the associated bundle $\mathcal{S}$.

15.5 The Standard Model

We refer to de Faria and de Melo [dFdM10], Chap. 9, for the description of the standard model representation and Lagrangian. A refined description of the (modified) standard model Lagrangian is also given in Connes-Marcolli [CM08], Chap. 9. For a coordinate-free presentation, one can also use Derdzinski [Der92].

The Yang-Mills theory underlying the standard model of elementary particles has gauge group $K = U(1) \times \mathrm{SU}(2) \times \mathrm{SU}(3)$. For $i = 1, 2, 3$, denote by $\det_i$ the determinant representations of these three groups on $\mathbb{C}$, V_i their standard representation on $\mathbb{C}^i$, $S = \mathrm{Res}_{\mathbb{C}/\mathbb{R}}\mathbb{C}^2$ the spinorial representation of $\mathrm{Spin}(3, 1)$ and V its standard representation. Let $G = K \times (V \rtimes \mathrm{Spin}(3, 1))$ be the full gauge group of gravity paired with matter. We denote by $\bar{S}$ and $\bar{V}_i$ the complex conjugated representations.

The representation of G underlying the standard model is given by defining:

1. the left-quark doublet representation

$$Q_L = V_1 \otimes V_2 \otimes V_3 \otimes S;$$

2. the Up-quark singlet representation

$$U_R = V_1 \otimes \det_2 \otimes \bar{V}_3 \otimes S;$$

3. the Down-quark singlet representation

$$D_R = V_1 \otimes \underset{2}{\det} \otimes \bar{V}_3 \otimes S;$$

4. the left lepton doublet representation

$$L_L = V_1 \otimes V_2 \otimes \underset{3}{\det} \otimes S;$$

5. the right-electron singlet representation

$$E_R = V_1 \otimes \underset{2}{\det} \otimes \underset{3}{\det} \otimes S;$$

6. the Higgs representation

$$H = V_1 \otimes V_2 \otimes \underset{3}{\det} \otimes S,$$

and admitting the experimental fact that there are three generations of quarks and leptons, the full standard model representation for matter particles in interaction is

$$V_{sm} := (Q_L \oplus U_R \oplus D_R \oplus L_L \oplus E_R)^{\oplus 3} \oplus H.$$

The bundle underlying the standard model is thus the Yang-Mills bundle

$$\boxed{\pi : C = \mathrm{Conn}_K(R) \times_M \Pi \underline{V}_{sm} \to M}$$

where $\underline{V}_{sm}$ is the bundle associated to a given principal bundle $R \times Q$ under

$$G = \big(\mathrm{U}(1) \times \mathrm{SU}(2) \times \mathrm{SU}(3)\big) \times \big(V \rtimes \mathrm{Spin}(3, 1)\big)$$

and to the representation V_{sm} by the associated bundle construction

$$\underline{V}_{sm} := (R \times Q) \times_G V_{sm}.$$

The standard model Lagrangian (before its Batalin-Vilkovisky homotopical Poisson reduction) is given by a combination of a classical Yang-Mills Lagrangian with matter and a Higgs/Yukawa component, which gives masses to some interacting and matter particles. It is thus essentially of the form

$$L(A, \psi, \varphi) = -\frac{1}{2}\langle F_A \wedge *F_A\rangle + \psi \not{D}_A \psi + V(A, \psi, \varphi),$$

where $V(A, \psi, \varphi) = L_H(\varphi) + L_{Hg}(A, \varphi) + L_{Hf}(\psi, \varphi)$ is the Higgs potential, with

- L_H the Higgs self-coupling potential;
- L_{Hg} the coupling of Higgs with gauge fields; and
- L_{Hf} the (Majorana) coupling of Higgs with fermions.

These additional Higgs couplings are made to give certain fermions and gauge fields a mass, by the spontaneous symmetry breaking mechanism. For more details on this mechanism, we refer to Derdzinski [Der92], Chap. 11, Connes-Marcolli [CM08], Chap. 9 and de Faria-de Melo [dFdM10], Chap. 9.

We remark that the CERN experiment at the LHC has shown in July 2012 the existence of the Higgs boson φ. This results gives a very robust experimental meaning to the standard model, which had been lacking for roughly thirty years.

References

[CM08] Connes, A., Marcolli, M.: Noncommutative Geometry, Quantum Fields and Motives. American Mathematical Society Colloquium Publications, vol. 55, p. 785. Am. Math. Soc., Providence (2008). ISBN 978-0-8218-4210-2

[Cos11] Costello, K.: Renormalization and Effective Field Theory. Mathematical Surveys and Monographs, vol. 170, p. 251. Am. Math. Soc., Providence (2011). ISBN 978-0-8218-5288-0

[Der92] Derdziński, A.: Geometry of the Standard Model of Elementary Particles. Texts and Monographs in Physics, p. 199. Springer, Berlin (1992). ISBN 3-540-54356-2

[dFdM10] de Faria, E., de Melo, W.: Mathematical Aspects of Quantum Field Theory. Cambridge Studies in Advanced Mathematics, vol. 127, p. 298. Cambridge University Press, Cambridge (2010). With a foreword by Dennis Sullivan. ISBN 978-0-521-11577-3

[HT92] Henneaux, M., Teitelboim, C.: Quantization of Gauge Systems, p. 520. Princeton University Press, Princeton (1992). ISBN 0-691-08775-X; 0-691-03769-8

[Pol98] Polchinski, J.: String Theory. Vol. II. Cambridge Monographs on Mathematical Physics, p. 531. Cambridge University Press, Cambridge (1998). Superstring theory and beyond. ISBN 0-521-63304-4

[Pol05] Polchinski, J.: String Theory. Vol. I. Cambridge Monographs on Mathematical Physics, p. 402. Cambridge University Press, Cambridge (2005). ISBN 0-521-63303-6; 978-0-521-67227-6; 0-521-67227-9. An introduction to the bosonic string. Reprint of the 2003 edition

Chapter 16
Variational Problems of Theoretical Physics

We will now present, sometimes in a more sketchy fashion, some important models of theoretical/mathematical physics. Here we have chosen to give only the superspace formulations of superfields, because of its evident mathematical elegance. We refer to the two IAS volumes [DEF+99] for more examples. Schwarz's books [Sch94b] and [Sch93] also discusses in more detail the relation between physics and topology, and Jost's book [Jos09] gives a nice introduction to the relation between geometry and physics.

16.1 Kaluza-Klein's Theory

The aim of Kaluza-Klein's theory is to combine general relativity on spacetime M and electromagnetism into only one "generally relativistic" theory on a $4+1$ dimensional space, locally given by a product $M \times S^1$.

Let M be a 4-dimensional manifold equipped with a principal U(1)-bundle $p : P \to M$. Fix a $U(1)$-invariant metric g on $U(1)$, which is equivalent to a quadratic form $g_U : \mathfrak{u}(1) \times \mathfrak{u}(1) \to \mathbb{R}$ on its Lie algebra. In addition, fix a U(1)-invariant metric

$$g_P : TP \times TP \to \mathbb{R}_P$$

on P. The length of a fiber of P (isomorphic to $U(1)$) for this metric is denoted by λ_P. The metric on P induces a Levi-Civita connexion on TP. This connection, being also an Ehresmann connection, induces a splitting of the canonical exact sequence

$$0 \to VP \longrightarrow TP \xrightarrow{dp} p^*TM \to 0,$$

and thus a decomposition $TP = HP \oplus VP$ where $HP \cong p^*TM$ and $VP = \mathrm{Ker}(dp)$ is the space of vertical tangent vectors. Recall that since a principal bundle is parallelizable, there is a canonical trivialization of the vertical tangent bundle

$$VP \cong \mathfrak{u}(1)_P := \mathfrak{u}(1) \times P.$$

F. Paugam, *Towards the Mathematics of Quantum Field Theory*,
Ergebnisse der Mathematik und ihrer Grenzgebiete. 3. Folge / A Series of Modern Surveys in Mathematics 59, DOI 10.1007/978-3-319-04564-1_16,

We assume that:

1. the restriction of the metric g_P on P to VP induces the given fixed metric on $U(1)$; and
2. the restriction of the metric g_P to $HP \cong p^*TM$ is induced by the pull-back of a Lorentzian metric g on M.

The Kaluza-Klein theory has underlying fiber bundle the bundle

$$\boxed{\pi : C = \mathrm{Sym}^2_{KK}(T^*P) \to M}$$

of such metrics on P. The Kaluza-Klein action functional

$$S : \Gamma(M, C) \to \mathbb{R}$$

is given by

$$S(g_P) := \int_P R_{scal}(g_P)\mathrm{vol}_{g_P},$$

where R_{scal} is the scalar curvature.

The scalar curvature then decomposes in coordinates as

$$R_{scal}(g_P) = p^*\left(R_{scal}(g) - \frac{\Lambda^2}{2}|F|^2 \right)$$

and if one integrates on the fibers, one gets the action

$$S(g_P) := \Lambda \int_M \left(R(g) - \frac{1}{\Lambda^2}|F|^2 \right) \mathrm{vol}(g).$$

If g is fixed and the action varies with respect to the primitive A of F, one gets Maxwell's theory. If one fixes F and varies the action with respect to g, one gets Einstein's equation for the Einstein-Hilbert action with an electromagnetic term.

16.2 Bosonic Gauge Theory

We refer to Freed's lectures on supersymmetry [Fre99]. Let M be an oriented manifold, G be a Lie group with Lie algebra $\mathfrak{g}$, and suppose given a G-invariant product $\langle \cdot, \cdot \rangle$ on $\mathfrak{g}$. Let (X, g) be a metric manifold on which G acts by isometries. Let $V : X \to \mathbb{R}$ be a G-invariant potential function. Suppose given a principal G-bundle P. The bundle underlying bosonic gauge theory is the bundle

$$\boxed{\pi : C = \mathrm{Conn}_G(P) \times (P \times_G X) \to M}$$

whose sections are pairs (A, φ) of a principal G-connection on P and of a section φ of the associated bundle $P \times_G X$.

The action functional $S = \Gamma(M, C) \to \mathbb{R}$ is given by

$$S(A, \varphi) = \int_M -\frac{1}{2}|F_A|^2 + \frac{1}{2}|d_A\varphi|^2 - \varphi^* V$$

where d_A is the covariant derivative associated to A.

16.3 Poisson Sigma Model

We describe the Poisson sigma model because it inspired Kontsevich's construction in [Kon03] of a deformation quantization formula for general Poisson manifolds (see the article by Cattaneo and Felder [CF01], and the survey article [CKTB05], Part III for more details on the Poisson sigma model underlying Kontsevich's ideas on deformation quantization).

Let (X, θ) be a Poisson manifold, i.e., a manifold equipped with a bivector $\theta \in \Gamma(X, \wedge^2 TX)$. Let M be a smooth surface. The bundle underlying this variational problem is the bundle

$$\boxed{\pi : C \to M}$$

whose sections are pairs (x, η) composed of a map $x : M \to X$ and of a differential form $\eta \in \Omega^1(M, x^*T^*X)$ with values in the pull-back by x of the cotangent bundle on X.

The action functional is given by

$$S(x, \eta) = \int_M \langle \eta \wedge Dx \rangle + \frac{1}{2}\langle \eta \wedge x^*\theta(\eta) \rangle$$

where:

- $Dx : TM \to x^*TX$ is the differential of x;
- θ is viewed as a morphism $\theta : T^*X \to TX$ so that $x^*\theta$ is a map $x^*\theta : x^*T^*X \to x^*TX$; and
- $\langle \cdot, \cdot \rangle : T^*X \times TX \to R_X$ is the standard duality pairing.

16.4 Chern-Simons Field Theory

Let M be a three dimensional manifold and G be a Lie group. Suppose given an invariant pairing $\langle \cdot, \cdot \rangle$ on the Lie algebra $\mathfrak{g}$ of G and a principal G-bundle P. The bundle underlying this variational problem is the bundle

$$\boxed{\pi : C = \mathrm{Conn}_G(P) \to M}$$

of principal G-connections on P. Such a connection is given by a G-invariant differential form

$$A \in \Omega^1(P, \mathfrak{g})^G.$$

The Chern-Simons action functional was introduced by Witten in [Wit89] to study topological invariants of three manifolds from a quantum field theoretical viewpoint. One says that it is a topological action functional because it does not depend on a metric and it is diffeomorphism invariant. It is given by

$$S(A) = \int_M \mathrm{Tr}\left(A \wedge dA + \frac{1}{3} A \wedge A \wedge A\right).$$

One can show that this action is classically equivalent (meaning has same equations of motions) to the Cartan formalism version of Euclidean general relativity in dimension 3. This shows that general relativity in dimension 3 is a purely topological theory.

16.5 Higher Chern-Simons Field Theory

We refer to [Sch11] for a complete modern treatment of this and other higher topological field theories. We will use here the formalism of non-abelian and differential cohomology presented in Sects. 10.3 and 10.4.

We first start with a simple example. Let $G = U(n)$ and $A = U(1)$ be smooth unitary groups, and consider the determinant morphism

$$\det : G = U(n) \to U(1) = A.$$

The induced morphism

$$B\det : BG \to BA$$

corresponds, for each smooth manifold M, to a morphism

$$c_1 : \underline{\mathrm{Hom}}(M, BG) \longrightarrow \underline{\mathrm{Hom}}(M, BA),$$

that may be thought of as a smooth geometric refinement of the first Chern class, that is with values in

$$\pi_0\big(\underline{\mathrm{Mor}}(M, BA)\big) \cong H^1\big(M, U(1)\big) \cong H^2(M, \mathbb{Z}),$$

where the last isomorphism is given by the boundary of the exponential exact sequence. This smooth map c_1 has a differential refinement

$$\hat{c}_1 : BU(n)_{conn} \to BU(1)_{conn}$$

that applies to the classifying spaces of principal bundles with invariant connections.

If M is a one-dimensional compact manifold, the holonomy integral

$$\int_M : \underline{\mathrm{Hom}}\big(M, BU(1)_{conn}\big) \longrightarrow U(1)$$

induces an action functional S_{c_1} by the diagram

$$\begin{array}{ccc} \underline{\mathrm{Hom}}\big(M, BU(n)_{conn}\big) & \xrightarrow{\hat{c}_1} & \underline{\mathrm{Hom}}\big(M, BU(1)_{conn}\big) \\ & \searrow^{e^{iS_{c_1}}} & \downarrow \scriptstyle{\int_M} \\ & & U(1) \end{array}$$

This thus gives a morphism

$$e^{iS_{c_1}} : \mathrm{BUN}_{conn}\big(M, U(n)\big) \to U(1)$$

from the moduli space of $U(n)$ bundles with connections on M, which may be used for a functional integral quantization. Note that only the exponential of the action functional is canonical in this setting.

The basic idea of higher Chern-Simons field theory is to take advantage of the abstract setting of non-abelian and differential cohomology to generalize the above example by using arbitrary higher groups and higher differential cohomology classes. Let G be a smooth ∞-group with delooping BG, and let $A = U(1)$ (it may actually be an arbitrarily deloopable ∞-group). Let

$$c : BG \to B^n A$$

be a non-abelian cohomology class in $H^n(BG, A)$, and suppose given a differential refinement

$$\hat{c} : BG_{conn} \to B^n A_{conn}.$$

If M is an n-dimensional compact manifold, the higher holonomy integral

$$\int_M : \underline{\mathrm{Hom}}\big(M, B^n U(1)_{conn}\big) \to U(1)$$

induces an action function S_c by the diagram

$$\begin{array}{ccc} \underline{\mathrm{Hom}}(M, BG_{conn}) & \xrightarrow{\hat{c}_1} & \underline{\mathrm{Hom}}\big(M, B^n U(1)_{conn}\big). \\ & \searrow^{e^{iS_c}} & \downarrow \scriptstyle{\int_M} \\ & & U(1) \end{array}$$

Definition 16.5.1 The above defined functional

$$e^{iS_c} : \mathrm{BUN}_{conn}(M, G) \to U(1)$$

on the moduli space of ∞-G-bundles with connections is called the *higher Chern-Simons functional* associated to the refined characteristic class

$$\hat{c} : BG_{conn} \to B^n U(1)_{conn}.$$

16.6 Supersymmetric Particle

We refer to Freed's lectures on supersymmetry [Fre99] for the two classical presentations (in component and using superspaces) of the supersymmetric particle.

Let (X, g) be the Minkowski space. The super bundle underlying the supersymmetric particle is the bundle

$$\boxed{\pi : C = X \times \mathbb{R}^{1|1} \to \mathbb{R}^{1|1}}$$

whose sections are morphisms

$$\varphi : \mathbb{R}^{1|1} \to X.$$

We introduce on $\mathbb{R}^{1|1}$ the three vector fields

$$\partial_t, \qquad D = \partial_\theta - \theta \partial_t, \quad \text{and} \quad \tau_Q = \partial_\theta + \theta \partial_t.$$

The action functional for the supersymmetric particle is the integral

$$S(\varphi) = \int_{\mathbb{R}^{1|1}} -\frac{1}{2} \langle D\varphi, \partial_t \varphi \rangle$$

with respect to the standard super-measure $|dt|d\theta$. The action is invariant with respect to ∂_t, D and τ_Q, so that it is called supersymmetric.

16.7 Superfields

We refer to the IAS volume [DF99], Chapter "Supersolutions", for the description of the following example.

Let M be ordinary Minkowski space, C a positive cone of time-like vectors in V, S a real representation of $\mathrm{Spin}(V)$, and

$$\Gamma : S^* \otimes S^* \to V$$

a pairing which is positive definite in the sense that $\Gamma(s^*, s^*) \in \bar{C}$ for $s^* \in S^*$ with $\Gamma(s^*, s^*) = 0$ only for $s^* = 0$. Associated to this, there is a morphism

$$\tilde{\Gamma} : S \otimes S \to V.$$

Definition 16.7.1 The *super translation group* is the supergroup with underlying space $V^s = V \times \Pi S^*$ and multiplication given by

$$\left(v_1, s_1^*\right)\left(v_2, s_2^*\right) = \left(v_1 + v_2 - \Gamma\left(s_1^*, s_2^*\right), s_1^* + s_2^*\right).$$

Its points with value in a super-commutative ring R is the even part of

$$\left(V + \Pi S^*\right) \otimes R.$$

The *super Minkowski space* is the underlying manifold of the super translation group. The *super Poincaré group* is the semi-direct product

$$P^s := V^s \rtimes \mathrm{Spin}(V).$$

The super Minkowski space is usually called the "N superspace" if S is the sum of N irreducible real spin representations of $\mathrm{Spin}(V)$.

The bundle underlying superfield gauge theory is the bundle

$$\boxed{\pi : C = \mathrm{Conn}_K(R) \times X \to M^s}$$

with X an ordinary manifold, K a super-group and R a super-K-principal bundle over M^s.

Here we only give two examples of a Lagrangian action functional on such a super-Minkowski space $M^{4|4}$ that is based on ordinary Minkowski space $\mathbb{R}^{3,1}$ with 4 super-symmetries (be careful, this notation could be misleading: it has nothing to do with super-affine space $\mathbb{A}^{4|4}$; the first 4 is for spacetime dimension 4 and the second 4 is for 4 supersymmetries, i.e., one has 4 copies of the standard real spin representation S of Spin(3, 1)).

First consider the case of a chiral superfield φ, that is a section of

$$\boxed{\pi : C = \mathbb{C} \times M^{4|4} \to M^{4|4},}$$

i.e., a supermap $\varphi : M^{4|4} \to \mathbb{C}$. We use the standard volume form on $S = \mathrm{Res}_{\mathbb{C}/\mathbb{R}}\mathbb{C}^2$ and on M^4 to get a supertranslation invariant volume form on $M^{4|4}$. The action functional is given by

$$S(\varphi) = \int_{M^{4|4}} \frac{1}{4}\bar{\varphi}\varphi.$$

Now let X be a Kähler manifold, i.e.,

1. a holomorphic manifold X,
2. equipped with a symplectic form $\omega \in \Omega_X^2$,

such that

$$g(v, w) = \omega(v, Jw)$$

is a Riemannian metric.

We assume given locally on X a Kähler potential, i.e., a real valued function K such that the Kähler form is

$$\omega = i\partial\bar{\partial}K.$$

The Lagrangian action functional applies to sections of the bundle

$$\boxed{\pi : C = X \times M^{4|4} \to M^{4|4},}$$

and is given by

$$S(\varphi) = \int_{M^{4|4}} \frac{1}{2} K(\bar{\varphi}, \varphi).$$

16.8 Bosonic Strings

We carefully inform the reader that the action given here is not supposed to be quantized through a functional integral. There is at the present time no proper definition of a hypothetic Lagrangian field theory (called M-theory) whose quantum field theoretic perturbative expansions would give the perturbative series defined *ad hoc* by string theorists to define quantum string propagation amplitudes.

We refer to Polchinski's book [Pol05] for a complete physical introduction to bosonic string theory. Let M be a two dimensional manifold and let

$$\mathrm{Sym}^2_{min}(T^*M) \to M$$

be the bundle of Minkowski metrics on M. Let (P, g_P) be a given space with a fixed metric. For example, this could be a Kaluza-Klein $U(1)$-fiber bundle (P, g_P) over ordinary spacetime. The fiber bundle underlying bosonic string theory on a fixed string topology M is the bundle

$$\boxed{\pi : C = P \times \mathrm{Sym}^2_{min}(T^*M) \to M}$$

whose sections are pairs (x, g) composed of a map $x : M \to P$ and a Lorentzian metric on M.

The Polyakov action functional $S : \Gamma(M, C) \to \mathbb{R}$ for bosonic strings is given by the formula

$$S(x, g) = \int_M \mathrm{Tr}(g^{-1} \circ x^* g_P) d\mu_g,$$

where:

- $g^{-1} : T^*M \to TM$ is the inverse of the metric map g;
- $x^* g_P : TM \to T^*M$ is the inverse image of $g_P : TP \to T^*P$; and
- the trace applies to $g^{-1} \circ x^* g_P \in \mathrm{End}(TM)$.

If one fixes x and varies g, the extremal points for this action are given by $g = x^* g_P$, but in a quantization process, it is important to allow "quantum" fluctuations of the metric g around these extremal points.

One can also add a potential function $V : P \to \mathbb{R}$ to the Lagrangian to get more general bosonic string action functionals.

16.9 Superstrings

The same remark as in Sect. 16.8 applies to this section. There are two ways to define classical superstrings. We refer to the introductory book [McM09] for a basic introduction to superstrings and to [Pol98] and [GSW88] for more specialized presentations.

In the Ramond-Neveu-Schwarz formalism, superstrings are a special case of superfields on a Lorentzian curve with N supersymmetries $M^{2|N}$. For $N = 1$, for example, one considers the super-space extension $M^{2|1}$ of the surface M that appears in the description of the Bosonic string in Sect. 16.8 with two fermionic variables that correspond to the real two dimensional real spinorial representation of $\mathrm{Spin}(1, 1)$. The strings are then given by superfields with values in, say, a complex target manifold X with given local Kähler potential K as in Sect. 16.7

$$x : M^{2|N} \to X.$$

So the fields of the theory are superfields on the supersymmetric extension of a given surface, meaning that the underlying bundle is

$$\boxed{\pi : C = X \times M^{2|N} \to M^{2|N}.}$$

In the Green-Schwarz formalism, superstrings are given by maps

$$x : M^2 \to X_s^N$$

from a surface to an N-super-Poincaré extension X_s^N of a given Lorentzian manifold X. In this case, the fields are classical fields on a surface but with values in a super-space, meaning that the underlying bundle is

$$\boxed{\pi : C = X_s^N \times M^2 \to M^2.}$$

In the two cases, the space of fields $\Gamma(M, C)$ is a super-space that can be quantized by a functional integral approach.

16.10 Supergravity

We refer to Lott's paper [Lot90] and to Egeileh's thesis [Ege07] for the superspace formulation of supergravity.

The fields of supergravity are sections of the supergravity bundle, fulfilling some torsion constraints. We only define here the underlying bundle, and warn the reader that constraints (which we won't describe here), play an essential role in the theory.

We need the definition of a Cartan super-connection on $M^{4|4}$ for the pair $(H, G) = (\mathrm{Spin}(3,1), P^s)$ composed of an ordinary spinor group and super-Poincaré group. It is given by:

1. a principal G-bundle Q on $M^{4|4}$;
2. a principal H-sub-bundle $P \subset Q$ of Q;
3. a section $s : M \to Q \times_G G/H$ of the associated bundle; and
4. a principal G-connection $A \in \Omega^1(Q, \mathfrak{g})$ on Q,

such that the pull-back s^*A induces an isomorphism

$$e_A : TM^{4|4} \cong Q \times_G V^s$$

called the Cartan coframe, and also sometimes the *supervielbein*. One has to assume moreover that the torsion of the given Cartan connection satisfy some first order constraint, for which we refer the reader to Lott's paper cited above.

The bundle underlying super-gravity theory is the bundle

$$\boxed{\pi : C = \mathrm{Conn}_{\mathrm{Cartan},tc}(Q, P, s) \to M^{4|4} = M}$$

whose sections are Cartan connections with torsion constraints.

The super-gravity action functional is then essentially the same as the Cartan version of the Einstein-Hilbert-Palatini action given in Sect. 14.4.2.

References

[CF01] Cattaneo, A.S., Felder, G.: Poisson sigma models and deformation quantization. Mod. Phys. Lett. A **16**(4–6), 179–189 (2001). doi:10.1142/S0217732301003255. Euroconference on brane new world and noncommutative geometry (Torino, 2000)

[CKTB05] Cattaneo, A., Keller, B., Torossian, C., Bruguières, A.: Déformation, Quantification, Théorie de Lie. Panoramas et Synthèses [Panoramas and Syntheses], vol. 20, p. 186. Société Mathématique de France, Paris (2005). ISBN 2-85629-183-X

[DEF+99] Deligne, P., Etingof, P., Freed, D.S., Jeffrey, L.C., Kazhdan, D., Morgan, J.W., Morrison, D.R., Witten, E. (eds.): Quantum Fields and Strings: A Course for Mathematicians. Vol. 1, 2. Am. Math. Soc., Providence (1999). Material from the special year on quantum field theory held at the Institute for Advanced Study, Princeton, NJ, 1996–1997. ISBN 0-8218-1198-3

[DF99] Deligne, P., Freed, D.S.: Classical field theory. In: Quantum Fields and Strings: A Course for Mathematicians, Vol. 1, 2 (Princeton, NJ, 1996/1997), pp. 137–225. Am. Math. Soc., Providence (1999)

[Ege07] Egeileh, M.: Géométrie des champs de Higgs, compactifications et supergravité. (2007, preprint)

[Fre99] Freed, D.S.: Five Lectures on Supersymmetry, p. 119. Am. Math. Soc., Providence (1999). ISBN 0-8218-1953-4

[GSW88] Green, M.B., Schwarz, J.H., Witten, E.: Superstring Theory. Vol. 1, 2nd edn. Cambridge Monographs on Mathematical Physics, p. 470. Cambridge University Press, Cambridge (1988). Introduction. ISBN 0-521-35752-7

[Jos09] Jost, J.: Geometry and Physics, p. 217. Springer, Berlin (2009). doi:10.1007/978-3-642-00541-1. ISBN 978-3-642-00540-4

[Kon03] Kontsevich, M.: Deformation quantization of Poisson manifolds. Lett. Math. Phys. **66**(3), 157–216 (2003). doi:10.1023/B:MATH.0000027508.00421.bf

[Lot90] Lott, J.: Torsion constraints in supergeometry. Commun. Math. Phys. **133**(3), 563–615 (1990)

[McM09] McMahon, D.: String Theory Demystified: A Self-Teaching Guide, p. 306. McGraw-Hill, New York (2009)

[Pol98] Polchinski, J.: String Theory. Vol. II. Cambridge Monographs on Mathematical Physics, p. 531. Cambridge University Press, Cambridge (1998). Superstring theory and beyond. ISBN 0-521-63304-4

[Pol05] Polchinski, J.: String Theory. Vol. I. Cambridge Monographs on Mathematical Physics, p. 402. Cambridge University Press, Cambridge (2005). ISBN 0-521-63303-6; 978-0-521-67227-6; 0-521-67227-9. An introduction to the bosonic string. Reprint of the 2003 edition

[Sch93] Schwarz, A.S.: Quantum Field Theory and Topology. Grundlehren der Mathematischen Wissenschaften [Fundamental Principles of Mathematical Sciences], vol. 307, p. 274. Springer, Berlin (1993). Translated from the 1989 Russian original by Eugene Yankowsky [E.M. Yankovskiĭ] and Silvio Levy. ISBN 3-540-54753-3

[Sch94b] Schwarz, A.S.: Topology for Physicists. Grundlehren der Mathematischen Wissenschaften [Fundamental Principles of Mathematical Sciences], vol. 308, p. 296. Springer, Berlin (1994). Translated from the Russian by Silvio Levy. ISBN 3-540-54754-1

[Sch11] Schreiber, U.: Differential cohomology in a cohesive ∞-topos. Nlab website (2011)

[Wit89] Witten, E.: Quantum field theory and the Jones polynomial. Commun. Math. Phys. **121**(3), 351–399 (1989)

Part III
Quantum Trajectories and Fields

Chapter 17
Quantum Mechanics

For historical reasons, we present the first quantum theories as they were understood by Heisenberg, Schrödinger, von Neumann and Dirac, among others. These quantization methods are essentially restricted to systems without interactions, and the most mathematically developed theory based on them is the deformation quantization program that says that for any Poisson manifold (M, π), the Poisson bracket on $\mathcal{O}_M$ can be extended to an associative multiplication operation $*$ on $\mathcal{O}_M[[\hbar]]$ such that one has

$$f * g = fg + \hbar\{f, g\} + O(\hbar^2).$$

This general result is due to Kontsevich [Kon03], and was found using functorial integral methods (applied to a convenient Poisson sigma model). The author's viewpoint is that the covariant methods, using functional integrals, have better functorial properties than the Hamiltonian methods, and generalize better to systems with (complicated) gauge symmetries, which are important in mathematical applications and in theoretical physics.

Note that these deformation quantization methods can be extended to perturbative covariant quantum field theory (following Fredenhagen and his collaborators). We refer to Chap. 22 for an introduction to this causal approach to perturbative quantum field theory. The deformation quantization methods also generalize to chiral/factorization algebras (see Beilinson-Drinfeld's book [BD04], Costello's book [Cos11] and Costello-Gwilliam's book [CG10]). We refer to Chaps. 21 and 24 for an introduction to these generalized deformation quantization problems.

A hike through the Hamiltonian methods is however useful to better understand the physicists' intuitions with quantum fields. The lazy reader can pass directly to Chap. 18. We will systematically use spectral theory, referring to Sect. 4.4 for some basic background.

17.1 Principles in von Neumann's Approach

We recall the von Neumann presentation of the Heisenberg/Schrödinger approach to quantum mechanics, that one can find in the still very modern reference [vN96].

F. Paugam, *Towards the Mathematics of Quantum Field Theory*,
Ergebnisse der Mathematik und ihrer Grenzgebiete. 3. Folge / A Series of Modern Surveys in Mathematics 59, DOI 10.1007/978-3-319-04564-1_17,

Definition 17.1.1 A *quantum mechanical system* is a tuple

$$(\mathcal{H}, \mathcal{A}, I, \mathrm{Hist}, H : \mathcal{H} \to \mathcal{H})$$

made of the following data:

1. An interval I of $\mathbb{R}$ that represents the physical *time parameter.*
2. A Hilbert space $\mathcal{H}$ that plays the role of the *phase space* T^*X of classical Hamiltonian systems, and whose norm 1 vectors are the physical states of the system.
3. A C*-algebra $\mathcal{A} \subset \mathcal{L}(H)$ of (not necessarily continuous) operators on $\mathcal{H}$ called the *observables* of the system.
4. Two sets of applications $\mathrm{Hist}_{\mathcal{H}} \subset \mathrm{Hom}(I, \mathcal{H})$ and $\mathrm{Hist}_{\mathcal{A}} \subset \mathrm{Hom}(I, \mathcal{A})$ respectively called the space of *possible evolutions* for the states and observables of the system (analogous to the space of histories of Hamiltonian mechanics).
5. A self-adjoint operator $H : \mathcal{H} \to \mathcal{H}$ called the *Hamiltonian* of the system.

If A is an observable, the elements of its spectrum $\lambda \in \mathrm{Sp}(A) \subset \mathbb{R}$ are the possible *measures* for the value of A, i.e., the real numbers that one can obtain by measuring A with a machine.

From a quantum mechanical system, one defines the space $\mathcal{T} \subset \mathrm{Hist}_{\mathcal{H}}$ of possible trajectories for the states of the system as the solutions of the *Schrödinger equation*

$$\varphi_{t_0} = \varphi \quad \text{and} \quad \frac{\hbar}{2i\pi}\frac{\partial \varphi_t}{\partial t} = -H\varphi_t.$$

This translates more generally to the evolution of observables $A \in \mathrm{Hist}_{\mathcal{A}}$ of the system by

$$i\hbar\frac{\partial A}{\partial t} = [H, A],$$

which corresponds to the naïve quantization of the equation

$$\frac{\partial a}{\partial t} = \{H, a\}$$

of Hamiltonian mechanics obtained by replacing the Poisson bracket by commutators (see Sect. 17.2).

One can think of quantum mechanics as being based on the following physical principles, which provide the link with experiment:

Principle 3

1. *If $A \in \mathcal{A} \subset \mathcal{L}(\mathcal{H})$ is an observable (self-adjoint operator) of the physical system and $[b, c] \subset \mathbb{R}$ is an interval, the probability of finding a number in the interval $[b, c]$ by measuring with a device the value of $A \in \mathcal{B}(\mathcal{H})$ on the system in the state $\varphi \in \mathcal{H}$ is*

$$\mathbb{P}(A, [b, c]) := \sqrt{\langle \varphi, \mathbb{1}_{[b,c]}(A)\varphi \rangle}.$$

2. *Two observables can be measured simultaneously if and only if they commute.*
3. *A measure of the observable A on a state φ changes the state φ in the state $A\varphi$.*
4. *A projector P is an observable that gives a true or false property for a state of the system.*

If R and S are two observables that don't commute, one associates to them the operator P of projection on the kernel M of the commutator $[R, S]$. The projector then corresponds to the property that the two observables are measurable simultaneously. If $M = \{0\}$, then R and S cannot be measured simultaneously. One can restrict to operators P and Q that satisfy the relation

$$[P, Q] = \frac{\hbar}{2i\pi}\mathrm{Id}.$$

One can then evaluate the dispersion of the measures of P and Q on a state φ by setting for $R = P, Q$,

$$d(R, \varphi) = \big\| R\varphi - \langle R\varphi, \varphi\rangle\varphi \big\|.$$

This shows the difference between the state $R\varphi$ obtained after the measure R and the state $\langle R\varphi, \varphi\rangle\varphi$ essentially equivalent to φ (the physical states being unitary). One then gets the Heisenberg uncertainty principle:

$$d(P, \varphi) \cdot d(Q, \varphi) \geq \frac{\hbar}{4\pi}$$

which means that the less the value of P on φ changes the state φ, the more the measure of the value of Q on φ changes the state φ of the given system. This can also be translated by saying that the more precisely you measure P, the less precisely you can measure Q.

17.2 Canonical Quantization of Heisenberg/Schrödinger

There is a strong analogy between quantum mechanical systems and classical Hamiltonian systems, given in Table 17.1.

If a classical Hamiltonian system $(P = T^*X, \pi, I, \mathrm{Hist}, H)$ is given, one can define a quantum mechanical system $(\mathcal{H}, \mathcal{A}, I, \mathrm{Hist}, H)$ by the following method:

1. Set $\mathcal{H} = L^2(X)$.
2. To every function $q \in \mathcal{C}^\infty(X)$ one associates the operator/observable $Q = m_q$ of multiplication by q in $\mathcal{H}$. The coordinate functions on $X = \mathbb{R}^n$ are denoted by q_i, and the corresponding operators by Q_i.
3. One associates to a vector field $v \in \Theta_X$ the operator $P_v(-) = \frac{\hbar}{2i\pi}d_v(-)$ of Lie derivative along the vector field v given by

$$d_v(f) = Df \circ v : X \to \mathbb{R}.$$

The derivations along coordinates are denoted by P_i.

Table 17.1 Analogy between classical hamiltonian systems and quantum mechanical systems

	CLASSICAL	QUANTUM
System	$(P, \pi, I, \text{Hist}, H)$	$(\mathcal{H}, \mathcal{A}, I, \text{Hist}, H)$
States	$x \in P$	$x \in \mathcal{H},\ \|x\| = 1$
Observables	$a \in C^\infty(P)$	$a \in \mathcal{A}$
Bracket	$\{a, b\}$	$[a, b]$
Hamiltonian	$H: X \to \mathbb{R}$	$H : \mathcal{H} \to \mathcal{H}$ self-adjoint $\pi : \mathcal{A} \to \mathcal{L}(\mathcal{H})$
Time evolution	Solution of the Hamiltonian vector field $\xi_H = \{H, \cdot\}$	$\sigma : \mathbb{R} \to \text{Aut}(\mathcal{A})$ $e^{itH}\pi(a)e^{-itH} = \pi(\sigma_t(a))$

4. More generally, one would like to have a Lie algebra morphism

$$\varphi : \big(\mathcal{A}, \{\cdot, \cdot\}\big) \to \big(\text{End}(\mathcal{H}), [\cdot, \cdot]\big),$$

where $\mathcal{A} \subset C^\infty(T^*X)$ is the algebra of interesting observables, which in particular contains:

- the classical Hamiltonian H;
- the coordinate functions q_i; and
- the derivatives with respect to the q_i, that one can also see as the applications $\dot{q_i}(x, \omega) = \omega(\dot{q}_{i\,x})$.

Such a Lie algebra morphism, called a quantization, would allow us to quantize all the useful functions on $P = T^*X$ and give back the above explicitly given constructions as particular cases, where we compute the bracket between observable quantities by setting

$$[A, B] = \frac{\hbar}{i}\{a, b\}.$$

This problem is clarified and solved by the theory called deformation quantization (see [Kon03] for the main result and Chap. 23 for an overview of the proof).

5. In particular, one gets the quantum Hamiltonian by replacing, when possible, the variables q_i and p_i on T^*X by the corresponding operators Q_i and P_i.

The probability amplitude that a state $f \in \mathcal{H}$ of the system is situated in a Borel set B of the space X is given by the value

$$\mathbb{A}_B(f) := \int_B |f|^2 d\mu.$$

One can thus also think of the quantum states in $\mathcal{H}$ as wave functions for some given particles.

17.3 Algebraic Canonical Quantization

We now turn to the algebraic approach to canonical quantization, which was invented in the fermionic case by Dirac [Dir82], and for which we also refer to Sect. 5.4 on Clifford algebras. This section can be seen as an algebraic explanation of the fact that the so-called "first canonical quantization" of classical Newtonian particles give the free field equations of field theory, which we will use later.

The quantization of the algebra of polynomial functions on the cotangent bundle $P = T^*X$ of a Riemannian manifold (X, g), that is the phase space for a particle $x : [0, 1] \to X$ moving in X, can be given by the algebra of differential operators $\mathcal{D}_X$. This algebra acts on $L^2(X)$ and has a filtration F whose graded algebra is the algebra

$$\mathrm{gr}^F \mathcal{D}_X = \mathcal{O}_P$$

of functions polynomial in the fiber on phase space. Moreover, the bracket on $\mathcal{D}_X$ induces the Poisson bracket on $\mathcal{O}_P$. The quantum Hamiltonian can thus be regarded as a differential operator $H \in \mathcal{D}_X$ acting on the state space $L^2(X)$.

In the case of a fermionic particle $x : \mathbb{R}^{0|1} \to X$, the phase space is the supermanifold $P = T^*[1]X$ with coordinate algebra $\mathcal{O}_P = \wedge^*\Theta_X$. The Clifford algebra $\mathrm{Cliff}(\Theta_X, g)$ gives a nice canonical quantization of $\mathcal{O}_P$, since it also has a filtration whose graded algebra is the algebra

$$\mathrm{gr}^F \mathrm{Cliff}(TX, g) = \wedge^*\Theta_X = \mathcal{O}_P$$

of functions on phase space. Moreover, the bracket on $\mathrm{Cliff}(\Theta_X, g)$ induces the Schouten-Nijenhuis super Poisson bracket on $\mathcal{O}_P$. The quantum Hamiltonian is then an element $H \in \mathrm{Cliff}(\Theta_X, g)$, and it acts naturally on the complex spinor representation S of the Clifford algebra, that is the state space of the quantum fermionic particle.

17.4 Weyl Quantization and Pseudo-Differential Calculus

Let X be a manifold equipped with a metric $g : TX \times_X TX \to \mathbb{R}_X$. One can think of (X, g) as a solution of Einstein's equations with g of Minkowski's signature. To perform the Weyl quantization, one needs a Fourier transform and a linear structure on spacetime. The easiest way to replace a curved spacetime by a flat one is to work in the tangent space of spacetime at a given point, which is the best linear approximation to spacetime. This procedure may be implemented locally by using the exponential map associated to the given metric. This remark is very important, since it shows that from a mathematical viewpoint,

it does not really makes sense to combine gravity and quantum mechanics: one of them lives on spacetime, and the other lives in the fiber of the tangent bundle of spacetime at a given point.

We denote by (x,q) the local coordinates on TX and (x,p) the local coordinates on T^*X. We denote by

$$\langle\cdot,\cdot\rangle : T^*X \times_X TX \to \mathbb{R}_X$$

the natural pairing given by

$$\big\langle (x,p),(x,q)\big\rangle = \big(x,p(q)\big).$$

One has a fiber-wise Fourier transform

$$\mathcal{F} : \mathcal{S}\big(T^*X\big) \to \mathcal{S}(TX)$$

on the Schwartz spaces (functions with compact support on the base X and rapid decay along the fibers equipped with their linear Lebesgue measure) given by $f \in \mathcal{S}(T^*X)$ by

$$\mathcal{F}(f)(x,q) = \int_{T_x^*X} e^{i\langle p,q\rangle} f(x,p)dp.$$

Given a symbol $H(x,p,q) \in \mathcal{S}(T^*X \times_X TX)$ (for quantization, this will typically be the Hamiltonian, that depends only on p et q and not on x) and a function $f(x,q)$ of $\mathcal{S}(TX)$, one can associate to it the product $H.(f \circ \pi) \in \mathcal{S}(T^*X \times_X TX)$ where $\pi : T^*X \times_X TX \to TX$ is the natural projection.

Definition 17.4.1 We denote by $L^2_{\infty,c}(TX)$ the space of functions on TX that are smooth with compact support on X and square integrable in the fibers of $TX \to X$. The operator $\hat{H}$ on $L^2_{\infty,c}(TX)$ associated to H is then given by the formula

$$(\hat{H}.f)(x,q) := \int_{T_xX}\int_{T_x^*X} e^{i\langle p,q-q_0\rangle} H(x,p,q).f(x,q_0)d\mu_g(q_0)d\mu_{g^*}(p).$$

This map

$$\begin{aligned} \mathcal{S}\big(T^*X \times_X TX\big) &\to \mathcal{L}\big(L^2_{\infty,c}(TX)\big) \\ H &\mapsto \hat{H} \end{aligned}$$

is called *Weyl quantization* or infinitesimal pseudo-differential calculus.

The (non-canonical) relation with ordinary pseudo-differential calculus is the following: if $U \subset \mathbb{R}^n$ is an open set and $h : T^*U \cong U \times (\mathbb{R}^n)^* \to \mathbb{R}$ is a symbol $h(x,p)$ on U, one associates to it the function

$$H : T^*U \times_U TU \cong U \times \big(\mathbb{R}^n\big)^* \times \mathbb{R}^n \to \mathbb{R}$$

by replacing x by q in $h(x,p)$, i.e., by the formula

$$H(x,p,q) := h(p,q).$$

This is obtained geometrically by identifying U with a subspace of the fiber of its tangent bundle $\mathbb{R}^n$, by the implicit function theorem (this space is a plane in $\mathbb{R}^n \times \mathbb{R}^n$). One gets an operator $\hat{H}$ on $L^2_{\infty,c}(TU)$ that is constant on U and thus gives an operator on the fibers $L^2(\mathbb{R}^n)$.

In practice, one often assumes $X = \mathbb{R}^n$, which gives $TX = \mathbb{R}^n \times \mathbb{R}^n$ and one can then consider the functions on TX that depend only on q in the coordinates (x, q) on TX. The choice of this decomposition is clearly not canonical, but Weyl quantization is often described as the map

$$\begin{aligned} \mathcal{S}(T^*X) &\to \mathcal{L}(L^2(X)) \\ H &\mapsto \hat{H}. \end{aligned}$$

With a convenient normalization of the Fourier transform, the Weyl quantization gives back the Heisenberg quantization described in Sect. 17.2. Indeed, a linear coordinate is a function $q_n : TX \to \mathbb{R}$ (a coordinate in the fiber of TX) that one can compose with the projection $\pi_T : T^*X \times_X TX \to TX$, which gives by quantization an operator

$$Q_n := \widehat{q_n \circ \pi_T}$$

in $\mathcal{C}^\infty(X, \mathcal{L}(L^2_{\infty,c}(TX)))$ that is simply the multiplication by the coordinate q_n in the fibers.

Given a vector field $\mathbf{v}_n \in \Theta_X = \Gamma(X, TX)$ (derivation in the x_n coordinate direction on X, and not on TX, this time: it is at this point that the identification $x_n = q_n$ between coordinates on X and on TX, used by physicists, is non-canonical), one can see this as the application $p_{n,h} : T^*X \to \mathbb{C}$, given by $p_{n,h}(x, p) = \frac{h}{i}\langle p, \mathbf{v}_n\rangle$, and composable with the projection $\pi_{T^*} : T^*X \times_X TX \to T^*X$, which gives by quantization an operator

$$P_n := \widehat{p_{n,h} \circ \pi_{T^*}}$$

in $\mathcal{C}^\infty(X, \mathcal{L}(\mathcal{S}(TX)))$ that is nothing else than $P_n = \frac{1}{2i\pi}\frac{\partial}{\partial q_n}$.

17.5 Quantization of the Harmonic Oscillator

The quantum harmonic oscillator is at the base of the free quantum field theory. Its resolution by the ladder method (creation and annihilation operators) is due to Dirac [Dir82].

The classical harmonic oscillator on $X = \mathbb{R}$ is given by the Hamiltonian system whose trajectories are maps from $\mathbb{R}$ into the Poisson manifold $P = T^*X \cong \mathbb{R} \times \mathbb{R}$ (with coordinates (q, p) called position and impulsion) and whose Hamiltonian is the function $H : P \to \mathbb{R}$ given by

$$H(q, p) = \frac{1}{2m}\left(p^2 + m^2\omega^2 q^2\right).$$

Recall that the classical solution of the Hamilton equations are then given by

$$\begin{cases} \frac{\partial q}{\partial t} = \frac{\partial H}{\partial p} = \frac{p}{m} \\ \frac{\partial p}{\partial t} = -\frac{\partial H}{\partial q} = m\omega^2 q, \end{cases}$$

which gives Newton's equations of motion

$$\frac{\partial^2 q}{\partial t^2} = \omega^2 q.$$

Canonical quantization of this system is obtained by working with the Hilbert space $\mathcal{H} = L^2(X)$ and replacing the position and impulsion by the operators $Q = m_q$ of multiplication by the function $q : X \to \mathbb{R}$ and $P = -i\hbar\partial_q$ of derivation with respect to q.

One then computes the spectrum of the Hamiltonian (whose elements are possible energy levels for the system)

$$H(Q, P) = \frac{1}{2m}(P^2 + m^2\omega^2 Q^2),$$

which also means that one solves the eigenvalue equation

$$H\psi = E\psi.$$

To find the solutions of this eigenvalue equation, one uses, following Dirac's ladder method, two auxiliary operators

$$a = \sqrt{\frac{m\omega}{2\hbar}}\left(Q + \frac{i}{m\omega}P\right)$$
$$a^\dagger = \sqrt{\frac{m\omega}{2\hbar}}\left(Q - \frac{i}{m\omega}P\right).$$

This gives the relations

$$x = \sqrt{\frac{\hbar}{2m\omega}}(a^\dagger + a)$$
$$p = i\sqrt{\frac{\hbar m\omega}{2}}(a^\dagger - a).$$

Theorem 17.5.1 *One can solve the eigenvalue equation for the operator H by using the operators a and $a^\dagger$. More precisely*:

1. *One has $H = \hbar\omega(a^\dagger a + 1/2)$.*
2. *The spectrum of $a^\dagger a$ is the set $\mathbb{N}$ of natural numbers and the spectrum of H is $\hbar\omega(\mathbb{N} + 1/2)$.*

3. *If ψ_0 is nonzero and such that $a\psi_0 = 0$, we define $\psi_n = (a^\dagger)^n \psi_0$ and $\psi_0 = |\emptyset\rangle$ is called the empty state. One then has*

$$H\psi_n = \hbar\omega(n + 1/2)\psi_n.$$

4. *One has the commutation relations*

$$\left[a, a^\dagger\right] = 1$$
$$\left[a^\dagger, H\right] = \hbar\omega a^\dagger$$
$$[a, H] = -\hbar\omega a.$$

Note that the empty state ψ_0 has as energy (eigenvalue) $\frac{\hbar\omega}{2}$ and is given by the function (solution of the differential equation in q of first order $a\psi_0 = 0$)

$$\psi_0(x) = \left(\frac{m\omega}{\pi\hbar}\right)^{\frac{1}{4}} e^{-\frac{m\omega}{2\hbar}x^2}.$$

17.6 Canonical Quantization of the Free Scalar Field

We refer to Folland's book [Fol08] for a very complete treatment of this issue.

The free scalar field of mass m is a function on Minkowski space $X = \mathbb{R}^{3,1}$ that is a solution of the Klein-Gordon equation

$$\left(\Box + \frac{m^2c^2}{\hbar^2}\right)\varphi = \left(\frac{1}{c^2}\partial_t^2 - \partial_x^2 + \frac{m^2c^2}{\hbar^2}\right)\varphi = 0.$$

The partial Fourier transform on the variable x of $\varphi(t,x)$, denoted by $(\mathcal{F}_x\varphi)(t,\xi)$, satisfies, at each point $\xi \in \mathbb{R}^3$, the differential equation of the classical harmonic oscillator

$$\left(\frac{1}{c^2}\partial_t^2 + \xi^2 + \frac{m^2c^2}{\hbar^2}\right)(\mathcal{F}_x\varphi)(t,\xi) = 0$$

whose classical Hamiltonian is the function $h(q,p) = p^2 + \omega^2 q^2$ with, for ξ fixed, $\omega^2 = \xi^2 + \frac{m^2c^4}{\hbar^2}$, $q = (\mathcal{F}_x\varphi)(t,\xi)$ and $p = \frac{\partial q}{\partial t}$.

The quantification of the free scalar field can essentially be done by replacing the classical harmonic oscillator by a quantum harmonic oscillator with Hamiltonian $H_\xi(Q,P) = P^2 + \omega^2 Q^2$ acting on $L^2(\mathbb{R}^3)$. We remark that this Hamiltonian depends on $\xi \in \mathbb{R}^3$, that is the Fourier dual variable to the position variable x in space.

There exists a nice way to formalize the above construction more canonically. One works in relativistic units, i.e., $c = \hbar = 1$. Let φ be a tempered distribution on $\mathbb{R}^4$ that solves the Klein-Gordon equation

$$(\Box + m^2)\varphi = 0.$$

Its Lorentz-covariant Fourier transform is a tempered distribution that satisfies $(-p^2+m^2)\hat{\varphi}=0$ so that it is supported on the twofold hyperboloid

$$X_m = \{p \in \mathbb{R}^{3,1} | p^2 = m^2\}$$

called the mass hyperboloid. One can decompose X_m into two components X_m^+ and X_m^- corresponding to the sign of the first coordinate p_0 (Fourier dual to time). One puts on X_m^+ the normalized measure

$$d\lambda = \frac{d^3\mathbf{p}}{(2\pi)^3\omega_{\mathbf{p}}}$$

with $\omega_{\mathbf{p}} = p_0 = \sqrt{|\mathbf{p}|^2+m^2}$. One then considers the Hilbert space

$$\mathcal{H} = L^2(X_m^+, \lambda).$$

Definition 17.6.1 The *total* (*resp. bosonic, resp. fermionic*) *Fock space* on $\mathcal{H}$, denoted by $\mathcal{F}(\mathcal{H})$ (resp. $\mathcal{F}_s(\mathcal{H})$, $\mathcal{F}_a(\mathcal{H})$), is the completion of the tensor algebra $\mathcal{F}^0(\mathcal{H})$ (resp. symmetric algebra $\mathcal{F}_s^0(\mathcal{H}) := \mathrm{Sym}_{\mathbb{C}}(\mathcal{H})$, resp. exterior algebra $\mathcal{F}_a^0(\mathcal{H}) := \wedge_{\mathbb{C}}\mathcal{H}$). The subspaces given by the algebraic direct sums $\mathcal{F}^0$, $\mathcal{F}_s^0$ and $\mathcal{F}_a^0$ are called *finite particle subspaces.*

One defines the number operator N on $\mathcal{F}^0(\mathcal{H})$ by

$$N = kI \quad \text{on } \otimes^k \mathcal{H}.$$

For $v \in \mathcal{H}$, one defines the operators B and $B^\dagger$ on $\mathcal{F}^0(\mathcal{H})$ by

$$B(v)(u_1 \otimes \cdots \otimes u_k) = \langle v|u_1\rangle . u_2 \otimes u_k,$$

$$B(v)^\dagger (u_1 \otimes \cdots \otimes u_k) = v \otimes u_1 \otimes \cdots \otimes u_k,$$

and one shows that these operators are adjoint for $v \in \mathcal{H}$ fixed. The operator $B(v)^\dagger$ does not preserve the symmetric subspace $\mathcal{F}_s^0(\mathcal{H})$ but $B(v)$ does preserve it. One also defines $A(v) = B(v)\sqrt{N}$ and denotes by $A^\dagger(v)$ the adjoint of the operator $A(v)$ on the bosonic Fock space $\mathcal{F}_s^0(\mathcal{H})$. One can also view it as the operator

$$A(v)^\dagger = P_s B(v)^\dagger \sqrt{N+1} P_s$$

with $P_s : \mathcal{F}^0(\mathcal{H}) \to \mathcal{F}^0(\mathcal{H})$ the projection on the bosonic subspace $\mathcal{F}_s^0(\mathcal{H})$.

One has the following commutation relations of operators on the bosonic Fock space: $[A(v), A(w)^\dagger] = \langle v|w\rangle I$ and $[A(v), A(w)] = [A(v)^\dagger, A(w)^\dagger] = 0$.

One has a natural map $R : \mathcal{S}(\mathbb{R}^4) \to \mathcal{H}$ defined by

$$Rf = \hat{f}_{|X_m^+}$$

with $\hat{f}(p) = \int e^{ip_\mu x^\mu} f(x) d^4x$ the Lorentz-covariant Fourier transform.

Definition 17.6.2 The *neutral free quantum scalar field of mass* m is the distribution on $\mathbb{R}^4$ valued in operators on the bosonic Fock finite particle space $\mathcal{F}_s^0(\mathcal{H})$ given by

$$\Phi(f) = \frac{1}{\sqrt{2}}\left[A(Rf) + A(Rf)^\dagger\right].$$

Note that Φ is a distribution solution of the Klein-Gordon equation, i.e., $\Phi((\Box + m^2)f) = 0$ for every $f \in \mathcal{S}(\mathbb{R}^4)$ because $(-p^2 + m^2)\hat{f} = 0$ on X_m^+.

For a charged scalar field (i.e. with complex values), it is necessary to use the two Hilbert spaces $\mathcal{H}_+ = L^2(X_m^+)$ and $\mathcal{H}_- = L^2(X_m^-)$ and the anti-unitary operator $C : \mathcal{H}_\pm \to \mathcal{H}_\mp$ given by $Cu(p) = u(-p)^*$. One defines the creation and annihilation operators for $v \in \mathcal{H}_-$ on $\mathcal{H}_+$ by

$$B(v) = A(Cv) \quad \text{and} \quad B(v)^\dagger = A(Cv)^\dagger.$$

Now here comes the trick. One considers as before the Fock space $\mathcal{F}_s^0(\mathcal{H}_+)$ and view it as defined over $\mathbb{R}$ by considering $\mathrm{Res}_{\mathbb{C}/\mathbb{R}}\mathcal{F}_s^0(\mathcal{H}_+)$. The operators $B(v)$ and $B(v)^\dagger$ are $\mathbb{R}$-linear because the operator $C : \mathcal{H}_\pm \to \mathcal{H}_\mp$ is, and that's why one takes the scalar restriction.

One then extends the scalars from $\mathbb{R}$ to $\mathbb{C}$ by applying $\cdot \otimes_{\mathbb{R}} \mathbb{C}$ to get a space that we will denote by $\mathcal{F}_s^0$, and that can be decomposed as a tensor product of two Fock spaces isomorphic to $\mathcal{F}_s^0(\mathcal{H}_+)$, and exchanged by an automorphism induced by complex conjugation. One then defines, for $f \in \mathcal{S}(\mathbb{R}^4)$, two maps $R_\pm : \mathcal{S}(\mathbb{R}^4) \to \mathcal{H}_\pm$ by

$$R_\pm f = \hat{f}_{|X_m^\pm}.$$

Definition 17.6.3 The *charged free scalar field of mass* m is the data of the two distributions on $\mathbb{R}^4$ with values in operators on the Fock space $\mathcal{F}_s^0$ defined by

$$\Phi(f) = \frac{1}{\sqrt{2}}\left[A(R_+f) + B(R_-f)^\dagger\right],$$

$$\Phi(f)^\dagger = \frac{1}{\sqrt{2}}\left[A(R_+f)^\dagger + B(R_-f)\right].$$

To quantize non-free fields, one uses perturbative methods by expressing the Hamiltonian $H = F + \lambda V$ as a sum of a free (quadratic) part, that can be quantized freely as before and a potential λV, with λ a formal variable. One then uses the free state space (Fock space) of the previous section and the action of some operators constructed from the free field using the potential. This method is purely perturbative (in that the parameter λ is purely formal). More on this can be found in Folland's book [Fol08].

We prefer to use covariant and functional integral methods, because:

- they are closer to geometry and allow a uniform treatment of fermionic and bosonic variables, starting from classical Lagrangian field theory; and
- they allow a very general treatment of gauge symmetries.

References

[BD04] Beilinson, A., Drinfeld, V.: Chiral Algebras. American Mathematical Society Colloquium Publications, vol. 51, p. 375. Am. Math. Soc., Providence (2004). ISBN 0-8218-3528-9

[CG10] Costello, K., Gwilliam, O.: Factorization algebras in perturbative quantum field theory (2010, preprint)

[Cos11] Costello, K.: Renormalization and Effective Field Theory. Mathematical Surveys and Monographs, vol. 170, p. 251. Am. Math. Soc., Providence (2011). ISBN 978-0-8218-5288-0

[Dir82] Dirac, P.A.M.: The Principles of Quantum Mechanics (International Series of Monographs on Physics). Oxford University Press, London (1982). ISBN 0198520115. http://www.amazon.ca/exec/obidos/redirect?tag=citeulike09-20&path=ASIN/0198520115

[Fol08] Folland, G.B.: Quantum Field Theory: A Tourist Guide for Mathematicians. Mathematical Surveys and Monographs, vol. 149, p. 325. Am. Math. Soc., Providence (2008). ISBN 978-0-8218-4705-3

[Kon03] Kontsevich, M.: Deformation quantization of Poisson manifolds. Lett. Math. Phys. **66**(3), 157–216 (2003). doi:10.1023/B:MATH.0000027508.00421.bf

[vN96] von Neumann, J.: Mathematical Foundations of Quantum Mechanics. Princeton Landmarks in Mathematics, p. 445. Princeton University Press, Princeton (1996). Translated from the German and with a preface by Robert T. Beyer, Twelfth printing, Princeton Paperbacks. ISBN 0-691-02893-1

Chapter 18
Mathematical Difficulties of Perturbative Functional Integrals

In this chapter, we authorize ourselves to use the physicists' notations

and some of their problematic computations,

describing where the mathematical problems are, so that one has a good idea of what needs to be done to solve them.

The author's viewpoint is that functional integrals (and particularly the nonperturbative renormalization group and Dyson-Schwinger equations, to be described in Chap. 20) give the description of interacting quantum fields that is presently the easiest point of entry for mathematicians into the modern applications of quantum field theory methods in mathematics.

The road to such a theory is tricky but paved by physicists, at least in each and all of the examples of physical relevance. It thus gives an interesting area of research for mathematicians. It seems impossible to explain quantum gauge fields to a mathematician without using this clumsy road. We will see later in the renormalization Chaps. 19, 20, 21 and 22 that one can eventually arrive at proper mathematical definitions in some interesting examples. We refer to Zeidler's books [Zei06, Zei09] and [Zei11] for a complete mathematical treatment of the quantum field theoretical methods presented in this chapter, and for an extended presentation of the literature. One may also use Connes-Marcolli's book [CM08] and Cartier and DeWitt-Morette's book [CDM06] for a smooth mathematical presentation of the perturbative expansions.

18.1 Perturbative Expansions in Finite Dimension

The main idea of perturbative methods in quantum field theory is to use explicit computations of Gaussian integrals in finite dimensions as definitions of the analogous Gaussian integrals in field theory. We thus review perturbative computations of Gaussian integrals in finite dimensions.

F. Paugam, *Towards the Mathematics of Quantum Field Theory*, Ergebnisse der Mathematik und ihrer Grenzgebiete. 3. Folge / A Series of Modern Surveys in Mathematics 59, DOI 10.1007/978-3-319-04564-1_18,

18.1.1 Gaussian Integrals with Source

To understand the notations used by physicists, one must understand their normalizations. One can however use other normalizations than those commonly used in physics to escape the infiniteness problem of normalization of Gaussian integrals in infinite dimensions. This is explained in detail in Cartier and DeWitt-Morette's book [CDM06] and expanded to the case of fields in J. LaChapelle's thesis [LaC06].

Here we essentially follow the excellent presentation of Chap. 1.2.2 of Zinn-Justin's book [ZJ93]. Let S be a quadratic form on $\mathbb{R}^d$ and $\Delta = S^{-1}$ its inverse (which we will later call the propagator or the kernel), that induces a quadratic form on $(\mathbb{R}^d)^\vee$. For $b \in (\mathbb{R}^d)^\vee$ a linear form, one defines

$$Z(S,b) = \int_{\mathbb{R}^d} \exp\left(-\frac{1}{2}S(x) + b \cdot x\right) dx.$$

The change of variable $x = y + \Delta \cdot b$ gives that this integral is

$$Z(S,b) = \exp\left(\frac{1}{2}\Delta(b)\right) \int_{\mathbb{R}^d} \exp\left(-\frac{1}{2}S(y)\right) = \exp\left(\frac{1}{2}\Delta(b)\right) Z(S,0).$$

The value of $Z(S,0)$ is obtained by diagonalizing from the dimension 1 value and gives

$$Z(S,0) = \frac{(2\pi)^{d/2}}{\sqrt{\det(S)}}.$$

Observe that the limit as d goes to infinity of $Z(S,b)$ makes no sense but the limit of the quotient $\frac{Z(S,b)}{Z(S,0)}$ can exist, because one has the formula

$$\frac{Z(S,b)}{Z(S,0)} = \exp\left(\frac{1}{2}\Delta(b)\right).$$

If F is a function on $\mathbb{R}^n$, we define

$$Z_F(S,b) := \int_{\mathbb{R}^d} F(x) d\mu_{S,b}(x)$$

where

$$d\mu_{S,b}(x) := \frac{\exp(-\frac{1}{2}S(x) + b \cdot x)}{Z(S,0)} d\mu(x)$$

is the Laplace transform of the Gaussian measure and the Gaussian mean value of F is given by

$$\langle F \rangle := Z_F(S,b)|_{b=0} = \frac{\int_{\mathbb{R}^n} F(x) e^{-\frac{1}{2}S(x)} dx}{Z(S,0)}.$$

Note in particular that

$$\langle \exp(b \cdot x) \rangle = \frac{Z(S,b)}{Z(S,0)}.$$

If we define $b \cdot x = \sum b_i \cdot x_i$, we can find the mean values of monomials by expanding $\exp(b \cdot x)$ as a power series:

$$\langle x_{k_1} \cdots x_{k_r} \rangle = \left[\frac{\partial}{\partial b_{k_1}} \cdots \frac{\partial}{\partial b_{k_r}} \frac{Z(S,b)}{Z(S,0)} \right] \Bigg|_{b=0} = \left[\frac{\partial}{\partial b_{k_1}} \cdots \frac{\partial}{\partial b_{k_r}} \exp\left(\frac{1}{2} \Delta(b) \right) \right] \Bigg|_{b=0}.$$

Each partial derivative $\frac{\partial}{\partial b_{k_i}}$ can either take a linear factor b_i out, or differentiate one that is already present. The result is zero if there are the same number of these two operations, so that

$$\langle x_{k_1} \cdots x_{k_r} \rangle = \sum_{\text{index pairings}} w_{i_{j_1}, i_{j_2}} \cdots w_{i_{j_{l-1}}, i_{j_l}}$$

with $w_{i,j}$ the coefficients of the matrix w. This formula is called Wick's formula.

More generally, if F is a power series in the x_i's, one finds the identity

$$\langle F \rangle = \left[F\left(\frac{\partial}{\partial b} \right) \exp\left(\frac{1}{2} \Delta(b) \right) \right] \Bigg|_{b=0}.$$

This is this mean value function that physicists generalize to infinite dimensional settings. The auxiliary linear form b is often denoted by J and called the external source.

There is even another way to look at this mean value, which allows us to replace the linear pairing $b \cdot x$ by an arbitrary function of x. This type of viewpoint may be useful when working on a non-linear space X, instead of $\mathbb{R}^d$. Consider the partition function as a (partially defined) function

$$Z : \mathcal{O}(\mathbb{R}^d) \to \mathbb{R}$$

on functions on $\mathbb{R}^d$ given for $S \in \mathcal{O}(\mathbb{R}^d)$ by the formula

$$Z(S) = \int_{\mathbb{R}^d} \exp(S(x)) dx.$$

This formula has a meaning even if the space parametrized by the variable x is nonlinear. The relation with the usual partition function Z_{lin} for $x \in \mathbb{R}^d$ is given by

$$Z_{lin}(b) = Z(S + b).$$

One may then view the mean value of $F \in \mathcal{O}(\mathbb{R}^d)$ with respect to the quadratic form S as the derivative

$$\langle F \rangle_S = \frac{1}{Z(S)} \frac{\delta Z}{\delta F}(S) := \frac{1}{Z(S)} \frac{d}{d\epsilon}\Bigg|_{\epsilon=0} Z(S + \epsilon F).$$

This means that the observable F may be seen as a tangent vector to the space of functions at the point given by S. This viewpoint is at the basis of Costello and Gwilliam's constructions [CG10].

18.1.2 The Defining Equation of the Fourier-Laplace Transform

Another take at the computation of Gaussian integrals is given by the use of their defining equation. The analog in quantum field theory of this equation is called the Dyson-Schwinger equation and is very practical for the definition of fermionic or graded functional integrals (whose variables are, for example, sections of an odd fiber bundle of spinors on spacetime or sections of a graded bundle obtained by applying the BV formalism to an ordinary action functional).

Let F be a function on $\mathbb{R}^n$ with rapid decay at infinity. One then has, for all $i = 1, \ldots, n$, the fundamental identity of integration by parts

$$\int_{\mathbb{R}^n} \frac{\partial}{\partial x_i} F(x) dx = 0.$$

Applying this to the definition of $Z(S, b)$, we get

$$Z_i(S, b) := \int_{\mathbb{R}^d} \frac{\partial}{\partial x_i} \exp\left(-\frac{1}{2}S(x) + b \cdot x\right) dx = 0,$$

and

$$Z_i(S, b) = \int_{\mathbb{R}^d} \frac{\partial}{\partial x_i}\left(-\frac{1}{2}S(x) + b \cdot x\right) \exp\left(-\frac{1}{2}S(x) + b \cdot x\right) dx.$$

The basic property of the Laplace transform is that

$$\int_{\mathbb{R}^d} F(x)G(x)\exp(b \cdot x)dx = F\left[\frac{\partial}{\partial b}\right] \int_{\mathbb{R}^d} G(x)\exp(b \cdot x)dx.$$

Applying this to the above computation, we find

$$Z_i(S, b) = \frac{\partial}{\partial x_i}\left(-\frac{1}{2}S(x) + b \cdot x\right)\left[\frac{\partial}{\partial b}\right] Z(S, b),$$

where the bracket means replacing a function (symbol) of x by the corresponding differential operator in the universal derivation $\frac{\partial}{\partial b}$. Finally, the defining equations for the Gaussian with source $Z(q, b)$ are given, for $i = 1, \ldots, d$, by

$$\left(-\frac{1}{2}\frac{\partial S}{\partial x_i} + b_i\right)\left[\frac{\partial}{\partial b}\right] Z(S, b) = 0.$$

Now if F is a function on $\mathbb{R}^n$, by applying the fundamental identity of integration by parts and the basic properties of the Laplace transform, we find the defining equations for the Gaussian mean value with source

$$Z_F(S,b) := \int_{\mathbb{R}^d} F(x) \frac{\exp(-\frac{1}{2}S(x) + b \cdot x)}{Z(S,0)} d\mu(x).$$

These are given, for $i = 1, \ldots, d$, and if one already has found $Z(S,b)$, by

$$\frac{\partial F}{\partial x_i}\left[\frac{\partial}{\partial b}\right]\frac{Z(S,b)}{Z(S,0)} + \left(-\frac{1}{2}\frac{\partial S}{\partial x_i} + b_i\right)\left[\frac{\partial}{\partial b}\right] Z_F(S,b) = 0.$$

18.1.3 Perturbative Gaussians

We refer to the introductory article [Phi01]. The aim of Feynman diagrams is to encode the combinatorics of Wick's formula for the value of the mean values

$$\langle F \rangle = \left[F\left(\frac{\partial}{\partial b}\right) \exp\left(\frac{1}{2}\Delta(b)\right)\right]\Big|_{b=0}$$

of functions for a Gaussian measure.

To compute these mean values in the case where F is a monomial, one expands the exponential $\exp(\frac{1}{2}\Delta(b))$ as a formal power series:

$$\exp\left(\frac{1}{2}\Delta(b)\right) = \sum_n \frac{\Delta(b)^n}{2^n n!} = \sum_n \frac{1}{2^n n!}\left(\sum_{i,j=1}^{d} \Delta_{i,j} b_i b_j\right)^n.$$

The terms of this series are homogeneous in the b_i of degree $2n$. If one differentiates such a monomial k times, and evaluates it at 0, one gets 0, except if $k = 2n$. It is thus enough to compute the value of $2n$ differentiations on a monomial of the form $(\sum_{i,j=1}^{d} \Delta_{i,j} b_i b_j)^n$.

The differentiation most frequently used in this computation is

$$\frac{\partial}{\partial b_k}\left(\frac{1}{2}\sum_{i,j=1}^{d} \Delta_{i,j} b_i b_j\right) = \sum_{i=1}^{d} \Delta_{i,j} b_i,$$

where one uses the symmetry of the matrix of the quadratic form Δ, the inverse of the quadratic form S. We put $\partial_i = \frac{\partial}{\partial b_i}$.

Let us perform this computation on some examples:

1. The term of order $n = 1$ gives

$$\langle x_1 x_2 \rangle = \partial_2 \partial_1 \left(\frac{1}{2}\sum_{i,j=1}^{d} \Delta_{i,j} b_i b_j\right) = \partial_2\left(\sum_j \Delta_{1,j} b_j\right) = \Delta_{1,2}$$

by using the symmetry of the matrix Δ, and

$$\langle x_1 x_1 \rangle = \partial_1 \partial_1 \left(\frac{1}{2} \sum_{i,j=1}^{d} \Delta_{i,j} b_i b_j \right) = \Delta_{1,1}.$$

2. The term of order $n = 2$ gives

$$\begin{aligned}
&\langle x_1 x_2 x_3 x_4 \rangle \\
&= \partial_4 \partial_3 \partial_2 \partial_1 \left(\frac{1}{2^2 2!} \sum_{i,j=1}^{d} \Delta_{i,j} b_i b_j \right)^2 \\
&= \partial_4 \partial_3 \partial_2 \frac{1}{2} \left(\sum \Delta_{i,j} b_i b_j \right) \left(\sum \Delta_{1,j} b_j \right) \\
&= \partial_4 \partial_3 \left[\left(\sum \Delta_{2,j} b_j \right) \left(\sum \Delta_{1,j} b_j \right) + \frac{1}{2} \left(\sum \Delta_{i,j} b_i b_j \right) \Delta_{1,2} \right] \\
&= \partial_4 \left[\Delta_{2,3} \left(\sum \Delta_{1,j} b_j \right) + \Delta_{1,3} \left(\sum \Delta_{2,j} b_j \right) + \Delta_{1,2} \left(\sum \Delta_{3,j} b_j \right) \right] \\
&= \Delta_{2,3} \Delta_{1,4} + \Delta_{2,4} \Delta_{1,3} + \Delta_{3,4} \Delta_{1,2}.
\end{aligned}$$

Similarly, one finds

$$\langle x_1 x_1 x_3 x_4 \rangle = \partial_4 \partial_3 \partial_1 \partial_1 \left(\frac{1}{2^2 2!} \sum_{i,j=1}^{d} \Delta_{i,j} b_i b_j \right)^2 = 2\Delta_{1,4} \Delta_{1,3} + \Delta_{3,4} \Delta_{1,1},$$

$$\langle x_1 x_1 x_1 x_4 \rangle = \partial_4 \partial_1 \partial_1 \partial_1 \left(\frac{1}{2^2 2!} \sum_{i,j=1}^{d} \Delta_{i,j} b_i b_j \right)^2 = 3\Delta_{1,4} \Delta_{1,1},$$

$$\langle x_1 x_1 x_4 x_4 \rangle = \partial_4 \partial_4 \partial_1 \partial_1 \left(\frac{1}{2^2 2!} \sum_{i,j=1}^{d} \Delta_{i,j} b_i b_j \right)^2 = 2\Delta_{1,4} \Delta_{1,4} + \Delta_{4,4} \Delta_{1,1},$$

$$\langle x_1 x_1 x_1 x_1 \rangle = \partial_1 \partial_1 \partial_1 \partial_1 \left(\frac{1}{2^2 2!} \sum_{i,j=1}^{d} \Delta_{i,j} b_i b_j \right)^2 = 3\Delta_{1,1} \Delta_{1,1}.$$

The combinatorics of these computations being quite complex, it is practical to represent each of the products that appear above by a graph, whose vertices correspond to indices of the coordinates x_i that appear in the functions whose mean values are being computed and each $\Delta_{i,j}$ becomes an edge between the i and the j vertex.

One thus obtains the diagrams:

$\langle x_1x_2x_3x_4\rangle =$ 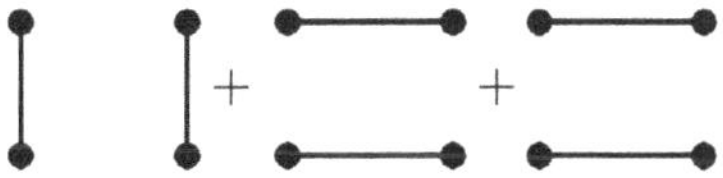

$\langle x_1x_1x_3x_4\rangle = 2.$ 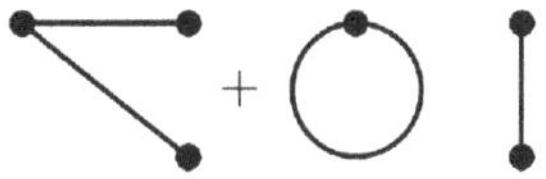

$\langle x_1x_1x_1x_4\rangle = 3.$ 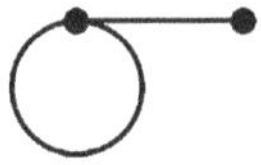

$\langle x_1x_1x_4x_4\rangle = 2.$ 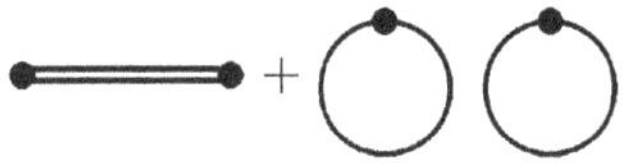

$\langle x_1x_1x_1x_1\rangle = 3.$ 

18.1.4 Finite Dimensional Feynman Rules

We retain the notations of the preceding sections and assume given a polynomial V in the coordinates $x_1, \dots, x_d$. The integrals whose generalizations are interesting for physics are of the type

$$Z(S,V) := \int_{\mathbb{R}^d} \exp\left(-\frac{1}{2}S(x) + \hbar V(x)\right)dx,$$

which one can rewrite in the form

$$Z(S,V) = \int_{\mathbb{R}^d} \exp\left(-\frac{1}{2}S(x)\right)\left[\sum_n \frac{1}{n!}(\hbar V(x))^n\right]dx.$$

One can evaluate these integrals as before by using the formal formula

$$Z(S,V) = Z(S,0)\exp\left(\hbar V\left(\frac{\partial}{\partial b}\right)\right)\exp\left(\frac{1}{2}\Delta(b)\right)\bigg|_{b=0},$$

whose power series expansion can be computed by using the preceding computations of the n-points function. The combinatorics of these computations is encoded in diagrams whose vertices correspond to the coefficients $a_{i_1,\dots,i_d}$ of the monomials of the polynomial $V = \sum_{i\in\mathbb{N}^d} a_i x^i$ and whose edges correspond to the coefficients $\Delta_{i,j}$ of the quadratic forms that appears in the Wick formula for these monomials.

For example, one can consider a polynomial of the form

$$V(x) = \sum_{i,j,k} v_{ijk}x_ix_jx_k$$

(a setting similar to the famous "φ^3" toy model) and compute

$$\begin{aligned} Z(S,V) &= \int_{\mathbb{R}^d} \exp\left(-\frac{1}{2}S(x) + \hbar \sum_{i,j,k} v_{ijk}x_i x_j x_k\right)dx \\ &= Z(S,0)\exp\left(\hbar \sum_{i,j,k} v_{ijk}\partial_i\partial_j\partial_k\right)\exp\left(\frac{1}{2}\Delta(b)\right)\Big|_{b=0}. \end{aligned}$$

Let us compute the terms of degree 2 in $\hbar$. These terms have 6 derivatives and their sum is

$$\sum_{i,j,k}\sum_{i',j',k'} v_{ijk}v_{i'j'k'}\partial_i\partial_j\partial_k\partial_{i'}\partial_{j'}\partial_{k'}\exp\left(\frac{1}{2}\Delta(b)\right)\Big|_{b=0}.$$

By Wick's theorem, one can rewrite this sum as

$$\sum_{i,j,k}\sum_{i',j',k'} \Delta_{i_1,i_2}\Delta_{i_3,i_4}\Delta_{i_5,i_6}v_{ijk}v_{i'j'k'}$$

where the internal sum is taken over all pairings $\{(i_1,i_2),(i_3,i_4),(i_5,i_6)\}$ of $\{i,j,k,i',j',k'\}$.

One associates to these pairings the graphs whose trivalent vertices correspond to the factors v_{ijk} and whose edges correspond to the $\Delta_{i,j}$. In this case, one gets exactly two distinct graphs called respectively the dumbbell and the theta:

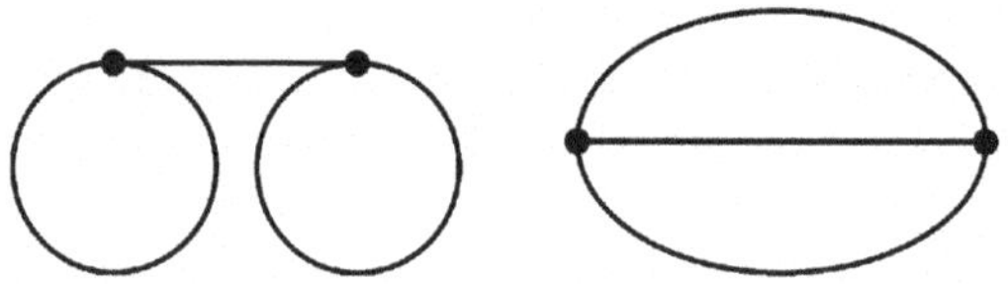

One then has to number the edges at each vertex to encode the corresponding monomial.

18.2 Feynman's Interpretation of Functional Integrals

We recall here the main idea of Feynman's sum over histories approach to quantum physics. Let M be a spacetime with its Lorentz metric g, let $\pi : C = M \times F \to M$ be a trivial linear bundle with fiber F, and $S : \Gamma(M,C) \to \mathbb{R}$ be a local action functional. Let $\iota_i : \Sigma_i \hookrightarrow M$, $i = 0,1$ be two space-like hyper-surfaces (that one can think of as the space at times t_0 and t_1). If we suppose that the action functional is of order 1 (first order derivatives only), one can specify a space of histories by choosing two sections $\varphi_i \in \Gamma(\Sigma_i, \iota_i^*C)$, $i = 0,1$ and saying that

$$H = H_{(\Sigma_0,\varphi_0),(\Sigma_1,\varphi_1)} := \left\{\varphi \in \Gamma(M,C),\ \varphi_{|\Sigma_0} = \varphi_0,\ \varphi_{|\Sigma_1} = \varphi_1\right\}.$$

We now suppose that $S(\varphi) = S_{free}(\varphi) + S_{int}(\varphi)$ is a sum of a free/quadratic part in the field, whose equations of motion give a linear partial differential equation

$$D\varphi = 0,$$

and of an interaction part that depends only on the values of the field and not of its derivatives.

One can think of the value $\varphi_0 : \Sigma_0 \to \iota_0^* C$ as specifying the value of the classical free field $\tilde{\varphi}_0$ on the spacetime region before the time corresponding to Σ_0, which is the starting time for the collision experiment. The value of $\tilde{\varphi}_0$ is measured experimentally in the preparation of the machine and its data is essentially equivalent to the datum of the pair (S_{free}, φ_0) composed of the free Lagrangian and the initial value for the experiment. One gets $\tilde{\varphi}_0$ from (S_{free}, φ_0) by solving the backward in time Cauchy problem for the hyperbolic linear partial differential equation of free motion

$$D\tilde{\varphi}_0 = 0, \qquad \varphi_{|\Sigma_0} = \varphi_0$$

with initial condition φ_0.

Similarly, the value $\varphi_1 : \Sigma_1 \to \iota_1^* C$ specifies the value of the classical free field $\tilde{\varphi}_1$ after the final time Σ_1 of the collision experiment, by solving the forward in time Cauchy problem for the free equation of motion. It is also measured after the experiment by the apparatus.

To sum up, what you put in the machine is a free field and what you get after the experiment is also a free field.

The quantum process essentially takes place between Σ_0 and Σ_1, and is composed of collisions and various particle creations and annihilations. The information that one can gather about the probabilistic property of this quantum process with respect to a given functional $A : H \to \mathbb{R}$, called an observable, is contained in Feynman's sum over history, also called the functional integral:

$$\langle \varphi_1 | A | \varphi_0 \rangle = \frac{\int_H A(\varphi) e^{\frac{i}{\hbar} S(\varphi)} [d\varphi]}{\int_H e^{\frac{i}{\hbar} S(\varphi)} [d\varphi]}.$$

One can understand this formal notation by taking a bounded box $B \subset M$ whose time-like boundaries are contained in the spacetime domain between Σ_0 and Σ_1, and that represent a room (or an apparatus' box) in which one performs the experiment. We now suppose that $M = \mathbb{R}^{3,1}$, B is a 0-centered square and $\Lambda = \mathbb{Z}^{3,1} \subset \mathbb{R}^{3,1}$ is the standard lattice. An observable is simple enough if it can be restricted to the space

$$\mathrm{Hom}(\Lambda \cap B, F) = F^{\Lambda \cap B} \subset \Gamma(M, C)$$

of functions on the finite set of points in $\Lambda \cap B$ with values in the fiber F of the bundle $\pi : C \to M$. For simple enough observables, one can make sense of the above formula by using ordinary integration theory on the finite dimensional linear space $F^{\Lambda \cap B}$ with its Lebesgue measure.

The main problem of renormalization theory is that if one makes the size L of the box B tend to infinity and the step ℓ of the lattice Λ tend to 0, one does not get a well defined limit. The task is to modify this limiting process to get well defined values for the above formal expression.

18.3 Functional Derivatives

In this book, we used well defined functional derivatives on the space of histories H of a field theory $(\pi : C \to M, S : \Gamma(M, C) \to \mathbb{R})$ along a given vector field $\mathbf{v} : H \to TH$. These kinds of functional derivatives are necessary to understand computations with fermionic variables.

We will now explain the analysts' approach to functional derivatives, even if it has some mathematical drawbacks. This will be necessary to properly understand the mathematical problems that appear in the manipulation of functional integrals for scalar fields (i.e., real valued functions).

Definition 18.3.1 Let $\mathcal{H}$ and $\mathcal{G}$ be two complete locally convex topological vector spaces, whose topology are given by a families of seminorms. If $F : \mathcal{H} \to \mathcal{G}$ is a functional, $f \in \mathcal{H}$ and $h \in \mathcal{H}$, one defines its *Gâteaux derivative*, if it exists, by

$$\frac{\delta F}{\delta \mathbf{h}}(f) := \lim_{\epsilon \to 0} \frac{F(f + \epsilon \cdot h) - F(f)}{\epsilon}.$$

One can apply this definition for example to functionals $F : \mathcal{H} \to \mathbb{R}$ on a topological space of fields $\mathcal{H} = \Gamma(M, C)$ of a bosonic field theory where $\pi : C = V \times M \to M$ is a trivial linear bundle on spacetime with fiber V, and this will suffice to understand the main difficulties. In particular, one can work with $\mathcal{H} = C_c^\infty(\mathbb{R}^n)$ or $\mathcal{H} = \mathcal{S}(\mathbb{R}^n)$ (equipped with their ordinary topologies).

One can also work with spaces of distributions. For example, the Gâteaux derivative of a functional $F : \mathcal{S}'(\mathbb{R}^n) \to \mathbb{R}$ along Dirac's δ_x distribution (that one often finds in physics books), if it exists, defines a map

$$\frac{\delta F}{\delta \boldsymbol{\delta}_x} : \mathcal{S}'(\mathbb{R}^n) \to \mathbb{R},$$

that associates for example to a locally integrable function φ a derivative

$$\frac{\delta F}{\delta \boldsymbol{\delta}_x}(\varphi) \in \mathbb{R}$$

of F at φ along the Dirac δ distribution, denoted by $\frac{\delta F(\varphi)}{\delta \varphi(x)}$ by physicists. For example, if $\varphi_0 \in \mathcal{S}(\mathbb{R}^n)$ is fixed, the corresponding functional $[\varphi_0](\varphi) := \int_{\mathbb{R}^n} \varphi_0(x)\varphi(x)dx$ on $\mathcal{S}(\mathbb{R}^n)$ can be extended to a functional of distributions $[\varphi_0] : \mathcal{S}'(\mathbb{R}^n) \to \mathbb{R}^n$ by setting

$$[\varphi_0](D) := D(\varphi_0).$$

One then has a well defined functional derivative for the functional $[\varphi_0]$ given by

$$\frac{\delta[\varphi_0]}{\delta\varphi(x)}(\varphi) := \frac{\delta[\varphi_0]}{\delta\boldsymbol{\delta}_x} = \varphi_0(x).$$

One can also consider the evaluation functional $F = \mathrm{ev}_x : \mathcal{S}(\mathbb{R}^n) \to \mathbb{R}$ given by $f \mapsto f(x)$. One then has

$$\frac{\delta F}{\delta \mathbf{h}}(f) := \lim_{\epsilon\to 0} \frac{F(f+\epsilon\cdot h) - F(f)}{\epsilon} = \lim_{\epsilon\to 0} \frac{f(x)+\epsilon\cdot h(x) - f(x)}{\epsilon} = h(x)$$

for every $f \in \mathcal{S}(\mathbb{R}^n)$, so that the functional derivative of F at f exists and is

$$\frac{\delta F}{\delta \mathbf{h}}(f) = F(\mathbf{h}).$$

Here we observe that the functional derivative has nothing in common with the derivative of distributions. One should also note that the function $F = \mathrm{ev}_x$ does not extend to tempered distributions and that one can't compute its functional derivative along δ_x, because $F(\delta_x) = \mathrm{ev}_x(\delta_x)$ isn't well defined. This is one of the explanation of the infinities that appear in the physicists' computations of functional integrals, leading to the use of renormalization methods.

For the rest of this section, we will assume that all our functionals extend to distributions and use the following definition of functional derivatives:

$$\frac{\delta F(f)}{\delta f(x)} := \frac{\delta F(f)}{\delta(\delta_x)} = \lim_{\epsilon\to 0} \frac{F(f+\epsilon\delta_x) - F(f)}{\epsilon}.$$

The most useful functional derivatives in field theory are given by the following proposition (that one can find in Folland's book [Fol08], p. 269).

Proposition 18.3.1 *One has the following computations*:

Functional	Formula $F(f)$	Functional derivative
Integral	$\int f(y)h(y)dy$	$\frac{\delta F(f)}{\delta f(x)} = h(x)$
Exponential integral	$\exp[\int f(y)h(y)dy]$	$\frac{\delta^J F(f)}{\delta f(x_1)\cdots\delta f(x_J)} = h(x_1)\cdots h(x_J)F(f)$
Quadratic integral	$\int\int f(x)K(x,y)g(y)dxdy$	$\frac{\delta F(f)}{\delta f(u)} = 2\int K(u,z)f(z)dz$
Quadratic exponential	$\exp(\int\int f(x)K(x,y)g(y)dxdy)$	$\frac{\delta F(f)}{\delta f(u)} = [2\int K(u,z)f(z)dz]F(f)$

Proof We will only give the proof for the two first integrals. The functional $F(f) = \int f(y)h(y)dy$ can be extended from $\mathcal{S}(\mathbb{R})$ to distributions f on $\mathcal{S}(\mathbb{R})$ by $F(f) = f(h)$: this is simply the evaluation of distributions at the given function h. It is a linear functional in the distribution f. One can thus compute $F(f+\epsilon\delta_x) = F(f) + \epsilon F(\delta_x)$ and get

$$\frac{\delta F(f)}{\delta(\delta_x)} = \delta_x(h) = h(x).$$

Similarly, for $F(f) = \exp[\int f(y)h(y)dy]$, one can extend it to distributions by $F(f) = \exp(f(h))$, yielding

$$\begin{aligned}
\frac{\delta F(f)}{\delta(\delta_x)} &= \lim_{\epsilon\to 0} \frac{\exp(f(h)+\epsilon\delta_x(h)) - \exp(f(h))}{\epsilon} \\
&= \lim_{\epsilon\to 0} \exp\big(f(h)\big) \cdot \frac{\exp(\epsilon\delta_x(h)) - 1}{\epsilon} \\
&= h(x)\exp\big(f(h)\big).
\end{aligned}$$

The other equalities are obtained by similar methods. □

18.4 Perturbative Definition of Functional Integrals

The main idea concerning perturbative expansions is to define the functional integrals by their expansions in terms of Feynman diagrams, using the analogy with finite dimensional Gaussian integrals. In this section, as in Chap. 18, we authorize ourselves to use the physicists' notations

and some of their problematic computations,

describing where the mathematical problems are, so that one has a good idea of what needs to be done to solve them.

One thus considers the action functional $S(\varphi)$ of a given system $\pi : C \to M$ without gauge symmetries as a perturbation of a free action functional $S_{free}(\varphi) = \int_M \langle \varphi, D\varphi \rangle$ that is quadratic in the fields with the linear equation $D\varphi = 0$ as Euler-Lagrange equation, by an interaction term $S_{int}(\varphi) = \int_M V(\varphi)$ with V a potential function. One writes

$$S(\varphi) = S_{free}(\varphi) + \lambda S_{int}(\varphi).$$

If Δ is the Schwartz kernel of D, also called the free propagator, one then defines

$$Z_{free}(J) = \int_H e^{\frac{i}{\hbar}(S_{free}(\varphi)+\int J(\varphi))}[d\varphi] := e^{-\frac{i}{\hbar}J\Delta J}.$$

If the kernel Δ is a smooth function, this free partition function satisfies the free Dyson-Schwinger equation

$$\left(\frac{\delta S_{free}}{\delta\varphi(x)}\left[-\frac{i}{\hbar}\frac{\delta}{\delta J}\right] + J(x)\right) \cdot Z_{free}(J) = 0$$

which is perfectly well defined in this case because there are no higher order functional derivatives in play. If Δ is only a distribution, one cannot apply the functional derivative to $Z_{free}(J)$ because it involves evaluating a distribution at the δ function. The solution of this problem is given by the regularization procedure for propagators.

One then formally writes

$$Z(J) = \int_H e^{\frac{i}{\hbar}(S(\varphi)+\int J(\varphi))}[d\varphi] = e^{\frac{i}{\hbar}S_{int}}\left[-\frac{i}{\hbar}\frac{\delta}{\delta J}\right] \cdot Z_{free}(J)$$

and permutes ordinary and functional integrals to finally define

$$\begin{aligned} Z(J) &:= \sum_{k=0}^{\infty} \frac{i^k}{k!\hbar^k} \underbrace{\int \cdots \int}_{k \text{ terms}} S_{int}^k \left[\frac{\delta}{\delta J}\right] e^{-\frac{i}{\hbar}J\Delta J} dx_1 \cdots dx_k \\ &= \sum_{k=0}^{\infty} \frac{i^k}{k!} \int J(x_1) \cdots J(x_k) G_k(x_1, \ldots, x_k) dx_1 \cdots dx_k. \end{aligned}$$

The terms that appear in this formal power series in $\hbar$ are integrals of some expressions that are ill defined because they involve the evaluation of distributions at δ functions. One can however check that this expression formally satisfies the Dyson-Schwinger equation

$$\left(\frac{\delta S}{\delta\varphi(x)}\left[-\frac{i}{\hbar}\frac{\delta}{\delta J(x)}\right] + J(x)\right) \cdot Z(J) = 0.$$

One then has to regularize, i.e., replace these distributions by usual families of functions that converge to them, compute everything with these regularized solutions and then pass to the distributional limit. The Green functions G_k that appear in the perturbative expansion can be described (or actually defined) by analogy with the usual Gaussian case as sums

$$G_k(x_1, \ldots, x_k) = \sum_{\Gamma} \int \frac{V(\Gamma)(p_1, \ldots, p_k)}{\sigma(\Gamma)} e^{i(x_1 p_1 + \cdots + x_k p_k)} \prod_i \frac{dp_i}{(2\pi)^d},$$

indexed by Feynman graphs that are not necessarily connected. The problem is that the rational fractions $V(\Gamma)(p_1, \ldots, p_k)$ are usually not integrable, so that one has to replace them by their renormalized values, which are obtained by adding to them some counter-terms (obtained by adding local counter terms to the Lagrangian), to make them convergent. This is the theory of renormalization.

Note that it is also interesting to consider the connected partition function defined by the equality

$$Z(J) =: e^{\frac{i}{\hbar}W(J)}$$

since its perturbative expansion is described by connected Feynman graphs, and so is combinatorially simpler. One actually uses the more simple 1PI partition function, also called the effective action $S_{eff}(\varphi)$. It is formally defined as a kind of Legendre transform by

$$S_{eff}(\varphi) = \left[J_E(\varphi) - W(J_E)\right]_{|J_E=J(\varphi)}$$

where J_E is chosen so that

$$\frac{\delta W}{\delta J}(J_E) = \varphi.$$

One can properly define it by taking the unrenormalized values

$$U(\Gamma)(\varphi) = \frac{1}{k!}\int_{\sum p_j=0} \hat{\varphi}(p_1)\cdots\hat{\varphi}(p_k)U\big(\Gamma(p_1,\ldots,p_k)\big)\prod_i \frac{dp_i}{(2\pi)^d},$$

and defining

$$\Gamma(\varphi) = S(\varphi) - \sum_{\Gamma\in 1\text{PI}} \frac{U(\Gamma)(\varphi)}{\sigma(\Gamma)}.$$

In this perturbative setting, where one defines functional integrals by their Feynman graph expansions, one has the equality

$$\int F(\varphi)e^{\frac{i}{\hbar}S(\varphi)}[d\varphi] = \int_{\text{tree level}} F(\varphi)e^{\frac{i}{\hbar}\Gamma(\varphi)}[d\varphi]$$

where the tree-level functional integral means that one only uses diagrams that are trees. This shows that the knowledge of the effective action is enough to determine all quantum observables.

18.5 Schwinger's Quantum Variational Principle

Another way of giving a mathematical meaning to the expectation value $\langle\varphi_1|A|\varphi_0\rangle$ is to consider the dual linear bundle $C^\dagger$ of C and to add to the action functional a source term

$$S_{source}(\varphi) = S(\varphi) + \int_M \langle J,\varphi\rangle(x)d^4x.$$

The aim is to make sense of the partition function

$$Z_A(J) = \langle\varphi_1|A|\varphi_0\rangle_J = \frac{\int_H A(\varphi)e^{\frac{i}{\hbar}S_{source}(\varphi)}[d\varphi]}{\int_H e^{iS(\varphi)}[d\varphi]},$$

whose value at $J=0$ gives the sum over histories. The main advantage of this new formal expression is that it replaces the problem of defining an integral on the space of histories by the problem of defining the functional differential equation on the dual space whose solution is $Z_A(J)$. This equation is called the Dyson-Schwinger equation and may be used to implicitly define the partition function, since this is usually how physicists using functional integrals proceed (see for example Zinn-Justin's book [ZJ93], Chap. 7 or Rivers' book [Riv90]).

The idea behind the Dyson-Schwinger approach can be explained quickly in the case of a scalar field theory $\pi : C = \mathbb{R} \times M \to M$ on spacetime. The space $\mathcal{O}_H$ of functions on H is equipped with a (partially defined) functional derivation $\frac{\delta}{\delta\varphi(x)}$ defined by

$$\frac{\delta F}{\delta\varphi(x)} = \lim_{t\to 0} \frac{F(\varphi + t\delta_x) - F(\varphi)}{t}.$$

The hypothetic functional integral $\int_H [d\varphi] : \mathcal{O}_H \to \mathbb{R}$ is supposed to satisfy the following formal equalities, that are inspired by ordinary properties of finite dimensional integrals and Fourier transforms:

$$\int_H [d\varphi] \frac{\delta}{\delta\varphi(x)} F(\varphi) = 0$$

and

$$\int_H [d\varphi] F(\varphi) G(\varphi) e^{\frac{i}{\hbar} J(\varphi)} = F\left[-\frac{i}{\hbar}\frac{\delta}{\delta J(x)}\right] \int_H G(\varphi) e^{\frac{i}{\hbar} J(\varphi)} [d\varphi].$$

Here, the expression $F[-\frac{i}{\hbar}\frac{\delta}{\delta J(x)}]$ is a functional differential operator obtained from $F(\varphi)$ by replacing φ by $-\frac{i}{\hbar}\frac{\delta}{\delta J(x)}$. Applying this to the (normalized) functional integral, one gets

$$\int_H \frac{\delta}{\delta\varphi(x)} e^{-\frac{i}{\hbar}[S(\varphi)+J(\varphi)]} [d\varphi] = 0,$$

and finally

$$\left(\frac{\delta S}{\delta\varphi(x)}\left[-\frac{i}{\hbar}\frac{\delta}{\delta J(x)}\right] + J(x)\right) \cdot Z(J) = 0$$

and more generally

$$\frac{\delta A}{\delta\varphi(x)}\left[-\frac{i}{\hbar}\frac{\delta}{\delta J(x)}\right] \cdot Z(J) + \left(\frac{\delta S}{\delta\varphi(x)}\left[-\frac{i}{\hbar}\frac{\delta}{\delta J(x)}\right] + J(x)\right) \cdot Z_A(J) = 0.$$

18.6 Theories with Gauge Freedom: Zinn-Justin's Equation

We refer to Henneaux and Teitelboim [HT92], Chaps. XVII and XVIII. The formalism of this section was first used by Zinn-Justin to study the renormalizability of Yang-Mills theory, and independently by Batalin-Vilkovisky. It is called the Zinn-Justin equation, or the quantum BV formalism in the physics literature.

Let $\pi : C \to M$ be a bundle and $S \in h(\mathcal{A})$ be a local action functional with $\mathcal{A} = \mathrm{Jet}(\mathcal{O}_C)$. In Sect. 13.4, we explained the methods of homological symplectic reduction in local field theory. Roughly speaking, the result of this procedure is a dg-$\mathcal{D}$-algebra $(\mathcal{A}^\bullet, D)$ equipped with a local bracket

$$\{\cdot,\cdot\}: \mathcal{A}^\bullet \boxtimes \mathcal{A}^\bullet \to \Delta_* \mathcal{A}^\bullet$$

and a particular element $S_{cm} \in h(\mathcal{A}^\bullet)$ called the classical master action, or also the BRST generator, such that

$$D = \{S_{cm}, \cdot\}.$$

This construction works in the most general context of a local field theory on a super-space. The space of classical observables is then given by the algebra

$$H^0(\mathcal{A}^\bullet, D).$$

Roughly speaking, the canonical version of the BV quantization (operator formalism) is given by a quantization of the graded algebra $\mathcal{A}^\bullet$, denoted by $\hat{\mathcal{A}}^\bullet$, that also gives a quantization of the BRST generator S_{cm} and thus a quantization $\hat{D}$ of the BRST differential. The algebra of quantum observables is then given by

$$H^0(\hat{\mathcal{A}}^\bullet, \hat{D}).$$

The functional integral quantization is a bit more involved. We denote by φ_i the fields and ghost variables and by φ_i^* the antifield variables. In the notation used by physicists, the duality between fields and antifields is given by

$$(\varphi^i, \varphi_i^*) = \delta^i_j.$$

This means that the antibracket is extended to the algebra $\mathcal{A}^\bullet$ by

$$(A, B) = \frac{\delta^R A}{\delta \varphi^i}\frac{\delta^L B}{\delta \varphi_i^*} - \frac{\delta^R A}{\delta \varphi_i^*}\frac{\delta^L B}{\delta \varphi^i}.$$

Here, the left and right functional derivatives are distributions defined by

$$\frac{d}{dt}F(\varphi + t\psi_\varphi) = \int_M \frac{\delta^L F}{\delta \varphi^a(x)}\psi_\varphi^a(x)dx = \int_M \psi_\varphi^a(x)\frac{\delta^R F}{\delta \varphi^a(x)}dx$$

for ψ_φ tangent vectors to the graded field variable φ.

One cannot naively use the functional integral (Dyson-Schwinger solution)

$$\int [d\varphi] \exp \frac{i}{\hbar} S_{cm}(\varphi, \varphi^* = 0)$$

as this would give infinite values because of the gauge freedom. The idea is to replace this by an expression

$$\int [d\varphi] \exp \frac{i}{\hbar} S_{qm}\left(\varphi, \frac{\delta \psi}{\delta \varphi}\right)$$

for some given functional $\psi(\varphi)$ of the fields variables called the gauge fixing. This gives a canonical transformation of the antibracket that does not change the corresponding differential graded algebra, but such that the corresponding equations of motion are well-posed, so that one can perform the usual perturbative quantization, as in the scalar field case.

The classical master action has to be expanded to a quantum master action $S_{qm} = S_{cm} + \sum_{n\geq 1} \hbar^n S_n$ that satisfies the quantum master equation

$$i\hbar \Delta S_{qm} - \frac{1}{2}(S_{qm}, S_{qm}) = 0,$$

where $\Delta = \frac{\delta^R}{\delta\varphi^i}\frac{\delta^R}{\delta\varphi_i^*}$ is the super Laplacian on $\mathcal{A}^\bullet$ (which is not well defined and needs to be regularized because it involves double functional derivatives). This equation guarantees that the above functional integral does not depend on the gauge fixing functional ψ. It is equivalent to the equation

$$\Delta \exp \frac{i}{\hbar} S_{qm} = 0.$$

If $A_0(\varphi^i) \in \mathcal{A}$ is a classical local gauge invariant function, a classical BRST extension is given by an element $A(\varphi, \varphi^*)$ of $\mathcal{A}^\bullet$ which is a solution of $(A, S_{cm}) = 0$. A quantum BRST extension of A is given by an element $\alpha \in \mathcal{A}^\bullet[\![\hbar]\!]$ of the form

$$\alpha = A + \sum_{n\geq 1} \hbar \alpha_i,$$

such that the quantum master observable equation

$$\Delta \alpha \exp \frac{i}{\hbar} S_{qm} = 0$$

is satisfied, the quantum master action S_{qm} being already fixed. This quantum master observable equation is again equivalent to

$$i\hbar\Delta\alpha - (\alpha, S_{qm}) = 0.$$

The function $A_\psi(\varphi) = A(\varphi, \frac{\delta\psi}{\delta\varphi})$ is called the gauge fixed version of A. The expectation value of A_0 is finally given by

$$\langle A_0 \rangle = \int \alpha\left(\varphi, \frac{\delta\psi}{\delta\varphi}\right) e^{\frac{i}{\hbar} S_{qm}(\varphi, \frac{\delta\psi}{\delta\varphi})} [d\varphi].$$

The main problem with this definition is that the BV Laplacian does not make sense because it involves second order functional derivatives. One of course also gets into trouble when writing down the Dyson-Schwinger equation for this functional integral since it involves higher order functional derivatives.

References

[CDM06] Cartier, P., DeWitt-Morette, C.: Functional Integration: Action and Symmetries. Cambridge Monographs on Mathematical Physics, p. 456. Cambridge University Press, Cambridge (2006). ISBN 978-0-521-86696-5; 0-521-86696-0. Appendix D contributed by Alexander Wurm

[CG10] Costello, K., Gwilliam, O.: Factorization algebras in perturbative quantum field theory (2010, preprint)

[CM08] Connes, A., Marcolli, M.: Noncommutative Geometry, Quantum Fields and Motives. American Mathematical Society Colloquium Publications, vol. 55, p. 785. Am. Math. Soc., Providence (2008). ISBN 978-0-8218-4210-2

[Fol08] Folland, G.B.: Quantum Field Theory: A Tourist Guide for Mathematicians. Mathematical Surveys and Monographs, vol. 149, p. 325. Am. Math. Soc., Providence (2008). ISBN 978-0-8218-4705-3

[HT92] Henneaux, M., Teitelboim, C.: Quantization of Gauge Systems, p. 520. Princeton University Press, Princeton (1992). ISBN 0-691-08775-X; 0-691-03769-8

[LaC06] LaChapelle, J.: Functional integration for quantum field theory. arXiv Mathematical Physics e-prints (2006). arXiv:math-ph/0610035

[Phi01] Phillips, T.: Finite dimensional Feynman diagrams. AMS (2001). http://www.ams.org/featurecolumn/archive/feynman1.html, 1–7

[Riv90] Rivers, R.J.: Path Integral Methods in Quantum Field Theory, 2nd edn. Cambridge Monographs on Mathematical Physics, p. 339. Cambridge University Press, Cambridge (1990). ISBN 0-521-25979-7; 0-521-36870-7

[Zei06] Zeidler, E.: Quantum Field Theory. I. Basics in Mathematics and Physics, p. 1020. Springer, Berlin (2006). doi:10.1007/978-3-540-34764-4. ISBN 978-3-540-34762-0; 3-540-34762-3. A bridge between mathematicians and physicists

[Zei09] Zeidler, E.: Quantum Field Theory. II. Quantum Electrodynamics, p. 1101. Springer, Berlin (2009). ISBN 978-3-540-85376-3. A bridge between mathematicians and physicists

[Zei11] Zeidler, E.: Quantum Field Theory. III. Gauge Theory, p. 1126. Springer, Heidelberg (2011). doi:10.1007/978-3-642-22421-8. ISBN 978-3-642-22420-1. A bridge between mathematicians and physicists

[ZJ93] Zinn-Justin, J.: Quantum Field Theory and Critical Phenomena, 2nd edn. International Series of Monographs on Physics, vol. 85, p. 996. Clarendon/Oxford University Press, London/New York (1993). Oxford Science Publications. ISBN 0-19-852053-0

Chapter 19
The Connes-Kreimer-van Suijlekom View of Renormalization

In this chapter we follow the Hopf algebra/Riemann-Hilbert description of renormalization, due to Connes and Kreimer (see [CK00] and [CK01]). This method is described in [CM08] and in [dM98] for theories without gauge symmetries. We also refer to [Mar10] for more informations on the relation between renormalization and motives.

We give here an overview of van Suijlekom's articles on this formalism (see [vS09a] and [vS09b]) for theories with gauge symmetries.

These constructions give a nice mathematical formulation of the BPHZ renormalization procedure in the dimensional regularization scheme, which is the main method for the renormalization of the standard model.

19.1 Diffeographism Groups for Gauge Theories

Suppose we are working with the classical BV action of a given gauge theory, given by an element S_{cm} in $h(\mathcal{A}^\bullet)$ for $\mathcal{A}^\bullet$ the BV dg-algebra of fields and antifields. As a graded $\mathcal{A}[\mathcal{D}]$-algebra, this algebra is generated by an even family

$$\Phi = \{\varphi^1, \ldots, \varphi^n, \varphi_1^*, \ldots, \varphi_n^*\}$$

of generators that will be called the fields. One actually assumes given a linear bundle $E^{tot} = E \oplus E^*$ over spacetime M with E and E^* both graded such that $\mathrm{Jet}(\mathcal{O}_{E^{tot}}) = \mathcal{A}^\bullet$ as a graded $\mathcal{A}[\mathcal{D}]$-algebra.

We assume that S_{cm} is the cohomology class of a polynomial Lagrangian density $\mathcal{L}$ (which is the case for Yang-Mills with matter).

We denote by R the set of monomials in $\mathcal{L}$, and decompose it into the free massless terms R_E, called the set of edges, and the interaction and mass terms R_V, called the set of vertices. Each edge in R_E will be represented by a different type of line, called the propagator, that corresponds to the free massless motion of the corresponding particle. Each vertex in R_V will be denoted by a vertex whose entering

F. Paugam, *Towards the Mathematics of Quantum Field Theory*,
Ergebnisse der Mathematik und ihrer Grenzgebiete. 3. Folge / A Series of Modern
Surveys in Mathematics 59, DOI 10.1007/978-3-319-04564-1_19,

edges are half lines of the same type as the line in R_E that corresponds to the given type of field.

The datum of the Lagrangian density is thus equivalent to the datum of the pair $R = (R_E, R_V)$ of sets of colored edges and vertices and of a map

$$\iota : R_E \coprod R_V \to \mathcal{A}^\bullet.$$

Indeed, one can retrieve the Lagrangian density by setting

$$\mathcal{L}_{free\text{-}massless} = \sum_{e \in R_E} \iota(e), \qquad \mathcal{L}_{int\text{-}mass}(\varphi) = \sum_{v \in R_V} \iota(v),$$

and finally

$$\mathcal{L}(\varphi) = \mathcal{L}_{free\text{-}massless} + \mathcal{L}_{int\text{-}mass}.$$

Example 19.1.1 The Lagrangian density of quantum electrodynamics is a $U(1)$-gauge invariant Yang-Mills Lagrangian (see Sect. 15.3) of the form

$$L(A, \psi) = -\frac{1}{2}\langle F_A \wedge *F_A \rangle + \bar{\psi} \not{D}_A \psi + \bar{\psi} m \psi$$

where A is the photon field given by an $i\mathbb{R} = \mathrm{Lie}(U(1))$-valued 1-form $A \in \Omega^1_M$ and a section $\psi : M \to S$ of the (assumed) trivial spinor bundle $M \times \mathbb{C}^2$ on flat spacetime that represents the electron field. Once a basis for the space of connections is fixed, the A-dependent Dirac operator decomposes as a sum of a Dirac propagator $i\bar{\psi} \not{D} \psi$ and an interaction term $-e\bar{\psi}\gamma \circ A\psi$. The space of vertices and edges for this Lagrangian are thus given by

$R_V = \{$  ,  $\}$; $R_E = \{$  ,  $\}$.

The corresponding monomials in $\mathcal{A}^\bullet$ are

$\iota($  $) = -e\bar{\psi}\gamma \circ A\psi$, $\iota($  $) = -m\bar{\psi}\psi$,

$\iota($  $) = i\bar{\psi}\gamma \circ d\psi$, $\iota($  $) = -dA * dA$,

with e and m the electric charge and mass of the electron (coupling constants), respectively.

Example 19.1.2 Similarly, the Lagrangian density of quantum chromodynamics SU(3)-gauge invariant Yang-Mills Lagrangian (see Sect. 15.3) of the form

$$L(A, \psi) = -\frac{1}{2}\langle F_A \wedge *F_A \rangle + \bar{\psi} \not{D}_A \psi + \bar{\psi} m \psi$$

where A is the quarks field, described by wiggly lines, and ψ is the gluon field, described by straight lines. In addition, one has the ghost fields ω and $\bar{\omega}$, indicated by dotted lines, as well as BRST antifields K_ψ, K_A, K_ω and $K_{\bar{\omega}}$. Between these fields, there are four interactions, three BRST-source terms, and a mass term for the quark. This leads to the following sets of vertices and edges,

$R_V =$ 

with the dashed lines representing the BRST-source terms, and

$R_E =$ 

Note that the dashed edges do not appear in R_E, i.e. the source terms do not propagate and in the following will not appear as internal edges of a Feynman graph.

Definition 19.1.1 A *Feynman graph* Γ is a pair composed of a set Γ_V of vertices each of which is an element in R_V and Γ_E of edges in R_E, and maps

$$\partial_0, \partial_1 : \Gamma_V \to \Gamma_E \cup \{1, 2, \ldots, N\},$$

that are compatible with the type of vertex and edge as parametrized by R_V and R_E, respectively. One excludes the case that ∂_0 and ∂_1 are both in $\{1, \ldots, N\}$. The set $\{1, \ldots, N\}$ labels the external lines, so that $\sum_j \operatorname{card}(\partial_j^{-1}(v)) = 1$ for all $v \in \{1, \ldots, N\}$. The set of *external lines* is $\Gamma_E^{ext} = \partial_0^{-1}(\{1, \ldots, N\}) \cup \partial_1^{-1}(\{1, \ldots, N\})$ and its complement Γ_E^{int} is the set of *internal lines*.

One regards the external lines Γ_E^{ext} as labeled $e_1, \ldots, e_N$ where $e_k = \partial_0^{-1}(k) \cup \partial_1^{-1}(k)$.

Definition 19.1.2 An *automorphism* of a Feynman graph Γ is given by a pair of isomorphisms $g_V : \Gamma_V \to \Gamma_V$ and $g_E : \Gamma_E \to \Gamma_E$ that is the identity on Γ_E^{ext} that are compatible with the boundary maps, meaning that for all $e \in \Gamma_E$,

$$\cup_j g_V\big(\partial_j(e)\big) = \cup_j \partial_j\big(g_E(e)\big).$$

Moreover, we require g_V and g_E to respect the type of vertex/edge in the set R. The automorphism group $\operatorname{Aut}(\Gamma)$ of Γ consists of all such automorphisms; its order is called the *symmetry factor* of Γ and is denoted by $\operatorname{Sym}(\Gamma)$.

Definition 19.1.3 A Feynman graph is called a *one-particle irreducible graph* (1PI) if it is not a tree and cannot be disconnected by removal of a single edge. The *residue* $\operatorname{res}(\Gamma)$ of a Feynman graph Γ is obtained by collapsing all its internal edges and vertices to a point.

If a Feynman graph Γ has two external lines, both corresponding to the same field, we would like to distinguish between propagators and mass terms. In more mathematical terms, since we have vertices of valence two, we would like to indicate whether a graph with two external lines corresponds to such a vertex, or to an edge. A graph Γ with two external lines is dressed with a bullet when it corresponds to a vertex, i.e. we write $\Gamma_\bullet$. The above correspondence between Feynman graphs and vertices/edges is given by the residue. For example, we have:

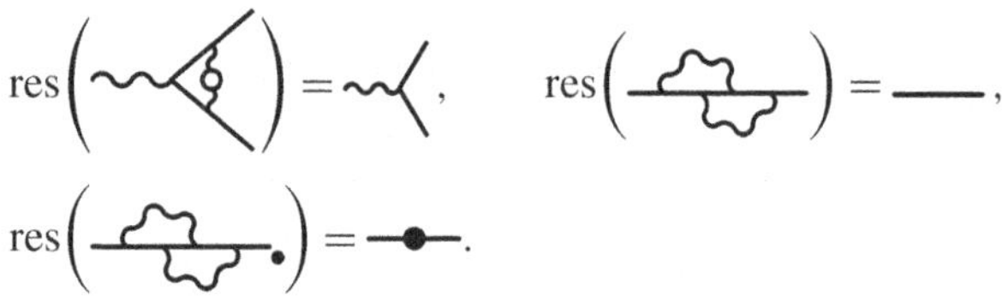

Definition 19.1.4 The *diffeographism group scheme* G is the affine group scheme whose algebra of functions A_G is the free commutative algebra generated by all 1PI graphs, with counit $\epsilon(\Gamma) = 0$ unless $\Gamma = \emptyset$, and $\epsilon(\emptyset) = 1$, coproduct

$$\Delta(\Gamma) = \Gamma \otimes 1 + 1 \otimes \Gamma + \sum_{\gamma \subsetneq \Gamma} \gamma \otimes \Gamma/\gamma,$$

and antipode given recursively by

$$S(\Gamma) = -\Gamma - \sum_{\gamma \subsetneq \Gamma} S(\gamma)\Gamma/\gamma.$$

In the above definition, the graph Γ/γ is obtained by contracting in Γ the connected components of the subgraph γ to the corresponding vertex/edge. If the connected component γ' of γ has two external lines, there are possibly two contributions corresponding to the valence two vertex and the edge; the sum involves the two terms $\gamma'_\bullet \otimes \Gamma/(\gamma' \to \bullet)$ and $\gamma' \otimes \Gamma/\gamma'$. To illustrate this with the QED example, one has

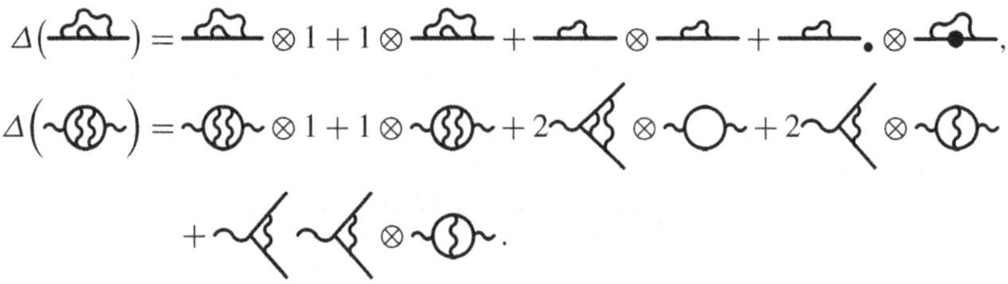

The Hopf algebra A_G is graded (meaning it has a grading that respects the product and coproduct) by the loop number $L(\Gamma) := b_1(\Gamma)$ of a graph. This gives a decomposition

$$A_G = \oplus_{n \in \mathbb{N}} A_G^n.$$

It has also a multi-grading indexed by the group $\mathbb{Z}^{(R_V)}$. For $R = R_E \coprod R_V$, one defines $m_{\Gamma,r}$ as the number of vertices/internal edges of type $r \in R$ appearing in Γ, and $n_{\gamma,r}$ the number of connected component of γ with residue r. For each $v \in R_V$, one defines the degree d_v by setting

$$d_v(\Gamma) = m_{\Gamma,v} - n_{\Gamma,v}.$$

The multi-degree indexed by R_V, given by

$$d : A_G \to \mathbb{Z}^{(R_V)},$$

is compatible with the Hopf algebra structure. This gives a decomposition

$$A_G = \oplus_{\alpha \in \mathbb{Z}^{(R_V)}} A_G^\alpha.$$

For $\alpha \in \mathbb{Z}^{(R_V)}$, we denote by $p_\alpha : A_G \to A_G^\alpha$ the corresponding projection.

19.2 The Riemann-Hilbert Correspondence

Let $X = \mathbb{P}^1(\mathbb{C}) - \{x_0, \dots, x_n\}$. A flat bundle on X is a locally free $\mathcal{O}_X$-module $\mathcal{F}$ with a connection

$$\nabla : \mathcal{F} \to \mathcal{F} \otimes \Omega_X^1$$

whose curvature is zero, i.e., such that $\nabla_1 \circ \nabla = 0$. Equivalently, this is a $\mathcal{D}_X$-module that is also a locally free $\mathcal{O}_X$-module.

Given a flat bundle $(\mathcal{F}, \nabla)$ its space of solutions (also called horizontal sections) is the locally constant bundle (local systems) on X (locally isomorphic to $\mathbb{C}_X^n$) given by

$$\mathcal{F}^\nabla = \{f \in \mathcal{F},\ \nabla f = 0\}.$$

Consider the functor

$$\begin{aligned} \mathrm{Sol} : \textsc{Flat Bundles} &\to \textsc{Local systems} \\ (\mathcal{F}, \nabla) &\to \mathcal{F}^\nabla. \end{aligned}$$

Note that a local system is equivalent to a representation of the fundamental group $\pi_1(X)$. One of Hilbert's problem is to show that it is essentially surjective, meaning that to every representation V of $\pi_1(X)$, one can associate a differential equation whose corresponding local system is given by the local system

$$\underline{V} := V \underset{\pi_1(X)}{\times} \tilde{X}$$

where $\tilde{X} \to X$ is the universal covering of X. Results in this direction are called Riemann-Hilbert correspondences.

Note that if $(\mathcal{F}, \nabla)$ is a flat bundle, and if $\gamma : x \to y$ is a path in X (whose homotopy class is an element in the fundamental groupoid $\Pi(X)$), the analytic continuation of solutions of $(\mathcal{F}, \nabla)$ along γ gives a linear map, called the monodromy transformation

$$M_\gamma : \mathcal{F}_x^\nabla \to \mathcal{F}_y^\nabla$$

between the two finite dimensional $\mathbb{C}$-vector spaces given by the fibers of the solution space at the two given points. If

$$\mathrm{Isom}\big(\mathcal{F}^\nabla\big) \to X \times X$$

is the algebraic groupoid whose fiber over (x, y) is the space of linear isomorphisms $\mathrm{Isom}(\mathcal{F}_x^\nabla, \mathcal{F}_y^\nabla)$, there is a natural groupoid morphism

$$M : \Pi(X) \to \mathrm{Isom}\big(\mathcal{F}^\nabla\big)$$

given by the monodromy representation.

A theorem due to Cartier and independently to Malgrange is that the Zariski-Closure of the image of the monodromy representation M in the relatively algebraic groupoid $\mathrm{Isom}(\mathcal{F}^\nabla)$ is equal to the differential Galois groupoid. To explain this result, one has to define the differential Galois groupoid of $(\mathcal{F}, \nabla)$. It is the groupoid whose fiber over a point $(x, y) \in X \times X$ is the space

$$\mathrm{Isom}(\omega_x, \omega_y)$$

between the fiber functors (i.e., symmetric monoidal functors between rigid symmetric monoidal categories, defined in Sect. 2.3; see Deligne [Del90] for more details on the general theory of Tannakian categories and their fundamental groups)

$$\omega_x, \omega_y : \big\langle(\mathcal{F}, \nabla)\big\rangle \to (\mathbb{C} - \text{VECTOR SPACES}, \otimes)$$

given by

$$\omega_{x,y}(\mathcal{G}, \nabla) = \mathcal{G}_{x,y}^\nabla,$$

where $\langle(\mathcal{F}, \nabla)\rangle$ denotes the sub-Tannakian category of the category of all flat bundles generated by the given flat bundle.

One can interpret the diffeographism groups of the Connes-Kreimer theory as some differential Galois groups associated to the defining equations of time-ordered exponentials (iterated integrals). This relation between differential Galois theory and the Connes-Kreimer theory is due to Connes and Marcolli and explained in detail in [CM08].

We will now explain a particular version of the Riemann-Hilbert correspondence that is directly related to the renormalization procedure in the dimensional regularization scheme.

Let $\gamma : C \to G$ be a loop with values in an arbitrary complex Lie group G, defined on a smooth simple curve $C \subset \mathbb{P}^1(\mathbb{C})$. Let $C_\pm$ be the two complements of

C in $\mathbb{P}^1(\mathbb{C})$, with $\infty \in C_-$. A Birkhoff decomposition of γ is a factorization of the form

$$\gamma(z) = \gamma_-(z)^{-1}\gamma(z), \quad (z \in C),$$

where $\gamma_\pm$ are (boundary values of) two holomorphic maps on $C_\pm$, respectively with values in G.

Definition 19.2.1 The value of $\gamma_+(z)$ at $z = 0$ is called the *renormalized value* of the given loop and the term $\gamma_-(z)$ is called the *counter-term* associated to the given loop.

Theorem 19.2.1 *Let G be a group scheme whose Hopf algebra is positively graded connected and commutative with degree zero component $\mathbb{C}$. Then any loop $\gamma : C \to G$ admits a Birkhoff decomposition.*

19.3 Connes-Kreimer's View of Renormalization

We refer to [CM08] for a complete overview of the Connes-Kreimer theory and for a description of its relations with Riemann-Hilbert correspondence and motivic Galois theory.

We consider here the effective action from Sect. 18.4. It is given as a formal Legendre transform of the connected partition function that is itself given by the logarithm of the full partition function. One expresses this effective action in terms of the unrenormalized values of the Feynman graphs of the given theory.

If S is the original (gauge fixed) action functional, one writes

$$\Gamma(\varphi) = S(\varphi) - \sum_{\Gamma \in 1\text{PI}} \frac{U(\Gamma)(\varphi)}{\text{Sym}(\Gamma)}$$

where

$$U(\Gamma)(\varphi) = \frac{1}{N!}\int_{\sum p_j = 0} \varphi(\hat{p}_1)\cdots\hat{\varphi}(p_N)U\big(\Gamma(p_1,\dots,p_N)\big)\prod_j \frac{dp_j}{(2\pi)^d},$$

where $U(\Gamma(p_1,\dots,p_N))$ actually denotes a multilinear distribution on the fields, called the bare value of the graph, and is specified by the Feynman rules of the theory. More precisely, one has

$$U\big(\Gamma(p_1,\dots,p_N)\big) = \int I_\Gamma(k_1,\dots,k_L,p_1,\dots,p_N)d^dk_1\cdots d^dk_L$$

where $L = b_1(\Gamma)$ is the loop number of the graph and $k_i \in \mathbb{R}^d$ are the momentum variables assigned to the internal edges. The rational fraction $I_\Gamma(k_1,\dots,k_L,p_1,\dots,p_N)$ is obtained by applying the Feynman rules. Roughly speaking, this is done by assigning:

- a fundamental solution (or its Fourier transform) of the linear differential operator that corresponds to each internal line (actually, one regularizes this solution to get a smooth function that approximate it),
- a momentum conservation rule to each vertex;
- a power of the coupling constant to each more than 3-valent vertex; and
- a factor m^2 to each mass 2-point vertex.

The rational fraction $I_\Gamma(k_1, \dots, k_L)$ is given by the product of all these terms, except some distributional terms. It is supposed to be (when regularized) a smooth function. The problem is that if one sends the regularization parameters to 0, one gets divergencies. It is thus necessary to modify these bare values $U(\Gamma)$ to get meaningful limits at the end. This is the aim of the BPHZ renormalization procedure.

Now if G is the diffeographism group of a given field theory, the bare values $U(\Gamma)(z)$ associated to the Feynman graphs give algebra morphisms from A_G to $\mathbb{C}$ and thus points of G with values in $\mathbb{C}$. If C is a curve around $0 \in \mathbb{P}^1(\mathbb{C})$, one can define a loop in G by $\gamma(z)(\Gamma) := U(\Gamma)(z)$.

Applying the Birkhoff decomposition Theorem 19.2.1 to the above loop in the diffeographism group, one gets finite values from the dimensionally regularized bare Feynman amplitudes. Connes and Kreimer showed in [CK00] that these are exactly the values computed by physicists by the BPHZ renormalization procedure.

19.4 Zinn-Justin's Equation and Slavnov-Taylor Identities

Recall that we defined the BV Laplacian, anti-bracket and quantum BV formalism in Sect. 18.6.

The Zinn-Justin equation, also called the quantum master equation, is the equation that must be satisfied by the deformed action functional S_{qm} associated to the classical master action S_{cm}, and the deformed observable A_{qm} associated to a classical observable $A_0 \in H^0(\mathcal{A}^\bullet)$, such that the corresponding functional integral with gauge fixing functional $\psi(\varphi)$

$$\langle A_0 \rangle = \int \alpha\left(\varphi, \frac{\delta\psi}{\delta\varphi}\right) e^{\frac{i}{\hbar} S_{qm}(\varphi, \frac{\delta\psi}{\delta\varphi})} [d\varphi]$$

makes sense and does not depend on the choice of the extensions S_{qm} and A_{qm} or on the gauge fixing functional. This is the computation of the effective action associated to this functional integral that is achieved purely in terms of Feynman graphs in this section. The quantum master equation is usually expressed in terms of relations between the renormalized Green functions called the Slavnov-Taylor identities. The aim of this section is to explain, following van Suijlekom's work, how these identities are encoded in the Connes-Kreimer setting.

Let $\mathcal{A}_R := \mathbb{C}[\![\lambda_{v_1}, \dots, \lambda_{v_k}]\!] \otimes \mathcal{A}^\bullet$, $k = |R_V|$, be the algebra obtained from the BV algebra by extension of scalars to the coupling constants algebra. We now suppose that all the interaction (and mass) terms of the Lagrangian density $\mathcal{L}$ of the given theory are affected with a corresponding coefficient that is now a formal variable among the λ_{v_i}. Let $A_G^{Green} \subset A_G$ be the Hopf sub-algebra generated by elements $p_\alpha(Y_v)$, $v \in R_V$ and $p_\alpha(G^e)$, $e \in R_E$, for $\alpha \in \mathbb{Z}^{R_V}$,

$$G^e = 1 - \sum_{\text{res}(\Gamma)=e} \frac{\Gamma}{\text{Sym}(\Gamma)},$$

$$G^v = 1 + \sum_{\text{res}(\Gamma)=v} \frac{\Gamma}{\text{Sym}(\Gamma)} \quad \text{and} \quad Y_v := \frac{G^v}{\prod_\varphi} \left(G^\varphi\right)^{N_\varphi(v)/2}$$

and

$$G^{\varphi^i} = G^e, \qquad G^{\varphi_i^*} = \left(G^e\right)^{-1}.$$

We denote by

$$G \longrightarrow\!\!\!\!\rightarrow G_{Green} = \text{Spec}\left(A_G^{Green}\right)$$

the corresponding quotient of the diffeomorphism group.

Theorem 19.4.1 *The algebra $\mathcal{A}_R$ is a comodule BV-algebra for the Hopf algebra A_G^{Green}.*

The BV-ideal $\mathcal{I}_{BV}$ in $\mathcal{A}_R$ is generated by the antibracket (S, S) (and corresponds to the classical BV equation $(S, S) = 0$). There is a natural Hopf ideal in A_G^{Green} such that the corresponding subgroup of diffeographisms

$$G_{ST} \subset G_{Green}$$

acts on $\mathcal{A}_R/\mathcal{I}_{BV}$. The corresponding group is given by the subgroup of G that fixes the ideal $\mathcal{I}$ by the above coaction. This takes into account the Slavnov-Taylor identities corresponding to the quantum master equation (here obtained by using only the classical master equation). To conclude geometrically, one has a diagram of diffeographism group actions on spaces

$$\begin{array}{ccccc} G_{ST} & \rightarrowtail & G_{Green} & \twoheadleftarrow & G \\ \curvearrowright & & \curvearrowright & & \\ \text{Spec}(\mathcal{A}_R/\mathcal{I}_{BV}) & \rightarrowtail & \text{Spec}(\mathcal{A}_R) & & \end{array}$$

and G_{ST} is defined as the subgroup of G_{Green} that fixes the subspace of $\text{Spec}(\mathcal{A}_R)$ that corresponds to the classical BV equation.

References

[CK00] Connes, A., Kreimer, D.: Renormalization in quantum field theory and the Riemann-Hilbert problem. I. The Hopf algebra structure of graphs and the main theorem. Commun. Math. Phys. **210**(1), 249–273 (2000). doi:10.1007/s002200050779

[CK01] Connes, A., Kreimer, D.: Renormalization in quantum field theory and the Riemann-Hilbert problem. II. The β-function, diffeomorphisms and the renormalization group. Commun. Math. Phys. **216**(1), 215–241 (2001). doi:10.1007/PL00005547

[CM08] Connes, A., Marcolli, M.: Noncommutative Geometry, Quantum Fields and Motives. American Mathematical Society Colloquium Publications, vol. 55, p. 785. Am. Math. Soc., Providence (2008). ISBN 978-0-8218-4210-2

[Del90] Deligne, P.: Catégories tannakiennes. In: The Grothendieck Festschrift, Vol. II, pp. 111–195. Birkhäuser Boston, Boston (1990)

[dM98] de Monvel, L.B.: Renormalisation (d'après Connes et Kreimer). Séminaire Bourbaki (1998)

[Mar10] Marcolli, M.: Feynman Motives, p. 220. World Scientific, Hackensack (2010). ISBN 978-981-4304-48-1; 981-4304-48-4

[vS09a] van Suijlekom, W.D.: Renormalization of gauge fields using Hopf algebras. In: Quantum Field Theory, pp. 137–154. Birkhäuser, Basel (2009). doi:10.1007/978-3-7643-8736-5_8

[vS09b] van Suijlekom, W.D.: The structure of renormalization Hopf algebras for gauge theories. I. Representing Feynman graphs on BV-algebras. Commun. Math. Phys. **290**(1), 291–319 (2009). doi:10.1007/s00220-009-0829-x

Chapter 20
Nonperturbative Quantum Field Theory

20.1 The Geometric Meaning of Perturbative Computations

In this section, we describe the geometric objects underlying the computations of physicists in perturbative quantum field theory. We will give more details on the various perturbative renormalization procedures in Chaps. 21 and 22.

In this section, we denote by X a smooth space and $\pi : \mathbb{R}_X := \mathbb{R} \times X \to X$ the trivial bundle. The jet space of $\mathbb{R}_X$ is the space

$$\mathrm{Jet}(\mathbb{R}_X/X) := \varprojlim \mathrm{Jet}^n(\mathbb{R}_X/X).$$

The algebra of functions

$$\mathcal{O}\big(\mathrm{Jet}(\mathbb{R}_X/X)\big) := \varinjlim \mathcal{O}\big(\mathrm{Jet}^n(\mathbb{R}_X/X)\big)$$

is thus naturally a filtered smooth algebra. The degree zero term of the filtration is

$$\mathcal{O}\big(\mathrm{Jet}^0(\mathbb{R}_X/X)\big) = \mathcal{O}(\mathbb{R}_X) = \mathcal{O}(X \times \mathbb{R}).$$

It is the smooth $\mathcal{O}(X)$-algebra associated to the bundle $\mathbb{R}_X$.

We assume given a dual affine connection

$$\nabla : \Omega^1_X \to \Omega^1_X \otimes \Omega^1_X,$$

where $\Omega^1_X \otimes \Omega^1_X$ denotes the module of sections so that we get a second order derivative

$$\nabla^2 : \mathcal{O}_X \to \mathrm{Sym}^2_{\mathcal{O}_X}\big(\Omega^1_X\big)$$

and more generally higher order derivatives ∇^n, that give an isomorphism

$$\mathrm{Jet}(\mathbb{R}_X/X) := \varprojlim_n \mathrm{Jet}^n(\mathbb{R}_X/X) \xrightarrow[\sim]{\nabla} \varprojlim_n \oplus_{k=0}^n \mathrm{Hom}_{\mathbb{R}_X}\big(\mathrm{Sym}^k(TX), \mathbb{R}_X\big)$$

of bundles over X.

F. Paugam, *Towards the Mathematics of Quantum Field Theory*,
Ergebnisse der Mathematik und ihrer Grenzgebiete. 3. Folge / A Series of Modern
Surveys in Mathematics 59, DOI 10.1007/978-3-319-04564-1_20,

Remark 20.1.1 If $X = \underline{\Gamma}(M, C)$ is the infinite dimensional space of fields of a given classical field theory, the bundle $\mathrm{Sym}^k(TX)$ has as fiber at $\varphi \in X$ the space

$$\mathrm{Sym}^k(TX)_\varphi \cong \Gamma\big(M^k, \varphi^* TC^k\big)_{S_k}.$$

Now we make the stronger assumption that the tangent bundle TX is trivialized by an isomorphism

$$TX \xrightarrow[\sim]{i} X \times V$$

with V a given vector space. This gives in particular an affine connection as above (the trivial one). This hypothesis is true if X is an affine space, i.e., a principal homogeneous space under a vector space V (if X itself is a vector space, one gets a canonical identification $X \cong V$). We then have an isomorphism

$$\mathrm{Jet}(\mathbb{R}_X/X) \xrightarrow{\sim} \varprojlim X \times \oplus_{k=0}^n \mathrm{Hom}_{\mathbb{R}}\big(\mathrm{Sym}^k(V), \mathbb{R}\big).$$

Now we are able to explain the geometric meaning of perturbative computations in quantum field theory. One works with the jet bundle $\mathrm{Jet}(\mathbb{R}_X)$ of functions on the space $X = \underline{\Gamma}(M, C)$ of fields around a given so-called background field φ_0, a solution of the Euler-Lagrange equation. If the bundle C is linear, one simply chooses $\varphi_0 = 0$. This situation is precisely the one postulated by Costello in his book [Cos11] and Epstein-Glaser in their original paper [EG73].

Now to explain fully the computations of perturbative renormalization, one also has to work on the bigger space Y of theories, given by $Y := \mathcal{O}(X) = \mathrm{Hom}(X, \mathbb{R})$, or of local theories, $Y = \mathcal{O}^{loc}(X)$. One then makes perturbative computations for functionals $F \in \mathcal{O}(Y)$ (i.e., functionals $F(S)$ of the action functionals S) around the Gaussian theory S_{free}, meaning we are working with $\mathrm{Jet}(\mathbb{R}_Y)$. This situation is exactly the one postulated by Borcherds [Bor10] (he worked on the space $\mathcal{O}^{loc}(X)$ of local Lagrangian densities, also sometimes called composite fields in the physics literature).

We now clearly see the interest of working in the setting of functorial analysis: it is much easier to deal with complicated spaces of functionals on functionals on functions, such as $\mathcal{O}(\mathcal{O}^{loc}(\underline{\Gamma}(M, C)))$, which are difficult to work with using functional analysis and topological vector spaces.

We refer to Chap. 22 for a detailed application of these methods to the Epstein-Glaser causal perturbative quantum field theory approach, and to Chap. 21 for Costello's description of the perturbative Wilson effective field theory for gauge theories.

We will first present in the next section the nonperturbative definition of a quantum field theory, whose perturbative expansion in our setting gives back Costello's effective quantum BV formalism.

20.2 Nonperturbative Quantum Field Theory

We will use the Wilson effective field theory approach to quantum field theory. We refer to Delamotte's survey [Del12] for a physical introduction to this formalism, which was worked out by Polchinski and Wetterich. This approach is called non-perturbative because it allows us to choose truncations that differ from the usual one, around the Gaussian and around a background field. We won't go into these technical details, and further refer to Gies' survey [Gie06] for complementary information (see also [Ros12] and [Paw07]).

Let $\pi : C \to M$ be a bundle and $X = \underline{\Gamma}(M, C)$ be its space of sections. Suppose given an action functional $S : X \to \mathbb{R}$. The Wetterich version of the renormalization group equation is a vector field on the space $Y = \mathcal{O}_X$ of theories that one wants to solve starting from the initial condition $S \in Y$. The definition of this vector field depends on an auxiliary function R_k of the scale parameter of the theory (a family of real numbers k), and looks like:

$$\partial_k S_k = \frac{1}{2}\tilde{\partial}_k \mathrm{Tr}\,\mathrm{Log}\big(S_k^{(2)} + R_k\big),$$

where $\tilde{\partial}_k := \frac{\partial R_k}{\partial k}\frac{\partial}{\partial R_k}$. The application of this operator means that the above equation is actually a vector field living on the space $Y \times Z$ of pairs (S, R) composed of an effective action S and a regularization operator R, that must be of the same nature as the second order field derivative of S_k. One wants to solve it using the initial condition

$$S_0(\varphi) = S(\varphi).$$

Recall that if $S : X \to \mathbb{R}$ is a functional, viewed as living in $\Gamma(X, \mathbb{R}_X)$, its jet is a section

$$j_\infty S : X \to \mathrm{Jet}(\mathbb{R}_X).$$

If we want to regard the second order derivative as a well defined expression, we need an affine connection on the space of fields, which may be induced by a (partially defined) metric on the space of fields by the Levi-Civita method. We thus assume given a metric

$$g_X : TX \times_X TX \to \mathbb{R}_X$$

and a compatible affine connection

$$\nabla : \Theta_X \to \Theta_X \otimes \Omega^1_X,$$

that is supposed to be *flat*. In this flat affine situation, we have an "identification"

$$\mathrm{Jet}(\mathbb{R}_X) \cong \prod_{n \geq 0} \mathrm{Hom}_{\mathbb{R}_X}\big(TX^{\otimes n}, \mathbb{R}_X\big)^{S_n}$$

of fiber bundles over X given by the symmetric powers of ∇, and we can define the second order derivative of S as a section

$$D^2 S \in \Gamma\big(X, \mathrm{Hom}_{\mathbb{R}_X}\big(TX^{\otimes 2}, \mathbb{R}_X\big)\big) \cong \Gamma\big(X, \mathrm{Hom}_{\mathbb{R}_X - sym}\big(TX, \mathrm{Hom}_{\mathbb{R}_X}(TX, \mathbb{R}_X)\big)\big).$$

Using the metric $g_X : TX \to \mathrm{Hom}_{\mathbb{R}_X}(TX, \mathbb{R}_X)$, or more precisely its (partially defined) inverse

$$g_X^{-1} : \mathrm{Hom}_{\mathbb{R}_X}(TX, \mathbb{R}_X) \to TX,$$

we can send $D^2 S$ to a section

$$g_X^{-1} \circ D^2 S \in \Gamma\big(X, \mathrm{Hom}_{\mathbb{R}_X}\big(\mathrm{End}(TX)\big)\big).$$

We also assume given another section

$$R \in \Gamma\big(X, \mathrm{Hom}_{\mathbb{R}_X}\big(TX^{\otimes 2}, \mathbb{R}_X\big)\big)$$

that is the regularization parameter. This allows us to define

$$g_X^{-1} \circ \big(D^2 S + R\big) \in \Gamma\big(X, \mathrm{Hom}_{\mathbb{R}_X}\big(\mathrm{End}(TX)\big)\big).$$

There is a partially defined logarithm morphism of bundles

$$\mathrm{Log} : \mathrm{End}(TX) \to \mathrm{End}(TX)$$

given by the usual power series

$$\mathrm{Log}(M) := \sum_{n \geq 0} (-1)^n \frac{(M - I)^{n+1}}{n+1}.$$

There is also a trace map

$$\mathrm{Tr} : \mathrm{End}(TX) \to \mathbb{R}_X,$$

given on the fibers by viewing an operator

$$T_{\varphi_0} X \to T_{\varphi_0} X$$

as a continuous (i.e., often, L^2) operator for the given metric g_{X,φ_0}. Applying this trace to the above section, we get

$$\mathrm{Tr}\big(\mathrm{Log}\big(g_X^{-1} \circ \big(D^2 S + R\big)\big)\big) \in \Gamma(X \times Z, \mathbb{R}_{X \times Z}).$$

One can then write down without coordinates the nonperturbative renormalization group equation

$$\partial_k S_k = \frac{1}{2} \tilde{\partial}_k \mathrm{Tr}\, \mathrm{Log}\big(g_X^{-1} \circ \big(D^2 S_k + R_k\big)\big),$$

where $\tilde{\partial}_k := \frac{\partial R_k}{\partial k}\frac{\partial}{\partial R_k}$. This equation is essentially equivalent to

$$\partial_k S_k = \frac{1}{2}\mathrm{Tr}\left(\partial_k R_k \left(D^2 S_k + R_k\right)^{-1}\right).$$

Note that the expression $\partial_k R_k (D^2 S_k + R_k)^{-1}$ in the above trace may be interpreted as a regularized version of the ill-defined nonperturbative propagator $(D^2 S)^{-1}$ corresponding to the solutions of the equations of motion. This interpretation is important in understanding the relation of the above equation to its perturbation around the Gaussian.

20.3 The Dyson-Schwinger Quantum Equations of Motion

One can obtain a regularized version of the Dyson-Schwinger equation by taking functional derivatives of the nonperturbative renormalization group equation

$$\partial_k S_k = \frac{1}{2}\mathrm{Tr}\left(\partial_k R_k \left(D^2 S_k + R_k\right)^{-1}\right).$$

Definition 20.3.1 The *regularized Dyson-Schwinger equations* are the equations

$$\frac{\delta}{\delta \mathbf{v}_1 \cdots \delta \mathbf{v}_n}\partial_k S_k = \frac{1}{2}\frac{\delta}{\delta \mathbf{v}_1 \cdots \delta \mathbf{v}_n}\mathrm{Tr}\left(\partial_k R_k \left(D^2 S_k + R_k\right)^{-1}\right)$$

derived from the nonperturbative renormalization group equation by functional derivation along a family $\{\mathbf{v}_i\}_{i\in\mathbb{N}}$ of vector fields on the space X of fields.

Usually, physicists use the ill-defined vector fields given by derivation along δ functions. This is well defined only when working with local functionals.

20.4 Nonperturbative Quantum Gauge Theory

We now use Sect. 13.5, where we described the Batalin-Vilkovisky gauge fixing procedure, to give a presentation of non-perturbative quantum gauge theories (which are explained in perturbative physical terms in [HT92], without a precise regularization procedure; see also [Bar00] for a perturbative formalism that takes care of anomalies). We refer to Gies' survey [Gie06] for a physical introduction to nonperturbative quantum gauge theory.

In the following, if X is a graded space, we define $\mathcal{O}(X) := \mathrm{Hom}_{\mathrm{PAR}}(X, \mathbb{A})$ where $\mathbb{A}$ is the *full* graded affine space whose points in a smooth graded algebra A are given by $\mathbb{A}(A) = A$.

We recall the notations of Sect. 13.5. Let $\mathcal{A}$ be the jet algebra of a bundle $C \to M$ and $S \in h(\mathcal{A})$ be a local action functional with gauge symmetries. Suppose given a Batalin-Vilkovisky algebra $\mathcal{A}_{BV}$ and a solution S_{cm} to the classical master equation.

Definition 20.4.1 A *nonperturbative quantization* of the gauge theory S is the datum of:

1. A graded gauge fixing in the sense of Definition 13.5.2, i.e., a symplectic quasi-isomorphism
$$i : \mathcal{A}_{BV}^{nm} \xrightarrow{\sim} \mathcal{A}_{BV}$$
and a graded Lagrangian subspace
$$\psi : L \hookrightarrow \underline{\mathrm{Spec}}_{\mathcal{D}}\big(\mathcal{A}_{BV}^{nm}\big),$$
that gives a gauge fixed action $S_\psi \in h(\mathcal{O}_L)$.
2. A family Γ_k of functionals $\Gamma_k : X_{BV}^{nm} \to \mathbb{R}$, where X_{BV}^{nm} is the graded space of solutions of the graded $\mathcal{D}$-algebra $\mathcal{A}_{BV}^{nm}$.
3. A family R_k of regularization variables
$$R_k : L \to \mathrm{Hom}_{\mathbb{R}_L}\big(TL^{\otimes 2}, \mathbb{R}_L\big)^{S_2},$$
such that $\Gamma_k(-,0)$ gives a solution to the nonperturbative renormalization group equation
$$\partial_k \Gamma_k(-,0) = \frac{1}{2}\mathrm{Tr}\big(\partial_k R_k \big(D^2\Gamma_k(-,0) + R_k\big)^{-1}\big),$$
with initial condition $\Gamma_0 = S_\psi$, and further satisfies the *Zinn-Justin equation*
$$\{\Gamma_k, \Gamma_k\} = \Delta(R_k),$$
where Δ is a functional of the regularization variable whose limit at infinity is 0.

We now describe the concrete implementation of the above construction using the original Batalin-Vilkovisky gauge fixing procedure. Let
$$\mathcal{A}_{fields} = \mathrm{Sym}_{\mathcal{A}}\big(\mathfrak{g}^{\circ}[-1] \oplus \mathfrak{a}_S[1] \oplus \mathfrak{l}_S\big)$$
be the algebra of field variables (with coordinates the ghosts, antighosts and their Lagrange multipliers). For simplicity, we will denote by
$$X_{BV} = \underline{\mathrm{Spec}}_{\mathcal{D}}(\mathcal{A}_{BV}) := T^*[-1]X_{fields}$$
the non-minimal Batalin-Vilkovisky graded symplectic space. We will denote by E_{BV} and E_{fields} the bundles over C whose jet algebras give the algebras $\mathcal{A}_{BV}$ and $\mathcal{A}_{fields}$, and $S_{cm} \in h(\mathcal{A}_{BV})$ the (extended) classical master solution, so that the associated dg-$\mathcal{D}$-structure is the usual one on X_{BV}.

We will also denote by $X_{fields} := \underline{\Gamma}(M, E_{fields})$ and $X_{BV} := \underline{\Gamma}(M, E_{BV})$ the associated spaces of sections. A morphism between these spaces is said to be local if it is a morphism of the underlying graded $\mathcal{D}$-spaces. One may identify the de Rham cohomologies of the jet algebras with algebras of functions on those spaces with values in $\mathbb{A}$, meaning that there are natural injections

$$h(\mathcal{A}_{fields}) \to \mathcal{O}(X_{fields}) \quad \text{and} \quad h(\mathcal{A}_{BV}) \to \mathcal{O}(X_{BV}).$$

There is a natural local projection morphism

$$\pi : X_{BV} \to X_{fields},$$

and we fix a local gauge fixing $\psi : X_{fields} \to \mathbb{A}$. This gives an associated local canonical transformation

$$f_\psi : X_{BV} \to X_{BV}$$

and Lagrangian subspace

$$L = X_{fields} \xrightarrow{f_\psi \circ 0} X_{BV}.$$

The gauge fixed action is the local functional

$$S_\psi := S_{cm} \circ f_\psi \circ 0 : X_{fields} \to \mathbb{R}.$$

If we solve the renormalization group equation perturbatively around a background field φ_0, finding a family of formal functionals: $\Gamma_k : \widehat{X_{BV}}_{\varphi_0} \to \mathbb{R}[[\hbar]]$, we may define observables as solutions to the associated regularized Dyson-Schwinger equations.

References

[Bar00] Barnich, G.: Classical and quantum aspects of the extended antifield formalism. arXiv High Energy Physics—Theory e-prints (2000). arXiv:hep-th/0011120

[Bor10] Borcherds, R.E.: Renormalization and quantum field theory. arXiv e-prints (2010). arXiv:1008.0129

[Cos11] Costello, K.: Renormalization and Effective Field Theory. Mathematical Surveys and Monographs, vol. 170, p. 251. Am. Math. Soc., Providence (2011). ISBN 978-0-8218-5288-0

[Del12] Delamotte, B.: An introduction to the nonperturbative renormalization group. In: Renormalization Group and Effective Field Theory Approaches to Many-Body Systems. Lecture Notes in Phys., vol. 852, pp. 49–132. Springer, Heidelberg (2012) (F-PARIS6-PMC)

[EG73] Epstein, H., Glaser, V.J.: The role of locality in perturbation theory. Ann. Henri Poincaré **19**(3), 211–295 (1973). oai:cds.cern.ch:880480

[Gie06] Gies, H.: Introduction to the functional RG and applications to gauge theories. arXiv High Energy Physics—Phenomenology e-prints (2006). arXiv:hep-ph/0611146

[HT92] Henneaux, M., Teitelboim, C.: Quantization of Gauge Systems, p. 520. Princeton University Press, Princeton (1992). ISBN 0-691-08775-X; 0-691-03769-8

[Paw07] Pawlowski, J.: Aspects of the functional renormalisation group. Ann. Phys. **322**, 2831–2915 (2007). doi:10.1016/j.aop.2007.01.007. arXiv:hep-th/0512261

[Ros12] Rosten, O.J.: Fundamentals of the exact renormalization group. Phys. Rep. **511**, 177–272 (2012). doi:10.1016/j.physrep.2011.12.003. arXiv:1003.1366

Chapter 21
Perturbative Renormalization à la Wilson

Here we briefly present the results of Costello's book [Cos11], which give a neat combination of the Batalin-Vilkovisky formalism with Wilson's effective field theory (see also [CG10] and [Gwi05] for applications to factorization algebras).

Note that Wilson's renormalization group method can also be applied in nonperturbative settings, where $\hbar$ is not a formal variable but a real number, as was explained in Chap. 20. This nonperturbative formalization allows the use of other types of truncation schemes, which sometimes have better analytic properties than the one presented here.

21.1 Energetic Effective Field Theory

One usually considers a (Euclidean) effective quantum field theory to be given by an effective action $S_{eff}(\Lambda,\varphi)$ that depends on a cut-off parameter Λ. Observables are then functionals

$$O:\mathcal{C}^\infty(M)_{\leq\Lambda}\to\mathbb{R}[\![\hbar]\!]$$

on the space of fields (here smooth functions) whose expression in the basis of eigenvalues for the Laplacian contain only eigenvalues smaller that the cut-off energy Λ. One clearly has an inclusion of observables at energy Λ' into observables at energy Λ for $\Lambda'\leq\Lambda$. The working hypothesis for Wilson's method is that the mean value of an observable O at energy Λ is given by the functional integral of O on the space $\mathcal{C}^\infty_{\leq\Lambda}$ (which is an ordinary integral since this space is finite dimensional)

$$\langle O\rangle:=\int_{\mathcal{C}^\infty_{\leq\Lambda}}O(\varphi)e^{S_{eff}(\Lambda,\varphi)}[d\varphi]$$

where $S_{eff}(\Lambda,\varphi)$ is a functional called the effective action. The equality between mean values for $\Lambda'\leq\Lambda$ induces an equality

$$S_{eff}(\Lambda,\varphi_L)=\hbar\log\int_{\varphi_H\in\mathcal{C}^\infty(M)_{[\Lambda',\Lambda]}}\exp\left(\frac{1}{\hbar}S_{eff}(\Lambda,\varphi_L+\varphi_H)\right).$$

F. Paugam, *Towards the Mathematics of Quantum Field Theory*,
Ergebnisse der Mathematik und ihrer Grenzgebiete. 3. Folge / A Series of Modern Surveys in Mathematics 59, DOI 10.1007/978-3-319-04564-1_21,

This equation is called the renormalization group equation. The relation of the effective action with the local action S (ordinary datum defining a classical field theory) is given by an ill-defined infinite dimensional functional integral

$$S_{eff}(\Lambda, \varphi_L) = \hbar \log\left(\int_{\varphi_H \in C^\infty_{[\Lambda,\infty]}} \exp\left(\frac{1}{\hbar} S(\varphi_L + \varphi_H)\right)\right).$$

A drawback of this energetic approach is that it is global on the manifold M because the eigenvalues of the Laplacian give global information. We will see in Sect. 21.4 how to use parametrices to give a localized version of this formalism.

21.2 Effective Propagators

We refer to Sect. 4.7 for an introduction to heat kernel regularizations. The starting point is to observe that, if $\Delta + m^2$ is the linear operator for the equations of motion of the free Euclidean scalar field, its inverse (free propagator) can be computed by using the heat kernel K_τ, that is the fundamental solution of the heat equation

$$\partial_\tau u(\tau, x) = \Delta u(\tau, x),$$

with Δ the Laplacian (here we denote τ the time variable, since one thinks of it as the proper time of a Euclidean particle). This amounts to describing $e^{-\tau(\Delta+m^2)}$ as the convolution by a function $K_\tau(x, y)$ (called the heat kernel) to obtain the propagator

$$P(x, y) = \frac{1}{\Delta + m^2} = \int_0^\infty e^{-\tau(\Delta+m^2)} := \int_0^\infty e^{-\tau m^2} K_\tau(x, y) d\tau.$$

The heat kernel can be interpreted as the mean value of a (Wiener) measure on the space of paths on M, meaning that one can give a precise meaning to the equality

$$K_\tau(x, y) = \int_{f:[0,1]\to M, f(0)=x, f(1)=y} \exp\left(-\int_0^\tau \|df\|^2\right).$$

The propagator can then be interpreted as the probability that a particle starts at x and ends at y along a given random path. This also permits us to regard Feynman graphs as random graphical paths followed by families of particles with prescribed interactions. It is also defined on a compact Riemannian manifold with the operator $\Delta_g + m^2$, where g is the metric.

Theorem 21.2.1 *The heat kernel has a small τ asymptotic expansion, in normal coordinates near a point (identify M with a small ball B), of the form*

$$K_\tau \equiv \tau^{-\dim(M)/2} e^{-\|x-y\|^2} \sum_{i \geq 0} \tau^i \Phi_i$$

with $\Phi_i \in \Gamma(B, M) \otimes \Gamma(B, M)$. *More precisely, if* K_τ^N *is the finite sum appearing in the asymptotic expansion, one has*

$$\| K_\tau - K_\tau^N \|_{C^l} = O\big(\tau^{N-\dim(M/2)-l}\big).$$

One can moreover define a weak notion of heat kernel for non-compact varieties, the use of which does not pose existential problems.

The effective propagator of the theory is then given by a cut-off of the heat kernel integral

$$P_{\epsilon,L}(x, y) := \int_\epsilon^L e^{-\tau m^2} K_\tau(x, y) d\tau.$$

21.3 The Renormalization Group Equation

Let $E \to M$ be a graded fiber bundle and $\mathcal{E} = \Gamma(M, E)$ be its space of sections (nuclear Fréchet space). We denote by $\mathcal{E}^{\otimes n}$ the space

$$\mathcal{E}^{\otimes n} := \Gamma\big(M^n, E^{\boxtimes n}\big).$$

Definition 21.3.1 A *functional* is an element of the space $\mathcal{O}(\mathcal{E})$ of formal power series on $\mathcal{E}$, given by

$$\mathcal{O}(\mathcal{E}) = \prod_{n \geq 0} \mathrm{Hom}\big(\mathcal{E}^{\otimes n}, \mathbb{R}\big)^{S_n},$$

where morphisms are supposed to be multilinear continuous maps. We denote by $\mathcal{O}^+(\mathcal{E})[\![\hbar]\!] \subset \mathcal{O}(\mathcal{E})[\![\hbar]\!]$ the space of functionals I that are at least cubic modulo $\hbar$. A *local functional* is an element of the space $\mathcal{O}^{loc}(\mathcal{E})$ of functionals that can be written

$$\Phi = \sum_n \Phi_n$$

with $\Phi_n : \mathcal{E}^{\otimes n} \to \mathbb{R}$ of the form

$$\Phi_n(e_1, \ldots, e_n) = \sum_{j=1}^k \int_M (D_{1,j} e_1) \cdots (D_{n,j} e_n) d\mu$$

with the $D_{i,j} : \mathcal{E} \to C_M^\infty$ some linear differential operators. One can identify these local functionals to global sections of the tensor product over $\mathcal{D}_M$ of the algebra of formal power series on jets by the maximal differential forms on M.

One has a natural algebra structure on $\mathcal{O}(\mathcal{E})$.

Let $I \in \mathcal{O}^+(\mathcal{E})[[\hbar]]$ be a functional and let

$$I = \sum_{i,k} \hbar^i I_{i,k}$$

with $I_{i,k}$ homogenous of degree k in $\mathcal{O}(\mathcal{E})$. We denote by $\mathrm{Sym}^n\mathcal{E}$ the space of S_n-invariants in $\mathcal{E}^{\boxtimes n}$.

Definition 21.3.2 Let $P \in \mathrm{Sym}^2\mathcal{E}$ and $I \in \mathcal{O}^+(\mathcal{E})[[\hbar]]$ be given. Let ∂_P be the contraction with P. The *P-renormalization group* evolution of I is given by the functional

$$W(P, I) = \hbar \log\bigl(\exp(\hbar\partial_P)\exp(I/\hbar)\bigr) \in \mathcal{O}^+[[\hbar]].$$

Example 21.3.1 If one starts with the standard local interaction functional $I \in \mathrm{Hom}(\mathcal{C}^\infty(M^3), \mathbb{R})$, given by

$$I = \int_M \varphi^3(x)dx = \int_{M^3} \varphi(x)\varphi(y)\varphi(z)\delta_{\Delta_{123}} dxdydz,$$

of the φ^3 theory (where $\Delta_{123} = M \subset M^3$ is the diagonal) and $P(x, y) \in \mathcal{C}^\infty(M \times M)$ is symmetric in x and y, the contraction of I by P is the distribution $\partial_P(I) \in \mathcal{C}^\infty(M)'$ defined as

$$\bigl\langle \partial_P(I), \varphi\bigr\rangle = \int_{M^3} \sigma P(x, y)\varphi(z)\delta_{\Delta_{123}} dxdydz,$$

where σ is the symmetrization operator.

21.4 Effective Theory with Interaction

Definition 21.4.1 Let D be a generalized Laplacian on a graded bundle $E \to M$. A *parametrix* for the operator D is a distributional section $P \in \mathrm{Hom}(\mathcal{E}^{\otimes 2}, \mathbb{R})$ of $E \boxtimes E \to M \times M$, which is symmetric, smooth away from the diagonal, and such that

$$(D \otimes 1)P - K_0 \in \Gamma_{\mathcal{C}^\infty}(M \times M, E \boxtimes E),$$

where K_0 is the time zero heat operator (δ_M for the Laplacian on flat space).

For example, if $P(0, L)$ is the heat kernel propagator, $P = P(0, L)$ is a parametrix. The difference $P - P'$ between two parametrices is a smooth function. Parametrices are partially ordered by their support.

Definition 21.4.2 A function $J \in \mathcal{O}(\mathcal{E})$ has *smooth first derivative* if the continuous linear map

$$\mathcal{E} \to \mathcal{O}(\mathcal{E}), \qquad \varphi \mapsto \frac{\partial J}{\partial \varphi}$$

extends to the space $\mathcal{E}'$ of distributional sections of $\mathcal{E}$.

Definition 21.4.3 A *quantum field theory* is a collection of functionals

$$I[P] \in \mathcal{O}^+(\mathcal{E})[[\hbar]]$$

indexed by parametrices, such that:

1. (Renormalization group equation) If P, P' are parametrices, then

$$W\big(P - P', I[P']\big) = I[P].$$

   This expression makes sense , because $P - P'$ is smooth in $\Gamma(M \times M, E \boxtimes E)$.
2. (Locality) For any (i, k), the support $\mathrm{supp}(I_{i,k}[P]) \subset M^k$ can be made as close as we like to the diagonal by making the parametrix P small (for any small neighborhood U of the diagonal, there exists a parametrix such that the support is in U for all $P' \leq P$).
3. The functionals $I[P]$ all have smooth first derivative.

One denotes by $\mathcal{T}^{(n)}$ the space of theories defined modulo $\hbar^{n+1}$.

One may also give a more concrete formulation of the above definition in terms of the effective propagator $P(\epsilon, L)$, and of the length scale effective interactions $I(L)$, using the renormalization group equation

$$W\big(P(\epsilon, L), I(\epsilon)\big) = I(L).$$

If $I[L]$ is a family of effective interactions in the length formulation,

$$I[P] = W\big(P - P(0, L), I[L]\big)$$

gives a quantum field theory in the parametrix sense.

We gave the parametrix formulation because it makes the comparison of Costello's work with the causal approach of Chap. 22 more transparent.

21.5 Feynman Graph Expansion of the Renormalization Group Equation

We now explain the relation of Costello's approach to the physicists' Feynman graph computations. This is not strictly necessary to prove the basic theorems, but may be helpful in understanding the relation with other approaches.

Definition 21.5.1 A *stable graph* is a graph γ, with external edges, and for each vertex v, an element $g(v) \in \mathbb{N}$ called the *genus* of the vertex, with the property that any vertex of genus 0 is at least trivalent.

If $P \in \mathrm{Sym}^2\mathcal{E}$ (a propagator) and $I \in \mathcal{O}^+(\mathcal{E})[\![\hbar]\!]$ (an interaction), for every stable graph γ, one defines

$$w_\gamma(P, I) \in \mathcal{O}(\mathcal{E})$$

by the following.

Let $T(\gamma)$ be the set of tails, $H(\gamma)$ be the set of half edges, $V(\gamma)$ the set of vertices and $E(\gamma)$ the set of internal edges. We denote by $b_1(\gamma)$ the first Betti number of the graph. The tensor product of interactions $I_{i,k}$ at vertices of γ with valency k and genus i define an element of

$$\mathrm{Hom}\big(\mathcal{E}^{\otimes H(\gamma)}, \mathbb{R}\big).$$

One can define an element of

$$\mathcal{E}^{\otimes 2E(\gamma)} \otimes \mathcal{E}^{\otimes T(\gamma)} \simeq \mathcal{E}^{\otimes H(\gamma)}$$

by associating to each internal edge a propagator and to each tail a field $\varphi \in X$. The contraction of these two elements give an element

$$w_\gamma(P, I)(\varphi) \in \mathbb{R}.$$

To say it differently, propagators give an element of $\mathcal{E}^{\otimes 2E(\gamma)}$ that can be contracted to get

$$w_\gamma(P, I) \in \mathrm{Hom}\big(\mathcal{E}^{\otimes T(\gamma)}, \mathbb{R}\big).$$

One can thus define

$$W(P, I) = \sum_\gamma \frac{1}{|\mathrm{Aut}(\gamma)|} \hbar^{b_1(\gamma)} w_\gamma(P, I) \in \mathcal{O}^+(\mathcal{E})[\![\hbar]\!].$$

Example 21.5.1 Take a scalar φ^3 theory with

$$I = I_{3,0} = \int_{M^3} \varphi_1(x)\varphi_2(y)\varphi_3(z)\delta_\Delta(x, y, z)d\mu_3 := \int_M \varphi_1(x)\varphi_2(x)\varphi_3(x)d\mu,$$

where $\Delta : M \subset M^3$ is the diagonal. The effective propagator $P(\epsilon, L)$ for the Laplacian is a distribution on M^2, that is given by a smooth function on M^2. One can also think of it as an element of the tensor product $C^\infty(M) \otimes C^\infty(M)$, i.e., of $\mathcal{E}^{\otimes 2}$. One associates to the vertex of valency 3 simply

$$I_{3,0} \in \mathrm{Hom}\big(\mathcal{E}^{\otimes 3}, \mathbb{R}\big).$$

If one has a graph with only one propagator and two vertices of valency 3 and genus 0, one associates to it the contraction of the tensor product

$$I_{3,0} \otimes I_{3,0} \in \mathrm{Hom}\big(\mathcal{E}^{\otimes 6}, \mathbb{R}\big)$$

with propagator $P = P(\epsilon, L) \in \mathcal{E}^{\otimes 2}$, that gives a morphism

$$w(P, \gamma) \in \operatorname{Hom}(\mathcal{E}^{\otimes 4}, \mathbb{R}).$$

The explicit expression for $w(P, \gamma)$ is given by

$$\begin{aligned} & w(P,\gamma)(\varphi_1,\dots,\varphi_6) \\ & := \sigma \int_{M^6} \varphi_1(x_1)\varphi_2(x_2)P(x_3,x_4)\varphi_5(x_5)\varphi_6(x_6)\delta_{\Delta_{123}}(x_i)\delta_{\Delta_{456}}(x_i)d\mu_6 \\ & := \sigma \int_{M^2} \varphi_1(x_1)\varphi_2(x_1)P(x_1,x_2)\varphi_5(x_2)\varphi_6(x_2)d\mu_2, \end{aligned}$$

where $\Delta_{123} : M \subset M^6$ and $\Delta_{456} : M \subset M^6$ are the diagonal maps into the indexed subspaces and σ is the symmetrization operator. Note that this new functional $w(P, \gamma)$ is not supported on the diagonal $\Delta_{123456} : M \subset M^6$, so that it is no longer a local functional.

One can associate to $P \in \operatorname{Sym}^2\mathcal{E}$ a differential operator

$$\partial_P : \mathcal{O}(\mathcal{E}) \to \mathcal{O}(\mathcal{E})$$

that gives a contraction in degree n. One then gets

$$W(P, I) = \hbar \log\big(\exp(\hbar\partial_P)\exp(I/\hbar)\big),$$

which is the usual equation for the definition of Feynman graphs.

21.6 The Quantization Theorem

Period variations are functions $f(t)$ on $]0, \infty[$ of the form

$$f(t) = \int_{\gamma_t} \omega_t$$

with ω a relative logarithmic differential form on a family of varieties $(X, D) \to U$ with normal crossing divisor defined on an open subset of the algebraic affine line $\mathbb{A}^1_{\mathbb{R}}$ that contains $]0, \infty[$. We denote by $\mathcal{P} = \mathcal{P}(]0, \infty[)$ the set of period variations. These are smooth functions of t. We denote by $\mathcal{P}_{\geq 0} \subset \mathcal{P}$ the sub-space of period variations that have a limit at 0.

Definition 21.6.1 A *renormalization scheme* is a choice of a complementary sub-space (singular parts) $\mathcal{P}_{<0}$ to $\mathcal{P}_{\geq 0}$ in $\mathcal{P}$, i.e., of a direct sum decomposition

$$\mathcal{P} = \mathcal{P}_{\geq 0} \oplus \mathcal{P}_{<0}.$$

The *singular part* of a period variation for this scheme is its projection

$$\mathrm{sing} : \mathcal{P} \to \mathcal{P}_{<0}.$$

Theorem 21.6.1 *The space $\mathcal{T}^{(n+1)} \to \mathcal{T}^{(n)}$ is canonically equipped with a principal bundle structure under the group $\mathcal{O}^{loc}(\mathcal{E})$ of local functionals. Moreover, $\mathcal{T}^{(0)}$ is canonically isomorphic to the space $\mathcal{O}^{+,loc}$ of local functionals that are at least cubic modulo $\hbar$.*

Proof Note first that the proof of the theorem uses a renormalization scheme, but that its statement doesn't. So let's fix a renormalization scheme. If $I[P]$ and $J[P]$ are two theories that coincide modulo $\hbar^{n+1}$, then

$$I_{0,*}[P] + \delta\hbar^{-(n+1)}\big(I[P] - J[P]\big) \in \mathcal{O}[\delta]/\delta^2$$

define a tangent vector to the space $\mathcal{T}^{(0)}$ of classical theories, that is canonically isomorphic to $\mathcal{O}^{+,loc}$. To prove the surjectivity, one needs to show that if

$$I[P] = \sum_{i,k} \hbar^i I_{i,k}[P] \in \mathcal{T}^{(n)}$$

is a theory defined modulo $\hbar^{n+1}$, there exist counter-terms

$$I^{CT}(\epsilon) = \sum_{i,k} \hbar^i I^{CT}_{i,k}(\epsilon) \in \mathcal{O}^{loc}[\![\hbar]\!] \otimes \mathcal{P}_{<0}$$

such that

$$I[P] := \lim_{\epsilon \to 0} W\big(P - P(0,\epsilon), I - I^{CT}(\epsilon)\big)$$

exists and defines a theory in $\mathcal{T}^{(n+1)}$. Recall that the renormalization group equation reads

$$W\big(P - P', I\big) = \hbar \log\big(\exp(\hbar\partial_{P-P'})\exp(I/\hbar)\big)$$

for parametrices P and P'. One has by construction

$$W\big(P - P', W\big(P' - P'', I\big[P''\big]\big)\big) = W\big(P - P'', I\big[P''\big]\big)$$

and decomposing in homogeneous components in $\hbar$ and $\mathcal{E}$, one also has

$$W\big(P - P', I\big)_{i,k} = W\big(P - P', I_{<(i,k)}\big)_{i,k} + I_{i,k}.$$

Suppose that the limit of

$$W_{<(i,k)}\left(P - P(0,\epsilon), I - \sum_{(r,s)<(i,k)} \hbar^r I^{CT}_{r,s}(\epsilon)\right)$$

as ϵ tends to 0 exists. Using the above relations, one has to set

$$I^{CT}_{(i,k)}(\epsilon) := \text{sing}_\epsilon\left(W_{(i,k)}\left(P - P(0,\epsilon), I - \sum_{(r,s)<(i,k)} \hbar^r I^{CT}_{(r,s)}(\epsilon)\right)\right).$$

One has to check that this expression does not depend on P. This can be proven in the case $P = P(0, L)$ by showing that derivative in L of what is inside the singular part is non-singular: it is given by graphs with vertices containing small degree expressions that are non-singular by induction and one just adds a propagator $P(\epsilon, L)$ between them that is also non-singular. Some more work is needed to prove the locality axiom, based on the asymptotic expansion of the heat kernel. □

21.7 The Effective Quantum BV Formalism

Definition 21.7.1 A *free BV theory* is given by:

1. A $\mathbb{Z}$-graded vector bundle E on M, whose graded space of global sections is denoted by $\mathcal{E}$.
2. An odd antisymmetric pairing of graded vector bundles

$$\langle\cdot,\cdot\rangle_{loc} : E \otimes E \to \text{Dens}(M)[-1]$$

   that is assumed to be fiber-wise non-degenerate.
3. A differential operator $Q : \mathcal{E} \to \mathcal{E}[1]$ of square zero that is anti-self-adjoint for the given pairing such that $H^*(\mathcal{E}, Q)$ is finite dimensional.
4. A solution $I \in \mathcal{O}^+_{loc}$ of the classical master equation

$$Q(I) + \frac{1}{2}\{I, I\} = 0.$$

A *gauge fixation* on a BV theory is the datum of an operator

$$Q^{GF} : \mathcal{E} \to \mathcal{E}[-1]$$

such that:

1. Q^{GF} is odd, of square zero and self-adjoint; and
2. the commutator

$$D := \left[Q, Q^{GF}\right]$$

   is a (generalized) Laplacian.

One thinks of Q^{GF} as a homological version of the ordinary gauge fixing, that gives a way to fix a Lagrangian subspace in $\mathcal{E}$ on which the quadratic form $\int \varphi D\varphi$ is non-degenerate.

Definition 21.7.2 A *parametrix* Φ for a BV theory is a symmetric distributional section of $(E \boxtimes E)[1]$ on $M \times M$ closed under the differential $Q \otimes 1 + 1 \otimes Q$, with proper support and such that $(D \otimes 1)\Phi - K_0$ is a smooth section of $E \boxtimes E$. The *propagator* of a parametrix is

$$P(\Phi) = \left(Q^{GF} \otimes 1\right)\Phi$$

and its *fake heat kernel* is

$$K(\Phi) = K_0 - (D \otimes 1)\Phi.$$

The difference between two parametrices is smooth because it is in the kernel of the elliptic operator D.

The effective BV Laplacian is defined as the insertion of $K(\Phi)$ by

$$\Delta_\Phi := -\partial_{K(\Phi)} : \mathcal{O} \to \mathcal{O}.$$

The ill-defined BV Laplacian would be

$$\Delta = \lim_{L \to 0} \Delta_{P(0,L)}.$$

One defines a bracket $\{\cdot, \cdot\}_\Phi$ on $\mathcal{O}$ by the formula

$$\{\alpha, \beta\}_\Phi = \Delta_\Phi(\alpha\beta) - (\Delta_\Phi\alpha)\beta - (-1)^{|\alpha|}\alpha\Delta_\Phi\beta$$

that measures the obstruction for Δ_Φ to be a graded-derivation.

Definition 21.7.3 A functional $I[\Phi] \in \mathcal{O}^+[\![\hbar]\!]$ satisfies the *scale Φ quantum master equation* if

$$QI[\Phi] + \left\{I[\Phi], I[\Phi]\right\}_\Phi + \hbar\Delta_\Phi I[\Phi] = 0.$$

One can also show that

$$[\partial_{P(\Psi)-P(\Phi)}, Q] = \partial_{K_\Psi} - \partial_{K_\Phi} = \Delta_\Phi - \Delta_\Psi$$

so that $\partial_{P(\Psi)-P(\Phi)}$ defines a homotopy operator between the BV Laplacians at different scales. Note that if $I[\Phi]$ satisfies the scale Φ quantum master equation, then

$$I[\Psi] = W\left(\Psi - \Phi, I[\Phi]\right)$$

satisfies the scale Ψ quantum master equation.

Definition 21.7.4 A *BV theory* is the datum of a family $I[\Phi] \in \mathcal{O}[\![\hbar]\!]$ such that:

1. the collection defines a theory in the classical sense;
2. the collection satisfies the quantum master equation.

We denote by $\mathcal{T}^n_{BV}$ the space of BV theories defined up to order n in $\hbar$.

By Theorem 21.6.1, one can always associate an effective theory to the gauge fixed local theory (which no longer has symmetries). However, there exist obstructions for this to be constructed and to also satisfy the quantum master equation. More precisely, if one looks at the application

$$\mathcal{T}_{BV}^{n+1} \longrightarrow \mathcal{T}_{BV}^{n}$$

between theories defined up to order $n+1$ in $\hbar$ and theories defined up to order n, the obstruction to its lifting is of a local cohomological nature, i.e., is given by a simplicial map

$$O_{n+1} : \mathcal{T}^n \longrightarrow \mathcal{O}_{loc}$$

for families of theories parametrized by $\Omega^*(\Delta^*)$.

More precisely, if $I[P]$ is a theory defined up to order $\hbar^{n+1}$, it satisfies quantum BV and $\tilde{I}[P]$ is a lift of this theory up to order $\hbar^{n+2}$, the obstruction for it to satisfy quantum BV is given by the formula

$$O_{n+1}[P] := \hbar^{-(n+1)}\left(Q(\tilde{I}) + \frac{1}{2}\{\tilde{I}, \tilde{I}\}_P + \hbar\Delta_P(\tilde{I})\right).$$

Setting $P = P(0, L)$, one may show that the limit as L tends to 0 of $O_{n+1}(P(0, L))$ exists and is a local functional. We refer to [Cos11] for a refined study of obstructions.

21.8 Observables and the Partition Function

We refer to [CG10] for the following considerations.

Let X denote the space of fields and $\mathcal{T}$ be the space of theories. One may view the partition function as a function

$$Z : \mathcal{T} \to \mathbb{R}$$

on the space of functionals (i.e., theories), given by the formal formula

$$Z(T) := \int_X e^{S(\varphi)+T(\varphi)}[d\varphi],$$

which makes sense when M is discrete. The advantage of this presentation is that we don't assume that T is given by a linear interaction of the form

$$T_J(\varphi) = \int_M J(x)\varphi(x)dx,$$

so that we don't have to assume that the space of fields is linear.

Now notice that if f is an observable, we have the equality

$$\int_X f(\varphi)e^{S(\varphi)}[d\varphi] = \frac{d}{d\epsilon}\bigg|_{\epsilon=0} \int e^{S(\varphi)+\epsilon f(\varphi)}[d\varphi] =: \frac{\delta Z}{\delta f}(0).$$

This implies that the mean value

$$\langle f\rangle_S := \frac{\int_X f(\varphi)e^{S(\varphi)}[d\varphi]}{Z(0)}$$

may be computed by the formula

$$\langle f\rangle_S := \frac{1}{Z(0)}\frac{\delta Z}{\delta f}(0).$$

The conclusion is that we may interpret observables f for a theory S as living in the tangent space

$$T_S\mathcal{T}$$

at S of the space of theories.

One may formulate the above formal computations properly by defining the mean value of an observable $f \in T_S\mathcal{T}$ (for a given decomposition $S = Q + I$ of S in a quadratic part Q and an interaction part I), by the formula

$$\langle f\rangle_I := \frac{d}{d\epsilon}\bigg|_{\epsilon=0} \left(\log e^{\hbar\partial_P} e^{(I+\epsilon f)/\hbar}\right) = \frac{d}{d\epsilon}\bigg|_{\epsilon=0} W(P, I+\epsilon f).$$

21.9 The Local Renormalization Group Action

We now consider theories on $\mathbb{R}^n$, which is not compact. One thus has to restrict to functionals with some decreasing conditions at infinity for the homogeneous components. Note that Hollands and Wald [HW10] also use a renormalization group in curved spacetime, by acting directly on the metric itself.

Consider the operation

$$R_l : \mathcal{S}(\mathbb{R}^n) \to \mathcal{S}(\mathbb{R}^n)$$

defined by

$$R_l(\varphi)(x) = l^{n/2-1}\varphi(lx).$$

If $I \in \mathcal{O}(\mathcal{S}(\mathbb{R}^n))$ is a functional on $\mathcal{S}(\mathbb{R}^n)$, one defines

$$R_l^*(I)(\varphi) := I\left(R_{l^{-1}}(\varphi)\right).$$

Definition 21.9.1 The *local renormalization group* is the action

$$\mathcal{RG}_l : \mathcal{T}^{(\infty)} \to \mathcal{T}^{(\infty)}$$

on the space of theories defined by

$$\mathcal{RG}_l(\{I_L\}) := \{R_l^*(I_{l^2L})\}.$$

One shows that the dependence on l is in $\mathbb{C}[\log l, l, l^{-1}]$.

Definition 21.9.2 A theory is called *valuable* if the terms of its effective action vary in $l^k(\log l)^r$ when $k \geq 0$ for $r \in \mathbb{Z}_{\geq 0}$. We denote by $\mathcal{R}^{(\infty)}$ the space of valuable theories. It is called *renormalizable* if it is valuable and, term by term in $\hbar$, it has only a finite number of deformations, i.e., the tangent space $T_{\{I_L\}}\mathcal{R}^{(n)}$ is finite dimensional.

Costello uses the above notion to give in [Cos11], Theorem 8.1.1, a complete classification of renormalizable scalar field theories on $\mathbb{R}^n$. In particular, he obtains the following theorem on $\mathbb{R}^4$.

Theorem 21.9.1 *Renormalizable scalar field theories on* $\mathbb{R}^4$, *invariant under* $\mathrm{SO}(4) \rtimes \mathbb{R}^4$ *and under* $\varphi \mapsto -\varphi$, *are in bijection with Lagrangians of the form*

$$\mathcal{L}(\varphi) = a\varphi D\varphi + b\varphi^4 + c\varphi^2$$

for $a, b, c \in \mathbb{R}[[\hbar]]$, *where* $a = -1/2$ *modulo* $\hbar$ *and* $b = 0$ *modulo* $\hbar$.

21.10 Application to Pure Yang-Mills Theory

In the Yang-Mills case of dimension 4, the graded space E is given by

$$\mathcal{E} = \Omega^0(M) \otimes \mathfrak{g}[1] \oplus \Omega^1(M) \otimes \mathfrak{g} \oplus \Omega^3(M) \otimes \mathfrak{g}[-1] \oplus \Omega^4(M) \otimes \mathfrak{g}[-2].$$

The elements of the four components of $\mathcal{E}$ are respectively called the ghosts, fields, antifields and antighosts and denoted by X, A, $A^\vee$ and $X^\vee$.

The differential is given by

$$\Omega^0(M) \otimes \mathfrak{g} \xrightarrow{d} \Omega^1(M) \otimes \mathfrak{g} \xrightarrow{d*d} \Omega^3(M) \otimes \mathfrak{g} \xrightarrow{d} \Omega^4(M) \otimes \mathfrak{g}.$$

Recall that the Yang-Mills field is in $\Omega^1(M) \otimes \mathfrak{g}$ and the gauge symmetries are in $\Omega^0(M) \otimes \mathfrak{g}$.

The pairing on $\Omega^*(M) \otimes \mathfrak{g}$ is given by

$$\langle \omega_1 \otimes E_1, \omega_2 \otimes E_2 \rangle := \int_M \omega_1 \wedge \omega_2 (E_1, E_2)_{\mathfrak{g}}.$$

The Yang-Mills action functional is given by

$$S(A) = \frac{1}{2}\int_M \left[\langle F_A, F_A \rangle + \langle F_A, *F_A \rangle\right].$$

The first term is topological and the second is the usual Yang-Mills functional.

The solution of the classical master equation is obtained by adding to the Lagrangian the ghost terms

$$\frac{1}{2}\langle [X, X], X^{\vee}\rangle$$

and the terms that give the action of the ghosts on fields

$$\langle dX + [X, A], A^{\vee}\rangle.$$

One has to use a first order formulation of Yang-Mills theory to have a gauge fixing (such that $D = [G, G^{GF}]$ is a Laplacian, and not a differential operator of degree 4). This theory has two fields: one in $\Omega^1(M) \otimes \mathfrak{g}$ and the other auto-dual in $\Omega^2_+(M) \otimes \mathfrak{g}$.

The problem of proving the existence of an effective BV theory is then reduced to a computation of a cohomological obstruction. We refer to Costello's book for more details and for applications to renormalization of pure Yang-Mills theory in dimension 4.

References

[CG10] Costello, K., Gwilliam, O.: Factorization algebras in perturbative quantum field theory (2010, preprint)

[Cos11] Costello, K.: Renormalization and Effective Field Theory. Mathematical Surveys and Monographs, vol. 170, p. 251. Am. Math. Soc., Providence (2011). ISBN 978-0-8218-5288-0

[Gwi05] Gwilliam, O.: Factorization algebras and free field theory. Thesis, Northwestern University (2005)

[HW10] Hollands, S., Wald, R.M.: Axiomatic quantum field theory in curved spacetime. Commun. Math. Phys. **293**, 85–125 (2010). doi:10.1007/s00220-009-0880-7. arXiv:0803.2003

Chapter 22
Causal Perturbative Quantum Field Theory

This section gives a short self-contained and coordinate-free presentation of causal perturbative quantum field theory in the spirit of the Bogoliubov-Epstein-Glaser renormalization method (see [EG73]), in the version developed on curved spacetime by Brunetti and Fredenhagen [BF09] (see also [Kel07] for more references and the relation with the Connes-Kreimer approach of Chap. 19), Brouder [Bro09], and Borcherds [Bor10] (see also [BBD+12]).

Note that these methods have been extended to gauge theories by Fredenhagen and Rejzner [FR11] (see also [Rej11]) and may also be applied in the Euclidean setting, thanks to the work of Keller [Kel09].

We adapt Keller's presentation to our functorial setting, which allows us to clarify the mathematical status of the space of fields underlying these computations, and to make precise the generality of the gauge fixing procedure.

22.1 Moyal Type Star Products and Hopf Algebras

Let M be a finite dimensional manifold and $\pi \in \wedge^2\Theta_M$ be a Poisson bracket. Let $\mathbb{R}_M \to M$ be the trivial bundle and $\mathrm{Jet}(\mathbb{R}_M) \to M$ be the associated jet bundle. We suppose given a *flat* affine connection ∇ on M that is compatible with the Poisson bracket, meaning that π is horizontal for ∇. The symmetric powers of ∇ give an isomorphism

$$\mathrm{Jet}(\mathbb{R}_M) \xrightarrow{\sim} \prod_{n\geq 0} \mathrm{Hom}_{\mathbb{R}_M}\left(TM^{\otimes n}, \mathbb{R}_M\right)^{S_n}.$$

Recall that the Poisson bracket of two functions f and g is defined by

$$\{f, g\} := \langle df \wedge dg, \pi\rangle.$$

The jet bundle $\mathrm{Jet}(\mathbb{R}_M) \to \mathbb{R}_M$ is a vector bundle. Consider the exterior product

$$\mathrm{Jet}(\mathbb{R}_M) \wedge \mathrm{Jet}(\mathbb{R}_M)$$

F. Paugam, *Towards the Mathematics of Quantum Field Theory*, 425
Ergebnisse der Mathematik und ihrer Grenzgebiete. 3. Folge / A Series of Modern Surveys in Mathematics 59, DOI 10.1007/978-3-319-04564-1_22,

over $\mathbb{R}_M$. One may generalize the above formula by computing

$$\langle D^n f \wedge D^n g, \pi^n \rangle.$$

This means that the exponential operator

$$P_\pi := \exp\left(\frac{i\hbar}{2}\pi\right) := \sum_{n\geq 0} \frac{1}{n!}\left(\frac{i\hbar}{2}\right)^n \langle D^n - \wedge D^n -, \pi^n \rangle,$$

where D^n is the projection of jets on their n-th component, and by convention,

$$\langle D^0 f \wedge D^0 g, \pi^0 \rangle := f \cdot g,$$

gives a well defined operator

$$\Gamma\big(M, \mathrm{Jet}(\mathbb{R}_M) \wedge \mathrm{Jet}(\mathbb{R}_M)\big) \to \mathcal{C}^\infty(M)[\![\hbar]\!].$$

Combining this operator with the jet map

$$j_\infty : \mathcal{C}^\infty(M) = \Gamma(M, \mathbb{R}_M) \to \Gamma\big(M, \mathrm{Jet}(\mathbb{R}_M)\big),$$

we get the Moyal star product

$$* : \mathcal{C}^\infty(M) \times \mathcal{C}^\infty(M) \to \mathcal{C}^\infty(M)[\![\hbar]\!].$$

Note that the associativity of this product holds only because the connection is flat and the given Poisson structure π is horizontal with respect to the given affine connection ∇. The $*$-product associated to a general Poisson bracket may be obtained in a similar way, but one has to work with the corresponding Poisson sigma model in the Batalin-Vilkovisky formalism, as described in Chap. 23.

The addition on the vector bundle TM induces a natural bundle coalgebra structure Δ on the power series algebra bundle $\mathrm{Jet}(\mathbb{R}_M)$, given fiberwisely by

$$\Delta(f)(v, w) := f(v + w).$$

Writing $\Delta(f) = \sum f_{(1)} \otimes f_{(2)}$ may be achieved by using the Taylor series expansion

$$\Delta(f)(v, w) = f(v + w) = \sum \frac{1}{n!}\big(D^n f\big)(v) \otimes w^n.$$

We may extend the Poisson pairing

$$\pi \in \Gamma\big(M, \mathrm{Hom}_{\mathbb{R}_M}\big(TM^{\otimes 2}, \mathbb{R}_M\big)\big)$$

to a bicharacter (see Sect. 6.2)

$$\psi_\pi \in \Gamma\big(M, \mathrm{Hom}_{\mathbb{R}_M}\big(\mathrm{Jet}(\mathbb{R}_M) \otimes \mathrm{Jet}(\mathbb{R}_M), \mathbb{R}_M[\![\hbar]\!]\big)\big),$$

which we may use to twist the bundle comodule algebra $(\mathrm{Jet}(\mathbb{R}_M), \otimes)$ under the bialgebra

$$\big(\mathrm{Jet}(\mathbb{R}_M), \otimes, \Delta\big).$$

This twist is given by an associative product

$$\overset{\hbar}{*} : \Gamma\big(M, \mathrm{Jet}(\mathbb{R}_M)\big) \times \Gamma\big(M, \mathrm{Jet}(\mathbb{R}_M)\big) \to \Gamma\big(M, \mathrm{Jet}(\mathbb{R}_M)\big)[\![\hbar]\!]$$

of the form

$$f \overset{\hbar}{*} g := \sum \psi_\pi(f_{(1)}, g_{(1)}) f_{(2)} g_{(2)}.$$

One may check that the product of two jets $j_\infty f$ and $j_\infty g$ is also a jet, which we may denote by $j_\infty(f * g)$. This thus gives by restriction a well defined $*$-product

$$\overset{\hbar}{*} : \mathcal{C}^\infty(M) \times \mathcal{C}^\infty(M) \to \mathcal{C}^\infty(M)[\![\hbar]\!],$$

that identifies, when correctly normalized, with the Moyal star product.

Note that the above reasoning can also be carried out in the setting where we replace M by a space of T of trajectories $T \subset \underline{\Gamma}(M, C)$ for a scalar field theory, and π is given by the Peierls-Dewitt bracket (see [Pei52, DeW03, DDM04]). The case of gauge theories can also be studied in a similar way in the setting of their homotopical Poisson reduction, described in Chap. 13. In these infinite dimensional cases, one has to carefully compute the insertions

$$\big\langle D^n f \wedge D^n g, \pi^n \big\rangle$$

because the Poisson bivector π is usually only a distributional bivector field. This leads to the regularization and renormalization problem.

22.2 Causal Star Products

We now use the formalism of Sect. 20.1 to give a geometric formulation of the Epstein-Glaser perturbative renormalization, inspired by Brunetti-Fredenhagen [BF09]. Our presentation using functors of points is however original, and adapted to more general situations. We denote by $X = \underline{\Gamma}(M, C)$ the space of fields of a quantum field theory, $TX \to X$ its tangent space, and $\mathbb{R}_X := \mathbb{R} \times X \to X$ the trivial bundle over X, whose sections are smooth functions on X. We assume given a flat affine connection ∇_X and a metric g_X on X that is flat for the given connection. We will actually assume that the metric is given at each $\varphi_0 \in X$ by the integration pairing

$$\int_M : \Gamma\big(M, \varphi_0^* TC\big) \times \Gamma\big(M, \varphi_0^* TC\big) \to \mathbb{R}$$

associated to a given density on M.

Definition 22.2.1 A regularized propagator is a section $P \in \Gamma(X, \mathrm{Sym}^2(TX))$ that is horizontal for ∇_X.

Note that the regularized propagators of physics usually do not depend on the variable $\varphi \in X$, and have constant value in the fiber of $\mathrm{Sym}^2(TX)$. This corresponds to the situation where X is a linear space, so that it carries a trivial affine connection, whose horizontal sections are simply the constant ones.

Using the trivialization

$$\mathrm{Jet}(\mathbb{R}_X/X) \cong \prod_{k\geq 0} \mathrm{Hom}_{\mathbb{R}_X}\left(\mathrm{Sym}^k(TX), \mathbb{R}_X\right)$$

induced by the affine connection on X, one can define a bundle algebra structure on $\mathrm{Jet}(\mathbb{R}_X)/X$, induced by the algebra structure on symmetric multilinear maps, which we will denote by $\otimes$.

Every functional $f : X \to \mathbb{R}$ thus defines a family

$$f^{(n)} = D^n f : X \to \mathrm{Hom}_{\mathbb{R}_X}\left(\mathrm{Sym}^n(TX), \mathbb{R}_X\right)$$

of distributions parametrized by X, whose value at $\varphi_0 \in X$ is a symmetric distribution

$$D^n_{\varphi_0} f : \otimes^n(T_{\varphi_0}X) := \Gamma\left(M^n, \varphi_0^* TC^n\right) \to \mathbb{R}.$$

From now on, we restrict to functionals whose derivatives are Schwarz distributions.

Definition 22.2.2 The *support of a functional* $f : X \to \mathbb{R}$ is defined by

$$\mathrm{supp}(f) := \overline{\cup_{\varphi_0 \in X} \mathrm{supp}\left(f^{(1)}_{\varphi_0}\right)}.$$

Definition 22.2.3 The star product associated to the regularized propagator is the operation

$$\begin{aligned} * : \mathcal{O}(X)[[\hbar]] \times \mathcal{O}(X)[[\hbar]] &\to \mathcal{O}(X)[[\hbar]] \\ (f, g) &\mapsto \left\langle j_\infty(f) \otimes j_\infty(g), \sum_{n\geq 0} \hbar^n P^{\otimes n} \right\rangle. \end{aligned}$$

We now want to define a coordinate-free version of the Epstein-Glaser renormalization scheme. The basic idea of this construction is to take a family P_λ of regularized propagators that tends to a distributional propagator

$$P \in \Gamma\left(X, \mathrm{Sym}^2_{dist}(TX)\right),$$

and to try to make sense of the above star product, at the limit. This is a way to formulate the renormalization problem, which we will now explain in more detail.

Definition 22.2.4 The bundle $\mathrm{Jet}_{dist}(\mathbb{R}_X/X)$ of *distributional jets* is defined by

$$\mathrm{Jet}_{dist}(\mathbb{R}_X/X) := \prod_{k\geq 0} \mathrm{Hom}_{\mathbb{R}_X}\big(\mathrm{Sym}^k\big(DTX^\vee\big), \mathbb{R}_X\big),$$

where $D \to X$ is the linear bundle over X given by

$$D := \underline{\Gamma}(M, \mathrm{Dens}_M) \times X \to X,$$

the duality is given at each point $\varphi_0 \in X$ by the finite dimensional bundle duality

$$T_{\varphi_0}X^\vee := \Gamma\big(M, \varphi_0^* TC^\vee\big)$$

over the base manifold M, and the product $DTX^\vee$ also denotes a fiberwise tensor product of linear bundles over M. We will also denote by

$$\otimes^k_{dist}(TX) := \mathrm{Hom}_{\mathbb{R}_X}\big(\otimes^k DTX^\vee, \mathbb{R}_X\big)$$

and

$$\mathrm{Sym}^k_{dist}(TX) := \mathrm{Hom}_{\mathbb{R}_X}\big(\mathrm{Sym}^k\big(DTX^\vee\big), \mathbb{R}_X\big)$$

the bundles of *distributional tensors* on the space X. We also denote by

$$\mathrm{Sym}^k_{dist}(\mathbb{R}_X) := \mathrm{Hom}_{\mathbb{R}_X}\big(\mathrm{Sym}^k(D), \mathbb{R}_X\big)$$

the bundle of *degree k distributional real numbers* on X, whose fiber at a point $\varphi_0 \in X$ is the space of real distributions on M^k.

Lemma 22.2.1 *There are natural monomorphisms*

$$\otimes^k(TX) \hookrightarrow \otimes^k_{dist}(TX), \qquad \mathrm{Sym}^k(TX) \hookrightarrow \mathrm{Sym}^k_{dist}(TX)$$

of bundles over X.

Proof At a given point φ_0, we have by definition

$$TX_{\varphi_0} = \Gamma\big(M, \varphi_0^* TC\big)$$

and

$$DTX^\vee_{\varphi_0} = \Gamma\big(M, \varphi_0^* TC^\vee \otimes \mathrm{Dens}_M\big).$$

The above monomorphisms are simply induced by the integration pairing

$$\int_M : TX_{\varphi_0} \times DTX_{\varphi_0}^{\vee} \longrightarrow \mathbb{R}$$

$$(\varphi, \psi d\mu) \longmapsto \int_M \langle \psi, \varphi \rangle d\mu,$$

by extending it to parametrized points. □

The definition of a coordinate-free Epstein-Glaser renormalization scheme involves the use of distributional propagators

$$P \in \Gamma\big(X, \otimes^2_{dist}(TX)\big),$$

that we may think of as the limit of a family of regularized propagators

$$P_\lambda \in \Gamma\big(X, \otimes^2(TX)\big).$$

Let us first construct the most naive notion of a 2-points time-ordered product. The notion of the Epstein-Glaser renormalization scheme will give a nice refinement of this construction. We also define the notion of a $*$-product, which is a non-symmetric version of the time-ordered product.

Proposition 22.2.1 *Let $P \in \Gamma(X, \otimes^2_{dist}(TX))$ be a distributionally valued section. The naive $*$-product associated to P is the binary operation*

$$*_2 : \mathcal{O}(X)[\![\hbar]\!] \times \mathcal{O}(X)[\![\hbar]\!] \to \mathcal{O}(X)[\![\hbar]\!]$$

$$(f, g) \mapsto \sum_{n \geq 0} \frac{i^n \hbar^n}{2^n n!} \Delta^*_{2n}\big(f^{(n)} \otimes P^{\otimes n} \otimes g^{(n)}\big)(1).$$

If $P \in \Gamma(X, \mathrm{Sym}^2_{dist}(TX))$ is symmetric, the above product is called the naive time-ordered product associated to P, and is denoted by τ_2. Its domain of definition is given by the set of pairs (f, g) such that for all $n \geq 0$, the wave front set of the distribution

$$f^{(n)} \otimes P^{\otimes n} \otimes g^{(n)}$$

is orthogonal to the conormal bundle $T^*_{\Delta_{2n}} M^{4n}$ of the diagonal $\Delta_{2n} : M^{2n} \hookrightarrow M^{4n}$.

In the case of a globally hyperbolic field theory on a Lorentzian spacetime, the Hadamard parametrix P (see [BF09]) has the good property of having well defined powers, because its wave front set is in a time-positive cone. The domain of definition of the associated $*$-product defined above is quite big and contains in particular multilocal (i.e., products of local) functionals. This $*$-product plays an important role in the Lorentzian causal Batalin-Vilkovisky formalism, to be discussed shortly in Sect. 22.3.

Definition 22.2.5 An *Epstein-Glaser renormalization scheme* is a tuple

$$\mathfrak{R} = \left(P, \leq, D, \{\mathrm{ext}_{2n}\}_{n\geq 0}\right)$$

composed of:

1. a section $P \in \Gamma(X, \mathrm{Sym}^2_{dist}(TX))$ called the *Epstein-Glaser parametrix*;
2. a partial ordering $\leq$ on M;
3. a domain of definition $D \subset \mathcal{O}(X)^2$;
4. a family of distributional extension maps

$$\mathrm{ext}_{2n} : \Gamma\left(X, \mathbb{R}^{2n}_{X,dist}\left(\Delta^c_{thin}\right)\right) \longrightarrow \Gamma\left(X, \mathbb{R}^{2n}_{X,dist}\right),$$

where $\Delta_{thin} \subset M^{2n}$ is the thin diagonal and $\mathbb{R}^{2n}_{X,dist}(\Delta^c_{thin})$ is the bundle on X whose fiber at a point $\varphi_0 \in X$ is the space of real valued distributions defined outside of the thin diagonal $\Delta_{thin} : M \hookrightarrow M^{2n}$.

The above data is further assumed to satisfy the following compatibility conditions, under the support condition $\mathrm{supp}(f) \leq \mathrm{supp}(g)$ on a pair $(f, g) \in D$:

1. For all $n \geq 0$, the wave front set of the distribution

$$f^{(n)} \otimes P^{\otimes n} \otimes g^{(n)}$$

is orthogonal to the conormal bundle $T^*_{\Delta_{2n}\backslash\Delta_{thin}} M^{4n}$ of the diagonal inclusion

$$\Delta_{2n\backslash\Delta_{thin}} : M^{2n}\backslash\Delta_{thin} = M^{2n}\backslash M \longrightarrow M^{4n}.$$

2. The pullback distribution

$$\Delta^*_{2n\backslash\Delta_{thin}}\left(f^{(n)} \otimes P^{\otimes n} \otimes g^{(n)}\right)$$

is in the domain of definition of the distributional extension map ext_{2n}.

Definition 22.2.6 The 2-point time-ordered product associated to an Epstein-Glaser renormalization scheme $\mathfrak{R}$ is the (partially defined) operation

$$\begin{aligned}\tau_2 : \mathcal{O}(X)[[\hbar]] \times \mathcal{O}(X)[[\hbar]] &\to \mathcal{O}(X)[[\hbar]]\\ (f, g) &\mapsto \sum_{n\geq 0} \frac{i^n \hbar^n}{2^n n!} \mathrm{ext}_{2n}\left(\Delta^*_{2n\backslash\Delta_{thin}}\left(f^{(n)} \otimes P^{\otimes n} \otimes g^{(n)}\right)\right)(1),\end{aligned}$$

with domain of definition $D[[\hbar]]$.

Example 22.2.1 The most simple renormalization scheme on an ordered spacetime is given by choosing the domain of definition to be the set of pairs (f, g) such that, for all $n \geq 0$, the wave front set of the distribution

$$f^{(n)} \otimes P^{\otimes n} \otimes g^{(n)}$$

is orthogonal to the conormal bundle $T^*_{\Delta_{2n}} M^{4n}$ of the full diagonal. The distribution pullbacks are then canonically extended to the thin diagonal, giving back the naive time-ordered product of Proposition 22.2.1. The problem is that this hypothesis is too strong in practice, when P is a true distribution, to allow both f and g to be local functionals, which is the case we are mostly interested in.

Proposition 22.2.2 *Let (M, g) be a metric manifold that is either globally hyperbolic (i.e., $M = \mathbb{R} \times M_0$, with $\mathbb{R}$ timelike) or Euclidean. Then there exists a natural Epstein-Glaser renormalization scheme $\mathfrak{R}(M, g)$ associated to scalar fields (i.e., functions) on M.*

Proof In both cases, one uses parametrices for the associated Laplacian Δ_g (which is elliptic in one case, and hyperbolic in the other). The partial ordering is given by equality in the Euclidean case and by time ordering in the Lorentzian case (two points are time-ordered if there is a Cauchy surface for the operator Δ_g, and a time-oriented path between them passing through this Cauchy surface, such that one is before it and the other after). The distribution extensions were described by Brunetti-Fredenhagen [BF09] and by Keller (see [Kel07] and [Kel09]). In the Euclidean case, a domain of definition that is already worked out is given by pairs of local functionals. In the Lorentzian case, there is a bigger domain of definition defined by the wave front set conditions we used, whose precise description can be found in [BF09]. □

We start with the central geometric lemma of the Epstein-Glaser method, whose proof can be found in [Ber04] (see also [BBD+12]).

Lemma 22.2.2 *Let $(M, \leq)$ be a partially ordered set, and define the relation $\nleq$ by*

$$x \nleq y \quad \Leftrightarrow \quad \text{not}(x \leq y).$$

Let N be a finite set and $I \subset N$ be a finite subset. Define $C_I \subset M^N$ by

$$C_I = \left\{x \in M^N, x_i \nleq x_j \text{ for all } i \in I \text{ and } j \in I^c\right\}.$$

Then there is a stratification

$$M^N \backslash \Delta_{thin} = \coprod_I C_I$$

of the complement of the thin diagonal $\Delta_{thin} : M \to M^N$, where I runs over proper subsets of N.

We now give, in our formalism, the Epstein-Glaser construction of n-point time-ordered products, that works uniformly in the Euclidean and Lorentzian setting. We are directly inspired here by the presentation of Keller [Kel09] and Brunetti-Fredenhagen [BF09].

Theorem 22.2.1 *Suppose given an Epstein-Glaser renormalization scheme from Proposition* 22.2.2.

$$\mathfrak{R} = \big(P, \leq, D, \{\mathrm{ext}_{2n}\}\big).$$

There is a natural family of maps

$$\tau_n : \mathcal{O}(X)[\![\hbar]\!]^n \to \mathcal{O}(X)[\![\hbar]\!],$$

called the renormalized n-point time-ordered products, that are associative, commutative, and with domain of definition given by the family (f_n) of functionals that are pairwise in D, i.e., such that $(f_i, f_j) \in D[\![\hbar]\!]$ for all $i \neq j$.

Proof This theorem is proved by induction on n. We already defined in Definition 22.2.6 the two point time-ordered product τ_2. Assume the maps τ_n are well defined for $n \leq N-1$. Let now $n = N$ and let $(f_n) \in \mathcal{O}(X)[\![\hbar]\!]^n$ be a family of functionals that are pairwise in D. Choose a partition of unity $\{\mathbb{1}_{C_I}\}$ for the stratification

$$M^n \backslash \Delta_{thin} = \coprod_I C_I$$

of Lemma 22.2.2. One may then define a section in $\Gamma(X, \mathbb{R}^{2n}_{X,dist}(\Delta^c_{thin}))$ from (f_n) by the formula

$$\tau_n(f_1, \ldots, f_n) := \sum_I \tau_2\big(\tau_I(\mathbb{1}_{C_I} f_I), \tau_{I^c}(\mathbb{1}_{C_{I^c}} f_{I^c})\big),$$

where $f_I = \otimes_{i \in I} f_i$. The above formula makes sense because the cardinals of I and I^c are both strictly smaller than n. □

The above theorem extends straightforwardly to vector valued fields, but since we have already described a similar result in Chap. 21, we shall not repeat the argument here.

We now define for the sake of completeness the S-matrix, which is often used by physicists.

Definition 22.2.7 The *time-ordered product* is the (partially defined) operation

$$._\tau : \mathcal{O}(X)[\![\hbar]\!] \times \mathcal{O}(X)[\![\hbar]\!] \to \mathcal{O}(X)[\![\hbar]\!]$$

induced by the family of compatible operations τ_n. The *Bogolyubov S-matrix* of a theory $S = S_0 + I$ with interaction part I (equipped with an Epstein-Glaser renormalization scheme for the free theory S_0) is the time-ordered exponential

$$\mathbb{S}_I := \exp_{._\tau}(I).$$

22.3 Causal Renormalization for Gauge Theories

We give here a sketch, in our functorial setting for spaces of fields, of the theory developed by Fredenhagen and Rejzner in [FR11] (see also [Rej11] for a more detailed presentation). It gives a causal analog of the effective quantum BV formalism, presented in Sect. 21.7.

Let $\pi : E_{fields} \to M$ be a gauge fixed BV theory, as in Sect. 13.5, with BV bundle

$$E_{BV} = T^*[-1]E_{fields},$$

and associated BV algebra $\mathcal{A}_{BV} := \mathrm{Jet}(\mathcal{O}_{E_{BV}})$. Choose a gauge fixed action $S_\psi \in h(\mathcal{A}_{BV})$ (of degree zero). We may view it as a functional

$$S_\psi : X_{BV} := \underline{\Gamma}(M, E_{BV}) \to \mathbb{R}$$

whose restriction to $X := X_{fields} := \underline{\Gamma}(M, E_{fields})$ via the zero section $0 : X \to T^*[-1]X$, also denoted by

$$S_\psi : X \to \mathbb{R},$$

has a non-degenerate second order derivative.

The choice of an Epstein-Glaser renormalization scheme for a parametrix of $D^2 S_\psi$ (assumed to exist) gives by renormalization a family of time-ordered products

$$\tau_n : \mathcal{O}(X)[[\hbar]]^n \longrightarrow \mathcal{O}(X)[[\hbar]].$$

These time-ordered products can be extended to the space X_{BV} by taking a family (f_n) of functionals, forming the associated gauge fixed functionals $(f_n \circ \psi \circ 0)$, and computing the time-ordered product on the space X. This gives a family of time-ordered products

$$\tau_n : \mathcal{O}(X_{BV})[[\hbar]]^n \longrightarrow \mathcal{O}(X)[[\hbar]].$$

Another way to extend this is to consider the class of functionals in $\mathcal{O}(X_{BV})$ given by the space

$$\mathcal{O}_{poly}(X_{BV}) := \Gamma\big(X, \mathrm{Sym}_{\mathbb{R}_X}(T[1]X)\big),$$

the partially defined map

$$\mathcal{O}_{poly}(X_{BV}) \to \mathcal{O}(X_{BV})$$

being given by insertion of tensors.

One may then extend the time-ordered product maps to

$$\tau_n : \mathcal{O}_{poly}(X_{BV})[[\hbar]]^n \longrightarrow \mathcal{O}_{poly}(X_{BV})[[\hbar]],$$

by simply using the usual product on the polynomial variables, and the time-ordered product on coefficients, which are functions on X.

The (polynomial) local functionals $A \in h(\mathcal{A}_{BV})$ are in $\mathcal{O}_{X_{BV}}$, so that we can now define their time-ordered product.

There is a natural Poisson bracket $\{-,-\}$ on $\mathcal{O}_{poly}(X_{BV})$ with domain of definition containing local functionals. This Poisson bracket is actually given by the natural symplectic structure on

$$X_{BV} = T^*[1]X_{fields}.$$

One may deform this Poisson bracket by defining a time-ordered odd Poisson bracket $\{-,-\}_\tau$ on $\mathcal{O}_{poly}(X_{BV})[[\hbar]]$. This extends naturally to functionals with good wave front set properties, and in particular to multilocal functionals, both in the Euclidean and Lorentzian setting.

Now we denote by $\{-,-\}_*$ the operation with two entries in $\mathcal{O}_{poly}(X_{BV})[[\hbar]]$ given by:

- the usual Poisson bracket $\{-,-\}$, in the Euclidean setting; and
- the $*$-product deformation of the usual Poisson bracket induced by a causal propagator (Hadamard parametrix)

$$P \in \Gamma\big(X, \otimes^2_{dist}(TX)\big)$$

(whose wave front set is concentrated in the positive cone).

The quantum master equation in this setting is then given by

$$\{e^I_\tau, S_0\}_* = 0,$$

where $S = S_0 + I$ is the decomposition of the action functional in its quadratic and interacting part. One may also define, in this setting, a renormalized anomalous Batalin-Vilkovisky Laplacian Δ_I that depends on the interaction term, so that the following quantum master equation

$$\frac{1}{2}\{S,S\}_\tau = \Delta_I(I)$$

is satisfied. It is a renormalized equation that looks like the equation used by physicists, but which is well defined (because Δ_I is renormalized) and that also works with an anomalous theory (the anomaly is included in the Δ_I term). For a related treatment of anomalous gauge theories, see [Bar00].

22.4 Deformation Quantization and Hopf Algebraic Structures

We refer to Chap. 6 for the necessary background on Hopf algebras. This section is inspired by Brouder's Hopf theoretic approach in [Bro09] to quantum field theory computations (see also [BFFO04] and [BBD+12]), and by Borcherds' formalization [Bor10] of perturbative functional integrals using Hopf algebraic methods.

22.4.1 A Finite Dimensional Toy Model

We refer to [Bro09] for this section. We present here the twisting methods from Sect. 6.2 on a simple example, which will be our main inspiration for the next section. We will use Sweedler's notation for coproducts:

$$\Delta(a) = \sum a_{(1)} \otimes a_{(2)}.$$

Let V be a finite dimensional vector space and $C = \mathrm{Sym}(V^*)$ be the algebra of polynomial functions on V, equipped with its commutative coproduct given by

$$\Delta_C(v^*) = v^* \otimes 1 + 1 \otimes v^*$$

on generators $v^* \in V^*$, and extended by algebra morphism. We may also interpret this coproduct as the one induced by addition on V, by the formula

$$\Delta_C\big(v^*(v)\big) = v^*(v_1 + v_2),$$

where v, v_1 and v_2 are in V and v^* in V^*. The coproduct Δ_C may be naturally extended to $\mathrm{Sym}(C) = \mathrm{Sym}(\mathrm{Sym}(V^*))$ by setting

$$\Delta(a) = \Delta_C(a)$$

for $a \in C = \mathrm{Sym}(V^*)$, and by extending this by algebra morphism

$$\Delta : \mathrm{Sym}(C) \to \mathrm{Sym}(C) \otimes \mathrm{Sym}(C).$$

This means in particular that

$$\Delta(a.b) = \sum (a_{(1)}.b_{(1)}) \otimes (a_{(2)}.b_{(2)})$$

for all $a, b \in C$. If D is a comodule over C, one also gets by the same formula a comodule algebra $(\mathrm{Sym}(D \otimes C), \Delta)$ over the bialgebra $(\mathrm{Sym}(C), \Delta)$. This is the type of object that will be used to define the deformation quantization in perturbative quantum field theory, through Drinfeld's twisting method (described in Sect. 6.2).

There is another coalgebra structure on $\mathrm{Sym}(C)$ given on generators $a \in C$ by

$$\delta(a) = a \otimes 1 + 1 \otimes a.$$

This corresponds to the additive group structure on the dual linear space C^*, by

$$\delta\big(a(a^*)\big) = a\big(a_1^* + a_2^*\big).$$

This coalgebra structure will be useful when defining the renormalization group. Note that there is a natural coaction of (C, Δ_C) on $(\mathrm{Sym}(C), \delta)$, given by a map

$$\Delta_C : \mathrm{Sym}(C) \to C \otimes \mathrm{Sym}(C)$$

induced by comultiplication on C by extension to an algebra morphism. At the level of functions on C^*, this coaction corresponds to the action of translations in the affine space V on the algebra of functions C on V, given by

$$\begin{aligned}(V,+)\times(C,+) &\to (C,+)\\ \big(w,a(v)\big) &\mapsto a(v+w).\end{aligned}$$

This action is clearly compatible with the addition of functions $a \in C$, so that one has a comodule structure over $(S(C),\delta)$.

If D is a comodule over C, given by a coaction map $\Delta_D : D \to C \otimes D$, we may also define a coaction

$$\Delta : \mathrm{Sym}(DC) \to \mathrm{Sym}(C) \otimes \mathrm{Sym}(DC)$$

of the coalgebra $(\mathrm{Sym}(C),\Delta)$ on the algebra $\mathrm{Sym}(DC)$ by extending the natural comodule structure

$$\Delta_{DC} : DC \to C \otimes DC$$

by algebra morphism.

If $\psi : V^* \otimes V^* \to \mathbb{R}$ is a bilinear form, it may be extended to a 2-cocycle on the bialgebras $C = \mathrm{Sym}(V^*)$ and $\mathrm{Sym}(C)$ by setting

$$\psi(ab,c) = \sum \psi(a,c_{(1)})\psi(b,c_{(2)}),$$

$$\psi(a,bc) = \sum \psi(a_{(1)},b)\psi(a_{(2)},c).$$

The associated twisted product (see Sect. 6.2) on the comodule algebra $\mathrm{Sym}(C)$ is defined by

$$a * b = \sum \psi(a_{(1)},b_{(1)})a_{(2)}b_{(2)},$$

where the Sweedler decomposition is written with respect to the $(\mathrm{Sym}(C),\delta)$-comodule structure map

$$\Delta : \mathrm{Sym}(C) \to C \otimes \mathrm{Sym}(C).$$

Morally, what we are doing here is a deformation quantization of the space C of functions on the affine space V, as a homogeneous space under the action of the group $(V,+)$.

Suppose now given a comodule D over C and define a coproduct δ on $\mathrm{Sym}(DC)$ by saying that DC is primitive. The coalgebra $(\mathrm{Sym}(DC),\delta)$ is a comodule under (C,Δ_C). In this abstract setting, the group G_{ren} of renormalizations is given by the group

$$G_{ren} := \mathrm{Aut}_{\mathrm{COMODCOALG}(C,\Delta_C)}\big(\mathrm{Sym}(DC),\delta\big)$$

of automorphisms of the coalgebra $\mathrm{Sym}(DC)$, as a comodule over the coalgebra C.

22.4.2 Hopf Algebraic Methods in Field Theory

We now describe how to generalize the methods of the previous section to the infinite dimensional setting of field theory. We use here [Bor10] and [BBD+12]. Throughout this section, we denote by normal letters the bundles, and by calligraphic letters their sheaves of sections.

Let $E \to M$ be a vector bundle, with associated jet bundle $\mathrm{Jet}(E) \to M$, and space of sections $\underline{\Gamma}(M, E)$ denoted by X. We denote by $\mathcal{E}$ the sheaf on M of sections of E, and $\mathrm{Jet}(\mathcal{E})$ the sheaf of sections of $\mathrm{Jet}(E)$. Let V be the jet bundle $\mathrm{Jet}(E)$ and denote by

$$V^* := \mathrm{Hom}_{\mathbb{R}_M}\big(\mathrm{Jet}(E), \mathbb{R}_M\big)$$

the dual bundle. Consider the bundle algebra

$$C = \mathrm{Sym}_{\mathbb{R}_M}\big(V^*\big)$$

whose sections $\mathcal{C}$ are polynomial Lagrangian functions on E, also called composite fields by physicists. The algebra $\mathcal{C}$ is equipped with a coproduct Δ_C given on generators by

$$\Delta_C(v) = v \otimes 1 + 1 \otimes v,$$

which gives the algebraic group structure on V corresponding to its natural addition, given by the fact that we started with a vector bundle E, equipped with an addition map

$$+ : E \times_M E \to E.$$

We define a coalgebra structure Δ on the sheaf $\mathrm{Sym}(\mathcal{C})$ by

$$\Delta(a) = \Delta_C(a)$$

for $a \in \mathcal{C}$ and extend this to

$$\Delta : \mathrm{Sym}(\mathcal{C}) \to \mathrm{Sym}(\mathcal{C}) \otimes \mathrm{Sym}(\mathcal{C})$$

by algebra morphism. Now consider the trivial comodule $\mathcal{D}$ over $\mathcal{C}$ given by the bundle $D = \mathrm{Dens}_M$ of densities on M. We equip the sheaf of algebras $\mathrm{Sym}(\mathcal{D}\mathcal{C}) := \mathrm{Sym}(\mathcal{D} \otimes \mathcal{C})$ with the coproduct δ given on its generators by

$$\delta(a) = a \otimes 1 + 1 \otimes a.$$

The space

$$Y = \mathcal{O}^{loc}(X) := \Gamma(M, \mathcal{D}\mathcal{C})$$

is the space of polynomial local Lagrangian densities on the space $X = \underline{\Gamma}(M, E)$ of fields. Morally, the coproduct δ corresponds to the additive group structure on the dual bundle $(DC)^*$. There is a natural coaction

$$\Delta_C : \mathrm{Sym}(DC) \to C \otimes \mathrm{Sym}(DC),$$

that corresponds to the action of the additive group $(V, +)$ on the space $(DC)^*$.

Definition 22.4.1 The renormalization group is the sheaf of groups

$$\mathcal{G}_{ren} := \mathrm{Aut}_{\mathrm{COALGCOMOD}(\mathcal{C},\Delta_{\mathcal{C}})}\big(\mathrm{Sym}(\mathcal{DC}),\delta\big)$$

of automorphisms of the $\mathcal{C}^\infty_M$-sheaf of coalgebras $\mathrm{Sym}(\mathcal{DC}) := \Gamma(-,\mathrm{Sym}(DC))$ that commute with the coaction of the sheaf of coalgebras $\mathcal{C} = \Gamma(-,C) := \Gamma(-,\mathrm{Sym}_{\mathbb{R}_M}(\mathrm{Jet}(E)^*))$ of local Lagrangian functions.

We carefully inform the reader that renormalizations are *not* given by algebra automorphisms of $\mathrm{Sym}(DC)$, so that the renormalization group is a very big group of automorphisms of a sheaf of $\mathcal{C}^\infty_M$-module coalgebras (not of bialgebras).

Up to now, we have been working with objects living on the manifold M, that are "local" objects in the physicists' sense. We now describe the main non-local Hopf algebraic structures that are used in perturbative quantum field theory.

Observe first that the algebra

$$\mathrm{Sym}_{\mathbb{R}}\big(\Gamma_c(M,DC)\big) := \oplus_{n\geq 0}\Gamma_c\big(M^n,\boxtimes^n DC\big)_{S_n}$$

is equipped with a natural coproduct δ that makes generators primitive. A coproduct δ similar to this one may be defined more intuitively on the dual algebra of formal functionals

$$\widehat{\mathcal{O}}(Y) := \prod_{n\geq 0}\mathrm{Hom}_{\mathbb{R}}\big(Y^{\hat{\otimes} n},\mathbb{R}\big)^{S_n} = \prod_{n\geq 0}\mathrm{Hom}_{\mathbb{R}}\big(\Gamma_c\big(M^n,(DC)^{\boxtimes n}\big),\mathbb{R}\big)^{S_n}$$

on the space $Y = \mathcal{O}^{loc}(X) := \Gamma(M,DC)$ of Lagrangian densities, by simply writing

$$\delta\big(F(\varphi_1,\ldots,\varphi_n)\big) := F(\varphi_1+\psi_1,\ldots,\varphi_n+\psi_n).$$

Note that this formula also extends naturally to the algebra $\widehat{\mathcal{T}}(Y)$ of noncommutative formal power series

$$\widehat{\mathcal{T}}(Y) := \prod_{n\geq 0}\mathrm{Hom}_{\mathbb{R}}\big(\Gamma\big(M^n,(DC)^{\boxtimes n}\big),\mathbb{R}\big).$$

The natural coaction map $\Delta_{DC} : DC \to DC \otimes C$ induces a coaction

$$\Delta : \mathrm{Sym}_{\mathbb{R}}\big(\Gamma_c(M,DC)\big) \to \mathrm{Sym}_{\mathbb{R}}\big(\Gamma_c(M,C)\big) \otimes \mathrm{Sym}_{\mathbb{R}}\big(\Gamma_c(M,DC)\big).$$

Definition 22.4.2 A propagator is a continuous bilinear map

$$G : \Gamma_c\big(M,DE^*\big) \otimes \Gamma_c\big(M,DE^*\big) \to \mathbb{C}.$$

A cut propagator is a propagator satisfying the following positive energy condition: at every point x of M, there is a partial order on the cotangent space defined by a proper closed convex cone, such that, if (p,q) is in the wave front set of G at some point $(x,y) \in M^2$, then $p \leq 0$ and $q \geq 0$. Also, as a distribution, G can be written as the boundary value of something in the algebra generated by smooth functions and powers and logarithm of polynomials. Moreover, if $x = y$, $p + q = 0$.

We may extend a propagator G to a continuous bilinear map

$$G : \Gamma_c(M, \mathrm{Jet}(E)^*) \otimes \Gamma_c(M, \mathrm{Jet}(E)^*) \to \mathcal{D}'(M^2) := \mathrm{Hom}_{\mathbb{R}}(\Gamma_c(M^2, D^{\boxtimes 2}), \mathbb{R})$$

by

$$G(s_\alpha^*, t_\beta^*) := (\partial^\alpha, \partial^\beta) \cdot G(s^* -, t^* -).$$

We extend this bilinear pairing by the bicharacter condition (see Examples 6.2.1 and 6.2.2) for the Hopf algebra $(\Gamma_c(M, C), \Delta_C)$ to a Hopf pairing

$$\psi_G : \Gamma_c(M, \mathcal{C}) \otimes \Gamma_c(M, \mathcal{C}) \to \mathcal{D}'(M^2),$$

and further through the bicharacter condition for the Hopf algebra $(\mathrm{Sym}_{\mathbb{R}}(\Gamma_c(M, C)), \Delta)$ to a Hopf pairing

$$\psi_G : \mathrm{Sym}^n_{\mathbb{R}}(\Gamma_c(M, C)) \otimes \mathrm{Sym}^m_{\mathbb{R}}(\Gamma_c(M, C)) \to \mathcal{D}(M^n \times M^m).$$

Finally, if we multiply the propagator by $\hbar$, we get a bicharacter ψ_G on the Hopf algebra $\mathrm{Sym}_{\mathbb{R}}(\Gamma_c(M, C))$ with values in the ring

$$\mathcal{R}[\![\hbar]\!] := \prod_{n \geq 0} \mathrm{Hom}_{\mathbb{R}}(\Gamma_c(M^n, \mathcal{D}^n), \mathbb{R})[\![\hbar]\!].$$

We may use this bicharacter to twist the comodule algebra $\mathrm{Sym}_{\mathbb{R}}(\Gamma_c(M, DC))$ under the bialgebra $(\mathrm{Sym}_{\mathbb{R}}(\Gamma_c(M, C)), \Delta)$, to get a partially defined $*$-product

$$f \overset{\hbar}{*} g := \sum \psi_G(f_{(1)}, g_{(1)}) f_{(2)} g_{(2)}.$$

This product is well defined only when one has a particular wave front set condition on f and g, because we used a distribution product to define it.

We will now describe Borcherds' main result.

Let L_F be a Lagrangian density and e^{iL_F} be the associated formal power series. Let $e^{iL_F}\mathrm{Sym}_{\mathbb{R}}(\Gamma_c(M, \mathcal{C}))$ be the free module of rank 1 generated by e^{iL_F}.

Definition 22.4.3 A Feynman measure associated to a cut propagator G is a continuous linear map $\omega : e^{iL_F}\mathrm{Sym}_{\mathbb{R}}(\Gamma_c(M, \mathcal{C})) \to \mathbb{C}$ satisfying:

- Smoothness on the diagonal: whenever $(p_1, \ldots, p_n)$ is in the wave front set of ω at the point $(x, \ldots, x)$ of the diagonal, then $p_1 + \cdots + p_n = 0$.
- Non-degeneracy: there is a smooth nowhere-vanishing function g such that $\omega(e^{iL_F} v)$ is $\int_M gv$ for $v \in \Gamma_c(M, \mathcal{D})$.
- Weak translation invariance: for A, B, with both sides interpreted as distributions,

$$\omega(AB) = \sum \omega(A') G(A'', B'') \omega(B'),$$

  whenever there is no element in the support of A that is $\leq$ some element of the support of B. Here, $\sum A' \otimes A''$ is given by the image of A under the coaction

$$\mathrm{Sym}_{\mathbb{R}}\Gamma_c(\mathcal{D}\mathcal{C}) \to \mathrm{Sym}_{\mathbb{R}}\Gamma_c(M, \mathcal{D}\mathcal{C}) \otimes \Gamma(M, \mathcal{C})$$

  induced by the coaction of $\mathcal{C}$ on $\mathcal{D}\mathcal{C}$.

The main theorem of Borcherds' paper [Bor10] is that Feynman measures exist and are unique up to renormalization.

Theorem 22.4.1 *There exists a Feynman measure ω associated to the cut propagator G and the renormalization group acts simply transitively on Feyman measures associated to G.*

The proof of the unicity result is given by induction, and uses the fact that there is a natural filtration of the renormalization group $\mathcal{G}_{ren}$ by the subgroups $\mathcal{G}_{ren,\geq n}$ that fix $\mathrm{Sym}^{\leq n}(\mathcal{DC})$, and such that $\mathcal{G}_{ren}$ is the limit of its quotients $\mathcal{G}_{ren}/\mathcal{G}_{ren,\geq n}$. This gives a decomposition of every element $g \in \mathcal{G}_{ren}$ as a product $g = \cdots g_n \cdots g_0$ with $g_k \in \mathcal{G}_{ren,\geq k}$.

The proof of the existence result uses complex analytic regularizations, à la Sato-Bernstein, to show that every cut propagator has an associated Feynman measure with values in meromorphic functions, and that any meromorphic Feynman measure can be made holomorphic by acting on it with a meromorphic renormalization.

References

[Bar00] Barnich, G.: Classical and quantum aspects of the extended antifield formalism. arXiv High Energy Physics—Theory e-prints (2000). arXiv:hep-th/0011120

[BBD+12] Bahns, D., Brouder, C., Dang, N.V., Fabretti, A., Paugam, F.: Introduction to Borcherds' renormalization and quantum field theory (2012, work in progress)

[Ber04] Bergbauer, C.: Epstein-Glaser renormalization, the Hopf algebra of rooted trees and the Fulton-MacPherson compactification of configuration spaces. Diplomarbeit, Freie Universität Berlin (2004)

[BF09] Brunetti, R., Fredenhagen, K.: Quantum field theory on curved backgrounds. arXiv (2009). arXiv:0901.2063

[BFFO04] Brouder, C., Fauser, B., Frabetti, A., Oeckl, R.: Quantum field theory and Hopf algebra cohomology. J. Phys. A **A37**, 5895–5927 (2004). doi:10.1088/0305-4470/37/22/014. arXiv:hep-th/0311253

[Bor10] Borcherds, R.E.: Renormalization and quantum field theory. arXiv e-prints (2010). arXiv:1008.0129

[Bro09] Brouder, C.: Quantum field theory meets Hopf algebra. Math. Nachr. **282**(12), 1664–1690 (2009). doi:10.1002/mana.200610828

[DDM04] DeWitt, B., DeWitt-Morette, C.: From the Peierls bracket to the Feynman functional integral. Ann. Phys. **314**(2), 448–463 (2004)

[DeW03] DeWitt, B.: The Global Approach to Quantum Field Theory. Vol. 1, 2. International Series of Monographs on Physics, vol. 114. Clarendon/Oxford University Press, Oxford/New York (2003). ISBN 0-19-851093-4

[EG73] Epstein, H., Glaser, V.J.: The role of locality in perturbation theory. Ann. Henri Poincaré **19**(3), 211–295 (1973). oai:cds.cern.ch:880480

[FR11] Fredenhagen, K., Rejzner, K.: Batalin-Vilkovisky formalism in perturbative algebraic quantum field theory. arXiv e-prints (2011). arXiv:1110.5232

[Kel07] Keller, K.: Dimensional regularization in position space and a forest formula for regularized Epstein-Glaser renormalization. Universität Hamburg, Hamburg (2007). Dissertation zur Erlangung des Doktorgrades des Department Physik des Universität Hamburg

[Kel09] Keller, K.J.: Euclidean Epstein-Glaser renormalization. J. Math. Phys. **50**(10), 103503 (2009). doi:10.1063/1.3202415. arXiv:0902.4789

[Pei52] Peierls, R.E.: The commutation laws of relativistic field theory. Proc. R. Soc. Lond. Ser. A **214**, 143–157 (1952)

[Rej11] Rejzner, K.: Batalin-Vilkovisky formalism in locally covariant field theory. arXiv e-prints (2011). arXiv:1111.5130

Chapter 23
Topological Deformation Quantizations

In this chapter, we give a very short description of Kontsevich's deformation quantization theorem and of his algebraic deformation quantization methods, that are based on deformation theory, described in Sect. 10.8.

23.1 Explicit Deformation Quantization of Poisson Manifolds

To illustrate the relation between homotopy and deformation, we briefly describe the modern approach to deformation quantization of Poisson manifolds, due to Kontsevich [Kon03] and Tamarkin [Tam98] (see also [Kon99] and [Kel05]). This approach is inspiring for the formulation of the quantization problem for factorization spaces, which will be discussed in Sect. 24.6. We refer to Sect. 10.8 for an introduction to the methods of deformation theory.

Suppose given a smooth Poisson manifold X, with Poisson algebra of functions $(\mathcal{C}^\infty(M), \{-,-\})$. A deformation quantization of X is given by a so-called associative *star-product*

$$*: \mathcal{C}^\infty(M)[[\hbar]] \times \mathcal{C}^\infty(M)[[\hbar]] \to \mathcal{C}^\infty(M)[[\hbar]]$$

that reduces to $\mathcal{C}^\infty(M)$ modulo $\hbar$. Such a product is given by extending the formula

$$f * g = \sum_{n=0}^{\infty} B_n(f,g)\hbar^n$$

to the formal series in $\mathcal{C}^\infty(M)$. The commutator returns the Poisson bracket via the formula

$$\{f,g\} = \frac{[f,g]}{\hbar} \quad \mathrm{mod}\ \hbar = B_1(f,g) - B_1(g,f).$$

F. Paugam, *Towards the Mathematics of Quantum Field Theory*,
Ergebnisse der Mathematik und ihrer Grenzgebiete. 3. Folge / A Series of Modern Surveys in Mathematics 59, DOI 10.1007/978-3-319-04564-1_23,

Associativity can be translated into the following equalities

$$\sum_{j+k=n} B_j\big(f, B_k(g,h)\big) = \sum_{j+k=n} B_j\big(B_k(f,g),h\big).$$

One usually asks that $*$ is defined by bidifferential operators $B_n : \mathcal{C}^\infty(M)^{\otimes 2} \to \mathcal{C}^\infty(M)$.

The simplest example of $*$-product is given by the Moyal $*$-product on $M = \mathbb{R}^2$. The Poisson bracket is given there by

$$\{f, g\} = \frac{\partial f}{\partial x_1}\frac{\partial g}{\partial x_2} - \frac{\partial f}{\partial x_2}\frac{\partial g}{\partial x_1},$$

for f and g two smooth functions on M. Kontsevich's constructions then yields the Moyal formal deformation given by

$$f * g = \sum_{n=0}^{\infty} \frac{\partial^n f}{\partial x_1^n}\frac{\partial^n g}{\partial x_2^n}\frac{\hbar^n}{n!}.$$

The explicit formula for the $*$-product for a general Poisson structure $\alpha = (\alpha^{ij})$ on $\mathbb{R}^n$ with variable coefficients is given by a formula of the form

$$f * g = \sum_{n=0}^{\infty} \hbar^n \sum_{\Gamma \in G_n} w_\Gamma B_\Gamma(f, g)$$

(where the sum is indexed by a certain class of graphs, and the terms $B_\Gamma(f, g)$ are written in terms of graph-indexed formulas involving partial derivatives of f, g and the matrix coefficient α^{ij} for the Poisson structure) which was derived by Kontsevich from the quantization of a Poisson sigma-model.

Let us mention more precisely the Cattaneo-Felder quantum field theoretic interpretation (see the original article [CF01] for the result and the survey article [CKTB05] for a refined presentation of the mathematical aspects) of Kontsevich's star product (Kontsevich actually found his formulas by computing functional integrals for this theory). We refer to Sect. 16.3 for a description of the classical Poisson sigma model. Let $D = \{z \in \mathbb{C}, |z| \leq 1\}$ be the Poincaré disc and (X, θ) be a Poisson manifold. Recall that fields of this theory are given by pairs (x, η) composed of a morphism $x : D \to X$ and a differential form $\eta \in \Omega^1(D, x^*T^*X)$. We fix three points 0, 1 and ∞ on the boundary of D. One may interpret a function $f : X \to \mathbb{R}$ as a pair of observables for this theory, given by the formulae

$$(x, \eta) \mapsto f\big(x(0)\big) \quad \text{and} \quad (x, \eta) \mapsto f\big(x(1)\big).$$

The Kontsevich formula for the *-product is formally given by the mean value (expressed by a functional integral)

$$(f * g)(y) := \frac{1}{Z(0)} \int_{x(\infty)=y} f\big(x(0)\big) g\big(x(1)\big) e^{\frac{i}{\hbar}S(x,\eta)}[dx d\eta].$$

Since the action functional has gauge symmetries, one needs to apply the Batalin-Vilkovisky formalism for gauge fixing. The detailed construction is provided in [CF01].

We will also describe in Sect. 24.2 some relations between the above construction and deformation quantization of chiral Lie algebras on the Poincaré disc, which may help to give a quantum field theoretic interpretation of Tamarkin's operadic proof [Tam98] of Kontsevich's theorem. This involves the study of the relation between $*$-products and deformation theory, which we will now discuss.

23.2 Deformation Theory and Quantization

Classically, the Hochschild complex of an algebra A is a positively graded complex $C^*(A, A)$ that gives a representative for the morphism complex

$$C^*(A, A) \cong \mathbb{R}\mathrm{Hom}_{A \otimes A^{op}}(A, A),$$

where the derived homomorphisms are computed in the differential graded setting. Its pth component, for $p \geq 0$, is the space of multilinear maps $\mathrm{Hom}(A^{\otimes n}, A)$. By definition, the differential of a p-cochain f is the $(p+1)$-cochain defined by

$$(-1)^p (df)(a_0, \dots, a_p) = a_0 f(a_1, \dots, a_p) - \sum_{i=0}^{p-1} (-1)^i f(a_0, \dots, a_i a_{i+1}, \dots, a_p)$$
$$+ (-1)^{p-1} f(a_0, \dots, a_{p-1}) a_p.$$

For example, first order deformations B_1 of the $*$-product (mod $\hbar^2$) satisfy the associativity equation

$$a B_1(b, c) - B_1(ab, c) + B_1(a, bc) - B_1(a, b)c = 0$$

which is the equation for a 2-cocycle in Hochschild cohomology. There is a natural dg-Lie bracket on $C^*(A, A)[1]$ given by

$$[f, g] = f \circ g - (-1)^{\deg(f).\deg(g)} g \circ f$$

where the (non-associative) product $\circ$ is defined as

$$(f \circ g)(a_0, \dots a_{k_1+k_2})$$
$$= \sum_{i=0}^{k_1} (-1)^{i k_2} f\big(a_0, \dots, a_{i-1}, g(a_i, \dots, a_{i+k_2}), a_{i+k_2+1}, \dots, a_{k_1+k_2}\big)$$

where $k_1 = \deg(f)$ and $k_2 = \deg(g)$.

Local cochains on $A = \mathcal{C}^\infty(M)$ are those defined by multilinear polydifferential operators, given in local coordinates by expressions of the form

$$(f_0, \dots, f_n) \mapsto \sum_{(\alpha_1,\dots,\alpha_n)} C^{(\alpha_1,\dots,\alpha_n)} \partial_{\alpha_1}(f_1) \cdots \partial_{\alpha_n}(f_n),$$

where $C^{(\alpha_1,\dots,\alpha_n)}$ are functions and the α_i are multi-indices. One shows that equivalence classes of $*$-products identify with solutions of the Mauer-Cartan equation

$$d(x) + \frac{1}{2}[x, x] = 0$$

of the dg-Lie algebra $C^*_{loc}(\mathcal{C}^\infty(M), \mathcal{C}^\infty(M))[1]$ given by local Hochschild cochains on $\mathcal{C}^\infty(M)$.

The space $H^0(X, \wedge^* \Theta_X)[1]$ of (shifted) multivector fields has a natural dg-Lie algebra structure with zero differential, and Lie bracket given by the Schouten-Nijenhuis bracket. A solution to the Mauer-Cartan equation in this dg-Lie algebra is precisely a Poisson structure:

$$\alpha \in H^0(X, \wedge^2 \Theta_X), \quad [\alpha, \alpha] = 0.$$

One may think of the complex $C^*_{loc}(\mathcal{C}^\infty(M), \mathcal{C}^\infty(M))[1]$ as the tangent complex to the formal derived moduli stack (that is simply defined by using Mauer-Cartan elements with values not only in classical infinitesimal Weil algebras but in infinitesimal differential graded parameters, see Sect. 10.8) associated to the given dg-Lie algebra. More concretely, it is the tangent space to the formal derived moduli stack of deformations of the dg-category $\mathrm{MOD}(A)$. There is an exact triangle

$$T_A \longrightarrow C^*(A, A)[1] \longrightarrow A[1]$$

where T_A is the tangent space to the space of deformations of A as an associative algebra, which explains for example why a first order deformation equivalence class of $*$-product as described explicitly above is given by an element in the second cohomology group $HH^2(A, A)$.

There is a natural, so-called Hochshild-Kostant-Rosenberg isomorphism

$$H^*\big(C^*_{loc}(\mathcal{C}^\infty(M), \mathcal{C}^\infty(M))\big) \cong H^0(X, \wedge^* \Theta_X)$$

with the space of multivector fields. Kontsevich's formality theorem says that the quasi-isomorphism

$$H^0(X, \wedge^* \Theta_X)[1] \to C^*_{loc}(\mathcal{C}^\infty(M), \mathcal{C}^\infty(M))[1],$$

explicitly given by the formula

$$\xi_0 \wedge \cdots \wedge \xi_n \mapsto \left((f_0, \dots, f_n) \mapsto \frac{1}{(n+1)!} \sum_{\sigma \in \Sigma_{n+1}} \mathrm{sng}(\sigma) \prod_{i=0}^{n} \xi_{\sigma_i}(f_i) \right),$$

for $n \geq 0$ and by

$$h \mapsto (1 \mapsto h),$$

for $n = -1$ and $h \in \mathcal{C}^{\infty}(M)$, may be made (homotopically) compatible with the dg-Lie structures on both sides. One then says that the considered dg-Lie algebra is *formal* (i.e., is quasi-isomorphic to its cohomology). Since a $\mathbb{R}[\hbar]/(\hbar^n)$-valued Mauer-Cartan element on the left is the same as a formal Poisson bracket up to order n, one gets that every formal deformation of the zero Poisson bracket corresponds to a $*$-product.

Part of the above reasoning can be translated into a purely geometric argument, as was done in [PTVV11]. This geometric translation also allows a nice treatment of the deformation quantization of Yang-Mills type gauge theories in any dimension (the deformation quantizations are only k-monoidal dg-categories, in this homotopical setting). The objects to be quantized are given by pairs (X, ω) composed of an Artin higher stack with a closed k-shifted symplectic 2-form ω (in the sense of Sect. 10.7) that induces an isomorphism

$$\Theta_{\omega} : \mathbb{T}_X \xrightarrow{\sim} \mathbb{L}_X[k]$$

between the tangent complex and the shifted cotangent complex of X.

A useful example of such a space is given by $X = BG$, equipped with the 2-shifted 2-form given by the natural G-invariant quadratic form

$$\omega : \mathfrak{g}[1] \wedge \mathfrak{g}[1] \cong \mathrm{Sym}^2(\mathfrak{g})[2] \to k[2].$$

The main result of loc. cit. implies that, if M is an oriented spacetime of dimension n, the above form induces a $2-n$-shifted 2-form ω on the space

$$X = \mathrm{BUNConn}_G(M) := \underline{\mathrm{Hom}}(M_{dR}, BG)$$

of G-bundles on M with flat connection. This is the space of trajectories of pure Yang-Mills theory. For $n = 4$, the symmetric monoidal category of such a space with 2-shifted 2-form may be deformed to a 2-monoidal dg-category.

This categorical approach also allows one to work in the algebraic or holomorphic setting, and to better understand obstructions. The Hochshild complex $C^*(A, A)$ identifies with the tangent space to the space of (tempered) deformations of dg-categories up to Morita equivalences at the derived category $D(A)$ of modules over A. All classes in Hochschild cohomology correspond to deformations of the dg-category $D(A)$, so that the deformation problem is unobstructed. Once Kontsevich's formality theorem is proved, the problem is to show that a given deformation of the category of modules is actually also a category of modules on an actual associative algebra. In the case of a smooth Poisson manifold, this is true, but it is false in the algebraic setting. The main idea of this geometric interpretation is that deforming categories of representations of algebras is easier than deforming algebras. This is also true when working with deformations of Hopf algebras. Once this setting is settled, the problem is to prove that a given $(k-1)$-monoidal dg-category is the category of modules over an E_k-algebra. We refer the reader to Francis work [Fra11] for the deformation theory of E_k-algebras and to Toen's work [Toe13] for the relation between deformation of monoidal dg-categories and shifted polyvectorfields.

To treat non-topological theories, one has to dive into the formalism of $\mathcal{D}$-geometry over the Ran space and factorization spaces, to be treated in Chap. 24. The type of objects that one obtains when forming the deformation quantization of a non-topological theory are chiral categories, which generalize n-monoidal dg-categories.

References

[CF01] Cattaneo, A.S., Felder, G.: Poisson sigma models and deformation quantization. Mod. Phys. Lett. A **16**(4–6), 179–189 (2001). doi:10.1142/S0217732301003255. Euroconference on brane new world and noncommutative geometry (Torino, 2000)

[CKTB05] Cattaneo, A., Keller, B., Torossian, C., Bruguières, A.: Déformation, Quantification, Théorie de Lie. Panoramas et Synthèses [Panoramas and Syntheses], vol. 20, p. 186. Société Mathématique de France, Paris (2005). ISBN 2-85629-183-X

[Fra11] Francis, J.: The tangent complex and Hochschild cohomology of E_n-rings. arXiv e-prints (2011). arXiv:1104.0181

[Kel05] Keller, B.: Deformation quantization after Kontsevich and Tamarkin. In: Déformation, Quantification, Théorie de Lie. Panor. Synthèses, vol. 20, pp. 19–62. Société Mathématique de France, Paris (2005)

[Kon99] Kontsevich, M.: Operads and motives in deformation quantization. Lett. Math. Phys. **48**, 35–72 (1999). arXiv:math.QA/9904055

[Kon03] Kontsevich, M.: Deformation quantization of Poisson manifolds. Lett. Math. Phys. **66**(3), 157–216 (2003). doi:10.1023/B:MATH.0000027508.00421.bf

[PTVV11] Pantev, T., Toen, B., Vaquie, M., Vezzosi, G.: Quantization and derived moduli spaces I: shifted symplectic structures. arXiv e-prints (2011). arXiv:1111.3209

[Tam98] Tamarkin, D.E.: Another proof of M. Kontsevich formality theorem. arXiv Mathematics e-prints (1998). arXiv:math/9803025

[Toe13] Toen, B.: Branes for ∞-operads. arXiv e-prints (2013). arXiv:1307.0405

Chapter 24
Factorization Spaces and Quantization

In this chapter, M denotes a smooth variety in a differentially convenient site (see Definition 3.1.3). We will be mostly interested in real, complex analytic, or smooth manifolds.

Factorization spaces give a geometric way to formalize the notion of operator product expansion, which is used by physicists to formalize quantum field theory in a locally covariant way (see e.g., Hollands-Wald [HW10] for their use in Lorentzian field theory on curved spacetime, and the Institute of Advanced Study semester notes [DEF+99], Volume 2, for their use in string theory). This approach was grounded by Beilinson-Drinfeld [BD04] in dimension 1 (see [Roz09, Gai99] and [FBZ04], Chap. 18, as additional references with concrete examples from conformal field theory). The generalization to a formal affine theory in higher dimension was developed by Francis-Gaitsgory [FG11]. The geometric approach through factorization spaces is presently in active development by Rozenblyum (see [Roz10, Roz11]), and Gaitsgory-Rozenblyum [GR11]. Our presentation of factorization spaces also owes pretty much to discussions with N. Rozenblyum on his ongoing project [Roz12]. Note however that the deformation quantization picture for factorization categories is original in our book.

All the chiral references work in the algebraic (or topological) setting, but we also need an analytic version of it. We will use the natural notion of a derived stack in $\mathcal{D}$-geometry, obtained as a functor of points on homotopical $\mathcal{D}$-algebras, following the general methods of Chap. 10. This formalism applies equally well to the algebraic or analytic setting, and we need to work with analytic spaces to get the tools of algebraic analysis of Chap. 11, which relate the general formalism to the usual computations of physicists, using distributions. One may also use manifolds with corners, based on the formalism described in Example 3.2.4 and the general differential calculus formalism of Sect. 2.5. We will use the general geometry of non-linear partial differential equations, described in Sect. 12.5.

There is another approach to factorization quantization that is due to Costello and Gwilliam [CG10] and uses the notion of factorization algebra. The relation of what we explain in this chapter with this approach is described in Remark 24.4.1. We have chosen the factorization space approach because it is directly related to our functor

F. Paugam, *Towards the Mathematics of Quantum Field Theory*,
Ergebnisse der Mathematik und ihrer Grenzgebiete. 3. Folge / A Series of Modern Surveys in Mathematics 59, DOI 10.1007/978-3-319-04564-1_24,

of points viewpoint of the geometry of spaces of fields, but the reader is strongly advised to refer to their work for an alternative approach, which may be accessible to a larger audience. Note that Costello and Gwilliam have a quite general existence theorem for quantizations of Euclidean field theories in their setting, that still needs to be adapted to ours, combining techniques of algebraic microlocal analysis from Chap. 11 with methods close to those used for causal perturbative quantum field theory in Chap. 22. We will partly explain this relation in Remarks 24.4.1 and 24.4.2, in a language that is directly adapted to the treatment of Lorentzian field theories, contrary to Costello and Gwilliam, that restrict their considerations to Euclidean (aka elliptic) field theories.

24.1 $\mathcal{D}$-Modules over the Ran Space

We begin our presentation of factorization methods by working with the perturbative theory of $\mathcal{D}$-spaces over the Ran space, which will play an important role in the perturbative theory. Recall that the Ran space is the derived stack defined by the ∞-colimit

$$\mathrm{Ran}(M) := \operatorname*{colim}_{I \twoheadrightarrow J} M^I$$

along diagonal embeddings. The category of $\mathcal{D}$-modules on the Ran space is the stable ∞-category

$$\mathcal{D}\big(\mathrm{Ran}(M)\big) := \mathrm{Mod}\big(\mathrm{Ran}(M)_{DR}\big)$$

of modules over $\mathrm{Ran}(M)_{DR}$. Note that by definition, one has

$$\mathcal{D}\big(\mathrm{Ran}(M)\big) \cong \lim_{I \twoheadrightarrow J} \mathcal{D}\big(M^I\big)$$

in the ∞-category of stable ∞-categories, so that a (right) $\mathcal{D}$-module on the Ran space may be seen as the datum of a compatible family $\{\mathcal{M}^I\}$ of $\mathcal{D}$-modules on M^I with a homotopy transitive system of homotopy equivalences

$$\Delta^* \mathcal{M}^I \cong \mathcal{M}^J.$$

The operation

$$\mathrm{union} : \mathrm{Ran}(M) \times \mathrm{Ran}(M) \to \mathrm{Ran}(M)$$

of union of subsets and

$$j : \big(\mathrm{Ran}(M) \times \mathrm{Ran}(M)\big)_{disj} \to \mathrm{Ran}(M) \times \mathrm{Ran}(M)$$

of inclusion of the disjoint pairs induce two monoidal structures

$$\otimes^* := \mathrm{union}_*(- \boxtimes -)$$

and

$$\otimes^{ch} := \mathrm{union}_* j_* j^*(- \boxtimes -),$$

respectively called the local and chiral monoidal structures. One also has the usual monoidal structure $\otimes^!$ on $\mathcal{D}(\mathrm{Ran}(M))$.

If $\mathcal{M}$ is a left $\mathcal{D}$-module on M, the associated de Rham functor

$$\mathrm{DR}_{\mathcal{M}} : \big(\mathcal{D}\big(\mathrm{Ran}(M)\big), \otimes^*\big) \to \big(\mathrm{DS}\textsc{h}\big(\mathrm{Ran}(M)\big), \otimes_{\mathbb{R}}\big)$$

is defined by

$$\mathrm{DR}_{\mathcal{M}}\big(\{\mathcal{M}^I\}\big) := \big\{\mathrm{DR}_{\mathcal{M}^{\boxtimes I}}\big(\mathcal{M}^I\big)\big\} := \big\{\mathcal{M}^I \otimes^{\mathbb{L}}_{\mathcal{D}_{M^I}} \mathcal{M}^{\boxtimes I}\big\}.$$

We will simply denote by DR the de Rham functor associated to $\mathcal{O}$. It plays an important role in the Euclidean case. Note that the de Rham functor associated to the $\mathcal{D}$-module of hyperfunctions $\mathcal{B}_M$ (or hyperfunctions with support in a positive time cone C on a Lorentzian spacetime M, denoted by $\mathcal{B}_C$) is more interesting in the case of hyperbolic spacetime.

The de Rham functor commutes with the main diagonal embedding, so that if $\mathcal{N}$ is a $\mathcal{D}$-module on M, then

$$\mathrm{DR}_{\mathcal{M}}\big(\Delta_*(\mathcal{N})\big) \cong \Delta_*\big(\mathrm{DR}_{\mathcal{M}}(\mathcal{N})\big).$$

This allows us to interpret the local functional calculus described in Sect. 12.2 directly on the Ran space.

24.2 Chiral Lie Algebras and Factorization Coalgebras

We now give a short summary of the theory of chiral algebras and their Koszul duality, due to Beilinson-Drinfeld [BD04] in dimension 1 and to Francis-Gaitsgory [FG11] in higher dimensions. This result allows us to relate formal affine factorization spaces (that are determined by their factorization coalgebra of "distributions") to Lie algebraic objects. It is also reminiscent of derived deformation theory, described in Sect. 10.8. The original property of this duality is that it is automatically an equivalence, because of the nilpotence property of the chiral tensor product on the Ran space. This may be compared with classical rational homotopy theory (see e.g., Quillen [Qui69]), where one needs to restrict the class of coalgebras in play to get an equivalence. This is avoided here because all factorization coalgebras already have the right nilpotence property.

Note that there is also a family version of this theory, which relates a class of Lie algebroids over a base $\mathcal{D}$-space Z/M_{DR} to factorization coalgebras over $Z^{(Ran)}/\mathrm{Ran}(M)_{DR}$. We only describe here the original theory, over a trivial base.

Definition 24.2.1 A chiral (Lie) algebra is a Lie algebra in $(\mathcal{D}(\mathrm{Ran}(M)), \otimes^{ch})$ that is supported on the main diagonal M. A local Lie algebra (also called a Lie*-algebra) is a Lie algebra in $(\mathcal{D}(\mathrm{Ran}(M)), \otimes^{*})$ that is supported on the main diagonal.

A chiral Lie algebra is thus given by a model

$$\mathcal{A} : (\mathrm{LIE}, \otimes) \longrightarrow (\mathcal{D}(\mathrm{Ran}(M)), \otimes^{ch})$$

of the symmetric linear theory of Lie algebras with values in $\mathcal{D}$-modules on the Ran space. We may think of this, in most cases, as the datum of a $\mathcal{D}$-module $\mathcal{A}$ on M together with bracket operations

$$[-, \cdots, -]_{i,ch} : \mathcal{A}^{\otimes^{ch} i} \to \mathcal{A},$$

that form a homotopy coherent L_∞-algebra. By adjunction, this may also be viewed as operations

$$[-, \cdots, -]_{ch} : j_* j^*(\mathcal{A}^{\boxtimes i}) \to \Delta_* \mathcal{A}.$$

The natural adjunction map $\mathrm{id} \to j_* j^*$ gives a morphism

$$(\mathcal{D}(\mathrm{Ran}(M)), \otimes^{ch}) \longrightarrow (\mathcal{D}(\mathrm{Ran}(M)), \otimes^{*}),$$

which allows us to associate a local Lie algebra to a chiral Lie algebra, by composition.

Remark 24.2.1 If one wants to have, on curves, an equivalence between commutative chiral algebras and commutative $\mathcal{D}$-algebras, it is necessary to add a unit to the Lie algebras in play, by defining unital chiral algebras. We refer to Beilinson-Drinfeld's book [BD04], Gaitsgory's survey [Gai99], Chap. 18 of Frenkel and Ben-Zvi's book [FBZ04], and Rozenblyum's lecture notes [Roz09] for more details on unital chiral algebra structures.

Locally constant chiral algebras give nice geometric ways to encode algebraic structures, and their homotopy versions. Let us illustrate this general fact with some simple examples.

Example 24.2.1 Let $M = \mathbb{R}$ be the real analytic affine line and A be an associative algebra. Let $\mathcal{A} := A \otimes \omega_{\mathbb{R}}$ be the associated right $\mathcal{D}$-module. The de Rham cohomology of $\mathcal{A}$ is the locally constant sheaf $\underline{A}$ on $\mathbb{R}$. There is a natural chiral bracket on $\mathcal{A}$ given by a morphism

$$\Delta^* j_* j^*(\mathcal{A} \boxtimes \mathcal{A}) \to \mathcal{A}$$

of $\mathcal{D}$-modules on $\mathbb{R}$. The left hand side is the $\mathcal{D}$-module $(\mathcal{A} \otimes^! \mathcal{A}) \oplus (\mathcal{A} \otimes^! \mathcal{A})$ with the two summands corresponding to two connected components of $\mathbb{R}^2 \backslash \Delta$. The

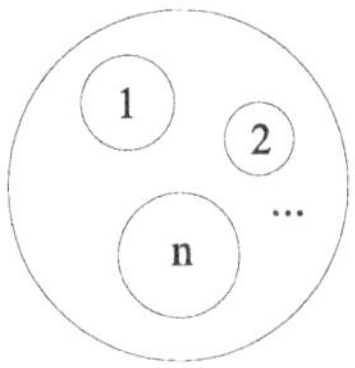

Fig. 24.1 The space of n-operations of the little discs operad

chiral product is the one induced by multiplication on one side and its opposite on the other. Its restriction to $\mathcal{A} \otimes^! \mathcal{A} = \Delta^*(\mathcal{A} \boxtimes \mathcal{A})$ gives the Lie bracket on $\mathcal{A}$. One may prove that locally constant chiral algebras on $\mathbb{R}$ correspond exactly to associative algebras up to homotopy, i.e., to E_1-algebras, via the above construction. More generally, locally constant chiral algebras on $\mathbb{R}^n$ are equivalent to E_n-algebras (see Fig. 24.1 for a drawing of the space of degree n-operations in E_2, given by families of n discs in a disc), as was proved by Lurie [Lur09c], Chap. 5 (he works in the topological setting, but the local constancy hypothesis allows a direct translation of his results to the chiral setting).

Example 24.2.2 The chiral algebra $\mathcal{A}$, associated in Example 24.2.1 with an associative algebra A, being locally constant, is in particular translation invariant. It thus descends to a chiral algebra $\mathcal{A}_{S^1}$ on $S^1 = \mathbb{R}/\mathbb{Z}$. One can show (see [Cos10] and [Lur09c] for the related construction in the setting of factorization algebras) that the chiral homology of $\mathcal{A}_{S^1}$ identifies with Hochshild homology of the underlying associative algebra A.

Example 24.2.3 A locally constant chiral algebra on the Poincaré half plane

$$\mathbb{H} := \{z \in \mathbb{C},\ \mathrm{Im}(z) > 0\}$$

is essentially the same as an E_2-algebra in complexes (recall that E_2-algebras are objects of a symmetric monoidal ∞-category equipped with two E_1-algebra structures that commute with each other up to homotopy). Its homology may be equipped with a Gerstenhaber structure, given by an action of the graded operad $G = H^*(E_2)$. This is essentially a graded commutative odd Poisson algebra. Note that Kontsevich's formality result over $\mathbb{R}$ from [Kon99] gives an explicit formula for the quasi-isomorphism of E_2 with $G = H^*(E_2)$, with coefficients in $\mathbb{R}$. A similar formula with coefficients in $\mathbb{C}$ is given by Drinfeld's associator [Dri87]. This formality result also gives an E_2 structure on the Hochshild cochain complex

$$\mathrm{CH}^*(A, A) := \mathbb{R}\mathrm{Hom}_{A \otimes A^{op}}(A, A)$$

of any associative algebra A, answering a conjecture of Deligne (whose validity is more generally known with coefficients in $\mathbb{Z}$, see [Lur09c], Sect. 6.1.4 for a proof conformal to our doctrinal approach to structures up to homotopy). One may intuitively understand the fact that this Hochshild complex has an E_2-algebra structure by thinking of it as a space of derived endomorphisms f of A, equipped with the

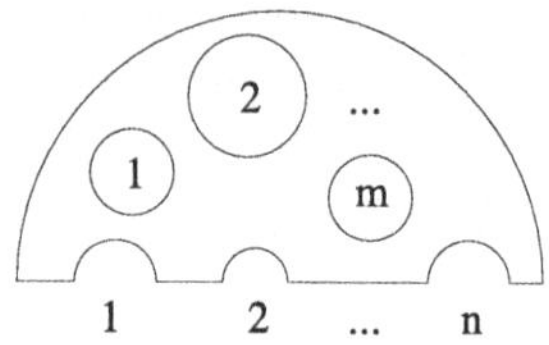

Fig. 24.2 The space of n-operations of the Swiss–Cheese operad

two operations

$$(f,g) \mapsto f \circ g \quad \text{and} \quad (f,g) \mapsto f *_A g.$$

These two operations get identified on the zero cohomology, that is simply the center $Z(A)$ of A, because of the Heckman-Hilton argument. However, they give two operations commuting up to homotopy on the full complex.

Example 24.2.4 If we work on the closure

$$\overline{\mathbb{H}} := \{z \in \mathbb{C},\ \mathrm{Im}(z) \geq 0\}$$

of the Poincaré half plane, which has boundary the real line $\mathbb{R}$, locally constant chiral algebras correspond to two sorted algebras: we have an E_2-algebra G coming from the $\mathbb{H}$ component and an E_1-algebra A coming from the $\mathbb{R}$ component, and they are linked by a morphism of E_2-algebras

$$f : G \to \mathrm{CH}^*(A, A).$$

The homology of these algebras form algebras over the Swiss-Cheese operad [Vor98] (see Fig. 24.2 for a drawing of the space of degree n operations of the Swiss-Cheese operad). These are pairs (G, A) composed of a Gerstenhaber algebra (i.e., an algebra over the homology $H^*(E_2)$ of the differential graded E_2 operad) and an associative algebra (i.e., an algebra over the operad $\mathrm{Ass} \cong H^*(E_1)$ of associative algebras) together with a map $f : G \to A$ between them satisfying compatibility conditions with the given structures. This situation is directly related to Tamarkin's operadic proof [Tam98] of Kontsevich's formality theorem. It actually gives a conceptual understanding of the relation between the Poisson sigma model, described in [CF01], and Tamarkin's work.

The main theorem of [FG11] is the following result, which tells us that the natural derived Koszul duality adjunction between LIE and COM for the chiral monoidal structure is an equivalence. This theorem essentially follows from the nilpotence property of the chiral tensor product, that makes all commutative coalgebras formal, in some sense.

Theorem 24.2.1 (Chiral Koszul Duality) *The natural pair of adjoint ∞-functors*

$$\mathrm{CE} : \mathrm{LIE}^{ch}\big(\mathcal{D}(\mathrm{Ran}(M))\big) \leftrightarrows \mathrm{COALG}^{ch}\big(\mathcal{D}(\mathrm{Ran}(M))\big) : \mathrm{Prim}[-1]$$

between Lie algebras and cocommutative coalgebras in $(\mathcal{D}(\mathrm{Ran}(M)), \otimes^{ch})$ is an equivalence of ∞-categories.

One may think of commutative factorization coalgebras as algebras of distributions on some distributional formal space.

Definition 24.2.2 Let $\mathcal{A}$ be a chiral commutative coalgebra. The distributional formal moduli space associated to $\mathcal{A}$ is the ∞-functor

$$\begin{aligned}\underline{\mathrm{Spec}}_{distrib}(\mathcal{A}) : \mathrm{COALG}^{ch}\big(\mathcal{D}\big(\mathrm{Ran}(M)\big)\big) &\to \mathrm{SH}_{\infty}\big(\mathrm{Ran}(M)\big)\\ \mathcal{B} &\mapsto \underline{\mathrm{Mor}}_{\mathrm{COALG}}(\mathcal{B}, \mathcal{A})\end{aligned}$$

with values in sheaves of ∞-groupoids on $\mathrm{Ran}(M)$.

One may think of the chiral Lie algebra $\mathrm{Prim}[-1]A$ geometrically as the tangent complex to the formal moduli space $\underline{\mathrm{Spec}}_{distrib}(\mathcal{A})$.

Definition 24.2.3 A commutative coalgebra given by the Chevalley-Eilenberg image of a chiral Lie algebra is called a factorization coalgebra on M.

Thus a factorization coalgebra on M is a particular kind of commutative chiral coalgebra in $\mathcal{D}(\mathrm{Ran}(M))$. Note that it almost never has support on the main diagonal, in contrast to chiral Lie algebras.

Definition 24.2.4 A chiral algebra is called *commutative* if the associated local bracket is trivial. A factorization coalgebra has *commutative factorization structure* if the associated chiral Lie algebra is commutative.

We now explain how to associate a chiral Lie algebra to a commutative $\mathcal{D}$-algebra. With the right notion of (higher dimensional) unital chiral algebra in hand, the obtained functor becomes an equivalence.

Proposition 24.2.1 *There is a functor between commutative derived $\mathcal{D}$-algebras and commutative chiral Lie algebras.*

Proof Let $\mathcal{A}^{\ell}$ be a commutative derived $\mathcal{D}$-algebra on M and $\mathcal{A}$ be the corresponding right $\mathcal{D}$-algebra. By definition, $\mathcal{A}$ is a model

$$\mathcal{A} : (\mathrm{COM}, \otimes) \longrightarrow \big(\mathrm{DMOD}\big(\mathcal{D}^{op}\big), \otimes^{!}\big)$$

of the commutative symmetric linear theory of commutative monoids into the monoidal ∞-category of right $\mathcal{D}$-modules. Using the natural morphisms

$$j_* j^*\big(\mathcal{A}^{\boxtimes n}\big) \longrightarrow \Delta_*\big(\mathcal{A}^{\otimes^! n}\big),$$

one can extend this to a chiral Lie algebra structure

$$\mathcal{A} : (\mathrm{LIE}, \otimes) \longrightarrow \big(\mathcal{D}\big(\mathrm{Ran}(M)\big), \otimes^{ch}\big).$$

Using the fact that the sequence

$$\mathcal{A}^{\boxtimes n} \to j_* j^*(\mathcal{A}^{\boxtimes n}) \to \Delta_*(\mathcal{A}^{\otimes^! n})$$

is a fibration sequence, we get that the obtained chiral algebra is commutative. □

Remark 24.2.2 Let us describe the $\mathcal{D}$-algebra of a free Euclidean scalar theory, given by a linear differential operator

$$\mathcal{D} \xrightarrow{.D} \mathcal{D} \twoheadrightarrow \mathcal{M}_D.$$

The corresponding $\mathcal{D}$-algebra is simply the differential graded $\mathcal{D}$-algebra

$$\mathbb{R}\mathrm{Sym}(\mathcal{M}_D) := \mathrm{Sym}\left(\left[\underset{1}{\mathcal{D}} \xrightarrow{.D} \underset{0}{\mathcal{D}}\right]\right),$$

and its solutions with values in a $\mathcal{D}$-algebra $\mathcal{A}$ identify with the derived solutions of $\mathcal{M}_D$ in the underlying $\mathcal{D}$-module. In this linear setting, one may take solutions of $\mathcal{M}_D$ with values in hyperfunctions, and if the operator is elliptic, one has an identification

$$\mathbb{R}\mathcal{H}om_{\mathcal{D}}(\mathcal{M}_D, \mathcal{O}_M) \xrightarrow{\sim} \mathbb{R}\mathcal{H}om_{\mathcal{D}}(\mathcal{M}_D, \mathcal{B}_M).$$

This isomorphism means that D induces an isomorphism

$$D : \mathcal{B}_M/\mathcal{A}_M \xrightarrow{\sim} \mathcal{B}_M/\mathcal{A}_M,$$

whose inverse H is called the parametrix of P. Microlocally, this means that microfunction solutions

$$\mathbb{R}\mathcal{H}om_{\mathcal{E}}(\mathcal{M}_D, \mathcal{C}_M) \cong 0$$

annihilate outside of the zero section of $T^*_M X$. If the operator is globally hyperbolic, the above statement is no longer true, but one may reformulate it.

Remark 24.2.3 The easiest way to have a formalism that works both in the hyperbolic and elliptic setting is to regard the parametrix H of a differential operator $D \in \mathcal{D}$ not as acting on a particular distributional space, but as a microdifferential operator (see Definition 11.5.4), given by a kernel on X^2 supported on the diagonal. This actually works for any operator, by viewing it as a microdifferential operator

$$\mathcal{E} \xrightarrow{.D} \mathcal{E}.$$

One may then invert D outside of the characteristic variety $S = \mathrm{char}(D) \subset T^*X$, yielding a microdifferential operator $H = D^{-1} \in \mathcal{E}$, that can be called the parametrix. This formulation has the advantage of generalizing directly to systems, the microdifferential inverse being replaced by a local homotopy between the complex of $\mathcal{E}$-modules in play and zero. In the physicist's language, the parametrix H is called the propagator, and it is usually represented by its kernel, living on X^2, with support on the diagonal. One can then make the microdifferential operator H act on the space

$$\mathcal{C}_{M^n\setminus S^n}$$

of microfunctions $f(x_1,\xi_1,\dots,x_n,\xi_n)$ defined outside of the characteristic manifold. The formulas appearing in the $*$-products of quantum field theory are given by pullbacks to the diagonal $\{(x_i,\xi_i,y_i,\eta_i),\ x_i=y_i,\ \xi_i=\eta_i\}$ of expressions of the form

$$g(y_1,\eta_1,\dots,y_n,\eta_n)H(x_1,\xi_1)\cdots H(x_n,\xi_n)f(x_1,\xi_1,\dots,x_n,\xi_n),$$

where H is the microlocal parametrix operator. These pullbacks can be defined only under some wave front set properties. The causal renormalization method, described in Chap. 22, gives a systematic way to deal with the problems that appear when the pullback is not well defined.

24.3 Operator Product Expansions

We now briefly describe (analytic) operator product expansions, by using an analytic version of Beilinson-Drinfeld's construction [BD04], Sect. 3.9. This gives a direct relation between chiral algebras and the physicists' operator product expansions. This definition looks natural because of recent results of Hollands and Kopper on φ^4 theory [HK11], in the setting developed by Hollands and Wald [HW10]. They obtained an analyticity result for operator product expansions.

Let M be an analytic manifold and $(\mathcal{A},[\cdot,\cdot]_{ch})$ be a chiral algebra on M. We also denote by $\mathcal{A}=\mathrm{CE}_{\otimes^{ch}}(\mathcal{A},[\cdot,\cdot]_{ch})$ the associated chiral coalgebra on $\mathrm{Ran}(M)$. Its restriction to M is the original $\mathcal{A}$.

For each surjection $\pi:I\twoheadrightarrow J$, we denote by $\Delta(\pi):M^J\to M^I$ the associated diagonal map and by $j(\pi):U^J\to M^I$ the inclusion of the complement open subset. By definition, there is a natural homotopy equivalence

$$\Delta(\pi)^!\mathcal{A}^I\cong\mathcal{A}^J.$$

We denote by $\hat{\Delta}(\pi)^!\mathcal{A}^I$ the restriction of $\mathcal{A}^I$ to the formal completion of M^I along the diagonal and by

$$\tilde{\Delta}(\pi)^!\mathcal{A}^I:=\hat{\Delta}(\pi)^!\mathcal{A}^I\otimes j_*\mathcal{O}_{U(\pi)}$$

its restriction to the pointed formal neighborhood of the diagonal.

By definition of a factorization coalgebra, the natural maps

$$j(\pi)^*\big(\mathcal{A}^I\big)\longrightarrow j(\pi)^*\big(\boxtimes_{i\in J}\mathcal{A}^{I_j}\big)$$

are equivalences.

Definition 24.3.1 Let $\pi : I \to \{1\}$ be the canonical projection. The I-th operator product expansion associated to $\mathcal{A}$ is the morphism

$$o_I : \mathcal{A}^{\ell \boxtimes I} \to j(\pi)_* j(\pi)^* \mathcal{A}^{\ell \boxtimes I} \xrightarrow{\sim} j(\pi)_* j(\pi)^* \mathcal{A}^{\ell I} \to \tilde{\Delta}^I_* \mathcal{A}^\ell.$$

24.4 Derived $\mathcal{D}$-Geometry of Factorization Spaces

We now recall from Sect. 12.5, for the reader's convenience, the following basic constructions of differential calculus on the Ran space in derived geometry.

Definition 24.4.1 The Ran space is the derived stack defined by the ∞-colimit

$$\mathrm{Ran}(M) := \operatorname*{colim}_{I \twoheadrightarrow J} M^I$$

along diagonal embeddings.

We will denote by $\mathrm{Ran}(M)_{DR}$ the associated de Rham space.

Let X be an Artin stack. We denote by $C = M \times X$ and $\pi : C \to M$ the natural projection. One may define the jet space

$$\pi_\infty : \mathrm{Jet}_{DR}(C) \to M_{DR}$$

as the push-forward $p_* C \to M_{DR}$ for the natural map $p : M \to M_{DR}$. We will also denote by $\mathrm{Jet}(C) := p^* p_* C$ the corresponding bundle over M with flat connection.

There is a natural inclusion

$$\underline{\Gamma}\big(M_{DR}, \mathrm{Jet}_{DR}(C)\big) \hookrightarrow \underline{\Gamma}\big(M, \mathrm{Jet}(C)\big)$$

that identifies horizontal sections of the jet bundle, and an isomorphism

$$j_\infty : \underline{\Gamma}(M, C) \xrightarrow{\sim} \underline{\Gamma}\big(M_{DR}, \mathrm{Jet}_{DR}(C)\big)$$

that sends a section to its infinite jet, that is horizontal for the canonical connection on $\mathrm{Jet}(C)$.

The above considerations may be extended by viewing M and C as derived stacks. Derived mapping spaces are not always isomorphic to the usual mapping spaces, but we will assume from now on that we are working only with these derived mapping spaces.

One may give, following Rozenblyum [Roz11], another description of the derived stack $\underline{\Gamma}(M, C)$, as flat sections of the multijet bundle

$$\mathrm{Jet}^{Ran}_{DR}(C) \to \mathrm{Ran}(M)_{DR}$$

over the de Rham space of the Ran space. This multijet space has the property of being a so-called factorization space, that we now define.

By definition, the category of derived stacks over $\mathrm{Ran}(M)$ is the ∞-limit of the categories of derived stacks over M^I with respect to the diagonal maps associated to surjections. We will denote it by $\mathrm{DS}\textsc{tacks}/\mathrm{Ran}(M)$.

Definition 24.4.2 A *factorization space* is a formal stack $Z^{(Ran)}$ over $\mathrm{Ran}(M)_{DR}$, i.e., a family of formal stacks

$$p_I : Z^{(I)} \to M^I$$

with a connection along M^I for each finite set I, that are compatible with pullbacks along diagonals and composition of surjections, up to coherent homotopy, together with compatible homotopy coherent factorization isomorphisms

$$Z^{(I)}_{|U_\alpha} \cong \left(\prod_{j\in J} Z^{I_j}\right)_{|U_\alpha},$$

for surjections $\alpha : I \twoheadrightarrow J$ where $U_\alpha \subset M^I$ is the subset

$$U_\alpha := \{x_i \neq x_{i'} \text{ if } \alpha(i) \neq \alpha(i')\}.$$

A factorization space $Z^{(Ran)}$ is called counital if in addition there are maps

$$M^{(I_1)} \times Z^{(I_2)} \to Z^{(I_1 \coprod I_2)}$$

that are compatible with factorization and restrictions to diagonals. We will denote by $\mathrm{DS}\textsc{tacks}_{fact}$ the ∞-category of factorization spaces and $\mathrm{DS}\textsc{tacks}^{cu}_{fact}$ the ∞-category of counital factorization spaces.

There is a natural forgetful functor

$$\begin{aligned}(\_)^{(1)} : \mathrm{DS}\textsc{tacks}^{cu}_{fact}/\mathrm{Ran}(M)_{DR} &\to \mathrm{DS}\textsc{tacks}/M_{DR}\\ Z^{(Ran)} &\mapsto Z^{(1)}\end{aligned}$$

from counital factorization spaces to $\mathcal{D}_M$-spaces (aka, spaces over M_{DR}).

Definition 24.4.3 The *multijet ∞-functor* is the right adjoint

$$(\_)^{(1)} : \mathrm{DS}\textsc{tacks}^{cu}_{fact}/\mathrm{Ran}(M)_{DR} \leftrightarrows \mathrm{DS}\textsc{tacks}/M_{DR} : \mathrm{Jet}^{Ran}_{DR}$$

to the forgetful functor. The *multijet space of a derived stack F over M_{DR}* is the counital factorization space

$$\mathrm{Jet}^{Ran}_{DR}(F) \to \mathrm{Ran}(M)_{DR}.$$

In particular, the *multijet space of a map* $\pi : C \to M$ is the multijet space of the derived stack

$$\mathrm{Jet}_{DR}(C) \to M_{DR}.$$

There is a natural isomorphism

$$\underline{\Gamma}(M, C) \xrightarrow{\sim} \underline{\Gamma}\big(\mathrm{Ran}(M)_{DR}, \mathrm{Jet}_{DR}^{Ran}(C)\big)$$

that identifies sections of π with horizontal sections of the multijet space. A given PDE on these sections, given by a closed subspace Z of $\mathrm{Jet}_{DR}(C)$, corresponds to a closed subspace $Z^{(Ran)}$ of the corresponding multijet space, that is also a counital factorization space.

The following proposition can be found in [Roz10], Proposition 5.1.1.

Proposition 24.4.1 *The multijet functor is given by the formula*

$$\mathrm{Jet}_{DR}^{Ran}(F)^I = (\pi_{2,DR})_*(\pi_{1,DR})^* F \underset{M_{DR}^I}{\times} M^I,$$

where π_1 and π_2 are the morphisms making the following canonical diagram

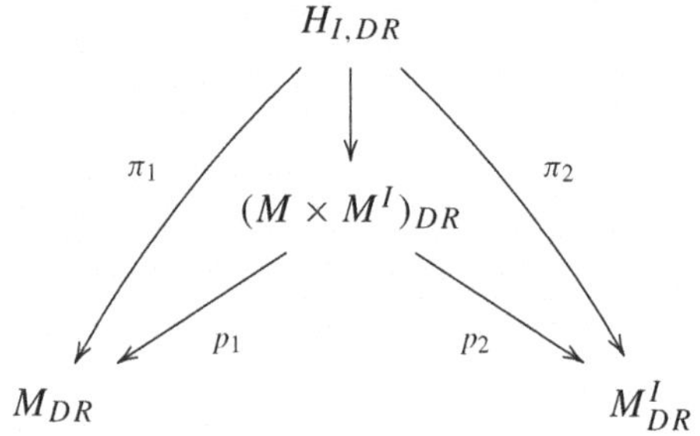

commutative and $H_I \subset M \times M^I$ is the incidence subspace, defined by

$$H_I := \big\{\big(x, \{x_i\}\big), x = x_i\big\}.$$

Definition 24.4.4 Let $Z^{(Ran)} \to \mathrm{Ran}(M)_{DR}$ be a factorization derived $\mathcal{D}$-stack over $\mathrm{Ran}(M)$. The chiral homology of Z is the derived stack

$$H(M, Z) := \underline{\Gamma}\big(\mathrm{Ran}(M)_{DR}, Z^{(Ran)}\big).$$

If Z is the derived $\mathcal{D}$-space $\mathbb{R}\underline{\mathrm{Sol}}_{\mathcal{D}}(\mathcal{I}_Z)$ of solutions of a PDE on $\underline{\Gamma}(M, C)$, and $Z^{(Ran)}$ is the corresponding multijet space, the chiral homology of Z identifies with the corresponding derived space $\mathbb{R}\underline{\mathrm{Sol}}(\mathcal{I}_Z)$, as a usual derived stack.

By definition, there is a natural evaluation morphism

$$\mathrm{ev} : H(M, Z) \times M \to Z.$$

Definition 24.4.5 Let $\mathcal{M}$ be a module over Z. The *localization* of $\mathcal{M}$ is the module over $H(M, Z)$ given by de Rham cohomology along M of $\text{ev}^*\mathcal{M}$.

Example 24.4.1 We present here an example due to Rozenblyum, that is related to the Witten genus construction. Let E be an elliptic curve and X be a derived Artin stack. Let $\pi : C = T^*X \times E \to E$ be the natural projection. Then $Z = \text{Jet}_{DR}(T^*X)$ is equipped with a local Poisson structure, given by a bivector

$$\pi \in \Theta_Z \wedge^* \Theta_Z$$

on the Ran space. The localization of Θ_Z identifies with the tangent complex

$$\mathbb{T}_{H(E,Z)} \cong \mathbb{T}_{\underline{\text{Hom}}(E,T^*X)}$$

and the symplectic structure on $\underline{\text{Hom}}(E, T^*X)$ defined by Costello [Cos10] and Pantev-Toen-Vaquie-Vezzosi [PTVV11] comes from the above bivector by localization.

Example 24.4.2 Let M be a smooth proper manifold and G be a compact Lie group, with an invariant pairing on its Lie algebra $\mathfrak{g}$. Let $\pi : C = \Gamma_M(M \times BG, M \times \text{Jet}^1 BG) \to M$ be the bundle whose sections are principal G-bundles equipped with a principal G-connection. Then $\text{Jet}(C) \to M_{DR}$ is the $\mathcal{D}$-bundle whose horizontal sections are principal G-bundles with a connection, and one may define a PDE $Z \subset \text{Jet}(C)$ by setting the curvature of the given connection to zero, in a derived sense, i.e., using the derived critical space of the curvature local functional. One then has

$$\mathbb{T}_{H(M,Z)} \cong \mathbb{T}_{\underline{\text{Hom}}(M_{DR},BG)}.$$

There is a natural local symplectic structure on Z induced by the given invariant pairing on the Lie algebra $\mathfrak{g}$ of G.

Remark 24.4.1 We want to present here a useful intuitive dictionary between Costello and Gwilliam's approach to the perturbative quantization [CG10] of factorization algebras and the presentation we give here. The heart of this dictionary is the fact that we work with sheaves on the Ran space, whereas Costello and Gwilliam work with cosheaves. The relation of their approach to ours is given by Lurie's non-abelian Poincaré duality theorem (see [Lur09c], Sect. 5.3 and [AFT12]), that sometimes allows the description of the space of compactly supported horizontal sections of a $\mathcal{D}$-space in a highly non-trivial way. Roughly speaking, if $(Z \to M_{DR}, s)$ is a pair composed of a $\mathcal{D}$-space over M and a horizontal section $s : M_{DR} \to Z$ of it, satisfying some additional *finiteness* and *connectivity* hypothesis, one may associate to it a space $\underline{\Gamma}_c(M_{DR}, Z)$ of compactly supported horizontal sections, and this space may be computed as the topological chiral homology

$$\underline{\Gamma}_c(M_{DR}, Z) \cong \int_M E^!$$

of a factorization M-cospace $E^!$ (essentially, a cosheaf on the Ran space satisfying an additional factorization property). Lurie's theorem, which applies to a locally compact topological space, probably extends directly to the semianalytic and subanalytic topoi (maybe with some overconvergence hypothesis), that we used to formalize algebraic analysis (see Example 3.2.3 and Sect. 11.3). We mention this result because it plays an important role in relating locally constant factorization algebras with algebras over some classical operads, to give a geometrical explanation of the usefulness of operads in the perturbative quantization of topological quantum field theories (illustrated by Cattaneo-Felder's paper [CF01] and Tamarkin's proof [Tam98] of Kontsevich's formality theorem).

Remark 24.4.2 Let $\mathcal{L}$ be an elliptic local Lie algebra with an invariant pairing $\psi : \mathcal{L} \otimes \mathcal{L} \to \omega_{\mathcal{D}}$. The Chevalley-Eilenberg commutative coalgebra $\mathrm{CE}_{\otimes^*}(\mathcal{L})$ for the local monoidal structure may be viewed as a chiral coalgebra via the natural transformation $\otimes^* \to \otimes^{ch}$. By [FG11], Proposition 6.1.2, the obtained factorization coalgebra is the Chevalley-Eilenberg algebra of the chiral envelope $U^{ch}(\mathcal{L})$ (this may actually be taken as a definition for the chiral envelope of a local Lie algebra). If $\mathcal{L}$ is dualizable for the $\otimes^*$ monoidal structure, with dual $\mathcal{L}^\circ$, the local invariant pairing ψ induces a Poisson bracket on the commutative $\mathcal{D}$-algebra $\mathcal{A} = \mathrm{CE}_{\otimes^*}(\mathcal{L}^\circ)$. By loc. cit., proof of Proposition 6.3.6, we know that

$$\mathrm{DR}_{\mathcal{O}}\big(\mathrm{CE}_{\otimes^*}(\mathcal{L})\big) \cong \mathrm{CE}_{\otimes}\big(\mathrm{DR}_{\mathcal{O}}(\mathcal{L})\big).$$

Since $\mathcal{L}$ is elliptic on M, we get that

$$\mathrm{DR}_{\mathcal{O}_{\mathrm{Ran}}}(\mathcal{A}) \longrightarrow \mathrm{DR}_{\mathcal{B}_{\mathrm{Ran}}}(\mathcal{A})$$

is a homotopy equivalence. Now, one may take the compactly supported sections

$$\Gamma_c\big(\mathrm{DR}_{\mathcal{B}_{\mathrm{Ran}}}(\mathcal{A})\big)$$

to construct a cosheaf on the Ran space, which is directly related, in the case of a commutative Lie algebra $\mathcal{L}$, to the commutative factorization algebras considered by Costello and Gwilliam [CG10].

24.5 Generalized Affine Grassmanians and Perturbative Quantization

We shall explain here Rozenblyum's generalization [Roz12] of the affine grassmanian quantization of Beilinson and Drinfeld [BD04]. This construction gives, in some sense, an analog in the chiral world of Lurie's non-abelian Poincaré-Verdier duality (see [Lur09c], Sect. 5.3). It also plays an important role, as we will see, in the geometric quantization of perturbative local field theories.

To every reasonable (in the sense of deformation theory, see [GR11]) pointed formal factorization $\mathcal{D}$-space Z on M (here pointed means, equipped with a horizontal section $s_0 \in \Gamma(M_{DR}, Z)$), one may naturally associate another formal factorization

$\mathcal{D}$-space Gr_Z on $\mathrm{Ran}(M)$ called the affine grassmanian $\mathcal{D}$-space of Z, with the following property: for $[S] \in \mathrm{Ran}(M)$ a point, we denote by D_S the union of the formal discs around points $s \in S \subset M$ and by D_S° the union of the corresponding punctured formal discs. The space Gr_Z has fiber at $[S]$ the fiber of the morphism

$$\underline{\Gamma}(D_S, Z) \longrightarrow \underline{\Gamma}\left(D_S^\circ, Z\right)$$

at the given fixed point s_0. The space Gr_Z is actually defined as the fiber of a morphism

$$Z^{(Ran)} \longrightarrow Z^{(0)}$$

at the given point s_0, where $Z^{(0)}$ is a punctured (aka local formal loop space) analog of the multijet construction. There is a relation between the chiral homology of Gr_Z and that of Z, that is valid only under some strong connectivity hypothesis (Z must be n-connected, where $n = \dim(M)$).

One may also think of this operation as a kind of geometric analog of the one dimensional twisted chiral envelope of Beilinson-Drinfeld [BD04], Sect. 3.7.10, and a twisted version of the higher dimensional chiral envelope of Francis-Gaitsgory [FG11], Sect. 6.4.

We now translate Costello and Gwilliam's definition from [CG10] of a perturbative quantum field to the setting of $\mathcal{D}$-geometry.

Definition 24.5.1 A *perturbative classical field theory* is the datum of a local (homotopy) Lie algebra $\mathcal{L}$ with invariant bilinear form ψ. A perturbative classical field theory is called *coherent* (resp. *holonomic*) if the $\mathcal{D}$-module underlying the Lie algebra $\mathcal{L}$ is coherent (resp. holonomic).

One denotes by $Z = B\mathcal{L}$ the formal $\mathcal{D}$-space with coalgebra of distributional coordinates the Chevalley-Eilenberg coalgebra $\mathrm{CE}_*(\mathcal{L})$. This coalgebra is equipped with a local Poisson bracket $\{\cdot,\cdot\}_\psi$ induced by the pairing ψ. If $\mathcal{L}$ is coherent, we may also define the algebra of coordinates on $B\mathcal{L}$ to be the Chevalley-Eilenberg algebra $\mathrm{CE}^*(\mathcal{L}^\circ)$.

Now if we apply the affine grassmanian construction to $Z = B\mathcal{L}$, we get a $\mathcal{D}$-space $\mathrm{Gr}_\mathcal{L}$, and the Poisson bracket $\{\cdot,\cdot\}_\mathcal{L}$ induces a line bundle $\mathcal{F}_\psi$ on $\mathrm{Gr}_\mathcal{L}$. The quantization of the formal Poisson $\mathcal{D}$-space $(B\mathcal{L}, \{\cdot,\cdot\}_\psi)$ is then given by the factorization algebra of global sections of this line bundle.

24.6 Categorical Quantization of Factorization Spaces

We now turn to the formal quantization question, to explain how to quantize $\mathcal{D}$-spaces that are not of the form $B\mathcal{L}$, for $\mathcal{L}$ a local Lie algebra equipped with an invariant pairing. The general approach that we give below may need some adaptations, in particular to obtain a well behaved notion of deformation theory (see

[PTVV11], and [GR11] for discussions of related issues). The main issue at stake here is to have good finiteness properties for the categories in play.

The basic idea of deformation theory of Poisson algebras, that we briefly discussed in Chap. 23, is that it is much easier to quantize the category of modules over a commutative algebra with a given Poisson structure, than to quantize the algebra itself. This idea was already clear in Kontsevich's formality approach to deformation quantization [Kon03].

Similarly, quantum group deformations, discussed in Chap. 6, were actually defined by Drinfeld in [Dri87] as deformation quantizations of braided monoidal categories. More generally, the article [PTVV11], briefly discussed in Chap. 23, also considers deformations of (the monoidal dg-category of modules over) symplectic derived stacks of maps between particular geometric stacks as n-monoidal dg-categories.

It is also clear in the homological mirror symmetry program of Kontsevich [Kon95] that the main objects at stake in mirror symmetry are some kinds of stable ∞-categories (or their more concrete analogs, given by A_∞-categories).

The above various quantization problems for categories may be restated in the setting of categorical factorization quantizations, using deformations of the stable ∞-category of modules (with additional finiteness conditions, such as perfectness, say) over a given factorization algebra, as a factorization category. The advantage of the factorization approach is that, even if it naturally generalizes the above quantization problems for algebraically structured categories, it allows us to give non-topological quantization of categories, with refined additional geometrical structures, encoded in factorization properties. This gives a neat combination of quantum field theory, in the general formulation given in this book (based on the geometry of non-linear partial differential equations), and deformation quantization of structured categories, which is at the basis of the above discussed (now standard) examples.

The precise relation with topological deformation quantizations of Chap. 23 is given by the fact that an n-monoidal differential graded category may be viewed as a locally constant factorization category over $\mathbb{R}^n$ (i.e., an E_n-algebra in the ∞-category of presentable stable ∞-categories enriched over dg-modules), and topological theories on a manifold (i.e., theories with a variable metric, that have a diffeomorphism gauge invariance) may be viewed as giving such factorization categories over $\mathbb{R}^n$.

We first define the notion of factorization categories, by simply adapting Definition 24.4.2 of factorization spaces. We need a notion of tensor product of presentable ∞-categories (these have small colimits and are generated by a small set of objects; see [Lur09c], Sect. 1.4.5 and Sect. 6.3). Note that the presentability property for module categories over factorization spaces may only be provable in very particular (for instance algebraic proper) cases, so that the following definition may need some modification to be adapted to smooth differential geometry.

Definition 24.6.1 A *factorization ∞-category* is a stack of stable presentable ∞-categories over $\mathrm{Ran}(M)_{DR}$, i.e., a family of stacks of ∞-categories

$$\mathcal{C}^{(I)} : \left(\mathrm{OPEN}_{M^I}^{op}, \tau\right) \longrightarrow {}^{\infty}\mathrm{CAT}_{sp}^{cont},$$

from the site of open subsets in M^I to the ∞-category of stable presentable ∞-categories (with continuous functors), with a categorical connection along M^I for each finite set I, i.e., an extension of $\mathcal{C}^{(I)}$ to

$$\mathcal{C}_{DR}^{(I)} \in \mathrm{SH}_{\infty\mathrm{CAT}_{sp}}\big(M_{DR}^I\big),$$

that are compatible with pullbacks along diagonals and composition of surjections, together with compatible homotopy coherent factorization equivalences

$$\mathcal{C}_{|U_\alpha}^{(I)} \cong \big(\boxtimes_{j\in J}\mathcal{C}^{(I_j)}\big)_{|U_\alpha},$$

for surjections $\alpha : I \twoheadrightarrow J$ where $U_\alpha \subset M^I$ is the subset

$$U_\alpha := \big\{x_i \neq x_{i'} \text{ if } \alpha(i) \neq \alpha\big(i'\big)\big\}.$$

Example 24.6.1 The category $\mathcal{D}(\mathrm{Ran}(M))$ is the most simple example of a factorization category. It is the factorization category of modules over the (factorization algebra associated to the) chiral algebra ω_M.

Proposition 24.6.1 *Let* $Z^{(Ran)}$ *be a geometric factorization space over* $\mathrm{Ran}(M)$. *The category* $\mathrm{MOD}(Z^{(Ran)})$ *is a factorization* ∞*-category.*

Proof This follows from the functoriality of module categories (see for example [BFN08]). □

Definition 24.6.2 A *categorical quantization* of a factorization category $\mathcal{C}$ is a factorization category $\mathcal{C}[\![\hbar]\!]$ flat over $\mathbb{R}[\![\hbar]\!]$, together with an equivalence

$$\mathcal{C} \cong \mathcal{C}' \quad \text{mod } \hbar.$$

If Z/M_{DR} is an Artin $\mathcal{D}$-space, a categorical quantization of Z is a categorical quantization of the factorization category $\mathrm{MOD}(Z^{(Ran)})$.

Example 24.6.2 To illustrate the above general definition, we give an example from Chap. 6. Let $\mathfrak{g}$ be a real Lie algebra and $U_\hbar(\mathfrak{g})$ be a quantum enveloping algebra of $U(\mathfrak{g})$. The category $\mathrm{MOD}(U_\hbar(\mathfrak{g}))$ is a braided monoidal (i.e., 2-monoidal) category that gives a formal deformation of the symmetric monoidal category $\mathrm{MOD}(U(\mathfrak{g}))$. Let $M = \mathbb{R}^2$ and consider the factorization categories

$$\mathcal{M}od\big(U_\hbar(\mathfrak{g})\big) \quad \text{and} \quad \mathcal{M}od\big(U(\mathfrak{g})\big)$$

on $\mathrm{Ran}(M)_{DR}$ associated to these two categorical E_2-algebras. Then the first is a categorical quantization of the second.

The following definition may be adapted by choosing a particular class of modules, like perfect modules, say. One may then formulate a general quantization problem, which is to find mild finiteness hypotheses under which,

there exists a categorical quantization
of an Artin Poisson $\mathcal{D}$-space as a factorization category.

If we know the existence of a categorical quantization $\mathcal{C}$ of an Artin Poisson $\mathcal{D}$-space Z (this is known in special cases), we may then ask if it has a formality property in the sense that its quantization bundle over $\mathbb{R}[\![\hbar]\!]$ has a natural splitting. This situation is quite analogous to the Hodge to de Rham spectral sequence degeneration, which is the conceptual heart of Hodge theory. Indeed, if X is a smooth complex projective variety, one always has a Hodge filtration on its de Rham cohomology, given, through the Rees construction, by a module over the affine line $\mathbb{A}^1$ that is equivariant under the action of GL_1. The Hodge to de Rham degeneration corresponds, in this case, to saying that there is a natural equivariant trivialization of this module (i.e., a splitting of the Hodge filtration).

References

[AFT12] Ayala, D., Francis, J., Tanaka, H.L.: Structured singular manifolds and factorization homology. arXiv e-prints (2012). arXiv:1206.5164

[BD04] Beilinson, A., Drinfeld, V.: Chiral Algebras. American Mathematical Society Colloquium Publications, vol. 51, p. 375. Am. Math. Soc., Providence (2004). ISBN 0-8218-3528-9

[BFN08] Ben-Zvi, D., Francis, J., Nadler, D.: Integral transforms and Drinfeld centers in derived algebraic geometry. arXiv e-prints (2008). arXiv:0805.0157

[CF01] Cattaneo, A.S., Felder, G.: Poisson sigma models and deformation quantization. Mod. Phys. Lett. A **16**(4–6), 179–189 (2001). doi:10.1142/S0217732301003255. Euroconference on brane new world and noncommutative geometry (Torino, 2000)

[CG10] Costello, K., Gwilliam, O.: Factorization algebras in perturbative quantum field theory (2010, preprint)

[Cos10] Costello, K.: A geometric construction of the Witten genus, I. In: Proceedings of the International Congress of Mathematicians. Volume II, pp. 942–959. Hindustan Book Agency, New Delhi (2010)

[DEF+99] Deligne, P., Etingof, P., Freed, D.S., Jeffrey, L.C., Kazhdan, D., Morgan, J.W., Morrison, D.R., Witten, E. (eds.): Quantum Fields and Strings: A Course for Mathematicians. Vol. 1, 2. Am. Math. Soc., Providence (1999). Material from the special year on quantum field theory held at the Institute for Advanced Study, Princeton, NJ, 1996–1997. ISBN 0-8218-1198-3

[Dri87] Drinfel'd, V.G.: Quantum groups. In: Proceedings of the International Congress of Mathematicians, Vol. 1, 2 (Berkeley, Calif., 1986), pp. 798–820. Am. Math. Soc., Providence (1987)

[FBZ04] Frenkel, E., Ben-Zvi, D.: Vertex Algebras and Algebraic Curves, 2nd edn. Mathematical Surveys and Monographs, vol. 88, p. 400. Am. Math. Soc., Providence (2004). ISBN 0-8218-3674-9

[FG11] Francis, J., Gaitsgory, D.: Chiral Koszul duality. arXiv e-prints (2011). arXiv:1103.5803

[Gai99] Gaitsgory, D.: Notes on 2D conformal field theory and string theory. In: Quantum Fields and Strings: A Course for Mathematicians, Vol. 1, 2 (Princeton, NJ, 1996/1997), pp. 1017–1089. Am. Math. Soc., Providence (1999)

[GR11] Gaitsgory, D., Rozenblyum, N.: Crystals and D-modules. arXiv e-prints (2011). arXiv: 1111.2087

[HK11] Hollands, S., Kopper, C.: The operator product expansion converges in perturbative field theory. arXiv (2011). arXiv:1105.3375

[HW10] Hollands, S., Wald, R.M.: Axiomatic quantum field theory in curved spacetime. Commun. Math. Phys. **293**, 85–125 (2010). doi:10.1007/s00220-009-0880-7. arXiv:0803.2003

[Kon95] Kontsevich, M.: Homological algebra of mirror symmetry. In: Proceedings of the International Congress of Mathematicians, Vol. 1, 2 (Zürich, 1994), pp. 120–139. Birkhäuser, Basel (1995)

[Kon99] Kontsevich, M.: Operads and motives in deformation quantization. Lett. Math. Phys. **48**, 35–72 (1999). arXiv:math.QA/9904055

[Kon03] Kontsevich, M.: Deformation quantization of Poisson manifolds. Lett. Math. Phys. **66**(3), 157–216 (2003). doi:10.1023/B:MATH.0000027508.00421.bf

[Lur09c] Lurie, J.: Higher algebra (2009, preprint)

[PTVV11] Pantev, T., Toen, B., Vaquie, M., Vezzosi, G.: Quantization and derived moduli spaces I: shifted symplectic structures. arXiv e-prints (2011). arXiv:1111.3209

[Qui69] Quillen, D.: Rational homotopy theory. Ann. Math. (2) **90**, 205–295 (1969)

[Roz09] Rozenblyum, N.: Introduction to chiral algebras (2009, preprint)

[Roz10] Rozenblyum, N.: Modules over a chiral algebra. arXiv e-prints (2010). arXiv:1010.1998

[Roz11] Rozenblyum, N.: Connections on conformal blocks. Thesis (2011)

[Roz12] Rozenblyum, N.: Continuous moduli functors and quantum field theory (2012, preprint)

[Tam98] Tamarkin, D.E.: Another proof of M. Kontsevich formality theorem. arXiv Mathematics e-prints (1998). arXiv:math/9803025

[Vor98] Voronov, A.A.: The Swiss-cheese operad. arXiv Mathematics e-prints (1998). arXiv: math/9807037

References

[AFT12] Ayala, D., Francis, J., Tanaka, H.L.: Structured singular manifolds and factorization homology. arXiv e-prints (2012). arXiv:1206.5164

[AGV73] Artin, M., Grothendieck, A., Verdier, J.-L.: Théorie des Topos et Cohomologie Étale des Schémas. Tome 3, Lecture Notes in Mathematics, vol. 305, p. 640. Springer, Berlin (1973). Séminaire de Géométrie Algébrique du Bois-Marie 1963–1964 (SGA 4), Dirigé par M. Artin, A. Grothendieck et J.L. Verdier. Avec la collaboration de P. Deligne et B. Saint-Donat

[And07] Andronikof, E.: Interview with Mikio Sato. Not. Am. Math. Soc. **54**(2), 243–245 (2007)

[Ara10] Ara, D.: The groupoidal analogue Theta˜ to Joyal's category Theta is a test category. arXiv e-prints (2010). arXiv:1012.4319

[Ara12] Ara, D.: Higher quasi-categories vs higher Rezk spaces. arXiv e-prints (2012). arXiv: 1206.4354

[Arn91] Arnol'd, V.I.: Mathematical Methods of Classical Mechanics. Graduate Texts in Mathematics, vol. 60, p. 516. Springer, New York (1991). Translated from the 1974 Russian original by K. Vogtmann and A. Weinstein. Corrected reprint of the second (1989) edition. ISBN 0-387-96890-3

[Ayo07] Ayoub, J.: Les Six Opérations de Grothendieck et Le Formalisme des Cycles Évanescents dans Le Monde Motivique. I (2007). Astérisque, 314:x+466 pp. (2008)

[Bar00] Barnich, G.: Classical and quantum aspects of the extended antifield formalism. arXiv High Energy Physics—Theory e-prints (2000). arXiv:hep-th/0011120

[Bar10] Barnich, G.: A note on gauge systems from the point of view of Lie algebroids. arXiv e-prints (2010). arXiv:1010.0899

[BBD+12] Bahns, D., Brouder, C., Dang, N.V., Fabretti, A., Paugam, F.: Introduction to Borcherds' renormalization and quantum field theory (2012, work in progress)

[BD86] Bunge, M., Dubuc, E.J.: Archimedian local C^∞-rings and models of synthetic differential geometry. Cah. Topol. Géom. Différ. Catég. **27**(3), 3–22 (1986)

[BD98] Baez, J.C., Dolan, J.: Categorification. arXiv Mathematics e-prints (1998). arXiv: math/9802029

[BD04] Beilinson, A., Drinfeld, V.: Chiral Algebras. American Mathematical Society Colloquium Publications, vol. 51, p. 375. Am. Math. Soc., Providence (2004). ISBN 0-8218-3528-9

[Ber71] Bernšteĭn, I.N.: Modules over a ring of differential operators. An investigation of the fundamental solutions of equations with constant coefficients. Funkc. Anal. Prilozh. **5**(2), 1–16 (1971)

F. Paugam, *Towards the Mathematics of Quantum Field Theory*,
Ergebnisse der Mathematik und ihrer Grenzgebiete. 3. Folge / A Series of Modern
Surveys in Mathematics 59, DOI 10.1007/978-3-319-04564-1,

[Ber90] Berkovich, V.G.: Spectral Theory and Analytic Geometry over Non-Archimedean Fields. Mathematical Surveys and Monographs, vol. 33, p. 169. Am. Math. Soc., Providence (1990). ISBN 0-8218-1534-2

[Ber01] Berndt, R.: An Introduction to Symplectic Geometry. Graduate Studies in Mathematics, vol. 26, p. 195. Am. Math. Soc., Providence (2001). Translated from the 1998 German original by Michael Klucznik. ISBN 0-8218-2056-7

[Ber04] Bergbauer, C.: Epstein-Glaser renormalization, the Hopf algebra of rooted trees and the Fulton-MacPherson compactification of configuration spaces. Diplomarbeit, Freie Universität Berlin (2004)

[Ber06] Bergner, J.E.: A survey of $(\infty, 1)$-categories. arXiv Mathematics e-prints (2006). arXiv:math/0610239

[Bes08] Besse, A.L.: Einstein Manifolds. Classics in Mathematics, p. 516. Springer, Berlin (2008). Reprint of the 1987 edition. ISBN 978-3-540-74120-6

[BF09] Brunetti, R., Fredenhagen, K.: Quantum field theory on curved backgrounds. arXiv (2009). arXiv:0901.2063

[BFFO04] Brouder, C., Fauser, B., Frabetti, A., Oeckl, R.: Quantum field theory and Hopf algebra cohomology. J. Phys. A **A37**, 5895–5927 (2004). doi:10.1088/0305-4470/37/22/014. arXiv:hep-th/0311253

[BFN08] Ben-Zvi, D., Francis, J., Nadler, D.: Integral transforms and Drinfeld centers in derived algebraic geometry. arXiv e-prints (2008). arXiv:0805.0157

[BGV92] Berline, N., Getzler, E., Vergne, M.: Heat Kernels and Dirac Operators. Grundlehren der Mathematischen Wissenschaften [Fundamental Principles of Mathematical Sciences], vol. 298, p. 369. Springer, Berlin (1992). ISBN 3-540-53340-0

[BH11] Baez, J.C., Hoffnung, A.E.: Convenient categories of smooth spaces. Trans. Am. Math. Soc. **363**(11), 5789–5825 (2011). doi:10.1090/S0002-9947-2011-05107-X

[BK72] Bousfield, A.K., Kan, D.M.: Homotopy Limits, Completions and Localizations. Lecture Notes in Mathematics, vol. 304, p. 348. Springer, Berlin (1972)

[BM88] Bierstone, E., Milman, P.D.: Semianalytic and subanalytic sets. Publ. Math. IHES **67**, 5–42 (1988)

[BM03] Berger, C., Moerdijk, I.: Axiomatic homotopy theory for operads. Comment. Math. Helv. **78**(4), 805–831 (2003). doi:10.1007/s00014-003-0772-y

[BN10] Ben-Zvi, D., Nadler, D.: Loop spaces and connections. arXiv e-prints (2010). arXiv:1002.3636

[BO78] Berthelot, P., Ogus, A.: Notes on Crystalline Cohomology, p. 243. Princeton University Press, Princeton (1978). ISBN 0-691-08218-9

[Bor10] Borcherds, R.E.: Renormalization and quantum field theory. arXiv e-prints (2010). arXiv:1008.0129

[Bou70] Bourbaki, N.: Éléments de Mathématique. Théorie des Ensembles, p. 349. Hermann, Paris (1970)

[BR12] Bergner, J.E., Rezk, C.: Comparison of models for (∞, n)-categories, I. arXiv e-prints (2012). arXiv:1204.2013

[Bro09] Brouder, C.: Quantum field theory meets Hopf algebra. Math. Nachr. **282**(12), 1664–1690 (2009). doi:10.1002/mana.200610828

[BS11] Barwick, C., Schommer-Pries, C.: On the unicity of the homotopy theory of higher categories. arXiv e-prints (2011). arXiv:1112.0040

[But07] Butterfield, J.: Philosophy of Physics, pp. 1–1369. Elsevier/North-Holland, Amsterdam (2007)

[BV81] Batalin, I.A., Vilkovisky, G.A.: Gauge algebra and quantization. Phys. Lett. B **102**(1), 27–31 (1981). doi:10.1016/0370-2693(81)90205-7

[BW05] Barr, M., Wells, C.: Toposes, Triples and Theories. Repr. Theory Appl. Categ. (2005). 12:x+288. Corrected reprint of the 1985 original [MR0771116]

[CB72] Choquet-Bruhat, Y.: Stabilité de solutions d'équations hyperboliques non linéaires. Application à l'espace-temps de Minkovski en relativité générale. C. R. Acad. Sci. Paris, Sér. A-B **274**, 843–845 (1972)

[CD09] Cisinski, D.-C., Déglise, F.: Local and stable homological algebra in Grothendieck Abelian categories. Homol. Homotopy Appl. **11**(1), 219–260 (2009)

[CDM06] Cartier, P., DeWitt-Morette, C.: Functional Integration: Action and Symmetries. Cambridge Monographs on Mathematical Physics, p. 456. Cambridge University Press, Cambridge (2006). ISBN 978-0-521-86696-5; 0-521-86696-0. Appendix D contributed by Alexander Wurm

[CE04] Calaque, D., Etingof, P.: Lectures on tensor categories. arXiv Mathematics e-prints (2004). arXiv:math/0401246

[CF01] Cattaneo, A.S., Felder, G.: Poisson sigma models and deformation quantization. Mod. Phys. Lett. A **16**(4–6), 179–189 (2001). doi:10.1142/S0217732301003255. Euroconference on brane new world and noncommutative geometry (Torino, 2000)

[CG10] Costello, K., Gwilliam, O.: Factorization algebras in perturbative quantum field theory (2010, preprint)

[CGM+06] Carinena, J.F., Gracia, X., Marmo, G., Martinez, E., Munoz-Lecanda, M., Roman-Roy, N.: Geometric Hamilton-Jacobi theory. arXiv Mathematical Physics e-prints (2006). arXiv:math-ph/0604063

[Cha01] Chaperon, M.: Jets, transersalité, singularités:petite introduction aux grandes idées de René Thom (2001, preprint)

[Che77] Chen, K.T.: Iterated path integrals. Bull. Am. Math. Soc. **83**(5), 831–879 (1977)

[Che97] Chevalley, C.: The Algebraic Theory of Spinors and Clifford Algebras, p. 214. Springer, Berlin (1997). Collected works, vol. 2. Edited and with a foreword by Pierre Cartier and Catherine Chevalley, with a postface by J.-P. Bourguignon. ISBN 3-540-57063-2

[Cis10] Cisinski, D.-C.: Catégories dérivables. Bull. Soc. Math. Fr. **138**(3), 317–393 (2010)

[CK00] Connes, A., Kreimer, D.: Renormalization in quantum field theory and the Riemann-Hilbert problem. I. The Hopf algebra structure of graphs and the main theorem. Commun. Math. Phys. **210**(1), 249–273 (2000). doi:10.1007/s002200050779

[CK01] Connes, A., Kreimer, D.: Renormalization in quantum field theory and the Riemann-Hilbert problem. II. The β-function, diffeomorphisms and the renormalization group. Commun. Math. Phys. **216**(1), 215–241 (2001). doi:10.1007/PL00005547

[CKTB05] Cattaneo, A., Keller, B., Torossian, C., Bruguières, A.: Déformation, Quantification, Théorie de Lie. Panoramas et Synthèses [Panoramas and Syntheses], vol. 20, p. 186. Société Mathématique de France, Paris (2005). ISBN 2-85629-183-X

[CL04] Cheng, E., Lauda, A.: Higher dimensional categories: an illustrated guidebook (2004, preprint)

[CM00] Chadozeau, A., Mistretta, E.: Le polynôme de Bernstein-Sato (2000, preprint)

[CM08] Connes, A., Marcolli, M.: Noncommutative Geometry, Quantum Fields and Motives. American Mathematical Society Colloquium Publications, vol. 55, p. 785. Am. Math. Soc., Providence (2008). ISBN 978-0-8218-4210-2

[Cos10] Costello, K.: A geometric construction of the Witten genus, I. In: Proceedings of the International Congress of Mathematicians. Volume II, pp. 942–959. Hindustan Book Agency, New Delhi (2010)

[Cos11] Costello, K.: Renormalization and Effective Field Theory. Mathematical Surveys and Monographs, vol. 170, p. 251. Am. Math. Soc., Providence (2011). ISBN 978-0-8218-5288-0

[DDM04] DeWitt, B., DeWitt-Morette, C.: From the Peierls bracket to the Feynman functional integral. Ann. Phys. **314**(2), 448–463 (2004)

[DEF+99] Deligne, P., Etingof, P., Freed, D.S., Jeffrey, L.C., Kazhdan, D., Morgan, J.W., Morrison, D.R., Witten, E. (eds.): Quantum Fields and Strings: A Course for Mathematicians. Vol. 1, 2. Am. Math. Soc., Providence (1999). Material from the special year on quantum field theory held at the Institute for Advanced Study, Princeton, NJ, 1996–1997. ISBN 0-8218-1198-3

[Del90] Deligne, P.: Catégories tannakiennes. In: The Grothendieck Festschrift, Vol. II, pp. 111–195. Birkhäuser Boston, Boston (1990)

[Del99] Deligne, P.: Notes on spinors. In: Quantum Fields and Strings: A Course for Mathematicians, Vol. 1, 2 (Princeton, NJ, 1996/1997), pp. 99–135. Am. Math. Soc., Providence (1999)

[Del12] Delamotte, B.: An introduction to the nonperturbative renormalization group. In: Renormalization Group and Effective Field Theory Approaches to Many-Body Systems. Lecture Notes in Phys., vol. 852, pp. 49–132. Springer, Heidelberg (2012) (F-PARIS6-PMC)

[Der92] Derdziński, A.: Geometry of the Standard Model of Elementary Particles. Texts and Monographs in Physics, p. 199. Springer, Berlin (1992). ISBN 3-540-54356-2

[DeW03] DeWitt, B.: The Global Approach to Quantum Field Theory. Vol. 1, 2. International Series of Monographs on Physics, vol. 114. Clarendon/Oxford University Press, Oxford/New York (2003). ISBN 0-19-851093-4

[DF99] Deligne, P., Freed, D.S.: Classical field theory. In: Quantum Fields and Strings: A Course for Mathematicians, Vol. 1, 2 (Princeton, NJ, 1996/1997), pp. 137–225. Am. Math. Soc., Providence (1999)

[dFdM10] de Faria, E., de Melo, W.: Mathematical Aspects of Quantum Field Theory. Cambridge Studies in Advanced Mathematics, vol. 127, p. 298. Cambridge University Press, Cambridge (2010). With a foreword by Dennis Sullivan. ISBN 978-0-521-11577-3

[DG62] Demazure, M., Grothendieck, A.: SGA 3: Schémas en Groupes. Lecture Notes in Mathematics, vol. 152. Springer, Berlin (1962)

[DHKS04] Dwyer, W.G., Hirschhorn, P.S., Kan, D.M., Smith, J.H.: Homotopy Limit Functors on Model Categories and Homotopical Categories. Mathematical Surveys and Monographs, vol. 113, p. 181. Am. Math. Soc., Providence (2004). ISBN 0-8218-3703-6

[Dir82] Dirac, P.A.M.: The Principles of Quantum Mechanics (International Series of Monographs on Physics). Oxford University Press, London (1982). ISBN 0198520115. http://www.amazon.ca/exec/obidos/redirect?tag=citeulike09-20&path=ASIN/0198520115

[DK80] Dwyer, W.G., Kan, D.M.: Simplicial localizations of categories. J. Pure Appl. Algebra **17**(3), 267–284 (1980). doi:10.1016/0022-4049(80)90049-3

[DK12] D'Agnolo, A., Kashiwara, M.: On a reconstruction theorem for holonomic systems. arXiv e-prints (2012). arXiv:1208.6104

[dM98] de Monvel, L.B.: Renormalisation (d'après Connes et Kreimer). Séminaire Bourbaki (1998)

[DM99] Deligne, P., Morgan, J.W.: Notes on supersymmetry (following Joseph Bernstein). In: Quantum Fields and Strings: A Course for Mathematicians, Vol. 1, 2 (Princeton, NJ, 1996/1997), pp. 41–97. Am. Math. Soc., Providence (1999)

[Dou66] Douady, A.: Le problème des modules pour les sous-espaces analytiques compacts d'un espace analytique donné. Ann. Inst. Fourier (Grenoble) **16**(fasc. 1), 1–95 (1966)

[Dri87] Drinfel'd, V.G.: Quantum groups. In: Proceedings of the International Congress of Mathematicians, Vol. 1, 2 (Berkeley, Calif., 1986), pp. 798–820. Am. Math. Soc., Providence (1987)

[DS95] Dwyer, W.G., Spaliński, J.: Homotopy theories and model categories. In: Handbook of Algebraic Topology, pp. 73–126. North-Holland, Amsterdam (1995). doi:10.1016/B978-044481779-2/50003-1

[DT83] Dubuc, E., Taubin, G.: Analytic rings. Cah. Topol. Géom. Différ. **24**(3), 225–265 (1983)

[Dub81] Dubuc, E.J.: C^∞-schemes. Am. J. Math. **103**(4), 683–690 (1981). doi:10.2307/2374046

[DZ94] Dubuc, E.J., Zilber, J.G.: On analytic models of synthetic differential geometry. Cah. Topol. Géom. Différ. Catég. **35**(1), 49–73 (1994)

[EG73] Epstein, H., Glaser, V.J.: The role of locality in perturbation theory. Ann. Henri Poincaré **19**(3), 211–295 (1973). oai:cds.cern.ch:880480

[Ege07] Egeileh, M.: Géométrie des champs de Higgs, compactifications et supergravité. (2007, preprint)

[Ehr81] Ehresmann, C.: Œuvres complètes et commentées. IV-1. Esquisses et complétions. Bull. Acad. Pol. Sci., Sér. Sci. Math. **29**(7–8):xxiv+403 (1981). Edited and with commentary in English by Andrée Charles Ehresmann

[Ehr82] Ehresmann, C.: Œuvres complètes et commentées. IV-2. Esquisses et structures monoïdales fermées. Cah. Topol. Géom. Différ. **23**(suppl. 2), 407–823 (1983). 1982. Edited and with commentary in English by Andrée Charles Ehresmann

[Eva98] Evans, L.C.: Partial Differential Equations. Graduate Studies in Mathematics, vol. 19, p. 662. Am. Math. Soc., Providence (1998). ISBN 0-8218-0772-2

[FBZ04] Frenkel, E., Ben-Zvi, D.: Vertex Algebras and Algebraic Curves, 2nd edn. Mathematical Surveys and Monographs, vol. 88, p. 400. Am. Math. Soc., Providence (2004). ISBN 0-8218-3674-9

[FG11] Francis, J., Gaitsgory, D.: Chiral Koszul duality. arXiv e-prints (2011). arXiv: 1103.5803

[FH90] Fisch, J.M.L., Henneaux, M.: Homological perturbation theory and the algebraic structure of the antifield-antibracket formalism for gauge theories. Commun. Math. Phys. **128**(3), 627–640 (1990)

[FHM02] Fausk, H., Hu, P., May, J.P.: Isomorphisms between left and right adjoints. arXiv Mathematics e-prints (2002). arXiv:math/0206079

[FLS02] Fulp, R., Lada, T., Stasheff, J.: Noether's variational theorem II and the BV formalism. arXiv (2002)

[Fol08] Folland, G.B.: Quantum Field Theory: A Tourist Guide for Mathematicians. Mathematical Surveys and Monographs, vol. 149, p. 325. Am. Math. Soc., Providence (2008). ISBN 978-0-8218-4705-3

[FR11] Fredenhagen, K., Rejzner, K.: Batalin-Vilkovisky formalism in perturbative algebraic quantum field theory. arXiv e-prints (2011). arXiv:1110.5232

[Fra11] Francis, J.: The tangent complex and Hochschild cohomology of E_n-rings. arXiv e-prints (2011). arXiv:1104.0181

[Fre99] Freed, D.S.: Five Lectures on Supersymmetry, p. 119. Am. Math. Soc., Providence (1999). ISBN 0-8218-1953-4

[Fre09] Fresse, B.: Modules over Operads and Functors. Lecture Notes in Mathematics, vol. 1967, p. 308. Springer, Berlin (2009). doi:10.1007/978-3-540-89056-0. ISBN 978-3-540-89055-3

[Gab96] Gabrielov, A.: Complements of subanalytic sets and existential formulas for analytic functions. Invent. Math. **125**(1), 1–12 (1996). doi:10.1007/s002220050066

[Gai99] Gaitsgory, D.: Notes on 2D conformal field theory and string theory. In: Quantum Fields and Strings: A Course for Mathematicians, Vol. 1, 2 (Princeton, NJ, 1996/1997), pp. 1017–1089. Am. Math. Soc., Providence (1999)

[Get83] Getzler, E.: Pseudodifferential operators on supermanifolds and the Atiyah-Singer index theorem. Commun. Math. Phys. **92**(2), 163–178 (1983)

[GG73] Golubitsky, M., Guillemin, V.: Stable Mappings and Their Singularities. Graduate Texts in Mathematics, vol. 14, p. 209. Springer, New York (1973)

[Gie06] Gies, H.: Introduction to the functional RG and applications to gauge theories. arXiv High Energy Physics—Phenomenology e-prints (2006). arXiv:hep-ph/0611146

[Gir71] Giraud, J.: Cohomologie Non Abélienne. Die Grundlehren der mathematischen Wissenschaften, vol. 179, p. 467. Springer, Berlin (1971)

[GJ99] Goerss, P.G., Jardine, J.F.: Simplicial Homotopy Theory. Progress in Mathematics, vol. 174, p. 510. Birkhäuser, Basel (1999). ISBN 3-7643-6064-X

[GR11] Gaitsgory, D., Rozenblyum, N.: Crystals and D-modules. arXiv e-prints (2011). arXiv:1111.2087

[Gro60] Grothendieck, A.: Éléments de géométrie algébrique. I. Le langage des schémas. Publ. Math. IHÉS **1–4**(4), 228 (1960)

[Gro67] Grothendieck, A.: Éléments de géométrie algébrique. IV. Étude locale des schémas et des morphismes de schémas. Publ. Math. IHÉS **32**, 361 (1967)

[Gro13] Grothendieck, A.: A la Poursuite des Champs (1983–2013). SMF, Maltsiniotis edition

[GS90] Guillemin, V., Sternberg, S.: Symplectic Techniques in Physics, 2nd edn., p. 468. Cambridge University Press, Cambridge (1990). ISBN 0-521-38990-9

[GSW88] Green, M.B., Schwarz, J.H., Witten, E.: Superstring Theory. Vol. 1, 2nd edn. Cambridge Monographs on Mathematical Physics, p. 470. Cambridge University Press, Cambridge (1988). Introduction. ISBN 0-521-35752-7

[GTZ09] Ginot, G., Tradler, T., Zeinalian, M.: A Chen model for mapping spaces and the surface product. arXiv e-prints (2009). arXiv:0905.2231

[Gwi05] Gwilliam, O.: Factorization algebras and free field theory. Thesis, Northwestern University (2005)

[GZ67] Gabriel, P., Zisman, M.: Calculus of Fractions and Homotopy Theory. Ergebnisse der Mathematik und ihrer Grenzgebiete, vol. 35, p. 168. Springer, New York (1967)

[Har70] Hartshorne, R.: Residues and Duality. Springer, Berlin (1970)

[Har77] Hartshorne, R.: Algebraic Geometry. Graduate Texts in Mathematics, vol. 52, p. 496. Springer, New York (1977). ISBN 0-387-90244-9

[Hin99] Hinich, V.: Deformations of homotopy algebras. arXiv Mathematics e-prints (1999). arXiv:math/9904145

[Hin01] Hinich, V.: DG coalgebras as formal stacks. J. Pure Appl. Algebra **162**(2–3), 209–250 (2001). doi:10.1016/S0022-4049(00)00121-3

[HK94] Huber, R., Knebusch, M.: On valuation spectra. In: Recent Advances in Real Algebraic Geometry and Quadratic Forms (Berkeley, CA, 1990/1991; San Francisco, CA, 1991). Contemp. Math., vol. 155, pp. 167–206. Am. Math. Soc., Providence (1994)

[HK11] Hollands, S., Kopper, C.: The operator product expansion converges in perturbative field theory. arXiv (2011). arXiv:1105.3375

[Hör03] Hörmander, L.: The Analysis of Linear Partial Differential Operators. i. Classics in Mathematics, p. 440. Springer, Berlin (2003). Distribution theory and Fourier analysis. Reprint of the second (1990) edition [Springer, Berlin; MR1065993 (91m:35001a)]. ISBN 3-540-00662-1

[Hör07] Hörmander, L.: The Analysis of Linear Partial Differential Operators. III. Classics in Mathematics, p. 525. Springer, Berlin (2007). Pseudo-differential operators. Reprint of the 1994 edition. ISBN 978-3-540-49937-4

[Hov99] Hovey, M.: Model Categories. Mathematical Surveys and Monographs, vol. 63, p. 209. Am. Math. Soc., Providence (1999). ISBN 0-8218-1359-5

[HT92] Henneaux, M., Teitelboim, C.: Quantization of Gauge Systems, p. 520. Princeton University Press, Princeton (1992). ISBN 0-691-08775-X; 0-691-03769-8

[Hub96] Huber, R.: Étale Cohomology of Rigid Analytic Varieties and Adic Spaces. Aspects of Mathematics, vol. 30, p. 450. Vieweg, Braunschweig (1996). ISBN 3-528-06794-2

[HW10] Hollands, S., Wald, R.M.: Axiomatic quantum field theory in curved spacetime. Commun. Math. Phys. **293**, 85–125 (2010). doi:10.1007/s00220-009-0880-7. arXiv:0803.2003

[Ill71] Illusie, L.: Complexe Cotangent et Déformations. I. Lecture Notes in Mathematics, vol. 239, p. 355. Springer, Berlin (1971)

[IZ99] Iglesias Zeimour, P.: Diffeology. Volume 1 of Web book (1999). http://math.huji.ac.il/~piz/Site/The Book/The Book.html

[Jan87] Jantzen, J.C.: Representations of Algebraic Groups, p. 443. Academic Press, Boston (1987). ISBN 0-12-380245-8

[Jar87] Jardine, J.F.: Simplicial presheaves. J. Pure Appl. Algebra **47**(1), 35–87 (1987). doi:10.1016/0022-4049(87)90100-9

[Jos09] Jost, J.: Geometry and Physics, p. 217. Springer, Berlin (2009). doi:10.1007/978-3-642-00541-1. ISBN 978-3-642-00540-4

[Joy07] Joyal, A.: Discs, duality and theta-categories (2007, preprint)

[Joy09] Joyce, D.: On manifolds with corners. arXiv e-prints (2009). arXiv:0910.3518

[Joy11] Joyal, A.: Notes on quasi-categories (2011, preprint)

[Kas95] Kassel, C.: Quantum Groups. Graduate Texts in Mathematics, vol. 155, p. 531. Springer, New York (1995). ISBN 0-387-94370-6

[Kas03] Kashiwara, M.: D-Modules and Microlocal Calculus. Translations of Mathematical Monographs, vol. 217, p. 254. Am. Math. Soc., Providence (2003). Translated from the 2000 Japanese original by Mutsumi Saito. Iwanami Series in Modern Mathematics. ISBN 0-8218-2766-9

[Ked10] Kedlaya, K.S.: Good formal structures for flat meromorphic connections, II: excellent schemes. arXiv e-prints (2010). arXiv:1001.0544

[Kel05] Keller, B.: Deformation quantization after Kontsevich and Tamarkin. In: Déformation, Quantification, Théorie de Lie. Panor. Synthèses, vol. 20, pp. 19–62. Société Mathématique de France, Paris (2005)

[Kel06] Keller, B.: A_∞-algebras, modules and functor categories. In: Trends in Representation Theory of Algebras and Related Topics. Contemp. Math., vol. 406, pp. 67–93. Am. Math. Soc., Providence (2006)

[Kel07] Keller, K.: Dimensional regularization in position space and a forest formula for regularized Epstein-Glaser renormalization. Universität Hamburg, Hamburg (2007). Dissertation zur Erlangung des Doktorgrades des Department Physik des Universität Hamburg

[Kel09] Keller, K.J.: Euclidean Epstein-Glaser renormalization. J. Math. Phys. **50**(10), 103503 (2009). doi:10.1063/1.3202415. arXiv:0902.4789

[KKK86] Kashiwara, M., Kawai, T., Kimura, T.: Foundations of Algebraic Analysis. Princeton Mathematical Series, vol. 37, p. 255. Princeton University Press, Princeton (1986). Translated from the Japanese by Goro Kato. ISBN 0-691-08413-0

[KM97] Kriegl, A., Michor, P.W.: The Convenient Setting of Global Analysis. Mathematical Surveys and Monographs, vol. 53, p. 618. Am. Math. Soc., Providence (1997). ISBN 0-8218-0780-3

[KM08] Kahn, B., Maltsiniotis, G.: Structures de dérivabilité. Adv. Math. **218**(4), 1286–1318 (2008)

[KMRT98] Knus, M.-A., Merkurjev, A., Rost, M., Tignol, J.-P.: The Book of Involutions, p. 593. Am. Math. Soc., Providence (1998). With a preface in French by J. Tits. ISBN 0-8218-0904-0

[KMS93] Kolář, I., Michor, P.W., Slovák, J.: Natural Operations in Differential Geometry, p. 434. Springer, Berlin (1993). ISBN 3-540-56235-4

[Koc06] Kock, A.: Synthetic Differential Geometry, 2nd edn. London Mathematical Society Lecture Note Series, vol. 333, p. 233. Cambridge University Press, Cambridge (2006). doi:10.1017/CBO9780511550812. ISBN 978-0-521-68738-6; 0-521-68738-1

[Kon95] Kontsevich, M.: Homological algebra of mirror symmetry. In: Proceedings of the International Congress of Mathematicians, Vol. 1, 2 (Zürich, 1994), pp. 120–139. Birkhäuser, Basel (1995)

[Kon99] Kontsevich, M.: Operads and motives in deformation quantization. Lett. Math. Phys. **48**, 35–72 (1999). arXiv:math.QA/9904055

[Kon03] Kontsevich, M.: Deformation quantization of Poisson manifolds. Lett. Math. Phys. **66**(3), 157–216 (2003). doi:10.1023/B:MATH.0000027508.00421.bf

[KRT97] Kassel, C., Rosso, M., Turaev, V.: Quantum Groups and Knot Invariants. Panoramas et Synthèses [Panoramas and Syntheses], vol. 5, p. 115. Société Mathématique de France, Paris (1997). ISBN 2-85629-055-8

[KS87] Kostant, B., Sternberg, S.: Symplectic reduction, BRS cohomology, and infinite-dimensional Clifford algebras. Ann. Phys. **176**(1), 49–113 (1987)

[KS90] Kashiwara, M., Schapira, P.: Sheaves on Manifolds. Grundlehren der Mathematischen Wissenschaften [Fundamental Principles of Mathematical Sciences], vol. 292, p. 512. Springer, Berlin (1990). With a chapter in French by Christian Houzel. ISBN 3-540-51861-4

[KS99] Kashiwara, M., Schapira, P.: Ind-sheaves, distributions, and microlocalization. In: Seminaire: Équations aux Dérivées Partielles, 1998–1999 (1999). Sémin. Équ. Dériv. Partielles, pages Exp. No. XXII, 13. École Polytech., Palaiseau

[KS01] Kashiwara, M., Schapira, P.: Ind-sheaves. Astérisque **271**, 136 (2001)

[KS06] Kashiwara, M., Schapira, P.: Categories and Sheaves. Grundlehren der Mathematischen Wissenschaften [Fundamental Principles of Mathematical Sciences], vol. 332, p. 497. Springer, Berlin (2006). ISBN 978-3-540-27949-5; 3-540-27949-0

[KSIW06] Kashiwara, M., Schapira, P., Ivorra, F., Waschkies, I.: Microlocalization of ind-sheaves. In: Studies in Lie Theory. Progr. Math., vol. 243, pp. 171–221. Birkhäuser Boston, Boston (2006). doi:10.1007/0-8176-4478-4_9

[KV98] Krasil'shchik, J., Verbovetsky, A.: Homological methods in equations of mathematical physics. arXiv (1998)

[LaC06] LaChapelle, J.: Functional integration for quantum field theory. arXiv Mathematical Physics e-prints (2006). arXiv:math-ph/0610035

[Lan93] Lang, S.: Algebra, 3rd edn., p. 914. Addison-Wesley, New York (1993). ISBN 0-201-55540-9

[Law79] Lawvere, F.W.: Categorical dynamics. In: Topos Theoretic Methods in Geometry. Various Publ. Ser., vol. 30, pp. 1–28. Aarhus Universitet, Aarhus (1979)

[Law04] Lawvere, F.W.: Functorial semantics of algebraic theories and some algebraic problems in the context of functorial semantics of algebraic theories. Theory Appl. Categ. **5**, 1–121 (2004). Reprinted from Proc. Natl. Acad. Sci. USA **50**, 869–872 (1963) [MR0158921] and Reports of the Midwest Category Seminar. II, pp. 41–61, Springer, Berlin (1968) [MR0231882]

[Law05] Lawvere, F.W.: An elementary theory of the category of sets (long version) with commentary. Theory Appl. Categ. **11**, 1–35 (2005). Reprinted and expanded from Proc. Natl. Acad. Sci. USA **52** (1964) [MR0172807]. With comments by the author and Colin McLarty

[Law06] Lawvere, F.W.: Adjointness in foundations. Theory Appl. Categ. **16**, 1–16 (2006). Reprinted from Dialectica **23** (1969)

[Lei04] Leinster, T.: Higher Operads, Higher Categories. Cambridge University Press, Cambridge (2004)

[LL66] Landau, L., Lifchitz, E.: Physique Théorique. Tome II: Théorie du Champ. Deuxième Édition Revue. Traduit du Russe Par Edouard Gloukhian, p. 451. Éditions Mir, Moscow (1966)

[LMB00] Laumon, G., Moret-Bailly, L.: Champs Algébriques. Ergebnisse der Mathematik und ihrer Grenzgebiete. 3. Folge. A Series of Modern Surveys in Mathematics [Results in Mathematics and Related Areas. 3rd Series. A Series of Modern Surveys in Mathematics], vol. 39, p. 208. Springer, Berlin (2000). ISBN 3-540-65761-4

[Łoj93] Łojasiewicz, S.: Sur la géométrie semi- et sous-analytique. Ann. Inst. Fourier (Grenoble) **43**(5), 1575–1595 (1993)

[Lot90] Lott, J.: Torsion constraints in supergeometry. Commun. Math. Phys. **133**(3), 563–615 (1990)

[Lur09] Lurie, J.: On the classification of topological field theories. arXiv e-prints (2009). arXiv:0905.0465

[Lur09a] Lurie, J.: Derived algebraic geometry. Doctoral thesis (2009, preprint)

[Lur09b] Lurie, J.: Derived algebraic geometry V: structured spaces. arXiv e-prints (2009). arXiv:0905.0459

[Lur09c] Lurie, J.: Higher algebra (2009, preprint)

[Lur09d] Lurie, J.: Higher Topos Theory. Annals of Mathematics Studies, vol. 170, p. 925. Princeton University Press, Princeton (2009). ISBN 978-0-691-14049-0; 0-691-14049-9

[Lur10] Lurie, J.: Moduli problems for ring spectra. In: Proceedings of the International Congress of Mathematicians. Volume II, pp. 1099–1125. Hindustan Book Agency, New Delhi (2010)

[Lur11] Lurie, J.: Derived algebraic geometry X: formal moduli problems (2011, preprint)

[LV12] Loday, J.-L., Vallette, B.: Algebraic Operads. Grundlehren der Mathematischen Wissenschaften [Fundamental Principles of Mathematical Sciences], vol. 346. Springer, Heidelberg (2012). ISBN 978-3-642-30361-6. xxiv+634 pp.

[Lyc93] Lychagin, V.: Quantizations of braided differential operators (1993, preprint)

[Maj95] Majid, S.: Foundations of Quantum Group Theory, p. 607. Cambridge University Press, Cambridge (1995). doi:10.1017/CBO9780511613104. ISBN 0-521-46032-8

[Mal10] Maltsiniotis, G.: Grothendieck ∞-groupoids, and still another definition of ∞-categories. arXiv e-prints (2010). arXiv:1009.2331

[Man97] Manin, Y.I.: Gauge Field Theory and Complex Geometry, 2nd edn. Grundlehren der Mathematischen Wissenschaften [Fundamental Principles of Mathematical Sciences], vol. 289, p. 346. Springer, Berlin (1997). Translated from the 1984 Russian original by N. Koblitz and J.R. King, with an appendix by Sergei Merkulov. ISBN 3-540-61378-1

[Mar10] Marcolli, M.: Feynman Motives, p. 220. World Scientific, Hackensack (2010). ISBN 978-981-4304-48-1; 981-4304-48-4

[Mat70] Mather, J.N.: Stability of C^∞ mappings. V. Transversality. Adv. Math. **4**, 301–336 (1970)

[McM09] McMahon, D.: String Theory Demystified: A Self-Teaching Guide, p. 306. McGraw-Hill, New York (2009)

[Mer10] Merkulov, S.A.: Wheeled pro(p)file of Batalin-Vilkovisky formalism. Commun. Math. Phys. **295**, 585–638 (2010). doi:10.1007/s00220-010-0987-x. arXiv:0804.2481

[ML98] Mac Lane, S.: Categories for the Working Mathematician, 2nd edn. Graduate Texts in Mathematics, vol. 5, p. 314. Springer, New York (1998). ISBN 0-387-98403-8

[Moc08] Mochizuki, T.: Good formal structure for meromorphic flat connections on smooth projective surfaces. arXiv e-prints (2008). arXiv:0803.1346

[MR91] Moerdijk, I., Reyes, G.E.: Models for Smooth Infinitesimal Analysis, p. 399. Springer, New York (1991). ISBN 0-387-97489-X

[MW74] Marsden, J., Weinstein, A.: Reduction of symplectic manifolds with symmetry. Rep. Math. Phys. **5**(1), 121–130 (1974)

[Nes03] Nestruev, J.: Smooth Manifolds and Observables. Graduate Texts in Mathematics, vol. 220, p. 222. Springer, New York (2003). Joint work of A.M. Astashov, A.B. Bocharov, S.V. Duzhin, A.B. Sossinsky, A.M. Vinogradov and M.M. Vinogradov. Translated from the 2000 Russian edition by A.B. Sossinsky, I.S. Krasil'schik and S.V. Duzhin. ISBN 0-387-95543-7

[Nis00] Nishimura, H.: Supersmooth topoi. Int. J. Theor. Phys. **39**(5), 1221–1231 (2000). doi:10.1023/A:1003630618417

[Pau11a] Paugam, F.: Histories and observables in covariant field theory. J. Geom. Phys. **61**(9), 1675–1702 (2011). doi:10.1016/j.geomphys.2010.11.002

[Pau11b] Paugam, F.: Homotopical Poisson reduction of gauge theories. In: Mathematical Foundations of Quantum Field Theory and Perturbative String Theory. Proc. Sympos. Pure Math., vol. 83, pp. 131–158. Am. Math. Soc., Providence (2011)

[Paw07] Pawlowski, J.: Aspects of the functional renormalisation group. Ann. Phys. **322**, 2831–2915 (2007). doi:10.1016/j.aop.2007.01.007. arXiv:hep-th/0512261

[Pei52] Peierls, R.E.: The commutation laws of relativistic field theory. Proc. R. Soc. Lond. Ser. A **214**, 143–157 (1952)

[Pen83] Penkov, I.B.: $\mathcal{D}$-Modules on supermanifolds. Invent. Math. **71**(3), 501–512 (1983). doi:10.1007/BF02095989

[Pen05] Penrose, R.: The Road to Reality, p. 1099. Knopf, New York (2005). A complete guide to the laws of the universe

[Phi01] Phillips, T.: Finite dimensional Feynman diagrams. AMS (2001). http://www.ams.org/featurecolumn/archive/feynman1.html, 1–7

[Pol98] Polchinski, J.: String Theory. Vol. II. Cambridge Monographs on Mathematical Physics, p. 531. Cambridge University Press, Cambridge (1998). Superstring theory and beyond. ISBN 0-521-63304-4

[Pol05] Polchinski, J.: String Theory. Vol. I. Cambridge Monographs on Mathematical Physics, p. 402. Cambridge University Press, Cambridge (2005). ISBN 0-521-63303-6; 978-0-521-67227-6; 0-521-67227-9. An introduction to the bosonic string. Reprint of the 2003 edition

[Pre07] Prelli, L.: Microlocalization of subanalytic sheaves. arXiv Mathematics e-prints (2007). arXiv:math/0702459

[Pri11] Pridham, J.P.: Notes characterising higher and derived stacks concretely. arXiv e-prints (2011). arXiv:1105.4853

[PTVV11] Pantev, T., Toen, B., Vaquie, M., Vezzosi, G.: Quantization and derived moduli spaces I: shifted symplectic structures. arXiv e-prints (2011). arXiv:1111.3209

[Qui67] Quillen, D.G.: Homotopical Algebra. Lecture Notes in Mathematics, vol. 43, p. 156. Springer, Berlin (1967)

[Qui69] Quillen, D.: Rational homotopy theory. Ann. Math. (2) **90**, 205–295 (1969)

[Qui70] Quillen, D.: On the (co-) homology of commutative rings. In: Applications of Categorical Algebra (Proc. Sympos. Pure Math., Vol. XVII, New York, 1968), pp. 65–87. Am. Math. Soc., Providence (1970)

[Rej11] Rejzner, K.: Batalin-Vilkovisky formalism in locally covariant field theory. arXiv e-prints (2011). arXiv:1111.5130

[Rez09] Rezk, C.: A cartesian presentation of weak n-categories. arXiv e-prints (2009). arXiv: 0901.3602

[Riv90] Rivers, R.J.: Path Integral Methods in Quantum Field Theory, 2nd edn. Cambridge Monographs on Mathematical Physics, p. 339. Cambridge University Press, Cambridge (1990). ISBN 0-521-25979-7; 0-521-36870-7

[Ros12] Rosten, O.J.: Fundamentals of the exact renormalization group. Phys. Rep. **511**, 177–272 (2012). doi:10.1016/j.physrep.2011.12.003. arXiv:1003.1366

[Roz09] Rozenblyum, N.: Introduction to chiral algebras (2009, preprint)

[Roz10] Rozenblyum, N.: Modules over a chiral algebra. arXiv e-prints (2010). arXiv: 1010.1998

[Roz11] Rozenblyum, N.: Connections on conformal blocks. Thesis (2011)

[Roz12] Rozenblyum, N.: Continuous moduli functors and quantum field theory (2012, preprint)

[RS80] Reed, M., Simon, B.: Methods of Modern Mathematical Physics. I, 2nd edn., p. 400. Academic Press/Harcourt Brace Jovanovich, New York/San Diego (1980). Functional analysis. ISBN 0-12-585050-6

[Sab09] Sabbah, C.: Introduction to Stokes structures. arXiv e-prints (2009). arXiv:0912.2762

[Sac09] Sachse, C.: Global analytic approach to super Teichmueller spaces. arXiv e-prints (2009). arXiv:0902.3289

[Sch66] Schwartz, L.: Théorie des Distributions. Publications de L'Institut de Mathématique de L'Université de Strasbourg, p. 420. Hermann, Paris (1966). No. IX–X. Nouvelle édition, entiérement corrigée, refondue et augmentée

[Sch93] Schwarz, A.S.: Quantum Field Theory and Topology. Grundlehren der Mathematischen Wissenschaften [Fundamental Principles of Mathematical Sciences], vol. 307, p. 274. Springer, Berlin (1993). Translated from the 1989 Russian original by Eugene Yankowsky [E.M. Yankovskiĭ] and Silvio Levy. ISBN 3-540-54753-3

[Sch94a] Schneiders, J.-P.: An introduction to $\mathcal{D}$-modules. Bull. Soc. R. Sci. Liège **63**(3–4), 223–295 (1994). Algebraic Analysis Meeting (Liège, 1993)

[Sch94b] Schwarz, A.S.: Topology for Physicists. Grundlehren der Mathematischen Wissenschaften [Fundamental Principles of Mathematical Sciences], vol. 308, p. 296. Springer, Berlin (1994). Translated from the Russian by Silvio Levy. ISBN 3-540-54754-1

[Sch07] Schapira, P.: Mikio Sato, a visionary of mathematics. Not. Am. Math. Soc. **54**(2), 243–245 (2007)

[Sch09a] Schapira, P.: Masaki Kashiwara and algebraic analysis. In: Algebraic Analysis and Around. Adv. Stud. Pure Math., vol. 54, pp. 1–9. Mathematical Society of Japan, Tokyo (2009)

[Sch09b] Schätz, F.: BFV-complex and higher homotopy structures. Commun. Math. Phys. **286**(2), 399–443 (2009). doi:10.1007/s00220-008-0705-0

[Sch10] Schapira, P.: Triangulated categories for the analysis. In: Triangulated Categories. London Math. Soc. Lecture Note Ser., vol. 375, pp. 371–388. Cambridge University Press, Cambridge (2010)

[Sch11] Schreiber, U.: Differential cohomology in a cohesive ∞-topos. Nlab website (2011)

[Sch12] Schapira, P.: Three lectures on algebraic microlocal analysis. arXiv e-prints (2012). arXiv:1206.1435

[Ser70] Serre, J.P.: Algèbre Locale—Multiplicités. Springer, Berlin (1970)

[SH98] Simpson, C., Hirschowitz, A.: Descente pour les n-champs (descent for n-stacks) (1998). arXiv:math.AG/9807049

[Sha97] Sharpe, R.W.: Differential Geometry. Graduate Texts in Mathematics, vol. 166, p. 421. Springer, New York (1997). Cartan's generalization of Klein's Erlangen program, with a foreword by S.S. Chern. ISBN 0-387-94732-9

[Sim01] Simpson, C.: Some properties of the theory of n-categories. arXiv Mathematics e-prints (2001). arXiv:math/0110273

[Sim10] Simpson, C.T.: Homotopy theory of higher categories. arXiv e-prints (2010). arXiv: 1001.4071

[SKK73] Sato, M., Kawai, T., Kashiwara, M.: Microfunctions and pseudo-differential equations. In: Hyperfunctions and Pseudo-Differential Equations (Proc. Conf., Katata, 1971; Dedicated to the Memory of André Martineau). Lecture Notes in Math., vol. 287, pp. 265–529. Springer, Berlin (1973)

[Smi66] Smith, J.W.: The de Rham theorem for general spaces. Tohoku Math. J. **18**, 115–137 (1966)

[Sou97] Souriau, J.-M.: Structure of Dynamical Systems. Progress in Mathematics, vol. 149, p. 406. Birkhäuser Boston, Boston (1997). A symplectic view of physics. Translated from the French by C.H. Cushman-de Vries. Translation edited and with a preface by R.H. Cushman and G.M. Tuynman. ISBN 0-8176-3695-1

[Spi08] Spivak, D.I.: Derived smooth manifolds. arXiv e-prints (2008). arXiv:0810.5174

[SSS09] Sati, H., Schreiber, U., Stasheff, J.: Differential twisted string and Fivebrane structures. arXiv e-prints (2009). arXiv:0910.4001

[Sta97] Stasheff, J.: Deformation theory and the Batalin-Vilkovisky master equation. In: Deformation Theory and Symplectic Geometry (Ascona, 1996). Math. Phys. Stud., vol. 20, pp. 271–284. Kluwer Academic, Dordrecht (1997)

[Sta98] Stasheff, J.: The (secret?) homological algebra of the Batalin-Vilkovisky approach. In: Secondary Calculus and Cohomological Physics (Moscow, 1997). Contemp. Math., vol. 219, pp. 195–210. Am. Math. Soc., Providence (1998)

[SW77] Sachs, R.K., Wu, H.H.: General Relativity for Mathematicians. Graduate Texts in Mathematics, vol. 48, p. 291. Springer, New York (1977). ISBN 0-387-90218-X

[Tam95] Tamsamani, Z.: On non-strict notions of n-category and n-groupoid via multisimplicial sets (1995). arXiv:math.AG/9512006

[Tam98] Tamarkin, D.E.: Another proof of M. Kontsevich formality theorem. arXiv Mathematics e-prints (1998). arXiv:math/9803025

[Tat57] Tate, J.: Homology of Noetherian rings and local rings. Ill. J. Math. **1**, 14–27 (1957)

[Tho56] Thom, R.: Un lemme sur les applications différentiables. Bol. Soc. Mat. Mexicana **1**, 59–71 (1956)

[Toe00] Toen, B.: Affine stacks (champs affines). arXiv Mathematics e-prints (2000). arXiv: math/0012219

[Toe02] Toen, B.: From homotopical algebra to homotopical algebraic geometry: lectures in Essen (2002). http://www.math.univ-toulouse.fr/~toen/essen.pdf

[Toe04] Toen, B.: Towards an axiomatization of the theory of higher categories. arXiv Mathematics e-prints (2004). arXiv:math/0409598

[Toe13] Toen, B.: Branes for ∞-operads. arXiv e-prints (2013). arXiv:1307.0405

[TV02] Toen, B., Vezzosi, G.: Homotopical algebraic geometry I: topos theory. arXiv Mathematics e-prints (2002). arXiv:math/0207028

[TV08] Toën, B., Vezzosi, G.: Homotopical Algebraic Geometry. II. Geometric Stacks and Applications. Mem. Am. Math. Soc. 193(902):x+224 (2008)

[TV09a] Toen, B., Vezzosi, G.: S^1-Equivariant simplicial algebras and de Rham theory. arXiv e-prints (2009). arXiv:0904.3256

[TV09b] Toën, B., Vaquié, M.: Au-dessous de Spec $\mathbb{Z}$. J. K-Theory **3**(3), 437–500 (2009). doi:10.1017/is008004027jkt048

[Val04] Vallette, B.: Dualite de Koszul des PROPs. arXiv Mathematics e-prints (2004). arXiv:math/0405057

[Vin01] Vinogradov, A.M.: Cohomological Analysis of Partial Differential Equations and Secondary Calculus. Translations of Mathematical Monographs, vol. 204, p. 247. Am. Math. Soc., Providence (2001). Translated from the Russian manuscript by Joseph Krasil'shchik. ISBN 0-8218-2922-X

[Vit09] Vitagliano, L.: The Lagrangian-Hamiltonian formalism for higher order field theories. arXiv e-prints (2009). arXiv:0905.4580

[Vit10] Vitagliano, L.: The Hamilton-Jacobi formalism for higher order field theories. arXiv e-prints (2010). arXiv:1003.5236

[Vit11] Vitagliano, L.: Hamilton-Jacobi diffieties. arXiv e-prints (2011). arXiv:1104.0162

[vN96] von Neumann, J.: Mathematical Foundations of Quantum Mechanics. Princeton Landmarks in Mathematics, p. 445. Princeton University Press, Princeton (1996). Translated from the German and with a preface by Robert T. Beyer, Twelfth printing, Princeton Paperbacks. ISBN 0-691-02893-1

[Vor98] Voronov, A.A.: The Swiss-cheese operad. arXiv Mathematics e-prints (1998). arXiv: math/9807037

[vS09a] van Suijlekom, W.D.: Renormalization of gauge fields using Hopf algebras. In: Quantum Field Theory, pp. 137–154. Birkhäuser, Basel (2009). doi:10.1007/978-3-7643-8736-5_8

[vS09b] van Suijlekom, W.D.: The structure of renormalization Hopf algebras for gauge theories. I. Representing Feynman graphs on BV-algebras. Commun. Math. Phys. **290**(1), 291–319 (2009). doi:10.1007/s00220-009-0829-x

[Wat79] Waterhouse, W.C.: Introduction to Affine Group Schemes, p. 164. Springer, New York (1979). ISBN 0-387-90421-2

[Wis10] Wise, D.K.: MacDowell-Mansouri gravity and Cartan geometry. Class. Quantum Gravity **27**(15), 155010 (2010). 26. doi:10.1088/0264-9381/27/15/155010

[Wit89] Witten, E.: Quantum field theory and the Jones polynomial. Commun. Math. Phys. **121**(3), 351–399 (1989)

[Wol84] Wolf, J.A.: Spaces of Constant Curvature, 5th edn., p. 412. Publish or Perish Inc., Houston (1984). ISBN 0-914098-07-1

[Zei06] Zeidler, E.: Quantum Field Theory. I. Basics in Mathematics and Physics, p. 1020. Springer, Berlin (2006). doi:10.1007/978-3-540-34764-4. ISBN 978-3-540-34762-0; 3-540-34762-3. A bridge between mathematicians and physicists

[Zei09] Zeidler, E.: Quantum Field Theory. II. Quantum Electrodynamics, p. 1101. Springer, Berlin (2009). ISBN 978-3-540-85376-3. A bridge between mathematicians and physicists

[Zei11] Zeidler, E.: Quantum Field Theory. III. Gauge Theory, p. 1126. Springer, Heidelberg (2011). doi:10.1007/978-3-642-22421-8. ISBN 978-3-642-22420-1. A bridge between mathematicians and physicists

[Zil90] Zilber, J.C.: Local analytic rings. Cah. Topol. Géom. Différ. Catég. **31**(1), 21–52 (1990)

[ZJ93] Zinn-Justin, J.: Quantum Field Theory and Critical Phenomena, 2nd edn. International Series of Monographs on Physics, vol. 85, p. 996. Clarendon/Oxford University Press, London/New York (1993). Oxford Science Publications. ISBN 0-19-852053-0

Index

F. Paugam, *Towards the Mathematics of Quantum Field Theory*,
Ergebnisse der Mathematik und ihrer Grenzgebiete. 3. Folge / A Series of Modern
Surveys in Mathematics 59, DOI 10.1007/978-3-319-04564-1,

The manufacturer's authorised representative in the EU is Springer Nature Customer Service Centre GmbH, Europaplatz 3, 69115 Heidelberg, Germany. If you have any concerns regarding our products, please contact ProductSafety@springernature.com

Printed and bound by CPI Group (UK) Ltd, Croydon, CR0 4YY
05/10/2026
02228422-0004